OOGENESIS

*Proceedings of a Symposium on Oogenesis held in
Baltimore, Maryland, in October, 1970.*

*Supported by a Contract between the Department of
Population Dynamics of the Johns Hopkins University
School of Hygiene and Public Health and the Center for
Population Research, National Institute of Child Health and
Human Development (Contract No. NIH70-782).*

OOGENESIS

Edited by

John D. Biggers *and* Allen W. Schuetz

University Park Press
BALTIMORE

Butterworths
LONDON

3/30/73

UNIVERSITY PARK PRESS
International Publishers in Science and Medicine
Chamber of Commerce Building
Baltimore, Maryland 21202

Published jointly by
UNIVERSITY PARK PRESS, BALTIMORE

and

BUTTERWORTH & CO. (PUBLISHERS) LTD., LONDON

ISBN 0–8391–0676–9 (University Park Press)
ISBN 0–0408–70369–5 (Butterworths)

LIBRARY OF CONGRESS CATALOGING IN PUBLICATION DATA
Main entry under title:

Oogenesis.

Proceedings of a symposium held in Baltimore, Md.,
in October, 1970, under the auspices of the Johns
Hopkins University School of Hygiene and Public Health.
1. Oogenesis—Congresses. I. Biggers, John D.,
1923–   ed. II. Schuetz, Allen W., 1936–   ed.
III. Johns Hopkins University. School of Hygiene and
Public Health.
QL965.064   599'.03'3   71–39292
ISBN 0–8391–0676–9

TABLE OF CONTENTS

Preface

1   Prologue 1

John D. Biggers
Allen W. Schuetz
School of Hygiene and Public Health
The Johns Hopkins University
Baltimore, Maryland

2   Comparative Studies on the Ultrastructure
of Mammalian Oocytes 5

Luciano Zamboni
Harbor General Hospital
Torrance, California

3   Changes of Some Cell Organelles
During Oogenesis in Mammals 47

Daniel Szollosi
School of Medicine
University of Washington
Seattle, Washington

4   The Kinetochore in Oocyte Maturation 65

Patricia G. Calarco
School of Medicine
University of Washington
Seattle, Washington

5   The Localization of Acid Phosphatase and the Uptake of
Horseradish Peroxidase in the Oocytes and
Follicle Cells of Mammals 87

Everett Anderson
The University of Massachusetts
Amherst, Massachusetts

6   Nuclear Structure and Function
    During Amphibian Oogenesis 119

        O. L. Miller, Jr.
        Barbara R. Beatty
        Barbara A. Hamkalo
        Oak Ridge National Laboratory
        Oak Ridge, Tennessee

7   Utilization of Genetic Information During Oogenesis 129

        Eric H. Davidson
        Barbara R. Hough
        The Rockefeller University
        New York, New York

8   Chromosomal Proteins 141

        Ru Chih C. Huang
        William Cieplinski
        The Johns Hopkins University
        Baltimore, Maryland

9   Ribonucleic Acid Synthesis During Oogenesis in *Xenopus laevis* 167

        Peter J. Ford
        MRC Epigenetics Research Group
        Institute of Animal Genetics
        Edinburgh, Scotland

10 Regulation of Ribosomal RNA Synthesis During Oogenesis
    of *Xenopus laevis* 193

        Marco Crippa
        Francesca Andronico
        C. N. R. Laboratory of Molecular Embryology
        Naples, Italy

        Glauco P. Tocchini-Valentini
        C. N. R. International Laboratory of Genetics and Biophysics
        Naples, Italy

11 Cytoplasmic DNA 215

        Igor B. Dawid
        Carnegie Institution of Washington
        Baltimore, Maryland

12   Protein Synthesis During Oocyte Maturation 227

    L. Dennis Smith
    Purdue University
    Lafayette, Indiana

13   Metabolism of the Oocyte 241

    John D. Biggers
    School of Hygiene and Public Health
    The Johns Hopkins University
    Baltimore, Maryland

14   Drosophila Oogenesis and Its Genetic Control 253

    Robert C. King
    Northwestern University
    Evanston, Illinois

15   Parthenogenesis and Heteroploidy in the Mammalian Egg 277

    R. A. Beatty
    University of Edinburgh
    Edinburgh, Scotland

16   Observations on the Behavior of Oogonia and Oocytes
in Tissue and Organ Culture 301

    Richard J. Blandau
    School of Medicine
    University of Washington
    Seattle, Washington

    D. Louise Odor
    Medical College of Virginia
    Virginia Commonwealth University
    Richmond, Virginia

17   Functional Interactions Between the Amphibian Oocyte
and General Ovarian Cells 321

    Antonie W. Blackler
    Cornell University
    Ithaca, New York

18   The Role of Protein Uptake in Vertebrate Oocyte Growth
      and Yolk Formation 339

          Robin A. Wallace
          Oak Ridge National Laboratory
          Oak Ridge, Tennessee

19   Follicle Growth in the Mouse Ovary 361

          Torben Pedersen
          The Finsen Institute
          Copenhagen, Denmark

20   Gonadotrophin-Induced Maturation of Mouse Graafian
      Follicles in Organ Culture 377

          T. G. Baker
          P. Neal
          University of Edinburgh
          Edinburgh, Scotland

21   Final Stages of Mammalian Oocyte Maturation 397

          Charles G. Thibault
          Institut National de la Recherche Agronomique
          Jouy-en-Josas, France

22   The Relation of Oocyte Maturation to Ovulation
      in Mammals 413

          Roger P. Donahue
          School of Medicine
          University of Washington
          Seattle, Washington

23   Maturation and Fertilization of Human Oocytes *in Vitro* 439

          Joseph F. Kennedy
          School of Medicine
          The Johns Hopkins University
          Baltimore, Maryland

24   Adenine Derivatives and Oocyte Maturation in Starfishes 459

          Haruo Kanatani
          University of Tokyo
          Tokyo, Japan

25 Hormones and Follicular Functions 479

Allen W. Schuetz
School of Hygiene and Public Health
The Johns Hopkins University
Baltimore, Maryland

26 Interaction Between Oocytes and Follicular Cells 513

A. V. Nalbandov
University of Illinois
Urbana, Illinois

Subject Index 525

Species Index 529

Author Index 533

PREFACE

In October, 1970, a Symposium on Oogenesis was held in Baltimore, Maryland, under the auspices of the Johns Hopkins University School of Hygiene and Public Health. The Symposium was supported by a Contract between the Department of Population Dynamics of the University and the Center for Population Dynamics, National Institute of Child Health and Human Development (Contract No. NIH 70-782).

An internationally drawn group of speakers was invited to cover many aspects of oogenesis in the animal kingdom, and their contributions are collected in this volume. We are indebted to all participants for their contributions to the Symposium.

The success of the Symposium and the production of this monograph have depended on the contributions of many people, to whom we are very grateful. The pages as seen in the book were prepared on an IBM MTSC typewriter. The typing and assembly of the final manuscript ready for photo-offset reproduction was skillfully done by Mrs. Jo Ann Sherbine and Mrs. Brenda Watson. Dr. Samuel Stern prepared the index. Mrs. L. Elaine Sheckells administered the budget, and Miss Reno Franks dealt with all secretarial work connected with the Symposium. We are indebted to these individuals for their major contributions.

We are also indebted to Drs. G. J. Marcus, A. R. Bellve, C. B. Ozias, M. Kramen, P. J. Olds, and R. A. Pedersen for checking manuscripts, and Mrs. Mary Thomas and Miss Beth Laube for technical assistance.

John D. Biggers
Allen W. Schuetz

PROLOGUE

John D. Biggers
Allen W. Schuetz

A single cell called the oocyte is the primary cellular link between the ongoing generation and the next. Generally, oogenesis can be regarded as the continuum of processes involved in the origin, growth and differentiation of the oocyte. The study of these processes, moreover, cannot be separated from considerations of the role of the oocyte in embryogenesis. Oogenesis, therefore, embraces all those cellular, molecular and physiological phenomena involved in producing a cell which is capable of expressing and maintaining the characteristics of the species. The study of these phenomena is the purpose of this symposium.

Why is it opportune to review the subject of oogenesis at the present time? Before this question can be answered it is necessary to discuss certain features of fertilization and early development.

At fertilization the information present in the mature ovum and spermatozoon is assembled into the fertilized egg. It is customary to associate fertilization with the transmission of heritable characters. At fertilization a set of paternal chromosomes is introduced into the ovum carrying a set of maternal chromosomes, thereby allowing the association of new sets of genes. In order to prevent the excessive accumulation of genetic material with the passage of generations the chromosome complement is reduced to one half during gametogenesis by a process called meiosis, from the Greek μειοῦν - to lessen. But meiosis has a much wider role than the reduction of the number of chromosomes from the diploid to the haploid condition. It is essential for the generation of genetic variation on which evolution depends through natural selection. While mutation produces variation by creating new genes, meiosis provides a mechanism for generating new combinations of genes within the gametes from the genes already in existence within a parent. In general, this generation of new combinations is accomplished by (1) the reassortment of the homologous pairs of chromosomes at the first meiotic division and (2) the process called crossing over, whereby segments of the chromatids from homologous pairs of chromosomes are exchanged to construct new chromosomes. The mechanisms involved in these chromosomal rearrangements during meiosis is, therefore, a major topic in the study of oogenesis.

Fertilization involves cell fusion, and this process is essential to ensure continuity of the cell line between generations. In a great many species the ovum is an enormous cell in comparison to the spermatozoon and is the dominant cell with respect to cytoplasmic mass. With fusion the ovum becomes the cellular vehicle in which its own genes interact with the genes of the fertilizing spermatozoon. Ample evidence suggests that the interaction of maternal and paternal genes is not immediately expressed and, in some species, it is delayed until the blastula stage of development. During this early period of development the changes that occur seem to be controlled by the ovum. The notion was first put forward in 1896 by E. B. Wilson that embryogenesis begins in oogenesis. Much more evidence now exists which suggests that the information controlling early embryonic development is programmed during oogenesis (see Davidson, 1968). The organelles and the chemical components of the oocyte as well as their functional significance in oogenesis and embryogenesis is therefore an area of major concern. Among the organelles the mitochondria are of particular interest in early development since they now appear to be separate carriers of genetic information and are capable of self-replication. In a considerable number of species the sperm contributes few, and sometimes no mitochondria at fertilization (Hersh, 1969). Thus, the transmission of mitochondria between generations may be entirely through the mother.

Apart from the transmission of developmental information between generations the ovum also carries nutrients necessary for the nurture of early development. The synthesis or accumulation of large quantities of yolk material in the oocytes is a major process in oogenesis, and it may require the development of extensive protein synthesizing machinery. Our present understanding of this process has primarily resulted from studies in non-mammalian species, particularly amphibian and avian species. In mammals, technical and methodological difficulties have markedly limited studies of these processes. The condition of viviparity in mammals, however, provides the opportunity for specialized mechanisms for the nutrition of the early embryo which is reflected in the differentiation and metabolism of the oocyte.

A different aspect of oogenesis concerns the regulation of oocyte growth and differentiation, and the synchronization of the release of mature ova at the site of fertilization at the optimum time. There is ample evidence that the regulation of certain aspects of oogenesis is under endocrine control; however, the manner in which the hormones act at the cellular level, and their relation to the stages of meiosis is poorly understood. Evidence is accumulating that small molecular messengers, like 1-methyladenine in starfish, are involved in the regulation and synchronization of many of these events. Likewise, the oocyte itself appears to play an active role in gonadogenesis, folliculogenesis and luteinization of the follicle.

Thus the study of oogenesis falls into three main areas:

    (1)   The mechanism of meiosis;

    (2)   Developmental aspects, concerned with the storage of information and the provision of a cellular vehicle and nutrients necessary for the development of a future individual;

    (3)   The endocrine regulation of oocyte maturation and release.

These three areas have been studied extensively in the past, but there has been a tendency for these topics to be studied independently with little exchange of information and ideas between investigators in the three areas. The last definitive reviews in the three areas were put together about ten years ago by Rhoades (1961) on meiosis, Raven (1961) on oogenesis and development and Zuckerman (1962) on the endocrine regulation. Many advances in the study of oogenesis and development have been made in the last decade because of the impact of cell and molecular biology. Often these advances have been facilitated by studying invertebrate and non-mammalian vertebrates, and only recently have the findings impinged on the study of oogenesis in mammals. It is now clear that the processes in mammals (including man) and lower organisms are similar in many ways.

A thorough understanding of oogenesis is of basic importance to man in several areas. These include the treatment of infertility in man and domestic animals, the control of excessive fertility, and the understanding of the effects of drugs on early development. Three questions need to be answered at the present time to guide future research in the applied fields, particularly those concerned with human fertility:

    (1)   How relevant are the discoveries on oogenesis in lower forms to studies of oogenesis in mammals?

    (2)   If the work on lower forms is relevant, what are the most useful species to study?

    (3)   If mammalian forms are uniquely different from the lower forms now being studied, how should mammalian oogenesis be investigated, and what new techniques need to be developed?

This Symposium was organized as a first step in the solution of these questions. Its purpose was to provide an up-to-date account of oogenesis in the several areas, and to encourage the exchange of ideas between the scientists involved in the three main areas of study. Major emphasis is given to the comparative approach, and to aspects of oogenesis not generally considered. The topics were chosen to stress the viewpoint that oogenesis is a major segment in the continuum of life, and that the oocyte is specialized by a process of differentiation to (1) undergo meiosis, (2) participate in cell fusion, and (3) accumulate and provide materials which govern the nature and nurture of the new individual produced when the ovum is fertilized. Although the questions have not been clearly answered, we hope the information gathered in this volume will help further discussion and promote useful work.

4

REFERENCES

Davidson, E. H. (1968) Gene activity in early development. New York, Academic Press.

Hersh, R. T. (1969) Mitochondrial genetics: A conjecture. *Science.* **166**, 402.

Raven, C. P. (1961) Oogenesis: The storage of developmental information. New York, Pergamon Press.

Rhoades, M. M. (1961) Meiosis. *In* "The cell". (J. Brachet and A. E. Mirsky, eds.) Vol. 3. New York, Academic Press.

Wilson, E. B. (1896) On cleavage and mosaic-work. *Arch. Entwicklungsmech Org.* **3**, 19.

Zuckerman, S. (1962) The Ovary. Vols. I and II. New York, Academic Press.

COMPARATIVE STUDIES ON THE ULTRASTRUCTURE OF MAMMALIAN OOCYTES

Luciano Zamboni

I.   Introduction
II.   Embryonic phase
        A.   Migration of primordial germ cells
        B.   Oogenesis
III.   Post-embryonic phase
        A.   Period of quiescence
        B.   Period of maturation
                1.   Follicle cell changes
                2.   Oocyte changes
IV.   Acknowledgments
V.   References

I. INTRODUCTION

The series of complex events that render the female germ cell capable of performing its ultimate function, conjugation with the spermatozoon, evolves over a long period which extends from the stage of oogonia differentiation in the embryonic ovary to the final maturation of the egg in the lumen of the Fallopian tube. The length of this period varies considerably among different species and for individual oocytes in the same animal. It is obviously shorter in those animals which have a brief gestation and attain rapid sexual maturity than in those, like primates, where both the gestation and pre-puberal periods are long. The life span is shortest for those oocytes which are ovulated early after the animal attains sexual maturity, and longest for those which are liberated from the follicles toward the end of the fertile period, shortly before the onset of menopause. In the human female, for example, the life span of individual germ cells varies from a minimum of about 12 years to a maximum of several decades.

The process of differentiation and maturation of the female germ cell may

suited
investi
not necessarily unimp
ace during the quiescent p

II. THE EMBRYONIC PHASE

The most salient phenomena of this phase are the migration of primordial germ cells to the genital ridges, and the process of oogenesis.

A. Migration of Primordial Germ Cells

The primordial germ cells, first recognizable in pre-somite embryos, arise in the entoderm of the yolk-sac stalk, the allantois, and the gut (Celestino da Costa, 1932; Debeyre, 1933; Witschi, 1948; Brambell, 1962). The cells are readily distinguishable because of their large size and round shape, the presence of numerous pseudopodial-like protrusions, and highly undifferentiated cytological characters which make them comparable to the blastomeres of eggs in advanced cleavage stages (Brambell, 1962). From their site of origin, the primordial germ cells migrate, mostly by means of active ameboid movements (Everett, 1943; Chiquoine, 1954) to the genital ridges (arrows, Fig. 2.1) located between the base of the mesentery and the Wolffian duct. With the arrival of the germ cells, the epithelium of the genital ridges, usually referred to as coelomic or germinal epithelium, begins to thicken and to proliferate actively (Fig. 2.2), becoming clearly differentiated from the underlying mesenchyme (Fig. 2.3). Due to the cellular multiplication, the region becomes markedly hypertrophic and soon assumes the form of a cylindrical ridge projecting ventrally into the splanchnocoele (Brambell 1927, 1962).

Upon arriving at the genital ridges, the primordial germ cells undergo morphological changes and lose some of those cytological characteristics, the pseudopodia for example, which had made their identification so easy during migration. At one time, this led some investigators to claim that the primordial germ cells all degenerate after reaching the genital ridges, and that the definitive germ cells later differentiate from a proliferating germinal epithelium (Kingery, 1917; Simkins, 1923, 1928; Hargitt, 1925), an hypothesis which was later abandoned following the acquisition of experimental evidence indicating that the primordial germ cells in the genital ridges not only give rise to all the definite germ cells, but also act as inductors in the development and future organization of the gonads (Bounoure, 1935, 1939; Nieuwkoop, 1949; Mintz, 1959; Zuckerman, 1960).

The primordial germ cells in the coelomic surface of the genital ridges can be easily recognized and distinguished from the germinal epithelium by electron microscopy. The germinal epithelium (Fig. 2.3) consists mostly of columnar cells with highly indented nuclei, coarsely aggregated and frequently marginated chromatin, and a relatively advanced organization of the cytoplasm with well

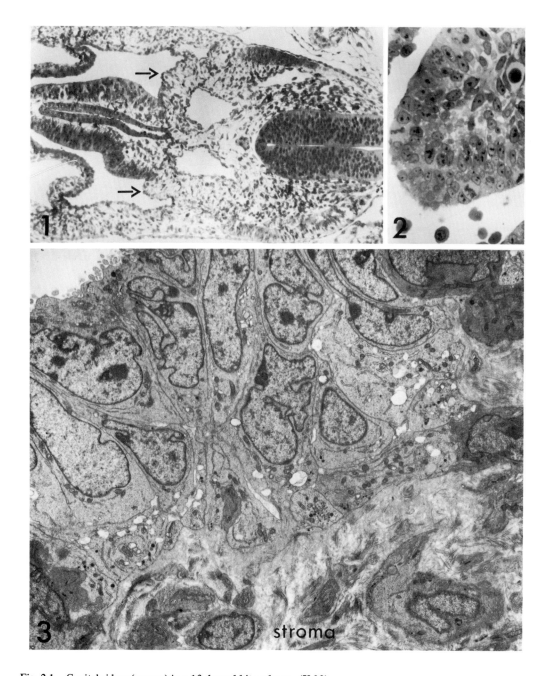

Fig. 2.1   Genital ridges (arrows) in a 13-day rabbit embryo   (X 90).

Fig. 2.2   Genital ridge of a rabbit embryo at the time of arrival of the primordial germ cells. Several cells of the coelomic (germinal) epithelium are in mitosis   (X 500).

Fig. 2.3   Coelomic (germinal) epithelium of an 11-day-old rabbit. The epithelial cells can be easily differentiated from the mesenchymal cells, from which they are separated by a thin basement membrane   (X 5,000).

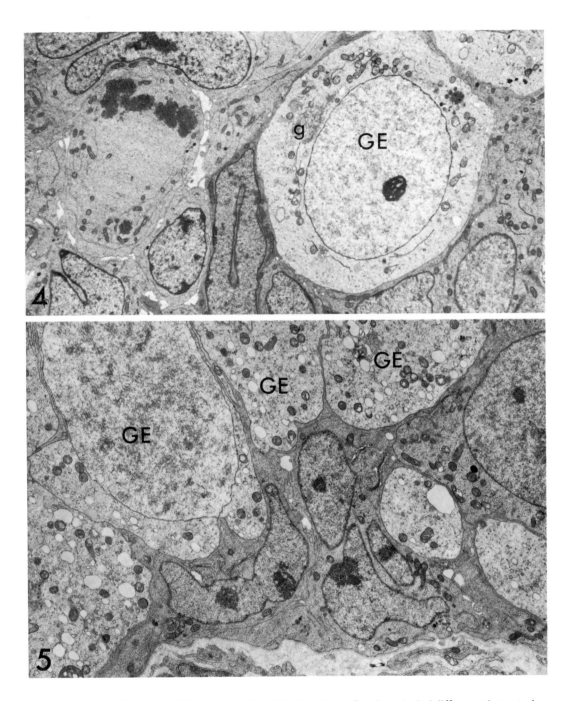

Figs. 2.4 and 2.5   Germ cells (GE) of a 4-day-old rabbit. Note the profound cytological differences between the germ cells and the surrounding cells from the germinal epithelium (see text). g, Golgi complex. (Micrograph courtesy of Dr. B. Gondos)   (Fig. 2.4 X 5,200) (Fig. 2.5 X 6,000).

differentiated organelles (Weakley, 1969a; Gondos, 1969a). The primordial germ cells are larger in size, have a round shape, and highly undifferentiated cytological characters (Figs. 2.4 and 2.5). The spheroidal and watery nucleus contains a finely granular and uniformly dispersed chromatin and one or two prominent nucleoli; the cytoplasm has only a limited number of organelles, mostly consisting of sparse mitochondria, a few elongated cisternae of the endoplasmic reticulum, and, occasionally, a small Golgi complex (Fig. 2.4). Fixation of the ovaries by vascular perfusion with glutaraldehyde permits recognition of the primordial germ cells also on the basis of their overall electron density which is much lower than that of the other cells of the germinal epithelium (Figs. 2.4 and 2.5).

B. Oogenesis

Upon arriving at the genital ridges, the primordial germ cells multiply actively giving rise to oogonia which soon leave the surface of the developing ovary by breaking through the basement membrane of the germinal epithelium. The oogonia of the first few generations reach the medullary region of the gonads where, together with epithelial cells which also migrated from the germinal epithelium, they give rise to the so-called medullary cords. These oogonia all degenerate, however, even though some may enter the first meiotic prophase and/or sporadically attempt follicular development (Kingsbury, 1913; Brambell, 1962). The oogonia of subsequent generations, those responsible for the development of the definitive ovarian cortex, leave the germinal epithelium but do not penetrate into the medulla as did those of the previous generations. Instead, they remain in the peripheral, or cortical region of the ovary (Fig. 2.6). Here they multiply actively, increasing rapidly in number. The fine morphology of the oogonia is essentially identical to that of the primordial germ cells (Odor and Blandau, 1969) and, thus, the character of these cells need not be described. However, it is important to mention that, like all mitotic cells, dividing oogonia consistently exhibit a pair of centrioles at either pole of their mitotic spindles (Fig. 2.7). This observation is of interest since, shortly after the end of mitotic proliferation, the centrioles apparently disappear, leaving the oocytes without centrioles during the meiotic divisions (Hertig and Adams, 1967; Zamboni, 1970); until the time of fertilization when typical centriolar structures are introduced into the activated egg by the spermatozoon (Stefanini, Oura, and Zamboni, 1969, 1970).

At the end of mitotic proliferation, the oogonia enter meiotic prophase (Figs. 2.8, 2.9 and 2.11) and differentiate into oocytes. Meiosis proceeds only through diplotene, at which time it is arrested, to be resumed in the adult ovary shortly before ovulation *(vide infra)*.

The fine morphology of meiotic oocytes has been studied in the rat (Franchi and Mandl, 1962), rabbit (Zamboni and Gondos, unpublished), cow (Baker and Franchi, 1967a), and human (Baker and Franchi, 1967b) and found to be essentially identical in these species. Prominent meiotic changes occur only in the nucleus, the organization of the cytoplasm remaining essentially unchanged throughout meiotic

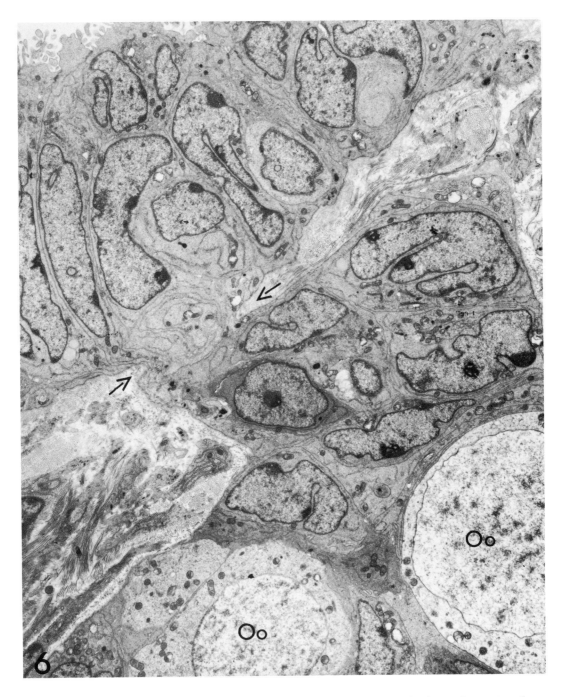

Fig. 2.6 Oogonia (Oo) and accompanying cells from the germinal epithelium in the ovarian cortex of an 11-day-old rabbit. The cells have streamed down from the surface of the ovary by breaking through the basement membrane (arrows). (Micrograph courtesy of Dr. B. Gondos) (X 4,900).

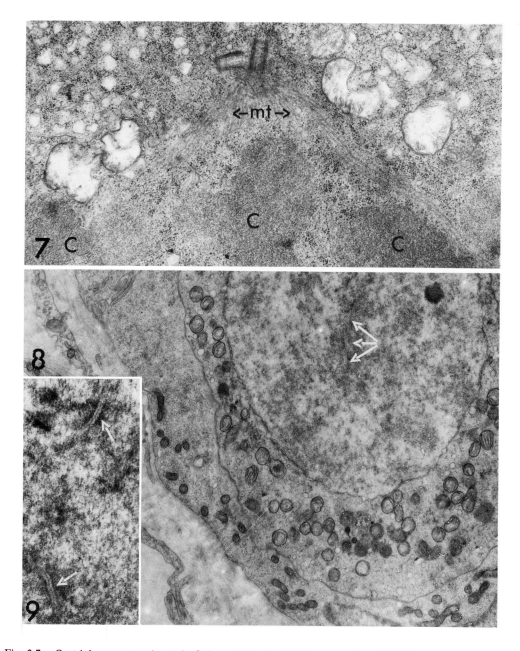

Fig. 2.7 Centrioles at one polar end of the spindle of a dividing oogonium in developing rabbit ovary. C, chromosomes. mt, microtubules (X 21,000).

Fig. 2.8 A meiotic oocyte at pachytene showing condensation of chromatin material around synaptinemal complexes (arrows). 11-day-old rabbit (X 7,300).

Fig. 2.9 Synaptinemal complexes (arrows) in the nucleus of meiotic oocyte at pachytene. 10-day-old rabbit (X 13,900).

prophase and being similar to that of oogonia. At pre-leptotene and leptotene, the chromatin condenses into isolated chromosomes which possess thin electron dense cores. At zygotene, the chromosomal patterns become much more evident and are frequently polarized in one nuclear hemisphere. The chromosomal cores are either single, as in leptotene, or associated in pairs, the latter form indicating that pairing of homologous chromosomes has begun. At pachytene, there is the transformation of the paired chromosomal cores into tripartite ribbons consisting of three linear, parallel structures separated from one another by translucent areas (arrows, Fig. 2.8 and 2.9). These structures, referred to as synaptinemal or axial complexes and first described by Moses (1956) in crayfish spermatocytes, indicate the fully synapsed condition of homologous chromosomes, and represent the core of each bivalent. Diplotene is characterized by reappearance of unpaired chromosomes similar to those in pre-leptotene and leptotene. In oocytes of the rat and rabbit in the resting stage, these chromosomes disappear and the chromatin assumes a "dictyate" form similar to that of interphase nuclei. In human oocytes, the diplotene chromosomal patterns persist throughout the length of the resting period.

The mitotic proliferation of the oogonia and the meiotic differentiation of the oocytes are accompanied by extensive, but not total loss of the germ cells (Fig. 2.10). In the rat, for example, four different waves of degeneration have been described (Beaumont and Mandl, 1962), the first and second affecting resting and dividing oogonia, the third oocytes in pachytene, and the fourth oocytes in diplotene. As a result, the number of normal germ cells, which had reached a peak around the seventeenth day of embryonic development, falls to its lowest value by the second day after birth.

The concept orginally proposed by Waldeyer (1870) and subsequently confirmed by the studies of de Winiwarter (1901, 1920) and de Winiwarter and Sainmont (1908, 1909), that oogenesis is completed at the end of embryonic development and that the ovary at birth has a finite stock of oocytes has been challenged in the past by some investigators who postulated that the definitive germ cells of the adult ovary are formed after birth and that oogonial differentiation continues throughout reproductive life (Kingery, 1917; Allen, 1923; Evans and Swezy, 1931). The massive degeneration of germ cells considered above was taken by these investigators as supporting evidence for their claim. Their hypothesis, however, had to be abandoned later since it proved inconsistent with the demonstration that in many species the total number of oocytes declines with age (see Franchi, Mandl and Zuckerman, 1962 for a review), and that *de novo* formation of oogonia does not occur in ovaries which have been experimentally depleted of germ cells (Zuckerman, 1951, 1956). Thus, in spite of the massive degeneration during the late stages of oogenesis, the number of oocytes left in the mammalian ovary at birth is sufficient to insure the normal reproductive activity of the adult animal.

A relatively high degree of synchronization appears to be a general feature of mammalian oogenesis. This is particularly evident in the rat (Beaumont and Mandl,

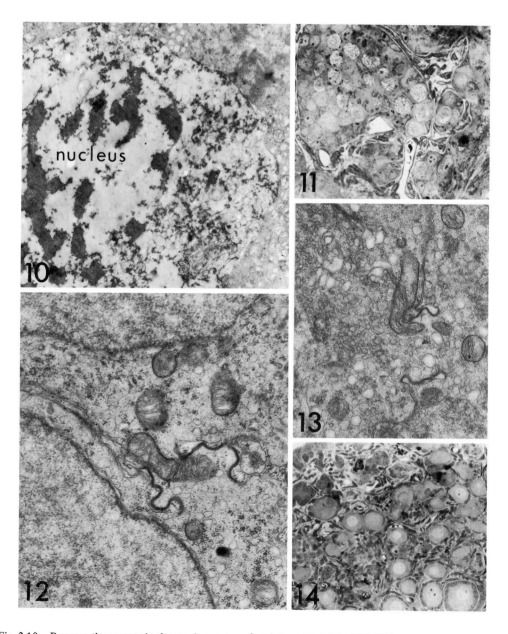

Fig. 2.10   Degenerating oocyte in the ovarian cortex of a 13-day-old rabbit   (X 7,900).

Fig. 2.11   A group of oocytes at the same stage of meiosis. Ovarian cortex of a 4-day-old rabbit   (X 450).

Figs. 2.12 and 2.13   Cytoplasmic bridges connecting germ cells in the ovaries of 5-day-old (Fig. 2.12) and 10-day-old (Fig. 2.13) rabbits. Through these bridges, cytoplasmic organelles appear to be exchanged from one cell to the other   (Fig. 2.12 X 21,300) (Fig 2.13 X 17,700).

Fig. 2.14   Ovarian cortex of a 19-day-old rabbit showing uniform population of unilaminar follicles   (X 450).

1962; Franchi and Mandl, 1962), mouse (Borum, 1961), guinea pig (Ioannou, 1964), hamster (Weakley, 1967), and rabbit (Peters, Levy and Crone, 1965; Teplitz and Ohno, 1963; Zamboni and Gondos, 1968; Gondos and Zamboni, 1969) where there is direct correspondence between stage of development and predominant stage (mitotic, meiotic, and degenerative) of cellular activity. Weakley (1967), Zamboni and Gondos (1968), and Gondos and Zamboni (1969) have recently shown that the synchronization of oogenesis is of a higher degree still since the germ cells mature in groups (Fig. 2.11), each of which consists of elements at an identical stage of differentiation. This synchronization is made possible by the presence of cytoplasmic bridges connecting the cells in each group to one another and bringing about a syncytial organization. These bridges (Figs. 2.12 and 2.13) are a common feature of mammalian oogensis having been observed in the rabbit (Zamboni and Gondos, 1968; Gondos and Zamboni, 1969), hamster (Weakley, 1967), rat (Franchi and Mandl, 1962), mouse (Odor and Blandau, 1969; Ruby, Dyer and Skalko, 1969), and human (Stegner, 1967). The intercellular bridges appear as cylindrical portions of cytoplasm limited by a plasma membrane which is thicker and more electron dense than the remainder of the cell membrane (Figs. 2.12 and 2.13). The cytoplasm of the bridge has a conventional population of organelles which appear to be freely exchanged from one cell to the other. These organelles consist mostly of mitochondria, elements of the endoplasmic reticulum, and ribosomes (Figs. 2.12 and 2.13). The frequency with which these bridges are found in the thin sections of embryonic ovarian tissue for electron microscopy indicates that their actual number is very high and that they form a network of intercellular connections which results in the organization of the germ cells into multiple syncytial groups. It has been proposed that the development of such a network is the result of a sequence of mitotic divisions characterized by incomplete cell separation and that the exchange of material and information between the connected cells contributes to the synchronous pattern of cell differentiation through mitosis and meiotic prophase (Zamboni and Gondos, 1968; Gondos and Zamboni, 1969), an hypothesis analogous to that formulated for the intercellular bridges between spermatogenetic cells in the seminiferous tubules (Fawcett, 1961), which are also organized as syncytia comprised of elements of identical age and stage of differentiation.

The oogonia which leave the surface of the ovary are accompanied by large numbers of other cells from the germinal epithelium (Okamoto, 1928; Brambell, 1962; Franchi, Mandl and Zuckerman, 1962). These cells, which are the precursors of the future follicle cells, closely surround groups of oogonia and oocytes (Figs. 2.6 and 2.8) from which they can easily be distinguished on the basis of their morphological characteristics. As seen with the electron microscope (Gondos, 1969b), they are smaller than the germ cells, and highly irregular in shape with profoundly indented nuclei and coarsely aggregated chromatin (Fig. 2.6). Their cytoplasmic organization is more differentiated than that of the germ cells: the endoplasmic reticulum and the Golgi complex are well developed and the mitochondria are numerous. Ribosomes and lipid droplets, signs of pronounced cellular activity, are also abundant.

The development of the unilaminar follicles and their final organization, i.e., a centrally located oocyte surrounded by a layer of flattened follicle cells (Fig. 2.14) is brought about by the interaction of several factors:

The disappearance of the intercellular bridges at the end of meiotic prophase leads to the isolation of individual oocytes which are no longer organized in syncytial groups.

The numerical ratio between follicle cells and oocytes makes it possible for each oocyte to become surrounded by several follicle cells.

The cellular and extracellular components of the ovarian stroma increase and grow between individual oocytes and surrounding follicle cells bringing about the isolation of the primary follicles and contributing to their typical spheroidal shape.

The importance of the above process for normal follicle development is demonstrated by the fact that alterations in these mechanisms may result in conspicuous abnormalities. The persistence of the cytoplasmic bridges, for example, may prevent separation of oocytes prior to the time they become surrounded by the follicle cells. This anomaly usually leads to the development of either polyovular follicles (Fig. 2.15) or, if followed by cytoplasmic fusion of conjoined cells, of polynucleated oocytes (Fig. 2.16).

The fine morphological characteristics of the components of the newly formed primary follicles are essentially identical to those of quiescent primary follicles in the adult ovary (Lanzavecchia and Mangioni, 1964; Weakley, 1966, 1967; Odor and Blandau, 1969). Thus, the description to be given in the following section applies also to the primary follicles in the ovarian cortex at the end of embryonic development.

III. POST-EMBRYONIC PHASE

The post-embryonic life of the mammalian oocyte consists of a long period of quiescence that spans the interval separating birth from sexual maturity, and a short maturative period, which precedes either ovulation or atresia *(vide infra)*.

During sexual immaturity, the ovarian cortex is crowded by numerous unilaminar follicles which are separated from one another by a thick connective tissue stroma consisting of fibroblasts, blood vessels and bundles of collagen and reticular fibers. During the period of sexual maturity, quiescent follicles and follicles in various stages of growth can be seen next to one another (Fig. 2.17). This provides an opportunity to study in detail the phenomena which accompany follicular growth and oocyte maturation. The fine morphological features of these phenomena have been investigated in follicular oocytes of the mouse (Yamada, Muta, Motomura and Koga, 1957; Wischnitzer, 1967; Weakley, 1968; Odor and Renninger, 1960; Szollosi, 1967; Weakley, 1968), hamster (Odor, 1965; Hadek, 1966; Szollosi, 1967; Weakley, 1968, 1969b), guinea pig (Adams and Hertig, 1964; Weakley, 1968), rabbit (Blanchette, 1961; Hadek, 1963, 1964; Zamboni and

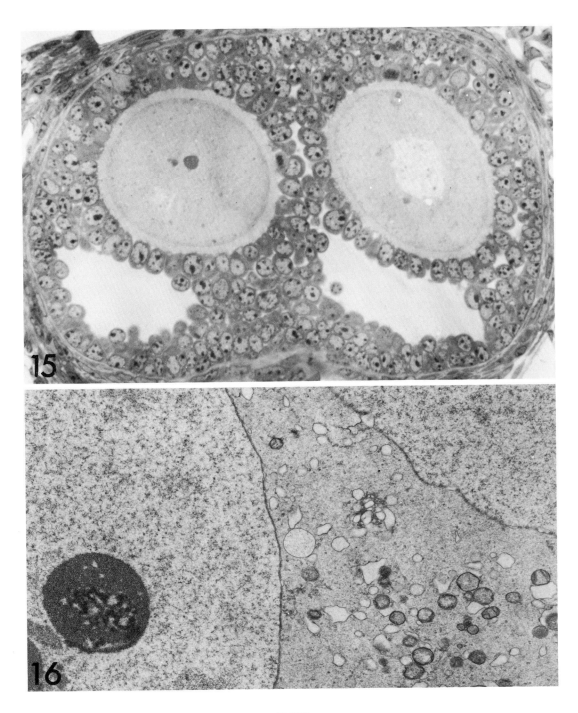

Fig. 2.15   Polyovular follicle in adult mouse ovary   (X 850).

Fig. 2.16   Binucleated oocyte in unilaminar follicle of adult human ovary   (X 13,500).

Mastroianni, 1966a), cat (Weakley, 1968), monkey (Hope, 1965), and human (Wartenberg and Stegner, 1960; Stegner and Wartenberg, 1961a, b; Adams and Hertig, 1965; Hertig and Adams, 1967; Baca and Zamboni, 1967; Hertig, 1968).

A. Period of Quiescence

During this period, no salient morphological changes can readily be demonstrated in the mammalian oocyte even though the cell is biochemically active (Oakberg, 1967; Baker, Beaumont and Franchi, 1969).

The quiescent oocyte in the center of the small follicle is surrounded by a single layer of flattened follicle cells resting on a thin basement membrane (Figs. 2.14, 2.18 and 2.19). The oocyte at this stage is a spherical cell which measures 50 to 70 μ in diameter. The cell outline is regular and the plasma membrane is smooth over most of the cell surface and is closely apposed to the cell membranes of the surrounding follicle cells (Figs. 2.18 and 2.19). Oocyte and follicle cells are zonally bound by intercellular junctions characterized by focally increased thickness and electron opacity of the two membranes. Occasionally, the very narrow intercellular slit separating the oocyte from the follicle cells may be irregularly dilated and occupied by a few microvillar extensions of the two cells.

The oocyte nucleus is large and of an overall spheroidal shape (Figs. 2.18 and 2.19). One or several nucleoli are usually apparent (Fig. 2.18). These are prominent and of the reticular type with the finely granular nucleolonema organized into anastomosing strands. As mentioned in the previous section, the organization of the chromatin varies among quiescent oocytes of different species. The "dictyate" nuclei of rodent oocytes show a uniform dispersion of the chromatin throughout the nucleoplasm, while the nuclei of primate oocytes are characterized by some degree of chromatin condensation into faint chromosomal patterns.

The structural organization of the cytoplasm is relatively simple. The majority of the organelles are clustered in a limited region around the nucleus (Figs. 2.18, 2.19, 2.22 and 2.23), probably as a consequence of the nuclear polarization of most of the oocyte metabolic activity at this stage. This concentration of organelles is evident even in histological preparations where the oocyte nucleus appears surrounded by a crescent-shaped body referred to in the past as Balbiani vitelline body (Raven, 1961) or yolk nucleus complex (Beams and Sheehan, 1941). Electron microscope studies (Hertig, 1968; Baca and Zamboni, 1967) have shown that the perinuclear organelles consist mostly of closely packed mitochondria, elements of the endoplasmic reticulum, lysosomes, a prominent Golgi complex, and, in human oocytes exclusively, annulate lamellae (Fig. 2.19).

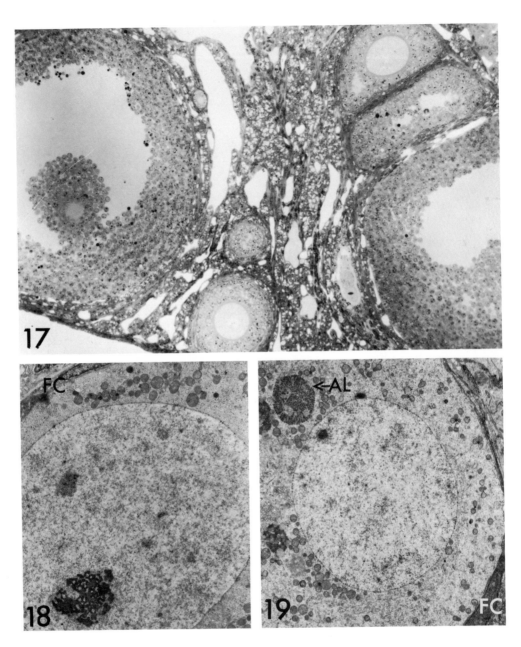

Fig. 2.17   Cortex of adult mouse ovary showing follicles in various stages of maturation   (X 200).

Fig. 2.18   Rabbit oocyte in unilaminar follicle. Note the vesicular nucleus and the prominent nucleolus. FC, follicle cells   (X 3,700).

Fig. 2.19   Human oocyte in unilaminar follicle. Most of the organelles are polarized in the perinuclear region. AL, annulate lamellae   (X 2,600).

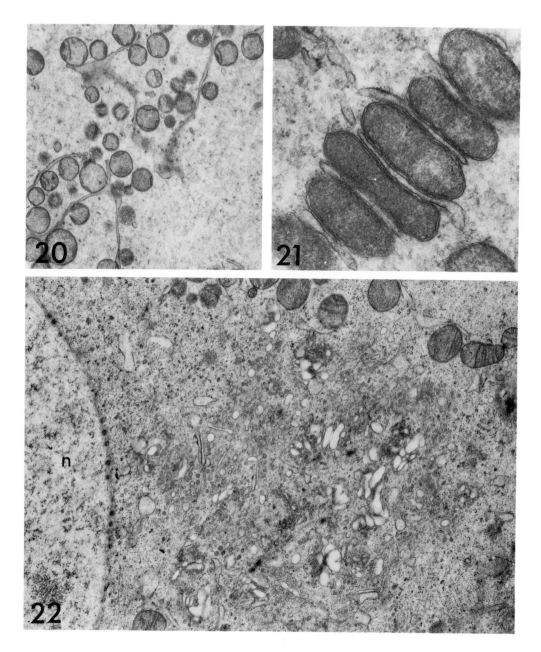

Figs. 2.20 and 2.21   These micrographs demonstrate the close association between mitochondria and cisternae of the endoplasmic reticulum in human (Fig. 2.20) and *Macaca mulatta* (Fig. 2.21) oocytes during the quiescent stage   (Fig. 2.20 X 11,300) (Fig. 2.21 X 22,600).

Fig. 2.22   The prominent Golgi complex of a quiescent oocyte of the rabbit. The complex consists of numerous microvesicles and microtubules which occupy an extensive area next to the nucleus (n). Notice also the close topographic association between mitochondria and endoplasmic reticulum cisternae   (X 7,500).

The mitochondria (Figs. 2.20, 2.21, 2.22) are spheroidal or slightly elongated, and provided with a limited number of cristae which either arch through the matrix or run parallel to the outer mitochondrial membranes (Adams and Hertig, 1964; Hope, 1965; Blanchette, 1961; Zamboni and Mastroianni, 1966; Baca and Zamboni, 1967). Individual mitochondria are closely surrounded by flattened and elongated cisternae of the rough endoplasmic reticulum (Figs. 2.20, 2.21, 2.22). This association is particularly pronounced in bovine oocytes (Senger and Saacke, 1970) where mitochondria are frequently hood- or basket-shaped and contain endoplasmic reticulum vesicles in the concavity delimited by their appendages.

With the exception of the annulate lamellae *(vide infra),* the Golgi complex (Figs. 2.22, 2.23 and 2.24) is the most prominent cytoplasmic component of the mammalian oocyte at this stage, not infrequently being so extended as to occupy up to two-thirds of the sectioned surface of the cell (Zamboni and Mastroianni, 1966a). Although the basic organization of the Golgi apparatus is essentially identical in oocytes of all the species studied, some differences may be noted. In the rabbit (Zamboni and Mastroianni, 1966a) and guinea pig (Adams and Hertig, 1964), it appears as an extensive collection of microtubules and microvesicles either devoid of any content, or containing sparse floccular material of low electron opacity (Fig. 2.22). In monkey oocytes (Zamboni, unpublished), the Golgi complex consists exclusively of elongated tubules with a relatively dense intraluminal content. At the periphery of the complex, these elements are associated with the elongated cisternae of the endoplasmic reticulum indicating that the two systems may be structurally and/or functionally continuous, as they frequently are in somatic cells (Fawcett, 1966). In human oocytes the apparatus is frequently associated with dense and straight fibrous or lamellar elements (Figs. 2.23 and 2.24) whose nature has yet to be determined (Hertig, 1968; Zamboni, unpublished). It is difficult to interpret the structural differences in the organization of the Golgi complex among oocytes of different species: they could be due to different stages of cellular activity, or they may represent actual species differences. A more exact understanding of the problem requires further studies which would have to be performed at equivalent stages of oocyte development and with standardized technical procedures.

Annulate lamellae are consistently present in the cytoplasm of human oocytes (Tardini, Vitali-Mazza and Mansani, 1960; Wartenberg and Stegner, 1960; Adams and Hertig, 1965; Hertig and Adams, 1967; Baker and Franchi, 1967; Baca and Zamboni, 1967). The presence of annulate lamellae in mammalian oocytes other than the human has been observed only by Weakley (1969b) in an atretic oocyte of the golden hamster. They have also been seen in the rabbit and human fertilized ova (Zamboni and Mastroianni, 1966b; Zamboni, Mishell, Bell and Baca, 1966). Although their number varies considerably, these structures frequently are so developed as to form stacks of up to 100 parallel, paired membranes limiting flattened cisternae 30 to 50 mμ in width (Figs. 2.25 and 2.26). The membranes of each unit are apposed against, or fused with one another at regularly spaced intervals. In cross section, the areas of apposition of the membranes appear as sites of increased electron density (Figs. 2.25 and 2.26) whereas in lamellae sectioned

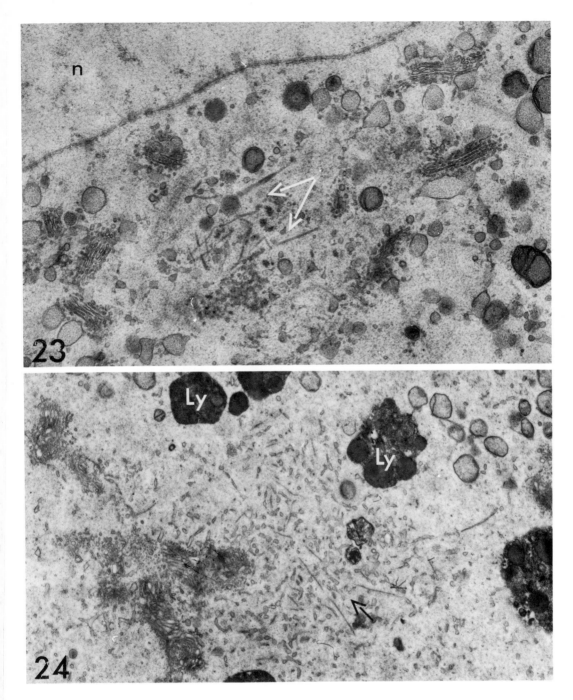

Figs. 2.23 and 2.24   The elements of the Golgi complexes of human quiescent oocytes are frequently associated with straight lamellae or fibrous structures (arrows) whose nature has not been determined. n, nucleus. ly, lysosomes   (Fig. 2.23 X 13,200) (Fig. 2.24 X 11,600).

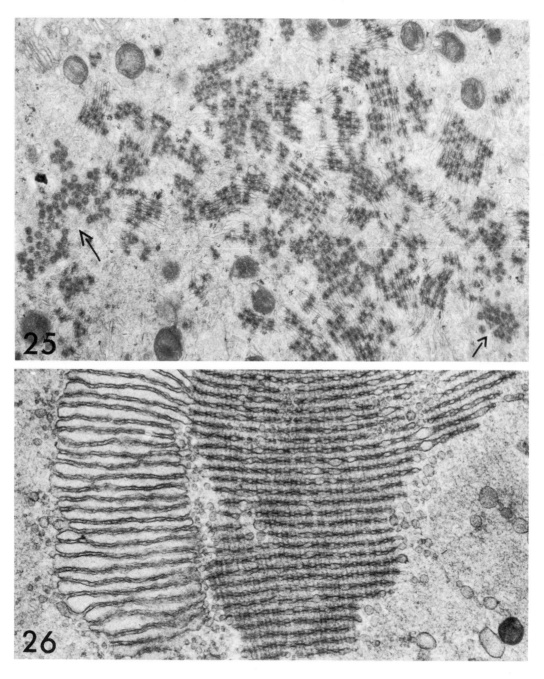

Figs. 2.25 and 2.26 Arrays of annulate lamellae in the cytoplasm of human quiescent oocytes. Arrows in Fig. 2.25 point to annuli of lamellae which have been cut tangentially. The stacks of membranes at the left of the annulate lamellae in Fig. 2.26 may represent an early stage of differentiation of these structures (Fig. 2.25 X 22,300) (Fig. 2.26 X 19,200).

Thus the study of oogenesis falls into three main areas:

(1) The mechanism of meiosis;

(2) Developmental aspects, concerned with the storage of information and the provision of a cellular vehicle and nutrients necessary for the development of a future individual;

(3) The endocrine regulation of oocyte maturation and release.

These three areas have been studied extensively in the past, but there has been a tendency for these topics to be studied independently with little exchange of information and ideas between investigators in the three areas. The last definitive reviews in the three areas were put together about ten years ago by Rhoades (1961) on meiosis, Raven (1961) on oogenesis and development and Zuckerman (1962) on the endocrine regulation. Many advances in the study of oogenesis and development have been made in the last decade because of the impact of cell and molecular biology. Often these advances have been facilitated by studying invertebrate and non-mammalian vertebrates, and only recently have the findings impinged on the study of oogenesis in mammals. It is now clear that the processes in mammals (including man) and lower organisms are similar in many ways.

A thorough understanding of oogenesis is of basic importance to man in several areas. These include the treatment of infertility in man and domestic animals, the control of excessive fertility, and the understanding of the effects of drugs on early development. Three questions need to be answered at the present time to guide future research in the applied fields, particularly those concerned with human fertility:

(1) How relevant are the discoveries on oogenesis in lower forms to studies of oogenesis in mammals?

(2) If the work on lower forms is relevant, what are the most useful species to study?

(3) If mammalian forms are uniquely different from the lower forms now being studied, how should mammalian oogenesis be investigated, and what new techniques need to be developed?

This Symposium was organized as a first step in the solution of these questions. Its purpose was to provide an up-to-date account of oogenesis in the several areas, and to encourage the exchange of ideas between the scientists involved in the three main areas of study. Major emphasis is given to the comparative approach, and to aspects of oogenesis not generally considered. The topics were chosen to stress the viewpoint that oogenesis is a major segment in the continuum of life, and that the oocyte is specialized by a process of differentiation to (1) undergo meiosis, (2) participate in cell fusion, and (3) accumulate and provide materials which govern the nature and nurture of the new individual produced when the ovum is fertilized. Although the questions have not been clearly answered, we hope the information gathered in this volume will help further discussion and promote useful work.

REFERENCES

Davidson, E. H. (1968) Gene activity in early development. New York, Academic Press.

Hersh, R. T. (1969) Mitochondrial genetics: A conjecture. *Science.* **166**, 402.

Raven, C. P. (1961) Oogenesis: The storage of developmental information. New York, Pergamon Press.

Rhoades, M. M. (1961) Meiosis. *In* "The cell". (J. Brachet and A. E. Mirsky, eds.) Vol. 3. New York, Academic Press.

Wilson, E. B. (1896) On cleavage and mosaic-work. *Arch. Entwicklungsmech Org.* **3**, 19.

Zuckerman, S. (1962) The Ovary. Vols. I and II. New York, Academic Press.

COMPARATIVE STUDIES ON THE ULTRASTRUCTURE OF MAMMALIAN OOCYTES

Luciano Zamboni

I. Introduction
II. Embryonic phase
    A. Migration of primordial germ cells
    B. Oogenesis
III. Post-embryonic phase
    A. Period of quiescence
    B. Period of maturation
        1. Follicle cell changes
        2. Oocyte changes
IV. Acknowledgments
V. References

I. INTRODUCTION

The series of complex events that render the female germ cell capable of performing its ultimate function, conjugation with the spermatozoon, evolves over a long period which extends from the stage of oogonia differentiation in the embryonic ovary to the final maturation of the egg in the lumen of the Fallopian tube. The length of this period varies considerably among different species and for individual oocytes in the same animal. It is obviously shorter in those animals which have a brief gestation and attain rapid sexual maturity than in those, like primates, where both the gestation and pre-puberal periods are long. The life span is shortest for those oocytes which are ovulated early after the animal attains sexual maturity, and longest for those which are liberated from the follicles toward the end of the fertile period, shortly before the onset of menopause. In the human female, for example, the life span of individual germ cells varies from a minimum of about 12 years to a maximum of several decades.

The process of differentiation and maturation of the female germ cell may be divided into an embryonic and postnatal phase, each consisting of periods of enhanced activity followed by intervals of quiescence. Only the active periods are suited for morphological studies, the presently available methods of morphological investigation being totally inadequate to provide information on the discrete, but not necessarily unimportant events that take place during the quiescent periods.

II. THE EMBRYONIC PHASE

The most salient phenomena of this phase are the migration of primordial germ cells to the genital ridges, and the process of oogenesis.

A. Migration of Primordial Germ Cells

The primordial germ cells, first recognizable in pre-somite embryos, arise in the entoderm of the yolk-sac stalk, the allantois, and the gut (Celestino da Costa, 1932; Debeyre, 1933; Witschi, 1948; Brambell, 1962). The cells are readily distinguishable because of their large size and round shape, the presence of numerous pseudopodial-like protrusions, and highly undifferentiated cytological characters which make them comparable to the blastomeres of eggs in advanced cleavage stages (Brambell, 1962). From their site of origin, the primordial germ cells migrate, mostly by means of active ameboid movements (Everett, 1943; Chiquoine, 1954) to the genital ridges (arrows, Fig. 2.1) located between the base of the mesentery and the Wolffian duct. With the arrival of the germ cells, the epithelium of the genital ridges, usually referred to as coelomic or germinal epithelium, begins to thicken and to proliferate actively (Fig. 2.2), becoming clearly differentiated from the underlying mesenchyme (Fig. 2.3). Due to the cellular multiplication, the region becomes markedly hypertrophic and soon assumes the form of a cylindrical ridge projecting ventrally into the splanchnocoele (Brambell 1927, 1962).

Upon arriving at the genital ridges, the primordial germ cells undergo morphological changes and lose some of those cytological characteristics, the pseudopodia for example, which had made their identification so easy during migration. At one time, this led some investigators to claim that the primordial germ cells all degenerate after reaching the genital ridges, and that the definitive germ cells later differentiate from a proliferating germinal epithelium (Kingery, 1917; Simkins, 1923, 1928; Hargitt, 1925), an hypothesis which was later abandoned following the acquisition of experimental evidence indicating that the primordial germ cells in the genital ridges not only give rise to all the definite germ cells, but also act as inductors in the development and future organization of the gonads (Bounoure, 1935, 1939; Nieuwkoop, 1949; Mintz, 1959; Zuckerman, 1960).

The primordial germ cells in the coelomic surface of the genital ridges can be easily recognized and distinguished from the germinal epithelium by electron microscopy. The germinal epithelium (Fig. 2.3) consists mostly of columnar cells with highly indented nuclei, coarsely aggregated and frequently marginated chromatin, and a relatively advanced organization of the cytoplasm with well along a tangential plane, they are seen as regularly spaced annuli about 1000 Å in diameter (arrows, Fig. 2.25). The precise origin and functional significance of these structures have not been fully elucidated (see Kessel, 1968, for an extensive review). The annulate lamellae are usually observed in cells characterized by active rates of growth and differentiation such as germ, embryonic, and neoplastic cells and are thought to be derived from the blebbing activity of the two leaflets of the nuclear membrane (Kessel, 1963), which are morphologically very similar to the annulate lamellae, or to represent a specialized form of endoplasmic reticulum (Yamamoto and Onozato 1965). These two hypotheses can easily be reconciled considering that the nuclear envelope is part of the endoplasmic reticulum system. The presently available information on the distribution and origin of the annulate lamellae, however, has contributed but little to our knowledge of the function of these structures which are tentatively considered to be involved in highly specialized synthetic processes of a yet undetermined nature (Kessel, 1968). Our understanding of the functional significance of these structures is rendered even more precarious by the yet unaccounted for absence of annulate lamellae in normal oocytes of most mammalian species.

B. Period Of Maturation

This is a period of relatively short duration which brings an oocyte to maturation and leads to the rupture of the follicle and ovulation. The changes which occur during this period involve follicle cells as well as oocytes.

1. Follicle cell changes. Onset of the stage of maturation is heralded by resumption of the mitotic activity of the follicle cells. The ensuing increase in follicle cell population and the accompanying secretion of *liquor folliculi* are the most important factors in the enlargement of the follicle. The results of autoradiographic studies (Böstrom and Odeblad, 1952; Zachariae, 1957), showing that the elaboration of *liquor folliculi* and zona pellucida constituents is related to the secretory activity of the follicle cells, are confirmed by the observation that the follicle cells during the phase of follicular growth exhibit morphological signs of enhanced secretory activity such as extensive ergastoplasmic networks, prominent Golgi complexes, and the presence of large numbers of cytoplasmic ribosomes. The continuous accumulation of *liquor folliculi* brings about the formation of the follicular antrum, whose further expansion accounts for the separation of the follicle cells closest to the oocyte (the cumulus oophorus) from the remaining follicle cells, as well as for the dislocation of the oocyte and surrounding cumulus which come to occupy an eccentric position in the cavity (Fig. 2.32).

In the past, little consideration has been paid to cystic bodies (Figs. 2.27 and 2.31), originally described by Call and Exner (1875) in rabbit follicles. In previous ultrastructural studies, these bodies have been described only twice, accurately by Motta (1965), and erroneously by Hadek (1969), who presented them as intracellular vacuoles or lacunae. In rabbit (Motta, 1965; Zamboni, unpublished) and human follicles (Zamboni, unpublished), the Call-Exner bodies appear as spheroidal

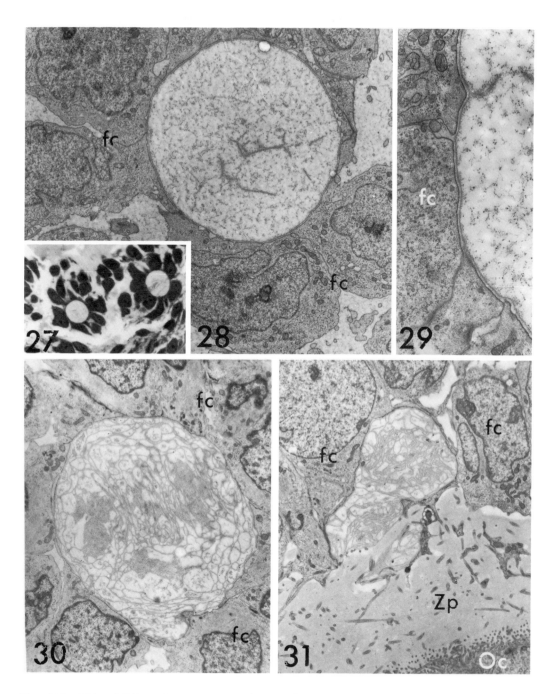

Figs. 2.27 - 2.31 Call-Exner bodies in ovarian follicles of rabbit (Figs. 2.27, 2.28, 2.29) and human ovaries (Figs. 2.30, 2.31). The wall of the Call-Exner body in Fig. 2.31 is open and the content of the cyst appears to be released in the follicular cavity adjacent to the zona pellucida (ZP). fc, follicle cells. Oc, oocyte (Fig. 2.27 X 350) (Fig. 2.28 X 7,500) (Fig. 2.29 X 15,700) (Fig. 2.30 X 3,100) (Fig. 2.31 X 3,900).

cysts surrounded by a corona of follicle cells bound to one another by tight intercellular junctions (Figs. 2.27, 2.28 and 2.30). A basement membrane-like lamina lines the periphery of the cyst running parallel and close to the plasma membrane of the follicle cells (Figs. 2.28, 2.29, 2.30). The lumen of the cyst is occupied by an irregular network of filamentous material, known to be PAS-positive, which in places blends with the basement membrane-like lamina (Figs. 2.28, 2.29, 2.30). It is conceivable that the reticular appearance of the material in the lumen of the Call-Exner bodies is due to precipitation of their content by the fixative, and that in the living state, the bodies contain follicular fluid and zona pellucida precursors which are collected in the cysts prior to release in the follicular antrum. This hypothesis is supported not only by the frequent observation of Call-Exner bodies with interrupted walls in the process of liberating their contents in the follicular cavity (Fig. 31), but also by the absence of Call-Exner bodies in large follicles with fully distended antra.

The formation of the zona pellucida is accompanied by increased surface activity of the cells of the cumulus oophorus resulting in the development of numerous and very long cytoplasmic projections which traverse the zona to reach the surface of the oocyte (Figs. 2.31 and 2.33). The expanded extremities of these projections, which in the living state are likely to be capable of ameboid movements, anchor on the oocyte surface by means of tight intercellular junctions (Figs. 2.34 and 2.35). In addition to this pattern of cumulus oophorus-oocyte relationship, observed in all species, the human oocyte exhibits an additional type of connection (Baca and Zamboni, 1967). This is characterized by profound indentations of the oocyte profile in which long and straight-running protrusions of the cumulus cells are accommodated (Fig. 2.36).

2. Oocyte changes. Oocyte changes have been studied in detail in numerous electron microscopic studies performed on follicular oocytes of a variety of species. No information is available, however, on the ultrastructure of the cytoplasmic and nuclear changes which occur during oocyte maturation *in vitro*. Yet, a knowledge of the phenomena which take place *in vitro* is important in view of recent interest in inducing mammalian oocyte maturation and fertilization in culture. The information to be obtained from studies performed on cultured oocytes would help establish whether the morphological pattern of maturation of oocytes *in vitro* is similar to that *in vivo* and, if not, what are the differences. For this purpose, the following review of the morphological changes of oocytes which mature in the ovarian follicle will be supplemented with a presentation of the fine morphology of human oocytes *in vitro*.*

---

*These oocytes (Zamboni and Smith, unpublished) were obtained from the ovaries of patients in reproductive age who underwent surgery for various gynecological disorders. Follicles of various sizes were punctured under a dissecting microscope, and the oocytes flushed from the follicular cavity and placed under paraffin oil in microdrops of Ham's F10 medium supplemented with 4 mg/ml albumin (Kennedy and Donahue, 1969). The cultures were maintained at 37°C in an atmosphere of 5% CO_2 in air. After 48 hours in culture, the oocytes were removed and processed for electron microscopy.

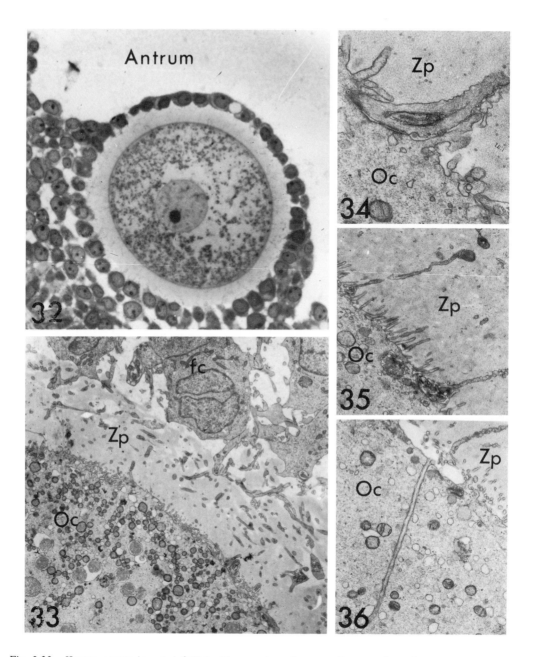

Fig. 2.32   Human oocyte in antral follicle. The oocyte and surrounding cumulus cells occupy an eccentric position in the follicular cavity. Note the diffuse distribution of the organelles throughout the oocyte cytoplasm   (X 500).

Figs. 2.33 - 2.36   The development of the cytoplasmic projections of the follicle cells (fc) and their pattern of attachment to the oocyte (Oc) is shown in micrographs of growing oocytes of the rabbit (Figs. 2.33 and 2.34), monkey (Fig. 2.35) and woman (Fig. 2.36). For explanation, see text. Zp, Zona pellucida   (Fig. 2.33 X 3,000) (Fig. 2.34 ,X 22,500) (Fig. 2.35 X 15,000) (Fig. 2.36 X 12,000).

The appearance of microvilli is one of the earliest changes to occur in the maturing oocyte (Figs. 2.33-2.37 and 2.41). The microvilli develop rapidly and uniformly over the whole surface of the oocyte but are consistently more numerous and elongated around the areas of contact with the cytoplasmic extensions of the cumulus cells, a phenomenon which testifies to the close functional relationship between the oocyte and the cumulus cells throughout the stage of maturation.

The organelles are no longer polarized in the perinuclear region (Fig. 2.32) and undergo profound structural changes. The mitochondria increase considerably in number and become uniformly distributed throughout the ooplasm while maintaining close association with the components of the endoplasmic reticulum (Fig. 2.37). The elements of the endoplasmic reticulum undergo pronounced development (Adams and Hertig, 1964; Zamboni and Mastroianni, 1966a; Baca and Zamboni, 1967; Weakley, 1968; Zamboni, 1970) and frequently assume the form of very elongated, tortuous, or concentrically arranged cisternae limited by ribosome-studded membranes and filled with highly dense material (Figs. 2.37 and 2.38). Obviously, the augmentation of the endoplasmic reticulum at this stage is functionally related to the high rate of protein synthesis known to take place during oocyte growth and maturation (Alfert, 1950; Hedberg, 1953).

The cytoplasm of rodent oocytes at this stage contains large numbers of fibrillar arrays (Szollosi, 1965; Weakley, 1968; Zamboni, 1970), each consisting of 10 to 20 parallel fibrils separated from one another by a distance of 150 - 200 Å. In cross section, a hexagonal crystalline pattern is evident. These fibrils have been seen to increase in number during oocyte meiotic maturation and are abundant in tubal ova (Zamboni, 1970). On the basis of morphological evidence suggesting their polysomal origin, it has been recently proposed that they may represent lattices of fibrillar RNA (Zamboni, 1970).

The most conspicuous cytoplasmic transformation of the maturing oocyte involves the Golgi complex (Figs. 2.39 and 2.40). As seen in follicular oocytes of the mouse (Zamboni, 1970), rat (Szollosi, 1967), hamster (Szollosi, 1967), guinea pig (Adams and Hertig, 1964), rabbit (Zamboni and Mastroianni, 1966a), and human (Baca and Zamboni, 1967)., and in human oocytes maintained in culture (Zamboni and Smith, unpublished), there is at first subdivision of the Golgi complex into multiple aggregates of vesicular and tubular elements which then migrate to the periphery of the cell (Figs. 2.39 and 2.40) becoming rather uniformly distributed in the cortical cytoplasm. There is evidence (Szollosi, 1967; Zamboni and Mastroianni, 1966a; Baca and Zamboni, 1967; Zamboni, 1970) that these changes are related to the synthesis and formation of those granules (Fig. 2.41) which were first observed by Austin (1956) in hamster eggs and which are commonly referred to as cortical granules in view of their similarity in morphology, distribution, and behavior to those of sea urchin eggs (Moser, 1939; Endo, 1952). Cortical granule synthesis in the mammalian oocyte seems to occur according to the following pattern (Figs. 2.39 and 2.40): the vesicles and tubules of the multiple Golgi complexes first become filled with a dense material and then coalesce to form larger vacuoles whose lumina

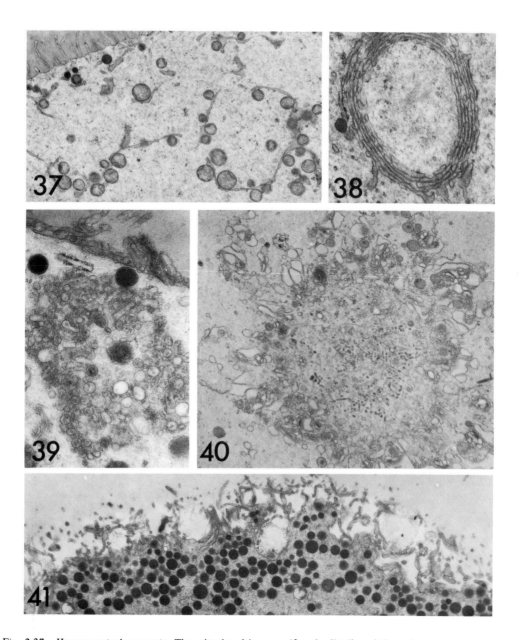

Fig. 2.37   Human maturing oocyte. The mitochondria are uniformly distributed throughout the ooplasm but maintain close association with the cisternae of the endoplasmic reticulum   (X 4,800).

Fig. 2.38   Concentric array of endoplasmic reticulum cisternae in maturing oocyte of the mouse   (X 9,900).

Figs. 2.39 and 2.40   Cortical granule synthesis in the multiple Golgi complexes of mouse (Fig. 2.39) and human oocytes (Fig. 2.40)   (Fig. 2.39 X 19,700) (Fig. 2.40 X 6,900).

Fig. 2.41   Human follicular oocyte after 48 hours in culture. The peripheral cytoplasm is crowded by a large number of cortical granules   (X 6,600).

are occupied by the fused contents of the individual vesicles. Condensation and progressive accumulation of the dense intraluminal content in these vacuoles brings about the formation of mature spheroidal granules 300 - 500 mμ in diameter, which subsequently migrate to the extreme periphery of the cell where they become organized in layers (Fig. 2.41). Cortical granule formation begins early during maturation and, in many species, numerous granules are already present in the cortical cytoplasm long before meiotic division is resumed. Human oocytes behave similarly *in vitro* (Zamboni and Smith, unpublished): they all show a large number of cortical granules and/or active cortical granule synthesis in the multiple Golgi complexes at the periphery of the cell (Fig. 2.41), whether or not they have entered meiotic division. The only known exception is the mouse oocyte where the number of cortical granules at the end of the follicular maturation is much lower than in oocytes of other species; a full complement of cortical granules is attained only at the end of the period of time spent by the unfertilized ovum in the oviduct shortly before fertilization (Zamboni, 1970).

In human oocytes maturing both *in vivo* and *in vitro,* an additional cytoplasmic change is the fragmentation and the disappearance of annulate lamellae in advanced maturative stages.

Condensation of the nucleolus is the first phenomenon to occur in the oocyte nucleus at resumption of meiosis. As seen in the mouse (Zamboni, 1970), rabbit (Zamboni and Mastroianni, 1966a), and human oocytes, the latter both *in vivo* (Baca and Zamboni, 1967) and *in vitro* (Zamboni and Smith, unpublished), the nucleolus loses its reticular pattern and becomes transformed into a highly compact and electron dense body (Figs. 2.42, 2.43 and 2.44) which is progressively reduced in size and disappears simultaneously with the fragmentation of the nuclear membrane. These changes are accompanied by the development of chromosomal patterns (Figs. 2.43 and 2.45) which are liberated in the ooplasm when the nuclear membrane is dismantled.

Metaphase 1 soon follows with the chromosomes becoming arranged on the equatorial plate of the meiotic spindle (Fig. 2.46). As mentioned previously, oocytes lack centrioles and thus divide meiotically in the absence of these structures. In the mouse the polar terminations of the spindle microtubules are associated with clusters of compactly arranged vesicles (Fig. 2.47) which conceivably correspond to the "granules" observed by Blandau (1945) and Odor and Blandau (1951) with the light microscope at the spindle poles of ova undergoing first and second meiotic division. In the rabbit, the spindle poles are associated with tubular profiles (Fig. 2.48) whose morphological characteristics resemble those of the endoplasmic reticulum (Zamboni and Mastroianni, 1966b). How these structures substitute, if they do, for the missing centrioles is an interesting but unsolved problem of cellular physiology.

The subsequent meiotic stages and the formation of the first polar body have been studied ultrastructurally in the mouse (Zamboni, 1970), rat (Odor and

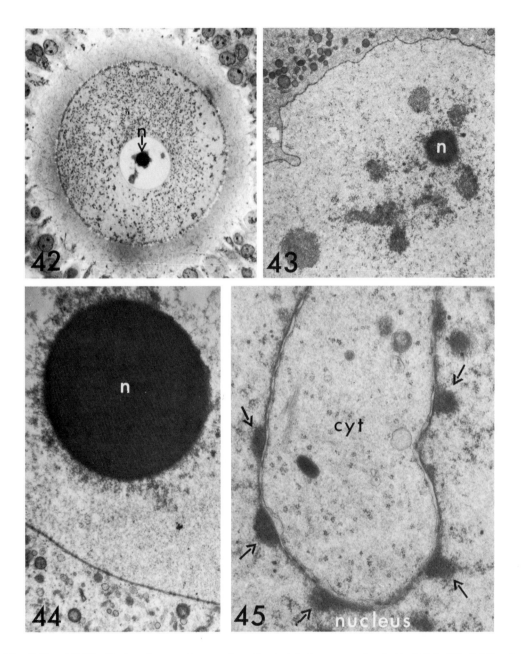

Figs. 2.42 and 2.43   Nuclear changes at resumption of meiosis in human oocytes, in the follicle (Fig. 2.42) and after 48 hours in culture (Fig. 2.43). The condensation of the nucleolus (n) and the development of chromosomal patterns are evident in both micrographs   (Fig. 2.42 X 475) (Fig. 2.43 X 5,500).

Figs. 2.44 and 2.45   Condensation of the nucleolus (Fig. 2.44) and development of chromosomal patterns (arrows, Fig. 2.45) in mouse oocytes at resumption of meiosis. Cyt, cytoplasm   (Fig. 2.44 X 7,000) (Fig. 2.45 X 20,500).

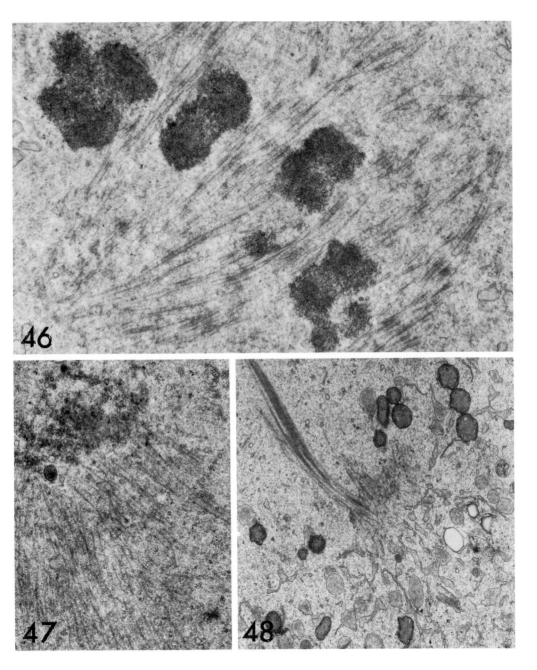

Fig. 2.46   Chromosomes in the metaphase of the first meiotic division in a human oocyte maintained for 48 hours in culture. Note the absence of centrioles at the spindle poles   (X 13,400).

Figs. 2.47 and 2.48   The polar ends of the microtubules of the meiotic spindle are associated with clusters of small vesicles in mouse oocytes (Fig. 2.47), and with elements resembling endoplasmic reticulum, in rabbit oocytes (Fig. 2.48)   (Fig. 2.47 X 16,000) (Fig. 2.48 X 15,300).

Renninger, 1960), rabbit (Zamboni and Mastroianni, 1966a) and human (Baca and Zamboni, 1967) and found to evolve in an essentially identical pattern in the various species. Similar changes also characterize the meiotic maturation of human oocytes *in vitro* (Zamboni and Smith, unpublished). Anaphase is associated with incipient and asymmetric cleavage of the oocyte cytoplasm. Cleavage is completed at telophase and culminates in the liberation into the perivitelline space of a small portion of the cell and of one-half of the chromosomes as a polar body (Figs. 2.49 and 2.50). The morphological characteristics of the polar body are obviously identical to those of the oocyte at the time of its liberation. This is an important point which helps differentiate, whenever necessary, the first polar body from the second. The first polar body, for example, contains microvilli and cortical granules (Figs. 2.49 and 2.51), elements which are absent from the second since it forms following penetration of the sperm into the ovum, i.e., after the liberation of the cortical granules and disappearance of the microvilli (Zamboni, Mishell, Bell and Baca, 1966). Another difference usually noted between first and second polar body concerns the organization of their nuclear complement. In the first polar body, this remains in the form of isolated chromosomes (Figs. 2.51 and 2.52) naked of any nuclear membrane (Odor and Renninger, 1960; Zamboni and Mastroianni, 1966a; Baca and Zamboni, 1967), whereas in the second a nucleus complete with a double membrane envelope is the norm (Zamboni, Mishell, Bell, and Baca, 1966; Stefanini, Ōura, and Zamboni, 1969). In this respect, the nuclear complement of the polar bodies behaves exactly as that of the ovum which remains in a chromosomal form at the end of the first meiotic division, and reverts to a nuclear (pronuclear) organization at the end of the second. The different behavior of the nuclear complement in the two polar bodies is difficult to evaluate. If the disappearance of chromosomal patterns and the reconstitution of the nuclear envelope in the nuclear complement of the egg is induced, or somehow conditioned by the penetration of the spermatozoon, it is not unreasonable to envision that the male gamete may influence also the future behavior of those chromosome halves which will be liberated into the second polar body.

In oocytes matured both *in vivo* and *in vitro,* there is a correlation between the localization of the oocyte chromosomes and the interval of time following polar body liberation (Figs. 2.49 and 2.50). The oocyte chromosomes remain close to the polar body for only a short time following completion of the first meiotic division (Zamboni and Mastroianni, 1966a; Baca and Zamboni, 1967). Then, they move progressively farther away from the polar body region so that, after ovulation, they occupy a position which is usually opposite that of the polar body (Stefanini, Ōura and Zamboni, 1969). The location of the oocyte chromosomes with respect to the polar body could be an important criterion for evaluating whether an oocyte attained maturation *in vitro* or was already mature before being placed in culture.

Completion of the first meiotic division just precedes, or occurs simultaneously with modifications of the relationship between the oocyte and the cells of the cumulus oophorus (Odor, 1960; Sotelo and Porter, 1959; Blandau, 1955; Zamboni and Mastroianni, 1966a; Zamboni, 1970). The cumulus cells pull away from the

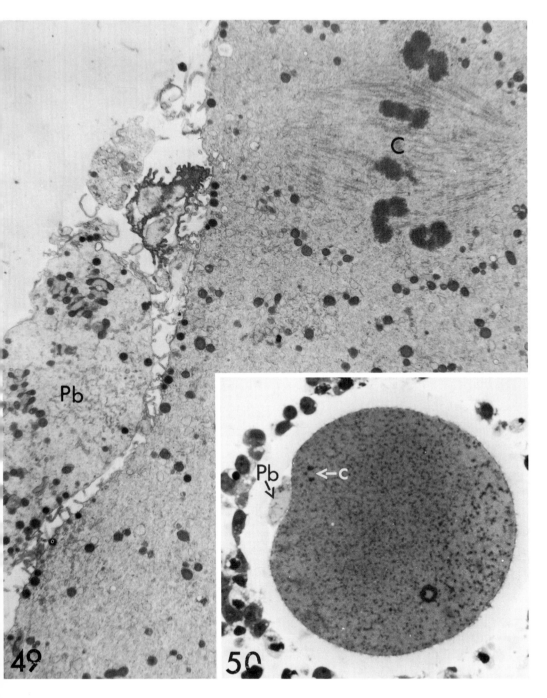

Figs. 2.49 and 2.50   These light and electron micrographs are from adjacent serial sections of a human oocyte found to be meiotically mature after 48 hours *in vitro*. The polar body (Pb) and the oocyte metaphase 2 chromosomes (c) are visible in both micrographs   (Fig. 2.49 X 6,500) (Fig. 2.50 X 570).

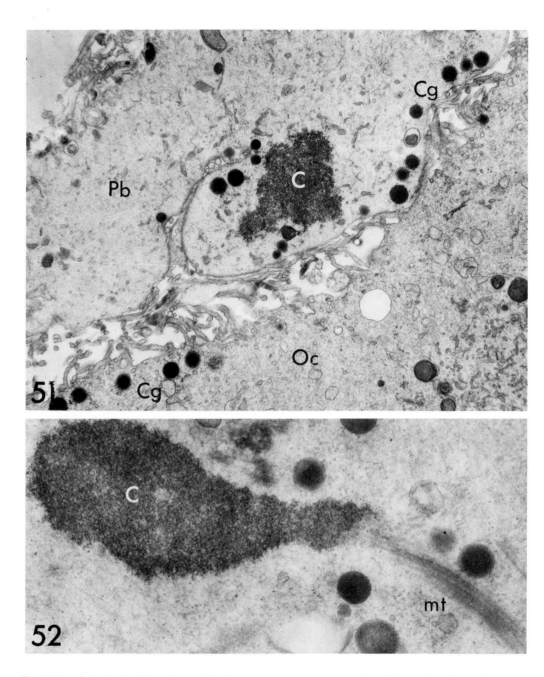

Fig. 2.51   Polar body (Pb) of a meiotically mature human oocyte after 48 hours in culture. Cg, cortical granules. c, chromosome   (X 8,900).

Fig. 2.52   Detail of polar body of a human oocyte after 48 hours in culture. The nuclear complement of the first polar body usually remains in the form of naked chromosomes (see also Fig. 2.51) which are often associated with the residual microtubules of the meiotic spindle (mt)   (X 27,800).

outer aspect of the zona pellucida, as well as from one another, forming a cellular investment which is considerably looser than that in the early stages of maturation. The cytoplasmic processes of the cumulus cells which traversed the zona pellucida to anchor on the oocyte surface decrease in number and then disappear. This change is brought about by retraction (Sotelo and Porter, 1959), as well as by degeneration, as indicated by the frequent presence of globular bodies of degenerated cytoplasm in the perivitelline space of mature oocytes (Zamboni and Mastroianni, 1966a; Zamboni, 1970). These modifications of the oocyte-cumulus cell relationship are a prelude to the process of egg denudation which is completed in the lumen of the fallopian tube after ovulation (Zamboni, 1970). Changes in oocyte-cumulus oophorus relationship similar to, but more pronounced than those described above are also observed when oocytes are maintained in culture. Here, however, it is more difficult to evaluate these changes since maintenance of oocytes *in vitro* brings about dissolution of the cumulus and swift denudation of the oocyte.

It must be remembered that some of the changes which have been described above are frequently the expression of atretic degeneration of the oocyte. It is known, in fact, that atresia may occur at any stage of follicular development and that, when it occurs in oocytes in medium-sized and large follicles, it is preceded by, or accompanied with changes such as completion of the first meiotic division and extrusion of the first polar body which are associated with normal maturation (for reviews see Ingram, 1962; Brambell, 1962; Branca, 1925; for ultrastructural studies on atretic follicles see Burkl and Thiel-Bartosh, 1967; Beltermann, 1965; Vasquez-Nin and Sotelo, 1967). Thus, the possibility that these changes may not represent normal phenomena associated with maturation but may be the expression of incipient atretic degeneration should always be considered. Unfortunately, this consideration is made only too rarely in morphological studies where it is automatically assumed that all the phenomena observed are part of the normal process of maturation of those oocytes which are destined to be ovulated. The same criticism can be made of the vast majority of studies on *in vitro* maturation of mammalian oocytes where the presence of a polar body and the particular configuration of the chromosomes are normally considered to be sufficient parameters to establish that an oocyte has "matured". In these studies, the possibility of dealing with atretic phenomena is not often considered. However, the chances of studying atretic oocytes are very great, especially in those species where only one egg is normally ovulated at each cycle. In the human female, for example, of approximately 500,000 oocytes present in the ovary at birth, only about 400 are ovulated, the remainder (more than 99%) undergoing atresia in various stages of their development.

In the monkey, the percentage of atresia is also very high. A simple histological survey of the cortex of the monkey ovary reveals that the vast majority of the follicles, regardless of their state of maturation, show signs of atretic degeneration (Fig. 2.53). This may help to explain the limited success of past experiments on maturation and fertilization *in vitro* of monkey oocytes and ova, as compared to others in which oocytes from other species were used. It is known, in fact that even

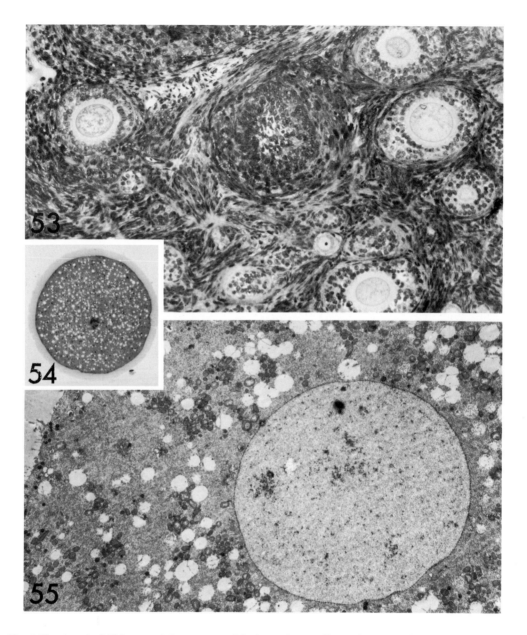

Fig. 2.53   Atretic follicles containing oocytes with clumped organelles in the ovary of a young adult *Macaca mulatta* monkey   (X 150).

Fig. 2.54   Monkey oocyte after 48 hours in culture. The oocyte cytoplasm exhibits diffuse vacuolization, a sign of advanced degeneration. The chromosomes (center) are in anaphase of the first meiotic division   (X 260).

Fig. 2.55   Monkey oocyte after 48 hours in culture. This oocyte shows degenerative signs identical to those of the oocyte in Fig. 2.54. The presence of a vesicular nucleus indicates that this oocyte has remained meiotically inactive throughout the culture period   (X 4,400).

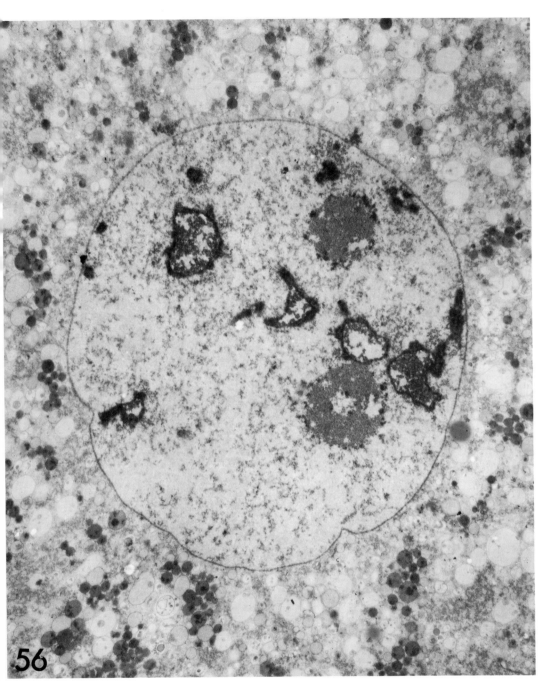

56

Fig. 2.56   Monkey oocyte after 48 hours in culture. The oocyte has not resumed meiosis and exhibits pronounced nuclear and cytoplasmic degeneration   (X 4,500).

though a certain percentage of monkey oocytes may reach meiotic maturation after 24 - 48 hours in culture (Edwards, 1962, 1965; Suzuki and Mastroianni, 1966), activation *in vitro* has not been obtained in this species (Suzuki and Mastroianni, 1968). This failure could be due to the fact that, in the monkey, the vast majority of follicular oocytes are earmarked for atresia and even though some may undergo meiotic maturation in culture, they are unable to proceed to fertilization.

This possibility seems to be supported by preliminary observations recently made in our laboratory on the ultrastructure of *Macaca mulatta* oocytes maintained *in vitro*. The oocytes (Zamboni and Smith, unpublished) were obtained from the ovaries of normally menstruating young adult monkeys at mid-cycle following the same technique described previously for human oocytes. Waymouth's medium supplemented with a 10% volume of monkey serum (Suzuki and Mastroianni, 1966) was used. The vast majority of the oocytes show regressive signs ranging from minute but diffuse vacuolization of the cytoplasm, detectable only by means of high resolution light microscopy (Fig. 2.54) or by electron microscopic examination (Fig. 2.55) to pronounced and widespread degeneration of cytoplasmic and nuclear components (Fig. 2.56). These signs of degeneration, very probably atretic in nature, were seen also in the few oocytes which were found to be meiotically mature at the end of the culture period (Zamboni and Smith, unpublished). These observations seem to indicate that monkey oocytes are less suited than those of other species for the experiments of maturation and fertilization *in vitro,* even though the reason for the higher rate of degeneration of monkey oocytes as compared, for example, to that of human oocytes remains unknown. The above findings also add evidence to the already known fact that incipient stages of atresia are frequently accompanied by the same changes which are the expression of normal oocyte maturation. Hence, the necessity of being extremely cautious in interpreting the morphological changes of oocytes maturing inside and outside follicles can never be overemphasized.

IV. ACKNOWLEDGMENTS

The author expresses his sincere gratitude to Miss Monica Karlsson, who helped in preparing the illustrations, and to Miss Elaine M. Jones, who typed the manuscript. The financial support of the U. S. P. H. S. (Research Grant HD-03341 and Research Contract 69-2220) and of the Ford Foundation is gratefully acknowledged.

V. REFERENCES

Adams, E. C. and Hertig, A. T. (1964) Studies on guinea pig oocytes. I. Electron microscope observations on the development of cytoplasmic organelles in oocytes of primordial and primary follicles. *J. Cell Biol.* **21**, 397.

Adams, E. C. and Hertig, A. T. (1965) Annulate lamellae in human oocytes in primordial and primary follicles. *J. Cell Biol.* **27**, 119A.

Alfert, M. (1950) A cytochemical study of oogenesis and cleavage in the mouse. *J. Cell Physiol.* **36**, 391.

Allen, E. (1923) Ovogenesis during sexual maturity. *Amer. J. Anat.* **31**, 439.

Austin, C. R. (1956) Cortical granules in hamster eggs. *Exp. Cell Res.* **10**, 533.

Baca, M. and Zamboni, L. (1967) The fine structure of human follicular oocytes. *J. Ultrastruct. Res.* **19**, 354.

Baker, T. G., Beaumont, H. M. and Franchi, L. L. (1969) The uptake of tritiated uridine and phenylalanine by the ovaries of rats and monkeys. *J. Cell. Sci.* **4**, 655.

Baker, T. G. and Franchi, L. L. (1967a) The fine structure of chromosomes in bovine primordial oocytes. *J. Reprod. Fertil.* **14**, 511.

Baker, T. G. and Franchi, L. L. (1967b) The fine structure of oogonia and oocytes in human ovaries. *J. Cell. Sci.* **2**, 213.

Beams, H. W. and Sheehan, J. R. (1941) The yolk nucleus complex of the human ovum. *Anat. Rec.* **81**, 545.

Beaumont, H. M. and Mandl, A. M. (1962) A quantitative and cytological study of oogonia and oocytes in the foetal and neonatal rat. *Proc. Roy. Soc. [Biol.]*. **155**, 557.

Beltermann, R. (1965) Elektronenmikroskopische Befunde bei beginnender Follikelatresie im Ovar der Maus. *Arch. Gynak.* **200**, 601.

Blanchette, E. J. (1961) A study of the fine structure of the rabbit primary oocyte. *J. Ultrastruct. Res.* **5**, 349.

Blandau, R. J. (1945) The first maturation division of the rat ovum. *Anat. Rec.* **92**, 449.

Blandau, R. J. (1955) Ovulation in the living albino rat. *Fertil. Steril.* **6**, 391.

Borum, K. (1961) Oogenesis in the mouse. *Exp. Cell Res.* **24**, 495.

Böstrum, H. and Odeblad, E. (1952) Autoradiographic observation on the uptake of S^{35} in the genital organs of the female rat and rabbit after injection of labeled sodium sulphate. *Acta Endocr.* **10**, 89.

Bounoure, L. (1935) Une preuve, experimentale du role du determinant germinal ches la grenouille rousse. *C. R. Acad. Sci., (Paris)*. **201**, 1223.

Bounoure, L. (1939) L'origine des Cellules Reproductrices et le Probléme de la Lignèe Germinale. Paris, Gauthiers-Villars.

Brambell, F. W. R. (1927) The development and morphology of the gonads of the mouse. I. The morphogenesis of the indifferent gonad and of the ovary. *Proc. Roy. Soc. [Biol.]*. **101**, 391.

Brambell, F. W. R. (1962) Ovarian changes. *In* "Marshall's Physiology of Reproduction." (A. S. Parks, ed.), p. 397. London, Longmans, Green & Co.

Branca, A. (1925) L'oocyte atresique et son involution. *Arch. Biol. (Paris).* **35**, 325.

Burkl, W. and Thiel-Bartosh, E. (1967) Elektronenmikroskopische Untersuchungen über die granulosa atresierender Tertiar follikel bei der Ratte. *Arch. Gynak.* **204**, 238.

Call, E. and Exner, S. (1875) Zur Kenntnis des Graafschen Follikels und der Corpus Luteum beim Kaninchen. *Sber. Akad. Wiss. Wien.* **71**, 321.

Celestino da Costa, A. (1932) L'état actuelle du probleme de l'origine des cellules sexuelles. *Bull. Ass. Anat.* **27**, 1.

Chiquoine, A. D. (1954) The identification, origin and migration of the primordial germ cells in the mouse embryo. *Anat. Rec.* **118**, 135.

Debeyre, A. (1933) Sur la presence de gonocytes chez un embryon humain au stade de la ligne primitive. *C. R. Ass. Anat. (Lisbonne).* **28**, 240.

de Winiwarter, H. (1901) Recherches sur l'ovogenese et l'organogenese de l'ovaire des mammiferes (lapin et homme). *Arch. Biol. (Paris).* **17**, 33.

de Winiwarter, H. (1920) Formation de la couche corticale definitive et origine des oeufs definitifs dans l'ovaire de chatte. *C. R. Soc. Biol. (Paris).* **83**, 1403.

de Winiwarter, H. and Sainmont, G. (1908) Uber die ausschliesslich postfötale Bildung der definitiven Eier bei der Katze. *Anat. Aus.* **32**, 613.

de Winiwarter, H. and Sainmont, G. (1909) Nouvelles recherches sur l'ovogenese et organogenese de l'ovaire des mammiferes (chat). *Arch. Biol. (Paris).* **24**, 165.

Edwards, R. G. (1962) Meiosis in ovarian oocytes of adult mammals. *Nature (London).* **196**, 446.

Edwards, R. G. (1965) Maturation *in vitro* of mouse, sheep, cow, pig, rhesus monkey, and human ovarian oocytes. *Nature (London).* **208**, 349.

Endo, Y. (1952) The role of the cortical granules in the formation of the fertilization membrane in eggs from Japanese sea urchins. *Exp. Cell Res.* **3**, 406.

Evans, H. M. and Swezy, O. (1931) Ovogenesis and the normal follicular cycle in adult mammalia. *Mem. Univ. Calif.* **9**, 119.

Everett, N. B. (1943) Observational and experimental evidence relating to the origin and differentiation of the definitive germ cells in mice. *J. Exp. Zool.* **92**, 49.

Fawcett, D. W. (1961) Intercellular bridges. *Exp. Cell Res. Suppl.* **8**, 174.

Fawcett, D. W. (1966) Atlas of Fine Structure. Philadelphia, Saunders.

Franchi, L. L. and Mandl, A. M. (1962) The ultrastructure of oogonia and oocytes in the foetal and neonatal rat. *Proc. Roy. Soc. [Biol.].* **157**, 99.

Franchi, L. L., Mandl, A. M. and Zuckerman, S. (1962) The development of the ovary and the process of oogenesis. *In* "The Ovary." (S. Zuckerman, ed.), Vol. 1, p. 1. New York, Academic Press.

Gondos, B. (1969a) Ultrastructure of the germinal epithelium during oogenesis in the rabbit. *J. Exp. Zool.* **172**, 465.

Gondos, B. (1969b) The ultrastructure of granulosa cells in the newborn rabbit ovary. *Anat. Rec.* **165**, 67.

Gondos, B. and Zamboni, L. (1969) Ovarian development: The functional importance of germ cell interconnections. *Fertil. Steril.* **20**, 176.

Hadek, R. (1963) Submicroscopic study on the cortical granules in the rabbit ovum. *J. Ultrastruct. Res.* **8**, 170.

Hadek, R. (1964) Submicroscope study on the cortical villi in the rabbit ovum. *J. Ultrastruct. Res.* **10**, 58.

Hadek, R. (1966) Cytoplasmic whorls in the golden hamster oocytes. *J. Cell Sci.* **1**, 281.

Hadek, R. (1969) Mammalian fertilization. "An Atlas of Ultrastructure." p. 248. New York, Academic Press.

Hargitt, G. T. (1925) The formation of the sex glands and germ cells of mammals. I. The origin of the germ cells in the albino rat. *J. Morph.* **40**, 517.

Hedberg, E. (1953) The chemical composition of the human ovarian oocyte. *Acta Endocr. (Copenhagen).* **14**, Suppl. 15, 1.

Hertig, A. T. (1968) The primary human oocyte: Some observations of the fine structure of Balbiani's vitelline body and the origin of the annulate lamellae. *Amer. J. Anat.* **122**, 107.

Hertig, A. T. and Adams, E. C. (1967) Studies on the human oocyte and its follicle. Ultrastructural and cytochemical observations on the primordial follicle stage. *J. Cell. Biol.* **34**, 647.

Hope, J. (1965) The fine structure of the developing follicle of the Rhesus ovary. *J. Ultrastruct. Res.* **12**, 592.

Ingram, D. L. (1962) Atresia. *In* "The Ovary". (S. Zuckerman, ed.), Vol. 1, p. 247. New York, Academic Press.

Ioannou, J. M. (1964) Oogenesis in the guinea pig. *J. Embryol. Exp. Morph.* **12**, 673.

Kennedy, J. F. and Donahue, R. P. (1969) Human oocytes: maturation in chemically defined media. *Science.* **164**, 1292.

Kessel, R. G. (1963) Electron microscope studies on the origin of annulate lamellae in oocytes of *Necturus. J. Cell. Biol.* **19**, 391.

Kessel, R. G. (1968) Annulate lamellae. *J. Ultrastruct. Res.* **25**, Suppl. 10.

Kingery, H. M. (1917) Oogenesis in the white mouse. *J. Morph.* **30**, 261.

Kingsbury, B. F. (1913) The morphogenesis of the mammalian ovary. *Felis domestica. Amer. J. Anat.* **16**, 59.

Lanzavecchia, G. and Mangioni, C. (1964) Étude de la structure et des constituants du follicule humain dans l'ovaire foetal. I. Le follicule primordial. *J. Microscopic.* **3**, 447.

Mintz, B. (1959) Continuity of the female germ cell line from embryo to adult. *Arch. Anat. Micro. Morph. Exp.* **48**, 155.

Moser, F. (1939) Studies on cortical layer response to stimulating agents in the *Arbacia* eggs. Response to insemination. *J. Exp. Zool.* **80**, 423.

Moses, M. J. (1956) Chromosomal structure in crayfish spermatocytes. *J. Biophys. Biochem. Cytol.* **2**, 215.

Motta, P. (1965) Sur l'ultrastructure des "Corps de Call et d'Exner" dans l'ovaire du lapin. *Z. Zellforsch.* **68**, 308.

Nieuwkoop, P. D. (1949) The present state of the problem of the "Kiembahn" in the vertebrates. *Experientia.* **5**, 308.

Oakberg, E. F. (1967) ^{3}H-uridine lability of mouse oocytes. *In* "Proceedings of the Colloquium on Physiology and Reproduction in Mammals." *Arch. Anat. Microsc. Morphol. Exp.* **56**, Suppl. 3-4.

Odor, L. D. (1960) Electron microscopic studies on ovarian oocytes and unfertilized tubal ova in the rat. *J. Biophys. Biochem. Cytol.* **7**, 567.

Odor, L. D. (1965) The ultrastructure of unilaminar follicles of the hamster ovary. *Amer. J. Anat.* **116**, 493.

Odor, L. D. and Blandau, R. J. (1951) Observations on the formation of the second polar body in the rat ovum. *Anat. Rec.* **110**, 329.

Odor, L. D. and Blandau, R. J. (1969) Ultrastructural studies on fetal and early postnatal mouse ovaries. II. Cytodifferentiation. *Amer. J. Anat.* **125**, 177.

Odor, L. D. and Renninger, D. R. (1960) Polar body formation in the rat oocyte as observed with the electron microscope. *Anat. Rec.* **137**, 13.

Okamoto, T. (1928) Uber den Ursprung des Follikelepithels des Eierstockes beim Hunde. *Folia Anat. Japon.* **6**, 689.

Peters, H., Levy, E. and Crone, M. (1965) Oogenesis in rabbits. *J. Exp. Zool.* **158**, 169.

Raven, C. P. (1961) Oogenesis. The storage of developmental information. New York, Pergamon Press, McMillan Co.

Ruby, J. R., Dyer, R. F. and Skalko, R. G. (1969) The occurrence of intercellular bridges during oogenesis in the mouse. *J. Morph.* **127**, 307.

Senger, P. L. and Saacke, R. G. (1970) Unusual mitochondria of the bovine oocyte. *J. Cell Biol.* **46**, 405.

Simkins, C. S. (1923) On the origin and migration of the so-called primordial germ cells in the mouse and the rat. *Acta. Zool. Stockholm.* **4**, 241.

Simkins, C. S. (1928) Origin of the sex cells in man. *Amer. J. Anat.* **41**, 249.

Sotelo, J. R. and Porter, K. R. (1959) An electron microscope study of the rat ovum. *J. Biophys. Biochem. Cytol.* **5**, 327.

Stefanini, M., Ōura, C. and Zamboni, L. (1969) Ultrastructure of fertilization in the mouse. 2. Penetration of sperm into the ovum. *J. Submicrosc. Cytol.* **1**, 1.

Stefanini, M., Ōura, C. and Zamboni, L. (1970) The fine morphology of activated mouse ova. *Proc. 7th Inter. Congr. Electron Microsc.* (P. Favard, ed.), **3**, 665.

Stegner, H. E. (1967) Die elektronenmikroskopische Struktur der Eizelle. *Erzbn. Anat. Entwicklungsgesch.* **39**, 6.

Stegner, H. E. and Wartenberg, H. (1961a) Elektronenmikroskopische und histopochemische Untersuchugen uber Struktur und Bildung der Zona pellucida Menschlicher Eizellen. *Z. Zellforsch.* **53**, 702.

Stegner, H. E. and Wartenberg, H. (1961b) Elektronenmikroskopische und histopochemische Befunde an menschlichen Eizellen. *Arch. Gynak.* **196**, 23.

Suzuki, S. and Mastroianni, L., Jr. (1966) Maturation of monkey ovarian follicular oocytes *in vitro. Amer. J. Obst. Gynec.* **96**, 723.

Suzuki, S. and Mastroianni, L., Jr. (1968) The fertilizability of *in vitro* cultured monkey ovarian follicular oocytes. *Fertil. Steril.* **19**, 500.

Szollosi, D. (1965) Development of "yolky substance" in some rodent eggs. *Anat. Rec.* **151**, 424.

Szollosi, D. (1967) Development of cortical granules and the cortical reaction in rat and hamster eggs. *Anat. Rec.* **159**, 431.

Tardini, A., Vitali-Mazza, L. and Mansani, F. E. (1960) Ultrastruttura dell'ovocita umano maturo. 1. Rapporti fra cellule della corona radiata, pellucida ed ovoplasma. *Arch. De Vecchi Anat. Pat.* **33**, 281.

Teplitz, R. and Ohno, S. (1963) Postnatal induction of ovogenesis in the rabbit *(Oryctolagus cuniculus). Exp. Cell Res.* **31**, 183.

Vasquez-Nin, G. H. and Sotelo, J. R. (1967) Electron microscope study of the atretic oocytes of the rat. *Z. Zellforsch.* **80**, 518.

Waldeyer, W. (1870) Eirstock und Ei. Leipzig, Enzelmann.

Wartenberg, H. and Stegner, H. E. (1960) Uber die elektronenmikroskopische Feinstruktur des Menschlichen Ovarialeis. *Z. Zellforsch.* **52**, 450.

Weakley, B. S. (1966) Electron microscopy of the oocyte and granulosa cells in the developing ovarian follicles of the golden hamster *(Mesocricetus auratus). J. Anat.* **100**, 503.

Weakley, B. S. (1967) Light and electron microscopy of developing germ cells and follicle cells in the ovary of the golden hamster: twenty-four hours before birth to eight days *postpartum. J. Anat.* **101**, 435.

Weakley, B. S. (1968) Comparison of cytoplasmic lamellae and membranous elements in the oocytes of five mammalian species. *Z. Zellforsch.* **85**, 109.

Weakley, B. S. (1969a) Differentiation of the surface epithelium of the hamster ovary. An electron microscopic study. *J. Anat.* **105**, 129.

Weakley, B. S. (1969b) Annulate lamellae in the oocyte of the golden hamster. *Z. Zellforsch.* **96**, 29.

Wischnitzer, S. (1967) Intramitochondrial transformations during oocyte maturation in the mouse. *J. Morph.* **121**, 29.

Witschi, E. (1948) Migration of the germ cells of human embryos from the yolk sac to the primitive gonadal folds. *Contr. Embryol. Carnegie Inst.* **32**, 67.

Yamada, E., Muta, T., Motomura, A. and Koga, H. (1957) The fine structure of the oocyte in the mouse ovary studied with electron microscope. *Kurume Med. J.* **4**, 148.

Yamamoto, K. and Onozato, H. (1965) Electron microscope study on the growing oocyte of the goldfish during the first growth phase. *Mem. Fac. Fisheries Hokkaido Univ.* **13**, 79.

Zachariae, F. (1957) Studies on the mechanism of ovulation: autoradiographic investigation on the uptake of radioactive sulfate (S^{35}) into the ovarian follicular mucopolysaccharides. *Acta Endocr.* **26**, 215.

Zamboni, L. (1970) Ultrastructure of mammalian oocytes and ova. *Biol. Reprod.* **2**, Suppl. 2, 44.

Zamboni, L. and Gondos, B. (1968) Intercellular bridges and synchronization of germ cell differentiation during oogenesis in the rabbit. *J. Cell Biol.* **36**, 276.

Zamboni, L. and Mastroianni, L., Jr. (1966a) Electron microscopic studies on rabbit ova. I. The follicular oocyte. *J. Ultrastruct. Res.* **14**, 95.

Zamboni, L. and Mastroianni, L., Jr. (1966b) Electron microscopic studies on rabbit ova. II. The penetrated tubal ovum. *J. Ultrastruct. Res.* **14**, 118.

Zamboni, L., Mishell, D. R., Jr., Bell, J. H. and Baca, M. (1966) Fine structure of the human ovum in the pronuclear stage. *J. Cell Biol.* **30**, 579.

Zuckerman, S. (1951) The number of oocytes in the mature ovary. *Rec. Progr. Hormone Res.* **6**, 63.

Zuckerman, S. (1956) The regenerative capacity of ovarian tissue. *Ciba Found. Coll. Ageing.* **2**, 31.

Zuckerman, S. (1960) Origin and development of oocytes in foetal and mature mammals. *In* "Sex Differentiation and Development". (C. R. Austin, ed.), 63. Mem. Soc. Endocrin. No. 7.

CHANGES OF SOME CELL ORGANELLES DURING OOGENESIS IN MAMMALS

Daniel Szollosi

I.   Introduction
II.   Vitellogenesis
III.   Centrioles
IV.   Mitochondria
V.   Acknowledgments
VI.   References

I. INTRODUCTION

Differentiation of mammalian oocytes has been studied by electron microscopy in various species. Most of these were reviewed and cited by Zamboni in this conference (see also Zamboni, 1970). From these studies it emerged that similar morphological changes take place during oocyte growth and maturation in eggs of most mammals although the details of these changes may be quite different. Instead of a chronological enumeration of the organelles found in oocytes and the description of their developmental changes, I would like to discuss certain problems in mammalian oogenesis--related to changes in cell organelles--which have escaped attention thus far and which may be of general importance. These are: (a) vitellogenesis, (b) centrioles, and (c) mitochondrial development. Most comments made relate to changes observed during oogenesis but in some cases postfertilization stages will be included for the sake of completeness.

II. VITELLOGENESIS

The development of yolk will be considered briefly here since it has been a neglected area in the formation of the mammalian ovum. A discussion is particularly necessary since it has been stated recently in basic textbooks of embryology that eggs of *Eutheria* (placental mammals) were devoid of yolk altogether (Balinsky,

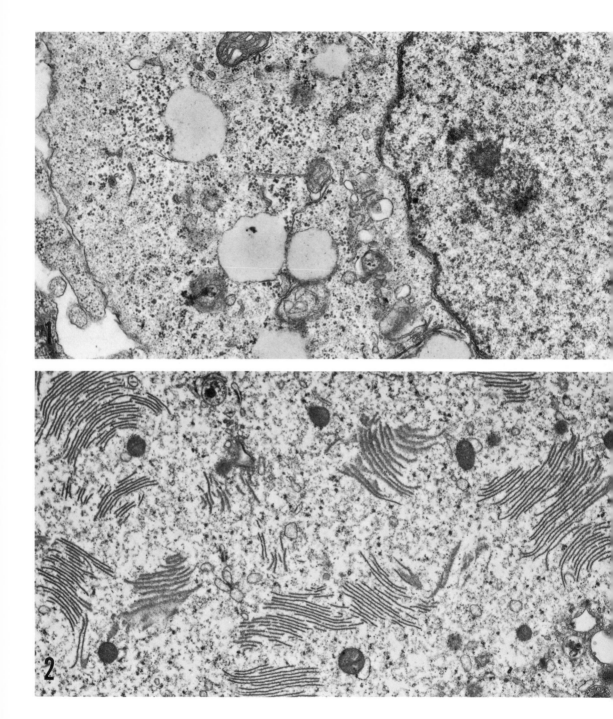

Fig. 3.1 Fat droplets and glycogen rosettes are numerous in a human oogonium   (X 25,300).

Fig. 3.2 In a secondary rat follicular oocyte "yolk plates" are closely packed   (X 10,100).

1965), and also because an inquiring investigator does not find a single entry on yolk in the only comprehensive book on the mammalian egg (Austin, 1961). Because of the varied use of the term "yolk", a definition must be given. It was suitably defined by Waddington (1962): "The reserve foodstuffs in eggs are often collectively referred to as 'yolk', but this is really a loose use of the word. Strictly speaking, the true yolk is only one part of the reserve; it consists of platelets or lumps of a protein-lipoid substance. Besides it, the reserve contains globules of more or less liquid fat, and granules of carbohydrate, which are usually in the form of glycogen."

In the loose sense of the definition, "yolk", that is fat droplets and glycogen granules, definitely exist ubiquitously in mammalian eggs (Fig. 3.1). The balance between the two varies in different eggs. In carnivores, for example, lipid droplets predominate with some glycogen dispersed in the cytoplasm (Szabo, 1967). In eggs of guinea pigs and rabbits, the frequency of lipid droplets decreases (Adams and Hertig, 1964) while in those of the rat, hamster, and several primates, there are relatively few.

Is true yolk to be found in mammalian eggs in the sense of Waddington's definition? The chemical composition of mammalian eggs has not been studied thoroughly. 12.5% of the dry weight of the zona-free mouse egg apparently is lipid (Loewenstein and Cohen, 1964), a percentage corresponding well to the levels contributed by lipid to the protoplasm of various sources (Giese, 1968). From these studies it cannot be said whether the lipid is free in the mouse egg or whether it is in platelets of protein-lipid, a requirement for it to be called yolk according to the accepted definition for this review.

In the cytoplasm of the mouse, rat, deer mouse and hamster eggs aggregations of fibrous material were described with a characteristic linear periodicity (Figs. 3.2, 3.3, 3.4) (Enders and Schlafke, 1965). The arrangement and the periodicity of this material differ slightly in eggs of the four rodents. The fibrous material was described most thoroughly in the hamster (Hadek, 1966; Weakley, 1966, 1967). In the mature rat and hamster eggs, the fibrous material forms large plates of parallel sheaths. Two (in the rat) or more (in the hamster) of the adjacent plates may be cross-linked, forming a prominent periodicity. In the rat the main periodicity measures approximately 35 nm (Fig. 3.3) while in the hamster it is 44.5 nm (Fig. 3.4). The periodicity is very prominent when the plates are sectioned in a grazing manner. If plates within hamster eggs are so sectioned, a second, finer periodicity appears, which measures 19.5 nm and which forms an angle with the more prominent periodicity of approximately 60° (Fig. 3.4). In the mouse eggs a bundle is formed of cross-linked fibrous material, with an approximately similar main periodicity.

The deposition of the fibrous material differs in the rat and hamster. In the rat, multiple stacks of fibrous material appear in cytoplasmic regions of small primary follicles (Fig. 3.5). The development is rapid and no indication can be found as to the source of the stacks. No other cell organelle appears to be involved in its developmental process. From the start, parallel stacks of the plates and the

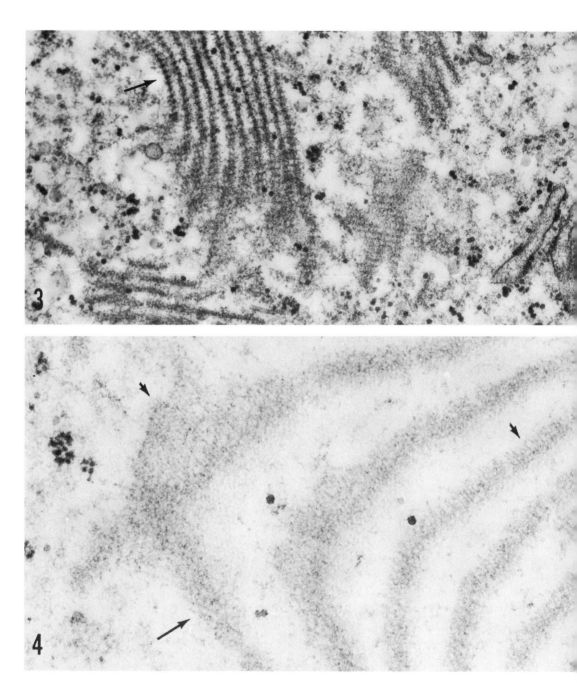

Fig. 3.3 The periodicity is prominent in "yolk plates" of rat oocytes when they are cross sectioned (long arrow) or viewed *en face* (X 48,300).

Fig. 3.4 Two types of periodicity are detected in "yolk plates" of hamster oocytes. The major repeat unit measures 44.5 nm (long arrow) and the smaller one 19.5 nm (short arrows) (X 63,100).

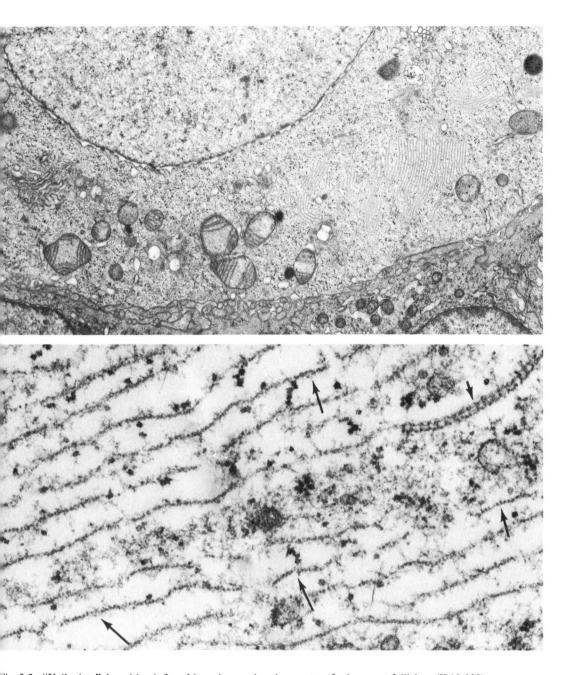

Fig. 3.5 "Yolk plate" deposition is found in various regions in oocytes of primary rat follicles (X 10,000).

Fig. 3.6 "Yolk plates" deposit as single lamellae in hamster oocytes. A faint periodicity of 19.5 nm is seen on the lamellae (long arrows). Two such lamellae appose to form the double plates with the larger 44.5 nm periodicity (short arrow) (X 48,600).

characteristic 35 nm periodicity are recognized. In the hamster the fibrous material develops later, subsequent to the deposition of the zona pellucida and usually in secondary follicles. At first, in contrast to the rat, long, single fibers appear with a slight zigzag pattern and fill a large part of the ooplasm (Fig. 3.6). Two such fibers later associate laterally to form the fibrous plates with the characteristic 44.5 nm periodicity. During enlargement of the oocytes more and more of the fibrous plates develop until the eggs reach their mature size. Subsequent to sperm penetration the fibrous plates are gradually reduced in number while in delayed blastocysts they disappear completely (Enders and Schlafke, 1965; see also discussion by Szollosi).

The development, reduction and disappearance of the fibrous material suggests that it may be utilized during the preimplantation period as an energy source and that it should be considered as yolk at least from the physiological point of view. Thus far, no attempt has been made to isolate the fibrous material and no chemical analysis has been performed on it. Results of fixation with osmium, permanganate and glutaraldehyde solutions suggest that the stacks of fibrous material may be proteinaceous primarily. The complex lipoprotein cannot be excluded on these grounds (Enders and Schlafke, 1965). The addition of calcium ions to osmium tetroxide solutions preserves to some extent the paracrystalline nature of these inclusions, while primary osmium fixation without calcium does not; but neither preserves the same degree of organization as the aldehyde solutions (Szollosi, 1967). Extraction procedures using a variety of solvents on differently fixed ovarian tissues further indicate that the stacks of fibrous material may be proteinaceous (Weakley, 1967). Recently it was suggested that the fibrillar arrays appear to form from polysomes: "At first the ribosomes become arranged in a curvilinear pattern and then they fuse giving rise to a typical fibril which may subsequently increase in length by addition of other ribosomal templates" (Zamboni, 1970). If the plates of fibrous material were composed of ribonucleoprotein they should be basophilic and should strongly absorb ultraviolet light at 260 nm. But the evenly spread weak basophilia and low level of UV absorption (Alfert, 1950; Flax, 1953; Austin, 1961) which is detected in the cytoplasm would seem to argue against the ribonucleoprotein nature of the filamentous material, which is densely packed in the egg cytoplasm. It should be pointed out also that in freshly ovulated mouse and rat eggs relatively few ribosomes could be located by electron microscopy.

The presence of the fibrous material has thus far been reported only in four species of myomorph rodents. May one assume that yolk exists in mammalian eggs in general in the sense of Waddington's definition? In various mammalian eggs morphological studies reported the existence of large vesicles with flocculent contents (Fig. 3.7a,b). Such is the case in the rabbit (Zamboni and Mastroianni, 1966a,b; Krauskopf, 1969a,b; Longo and Anderson, 1969; Gulyas, 1970, and personal communication), pig-tail monkey (Fig. 3.8), ferret (Hadek, 1969) and sheep (Tervit and Dickson, personal communication). The flocculent, membrane-bounded material is found most frequently in rabbit eggs, the most extensively studied species. Occasionally small clusters of ribosomes are attached to the membrane limiting the flocculent material, indicating that it is a portion of the rough endoplasmic reticulum system (Fig. 3.7a,b) (Zamboni and Mastroianni,

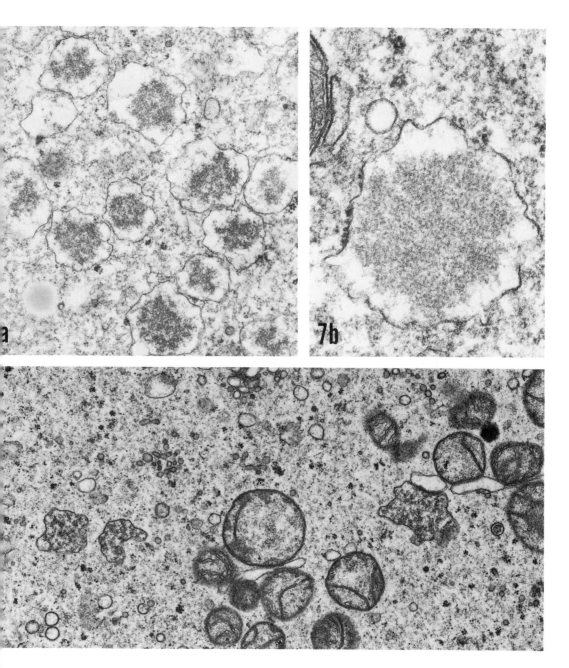

Fig. 3.7a, b Membrane-bounded vesicles with a flocculent content represent "yolk" in rabbit oocytes. Ribosomes are attached to some segments of the membrane. (Original electron micrographs kindly provided by Dr. B. Gulyas) (Fig. 3.7a X 22,600) (Fig. 3.7b X 59,000).

Fig. 3.8 Membrane-bounded vesicles with flocculent material are also found in oocytes of the pig-tail monkey (X 23,000).

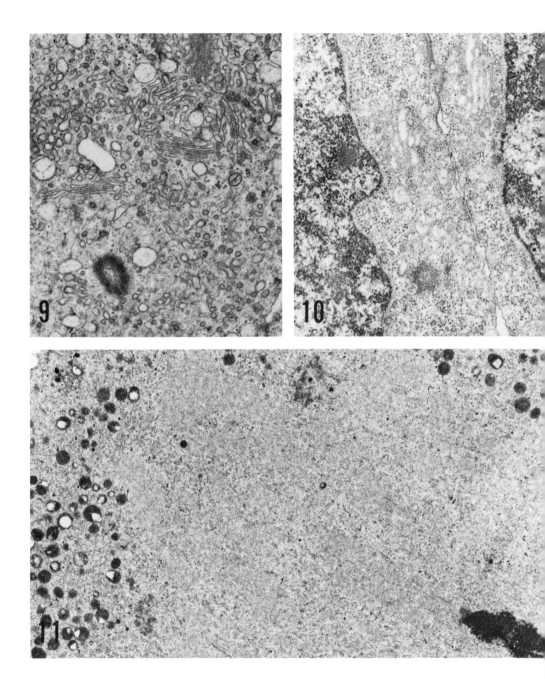

Fig. 3.9   A centriole is present in hamster primary oocytes in the dictyate stage   (X 23,000).

Fig. 3.10   A centriole is found in blastomeres of a 16-cell rat morula   (X 36,500).

Fig. 3.11   At the spindle pole of the first meiotic metaphase, two filamentous bodies (diffuse satellites) are detected   (X 8,700).

966a,b; Krauskopf, 1969a,b). These studies suggested that the vesicles containing the flocculent material might also represent yolk.

During cleavage new granules develop in rat and mouse eggs which were interpreted as yolk (Enders and Schlafke, 1965). The granules are complex, showing regions of differential electron density, and it was suggested that they may be composed primarily of lipid, phospholipid and glycogen particles. Although the identification of the substances composing the granules may be correct, one would expect that the material representing yolk would be deposited during some segment of oogenesis rather than during cleavage stages (four to eight cells) so that the deposited storage materials might be available to support the ovum's metabolic activities as soon as it reaches the oviduct.

III. CENTRIOLES

In contrast to most animal cells, mature mammalian oocytes lack centrioles (Szollosi, 1970b; Szollosi, Calarco and Donahue, unpublished; Hertig and Adams, 1967; Zamboni, 1970, and this conference). From serial sections of the first and second meiotic spindles, it is quite clear that at the spindle poles no centrioles are present. During oogonial divisions and in oocytes having reached zygotene and pachytene stages of the meiotic prophase at or near the time of birth, centrioles with the usual morphology are present (Fig. 3.9). Centrioles could not be localized during any of the latter phases of oocyte development. Centrioles are also absent after sperm penetration and in early cleavage stages (Szollosi, unpublished; Zamboni, personal communication). However, centrioles were detected anew in morulae and young blastocysts (Fig. 3.10). Apparently the morphological integrity of the centrioles is not necessary for the meiotic divisions nor for the first few cleavage divisions. These observations raised the possibility that the most important function of centrioles may be in formation of cilia and flagella rather than in cell division.

Although the spindle poles lack centrioles and asters, the microtubules constituting the spindle apparatus do not end randomly at the spindle pole region. The spindle apparatus forms a barrel-shaped structure and the fibers do not converge towards a single focal point as in mitosis. Instead, the microtubules converge on and apparently anchor at several small foci at the periphery of the spindle pole. Careful examination of serial sections of this region discloses the presence of a large number of small filamentous bodies which are reminiscent of centriolar satellites (Fig. 3.11). The filamentous bodies are more diffuse than satellites previously described, occasionally showing one or more condensed regions (Fig. 3.12) (Szollosi, 1970b). In higher plants, which also lack centrioles, no similar structures were detected. The exact number of diffuse satellites could not be determined with certainty, but in the mouse meiotic spindle there are at least up to a dozen at each pole (Szollosi, Calarco and Donahue, unpublished). The adjacent satellite-like bodies are independent of each other. Chromosomes, therefore, can easily separate from the spindle during aging of eggs in the oviduct, leading in the long run to the formation of micronuclei (Szollosi, 1970b). A small Golgi complex comprised of 2-3 short parallel lamellae, coated vesicles budding off the terminal portions of the Golgi sacks, and a group of

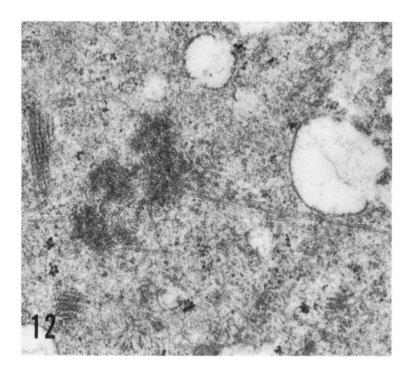

Fig. 3.12 The filamentous bodies demonstrate several condensed regions into which microtubules project. (⨉ 47,000).

small granules and vesicles are frequently found in close proximity to the filamentous bodies (diffuse satellites) (Fig. 3.13). It seems that both the continuous fibers and the kinetochore fibers may project towards the same focal point and may be anchored at the same diffuse satellite (Szollosi, 1964). Most kinetochore fibers originating at one particular chromosome project to one center. Only occasionally is there a microtubule which, making an acute angle at the kinetochore, does not run parallel to the other microtubules. In such a case the particular microtubule may project to an adjacent center or it may end freely, not being anchored at all.

The differences reported between the spindle poles of plant cells, mammalian oocytes, and early cleavage stages raise interesting questions which presently cannot be adequately answered. It should be noted that a faintly filamentous material has been reported to be present at other sites from which microtubules originate. Such are the postacrosomal ring in mammalian and other spermatids, the kinetochores satellites and others. The filamentous material may represent a local precursor pool within which the assembly of microtubules is possible.

The accumulation of an amorphous material, very similar to the filamentous bodies, at the spindle poles of a marine diatom during its meiotic and mitotic division, has been named the "paracentrosome", a neutral term for the material

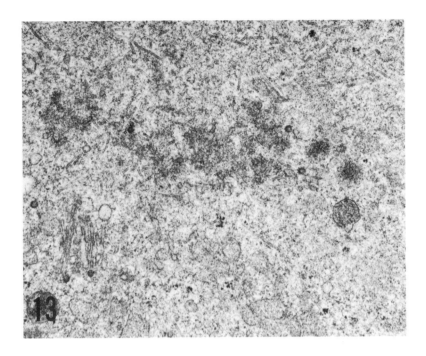

Fig. 3.13   A small Golgi unit can be detected in the proximity of the filamentous bodies. Peripherally, coated vesicles bud off the Golgi apparatus.   (X 31,000).

found at the "centrosomal" region (Manton, Kowallik, and Stosch, 1970). It reportedly represents a reserve substance from which morphologically more elaborate structures will be formed. Adjacent to the "paracentrosome" two centrioles are present, each in the process of forming a flagellum. In the case of the diatom *Lithodesmium undulatum,* then, the structures involved in spindle and flagellum formation can be separated and recognized as independent entities, further supporting the interpretation drawn from the studies on the maturation spindle in the mouse.

IV. MITOCHONDRIA

During gametogenesis extensive morphological transformations occur in the mitochondria of both male and female cell lines (Enders and Schlafke, 1965; Wischnitzer, 1967; Calarco and Brown, 1969; Anderson, Condon, and Sharp, 1970). Mitochondria of oogonia and young oocytes lack the typical shelf-like cristae, although such are present in the primordial germ cells during their motile, migratory period. The cristae become more tubular in nature with an enlarged lumen separating the two leaves of the cristae (Fig. 3.14). In the most extreme case, the mouse, the separating halves of the cristae displace neighboring cristae into an orientation parallel to the external membranes (Wischnitzer, 1967; Calarco and

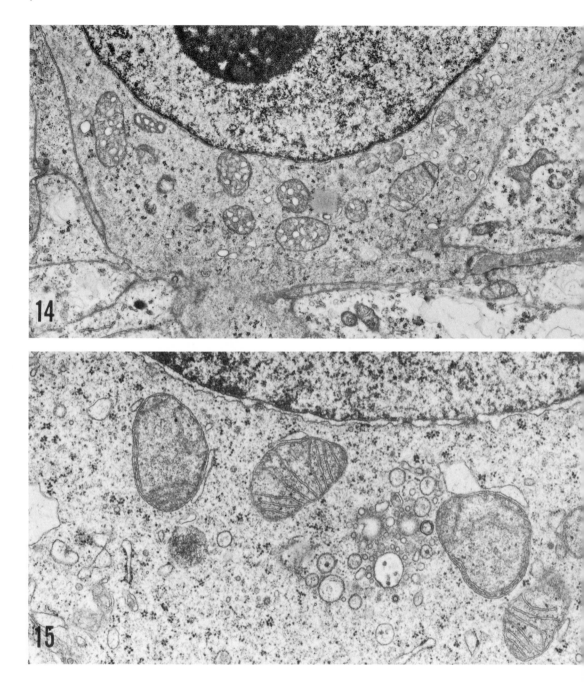

Fig. 3.14 The two components of cristae separate from each other and give the mitochondria a vesiculated appearance in human oogonia (X 15,645).

Fig. 3.15 Cristae are displaced peripherally in some mitochondria in primary rat oocytes, while in others they are organized in a shelf-like manner (X 26,373).

Brown, 1969). The cristae are displaced peripherally (Fig. 3.15) in the mitochondria of most mammalian oocytes. The mitochondrial matrix becomes very dense. The mitochondria maintain this configuration at the time of ovulation, fertilization and early cleavage stages. In morulae a reorganization of mitochondria starts to become detectable. The mitochondria elongate and swell and the shelf-like arrangement of cristae reappears, oriented transversely to the long axis of the mitochondria. The two leaflets of the cristae do not become parallel, however, and irregular intracristal spaces remain.

The separation of the cristae is maintained until day 14 development in the rat heart, when they then become similar to mitochondria of a large number of differentiated tissues. These contain several cristae with parallel membranes. The antimycin-sensitive DPNH oxidase and mitochondrial ATPase levels show that their specific activities are low until day 10 and then increase, reaching a maximum value at day 12 and leveling off thereafter (Mackler, 1970). The rise of respiratory enzyme activity corresponds to an increase in the number of cristae and elimination of intracristal spaces. These data were strengthened by feeding pregnant rats on a galactoflavin-augmented riboflavin-deficient diet or on a diet supplemented with chloramphenicol which significantly lowered the levels of activity of the electron transport system and which maintained the vacuolated mitochondrial morphology. Although the experiments did not include preimplantation stages of development, the mitochondrial structure suggests that, in the rat, oxidative metabolism may be relatively unimportant prior to day 11 of development.

A peculiar mitochondrial morphology was described in bovine ovarian oocytes (Senger and Saacke, 1970), in oocytes of goat and sheep (Saacke, personal communication), and in sheep blastomeres (2-8 cell stage) (Tervit and Dickson, personal communication). A linear or, more frequently, a curved appendage extends from the surface of the mitochondrion composed of its outer and inner membranes (Fig. 3.16a,b). The curved appendage forms, in a three-dimensional reconstruction, a hood over a portion of the mitochondrion. The functional significance of this modification is not known.

In oocytes and in preimplantation stages of all mammals studied a close morphological relationship is consistently observed between mitochondria and the rough endoplasmic reticulum. This morphological association occurs in primary follicles and is maintained in actively enlarging secondary follicles. At the point where the two organelles approximate most closely, polysomal complexes are lacking (Enders and Schlafke, 1965). During early phases of oocyte growth, an electron-dense, fibrillar material develops in association with the ER-mitochondrial complexes, while later almost all of the mitochondria cluster around this electron dense substance (Fig. 3.17). The ER-mitochondrial complexes are gigantic in primates, although the fibrillar, electron-dense matrix is relatively small and focal (Wartenberg and Stegner, 1960; Baker and Franchi, 1967). Long, flattened cisternae are associated with several, and often with large numbers of mitochondria. One possible interpretation of these associational complexes may be that the mitochondria provide the rough endoplasmic reticulum with the high energy

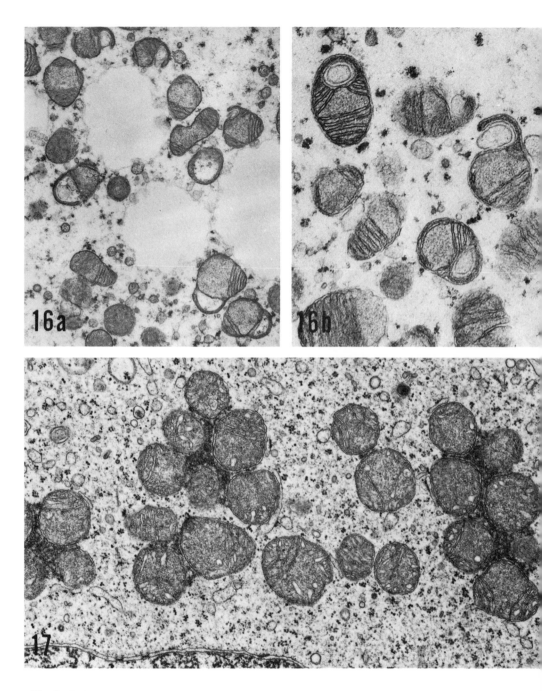

Fig. 3.16a, b Bovine mitochondria show the development of hood-shaped extensions. (By kind permission of Drs. Senger and Saacke)   (Fig. 3.16a X 8,300) (Fig. 3.16b X 25,000).

Fig. 3.17 An electron dense, fibrillar material is deposited in the cytoplasm around which mitochondria cluster. Hamster primary follicle   (X 20,100).

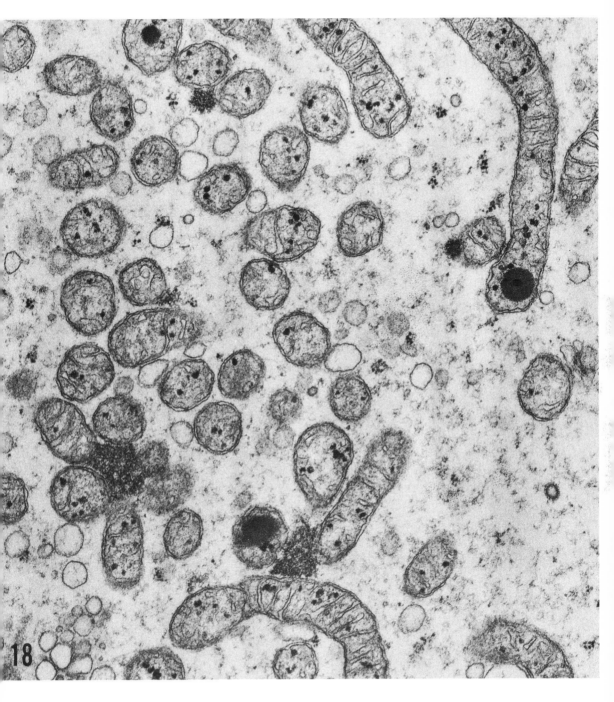

Fig. 3.18 In oocytes of secondary follicles of the spider monkey electron dense granules make their appearance and, in some, condense to form large spherical deposits. Between the mitochondria the fibrillar material can also be detected (X 44,200).

required for some synthetic activity, perhaps vitellogenesis. It has been suggested that the fibrillar material may be a substance playing a role in the genesis of mitochondria (Andre, 1962; Szollosi, 1969).

In oocytes of secondary follicles in the spider monkey, mitochondrial matrix granules develop, which aggregate often into a single, large electron dense deposit. Superficially this material appears very similar to the intramitochondrial yolk granules of Amphibia (Ward, 1962), even though it is homogeneous and lacks crystalline structure (Fig. 3.18). This interpretation was ruled out, however, by the consistent absence of such deposits in large tertiary follicles within the same ovaries.

Prior to ovulation the obvious ER-mitochondrial complexes disperse. However, in most species, a small agranular ER saccule apparently remains associated with every mitochondrion (Calarco and Brown, 1969; Szollosi, 1970a; Senger and Saacke, 1970). The small saccules may be remnants of the earlier, large, rough ER. During cleavage stages this association is maintained, and in the eight cell stage, when a rapid increase of cytoplasmic polysomes is observed (Szollosi, 1970a), the mitochondrial-associated membranes are the first membranes next to the nuclear envelope to become extensively studded with ribosomes.

V. ACKNOWLEDGMENTS

Supported by USPHS grants HD-03752 and GM-16598 from the National Institutes of Health.

VI. REFERENCES

Adams, E. C. and Hertig, A. T. (1964) Studies on guinea pig oocytes. I. Electron microscopic observations on the development of cytoplasmic organelles in oocytes of primordial and primary follicles. *J. Cell Biol.* **21**, 397.

Alfert, M. (1950) A cytochemical study of oogenesis and cleavage in the mouse. *J. Cell Physiol.* **36**, 381.

Anderson, E., Condon, W. and Sharp, D. (1970) A study of oogenesis and early embryogenesis in the rabbit, *Oryctolagus cuniculus,* with special reference to the structural changes of mitochondria. *J. Morph.* **130**, 67.

Andre, J. (1962) Contribution a la connaissance du chondriome. *J. Ultrastruct. Res.* Suppl. 3.

Austin, C. R. (1961) The Mammalian Egg. Oxford, Blackwell and Company.

Baker, T. G. and Franchi, L. L. (1967) The fine structure of oogonia and oocytes in human ovaries. *J. Cell Sci.* **2**, 213.

Balinsky, B. T. (1965) An Introduction to Embryology. 2nd edition, p. 273. Philadelphia and London, W. B. Saunders Company.

Calarco, P.and Brown, H. (1969) An ultrastructural and cytochemical study of preimplantation development of the mouse. *J. Exp. Zool.* **171**, 253.

Enders, A. C. and Schlafke, S. J. (1965) The fine structure of the blastocyst: Some comparative studies. *In* "Preimplantation Stages of Pregnancy". (G. E. W. Wolstenholme and M. O'Connor, eds.), p. 29. London, J. & A. Churchill, Ltd.

Flax, M. H. (1953) Ribose nucleic acid and protein during oogenesis and early embryonic development in the mouse. Columbia University Doctoral Dissertation No. 6615.

Giese, A. (1968) Cell Physiology. 3rd edition, p. 55. Philadelphia, W. B. Saunders Company.

Gulyas, B. (1970) The rabbit zygote: Formation of annulate lamellae. *J. Ultrastruct. Res.* (In press).

Hadek, R. (1966) Cytoplasmic whorls in the golden hamster oocyte. *J. Cell Sci.* **1**, 281.

Hadek, R. (1969) Mammalian Fertilization. New York and London, Academic Press.

Hertig, A. T. (1968) The primary human oocyte: Some observations on the fine structure of Balbiani's vitelline body and the origin of the annulate lamellae. *Amer. J. Anat.* **122**, 107.

Hertig, A. T. and Adams, E. (1967) Studies on the human oocyte and its follicle. Ultrastructural and cytochemical observations on the primordial follicle stage. *J. Cell Biol.* **34**, 647.

Krauskopf, C. (1969a) Elektronenmikroskopische Untersuchungen uber die Struktur der Oocyte und der 2-Zellstadiums beim Kaninchen. I. Oocyte. *Z. Zellforsch.* **92**, 275.

Krauskopf, C. (1969b) Elektronenmikroskopische Untersuchungen uber die Struktur der Oocyte und des 2-Zellstadiums beim Kaninchen. II. Blastomeren. *Z. Zellforsch.* **92**, 296.

Loewenstein, J. E. and Cohen, A. I. (1964) Dry mass, lipid content and protein content of intact and zona-free mouse ovum. *J. Embryol. Exp. Morph.* **12**, 113.

Longo, F. J. and Anderson, E. (1969) Cytological events leading to the formation of the two-cell stage in the rabbit: Association of the maternally and paternally derived genomes. *J. Ultrastruct. Res.* **29**, 86.

Mackler, B. (1970) Studies of mitochondrial energy systems during embryogenesis in the rat. *In* "Symposium on Metabolic Pathways in Mammalian Embryos During Organogenesis and Its Modification by Drugs". Berlin, Free University.

Manton, I., Kowallik, K. and Stosch, H. A. von (1970) Observations on the fine structure and development of the spindle at mitosis and meiosis in a marine centric diatom *(Lithodesmium undulatum)*. IV. The second meiotic division and conclusions. *J. Cell Sci.* **7**, 407.

Senger, P. L. and Saacke, R. G. (1970) Unusual mitochondria of the bovine oocyte. *J. Cell Biol.* **46**, 405.

Szabo, P. L. (1967) Ultrastructure of the developing dog oocyte. *Anat. Rec.* **157**, 330.

Szollosi, D. (1964) Structure and function of centrioles and their satellites from the jellyfish, *Phialidium gregarium. J. Cell Biol.* **21**, 465.

Szollosi, D. (1967) Fixation procedures of embryonal tissues for electron microscopy. *In* "Methods in Developmental Biology". (F. H. Wilt and N. K. Wessels, eds.) New York, Thomas Y. Crowell and Company.

Szollosi, D. (1969) Mitochondrial-rough endoplasmic reticulum complexes in maturing oocytes and spermatocytes. *J. Cell Biol.* **43**, 143a.

Szollosi, D. (1970a) Nucleoli and ribonucleoprotein particles in the preimplantation conceptus of the rat and mouse. *In* "The Biology of the Blastocyst". (R. J. Blandau, ed.) Chicago, The University of Chicago Press.

Szollosi, D. (1970b) Morphological changes in mouse eggs due to aging in the Fallopian tube. *Amer. J. Anat.* (In press).

Waddington, C. H. (1962) Principles of Embryology. p. 36. London, George Allen and Unwin, Ltd.

Ward, R. T. (1962) The origin of protein and fatty yolk in *Rana pipiens.* II. Electron microscopical and cytochemical observations of young and mature oocytes. *J. Cell Biol.* **14**, 309.

Wartenberg, H. and Stegner, H. E. (1960) Uber die electronenmikroskopische Feinstructur des menschlichen Ovarialeies. *Z. Zellforsch.* **52**, 450.

Weakley, B. S. (1966) Electron microscopy of the oocyte and granulosa cells in the developing ovarian follicles of the golden hamster *(Mesocricetus auratus). J. Anat.* **100**, 503.

Weakley, B. S. (1967) Investigations into the structure and fixation properties of cytoplasmic lamellae in the hamster oocyte. *Z. Zellforsch.* **81**, 91.

Wischnitzer, S. (1967) Intramitochondrial transformations during oocyte maturation in the mouse. *J. Morph.* **121**, 29.

Zamboni, L. (1970) Ultrastructure of mammalian oocytes and ova. *Biol. Reprod.* Suppl. **2**, 44.

Zamboni, L. and Mastroianni, L., Jr. (1966a) Electron microscopic studies on rabbit ova. I. The follicular oocyte. *J. Ultrastruct. Res.* **14**, 95.

Zamboni, L. and Mastroianni, L., Jr. (1966b) Electron microscopic studies on rabbit ova. II. The penetrated tubal ovum. *J. Ultrastruct. Res.* **14**, 118.

4

THE KINETOCHORE IN OOCYTE MATURATION

Patricia G. Calarco

I.   Introduction
II.   Materials and methods
III.   Results and discussion
        A.   Dictyate oocytes
        B.   Chromatin condensation
        C.   Circularly arranged bivalents
        D.   Prometaphase I
        E.   Metaphase I
        F.   Anaphase I
        G.   Telophase I
        H.   Chromatin mass
        I.   Prometaphase II
IV.   Summary
V.   Acknowledgments
VI.   References

I. INTRODUCTION

Meiotic maturation in the mouse begins when oogonia enter prophase about 8 days before birth. By approximately 5 days after birth primary oocytes have reached the diplotene stage and arrest at this point. The chromatin progresses to a noncondensed state and a large nucleolus is present. This stage is known as the dictyate stage and persists until the oocyte is stimulated to continue its maturation to metaphase II. The oocyte is then ovulated and remains arrested at metaphase II unless fertilization occurs.

The ultrastructural investigation of maturation in mammals has been somewhat handicapped because the oocyte is relatively inaccessible in the ovarian follicle, the number of eggs that can be studied is small, and the correct sequence of observed stages is subject to interpretation. Use of oocytes maturing *in vitro* (see chapter by R. P. Donahue) in large part overcomes these difficulties since the oocytes are more readily obtained, form a fairly synchronous population, and large numbers can be collected at one time. However, the sequence and timing of *in vitro* maturation has only been reported for the mouse (Donahue, 1968). The *in vitro* cultivation of the mouse oocyte, then, provides a good system in which to study maturation, from dictyate arrest to metaphase II arrest, at the fine structural level.

The kinetochore (centromere) is that structure on the chromosome which serves as the place of spindle filament attachment. The kinetochore functions in chromosome movement during meiosis (and mitosis) and may function additionally in the organization of the noncontinuous, chromosomal elements of the spindle (Inoue, 1964; Bajer and Mole-Bajer, 1969).

The fine structure of the mitotic kinetochore has been variously described as one or more coarse filaments each composed of two fibrils (Brinkley and Stubblefield, 1966), a crescentic structure (Robbins and Gonatas, 1964), a plate or disc-shaped structure (Krishan and Buck, 1965; Jokelainen, 1967), and as one or more dense bands consisting of two subunits (Krishan, 1968). The origin of the kinetochore and its chemical composition are not known with certainty although it has been suggested that the kinetochore is a region of the chromosome persisting in an extended state (Brinkley and Stubblefield, 1966). For further information on the mitotic kinetochore the reader is directed to the comprehensive article by Brinkley and Stubblefield (1970).

The meiotic kinetochore has been less extensively investigated. Because homologs are separated at the first meiotic division, while the chromatids of each homolog remain together, Darlington (1937) proposed that the metaphase I kinetochore of each homolog was undivided and this was supported by the observation of undivided kinetochores in metaphase I pigeon spermatocytes (Nebel and Coulon, 1962). However, two kinetochores per anaphase I homolog have been reported in light microscope studies on plants (Lima de Faria, 1958) and ultrastructural studies on *Urechis* eggs (Luykx, 1965).

The purpose of this article will be to discuss ultrastructural aspects of the kinetochore within the framework of maturation.

II. MATERIALS AND METHODS

Follicular oocytes were released from ovaries of CF-1 (Carworth, Inc.) female mice 8-12 weeks old, and cultured in a chemically defined Krebs-Ringer salt solution (pH 7.2) at 37°C in 5% CO_2 (Donahue, 1968). When oocytes are released from their follicles in this manner, maturation begins almost immediately and continues until metaphase II is reached between 11 and 17 hours of culture. In collaboration with Dr. Donahue, who cultured the oocytes, samples of approximately 30 eggs were taken at timed intervals during the *in vitro* maturation period. Ten of these 30 oocytes were scored by light microscopy for their stage of nuclear progression and the remaining 20 prepared for electron microscopy.

The fixation procedure generally used was a 60-minute fixation at room temperature in 3% glutaraldehyde in 0.1 M phosphate buffer, followed by a 15-minute fixation in 2% osmium tetroxide in 0.1 M phosphate buffer. After dehydration in a graded series of ethanols, eggs were embedded in Epon 812 by a modification of the procedure described by Luft (1961). Embeddings were made in a thin layer of Epon and it was possible to determine the stage of maturation of some of the embedded oocytes with the light microscope, thus facilitating the selection of eggs for ultrastructural study.

An alternating series of thick and thin sections were cut with a diamond knife on a Reichert microtome. Contrast of thin sections was enhanced by treating the sections with uranyl acetate (Watson, 1958) followed by lead citrate (Reynolds, 1963). Grids were examined in a Philips 200 electron microscope. Thick sections mounted on glass slides were stained with a combination of methylene blue and azure II (Richardson, Jarett and Finke, 1960).

III. RESULTS AND DISCUSSION

A. Dictyate Oocytes

Immediately after release from their follicles dictyate oocytes exhibit an intact germinal vesicle with a slightly undulated nuclear membrane. The nucleolus is spherical and agranular. Fine strands of chromatin and occasional areas of more condensed chromatin are seen (Fig. 4.1). Pardue and Gall (1970) have shown that labeled mouse satellite DNA hybridizes with the centromeric heterochromatin on condensed mouse chromosomes in both mitosis and meiosis. Furthermore, in mouse interphase nuclei, satellite DNA binds to the heterochromatic chromocenters suggesting that centromeric areas (kinetochores) are also located in the chromocenters. One might, therefore, expect the kinetochore to first appear near or within heterochromatic chromocenters. Areas which may correspond to these chromocenters are seen, but kinetochores are not observed in the intact germinal vesicle. Pores are present in the nuclear envelope and a few microtubules are seen in the cytoplasm in close proximity to the nuclear envelope.

Oocytes cultured for 30 min, which is approximately 1.5 hours after their release from the ovary, have a more convoluted nuclear membrane but differ little from dictyate oocytes. Cytoplasmic microtubules are seen in greater numbers and often are directed toward the nuclear membrane (Fig. 4.2). Nuclear pores are still present.

B. Chromatin Condensation

In the earliest stages of chromatin condensation, electron-dense granules are

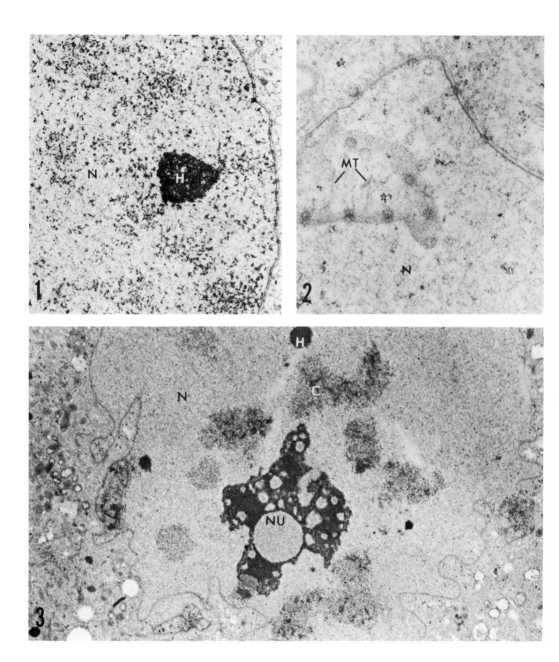

Fig. 4.1   Electron micrograph of a section through a dictyate oocyte. N, nucleus. H, heterochromatin   (X 9,400).

Fig. 4.2   Oocyte after 0.5 hour culture. MT, microtubule. N, nucleus   (X 30,400).

Fig. 4.3   Early stage of chromatin condensation. N, nucleus. NU, nucleolus. H, heterochromatin. C, chromosome   (X 6,600).

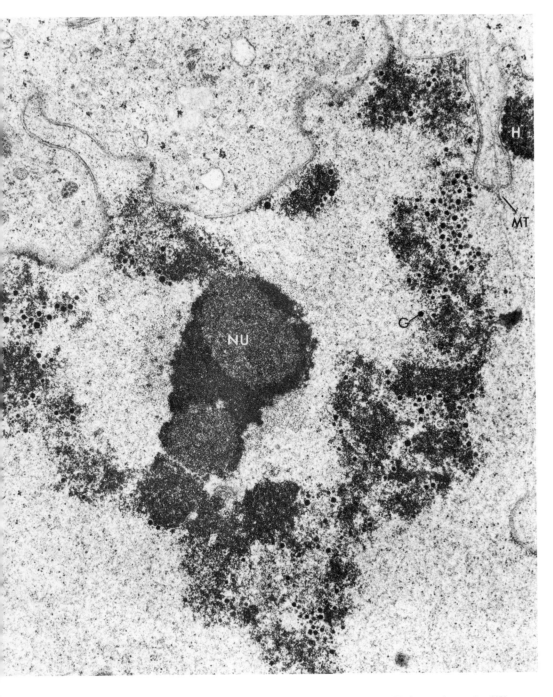

Fig. 4.4 Later stage of chromatin condensation. MT, microtubules. G, granules. H, heterochromatin. NU, nucleolus. C, chromosome (X 19,400).

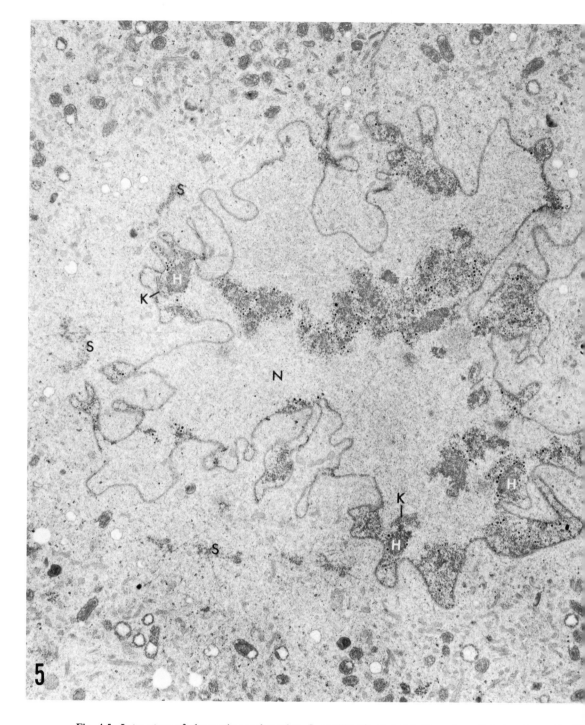

Fig. 4.5 Later stage of chromatin condensation. S, satellite-like material. N, nucleus. H, heterochromatin. K, kinetochore (X 7,400).

en throughout the chromosomes (Fig. 4.3). Several areas of more condensed iromatin, perhaps corresponding to the centromeric heterochromatin, are seen but ɔ kinetochores are visible. The nucleolus presents a reticulated appearance and is osely associated with areas of condensing chromatin. Chromosomes appear to ɔndense with part of the chromosome in contact with the convoluted nuclear ivelope. Woolam, Miller and Ford (1967) reported that pachytene chromosomes in iouse spermatocytes are attached to the nuclear membrane both at their eterochromatic (centromere) ends and their distal ends. The attachment of ɔndensing chromosomes to the nuclear envelope has also been reported in animal iitosis (Brinkley and Stubblefield, 1970) and in plants (Sparvoli, Gay and Kaufman, 964) and suggests that a chromosome-nuclear membrane association may be idespread. The nuclear envelope appears to be unbroken although pores are no ɔnger present. Microtubules, interpreted to be future spindle elements, are seen nly outside the nucleus and often are directed toward it. Very rarely, small ytoplasmic densities from which microtubules emanate are observed near the ucleus.

In later stages of chromatin condensation, the association of large lectron-dense granules with chromosomes is more clearly seen. These granules ɔsemble perichromatin granules (Watson, 1962) but are larger, reaching a maximum iameter of 90 nm (Figs. 4.4, 4.5). The nucleolus can still be recognized as fibrillar ɔnes of different densities (Fig. 4.4). Recognizable kinetochores are seen at this tage often in close proximity to heterochromatic regions (Figs. 4.5, 4.6). The uclear envelope (NE) is not continuous, for microtubules cross from the cytoplasm ɪto the nucleus and are in contact with all observed kinetochores. In addition, the JE is highly convoluted and in some regions folds back upon itself with ucleoplasmic surfaces apposed. This folding process eventually results in the ɔrmation of nuclear envelope doublets (Figs. 4.8, 4.10). These quadruple iembrane complexes often have ribosomes on their nonapposed cytoplasmic ɪrfaces (Szollosi and Calarco, 1970). Cytoplasmic areas of medium electron density ɪhich appear to be focal origins of microtubules are seen in the vicinity of the uclear envelope (Figs. 4.5, 4.7) and may be analogous to pericentriolar satellites Szollosi, 1964). No centrioles were observed in any of the maturation stages of the iouse.

ᴄ. Circularly Arranged Bivalents

After condensation of bivalents is completed, the chromosomes appear to be ircularly oriented in the light microscope. In the electron microscope circularly ɔriented bivalents are seen close to the inner margin of the nuclear envelope (Fig. ᵻ.9). Terminalized chiasmata are suggested by the V shape of some chromosomes. ᵀhe V shape at this stage is "tight", that is, centromere repulsion has not yet begun. ᵀhe point of the V is usually oriented toward the nuclear membrane and the elocentric kinetochores at the tips of the V usually face toward the inner area of ʰe nuclear region. A few denser regions of chromatin (perhaps heterochromatic) ʲan still be discerned on some chromosomes.

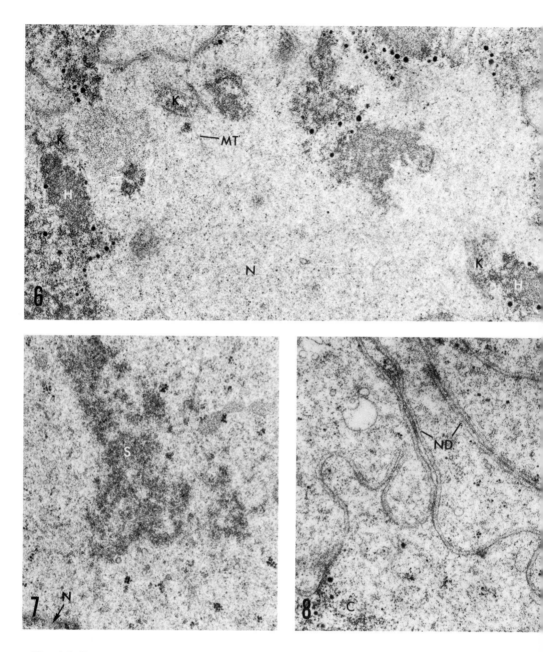

Fig. 4.6 Enlargement of a portion of Fig. 4.5. N, nucleus. H, heterochromatin. K, kinetochore. MT microtubule (X 15, 861).

Fig. 4.7 "Satellite-like" mass (S) of medium electron density near the nucleus (N) of a late condensing chromatin stage oocyte. Note the associated microtubules (X 28,200).

Fig. 4.8 Nuclear envelope doublets (ND) formed in the breakdown of the nucleus. C, chromatin (X 28,200).

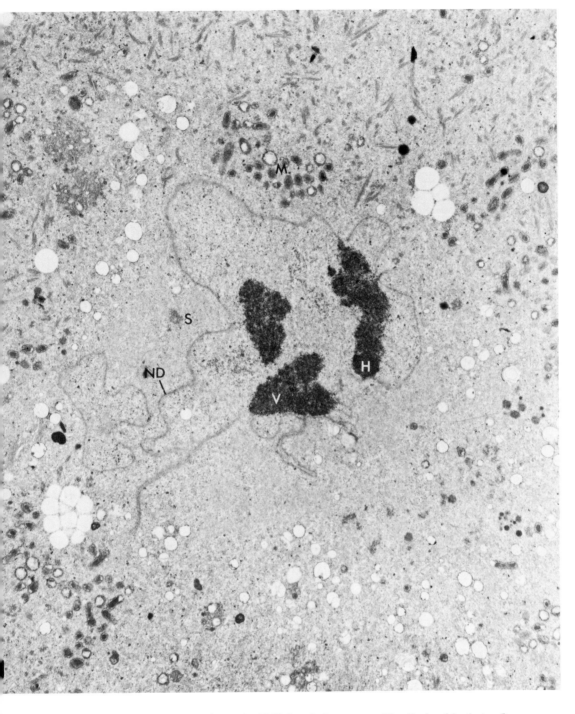

Fig. 4.9 Circular chromatin stage. H, heterochromatin. V, V-shaped chromosome. M, mitochondria cluster. S, atellite. ND, nuclear envelope doublet (X 8,850).

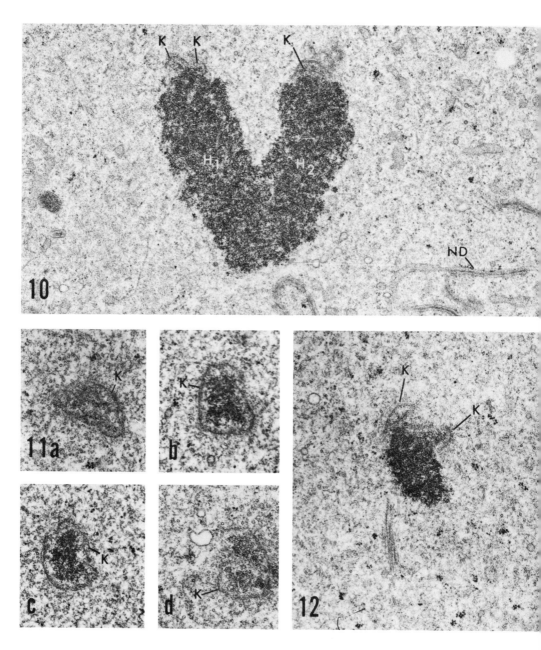

Fig. 4.10 Chromosome bivalent showing kinetochores (K) associated with two homologs (H$_1$ and H$_2$). ND nuclear envelope doublets (X 17,650).

Fig. 4.11 Cross sections through prometaphase I kinetochores (K). Darker material in the center is chromatin. Note the associated microtubules (Fig. 4.11a-c X 26,400) (Fig. 4.11d X 28,160).

Fig. 4.12 Prometaphase I section through end of one telocentric homolog showing the two kinetochore (K) (X 17,240).

The appearance of a wider V is interpreted to mean that centromere repulsion is begun (Fig. 4.10). Eventually the kinetochores on one arm of the V (one homolog) will come to lie 180° away from the kinetochores on the other arm of the (other homolog). Electron micrographs suggest that two kinetochores are present in each homolog, i.e., one per chromatid (Fig. 4.10). Therefore, the kinetochores of meiosis I are physically, if not functionally, divided. This agrees with earlier reports of two kinetochores per meiotic chromosome in *Urechis* eggs at anaphase I (Luykx, 1965). The former nuclear envelope can be detected only as nuclear envelope doublets which in favorable sections appear to break down into cisternae indistinguishable from those of the endoplasmic reticulum. "Satellites" are rarely seen in the immediate vicinity of the chromosomes. The orientation of spindle microtubules appears to be random at this time.

Prometaphase I

During prometaphase I, bivalents become oriented on the spindle. Cross-sections through kinetochore regions reveal some ellipsoidal and circular profiles (Fig. 4.11a-d). The wall of the circle is approximately 30-36 nm in diameter and favorable sections suggest it is composed of two dense layers separated by a lighter space (Fig. 4.11b). Cross sections of microtubules are usually seen surrounding the circular kinetochores, which suggests that the section is distal to the point at which microtubules attach to the kinetochores.

Journey and Whaley (1970) reported circular, band-like kinetochores encircling mitotic prometaphase sister chromosomes in Chinese hamster fibroblasts treated with vincristine sulfate and colchicine. They suggested that ring-like kinetochores might be a stage in kinetochore development and could be the structural basis for the firm attachment of chromatids in the kinetochore region. Since in the present study the kinetochores on one homolog appear separate prior to metaphase I, circular profiles are not considered to be of undivided kinetochores, but rather of kinetochores of single chromatids. Other sections closer to a longitudinal plane show two separate kinetochores on the two telocentric chromatids of one homolog (Fig. 4.12).

On the basis of serial sections through kinetochore regions of metaphase Chinese hamster fibroblasts, Brinkley and Stubblefield (1970) suggested that each mitotic sister kinetochore was composed of two kinetochore filaments, and circular profiles represented the two separate kinetochore filaments of one sister kinetochore. The axial or central element of their kinetochore filaments was reported to be 15-24 nm wide. The observation of circular kinetochores through 3 serial sections (180-300 nm total thickness) in this study suggests that the meiotic kinetochore is wider than a 15-24 nm filament. The circular profiles are more consistent with transverse sections of a disc-shaped kinetochore.

Metaphase I

At metaphase I the chromosomes are co-oriented on a broad spindle

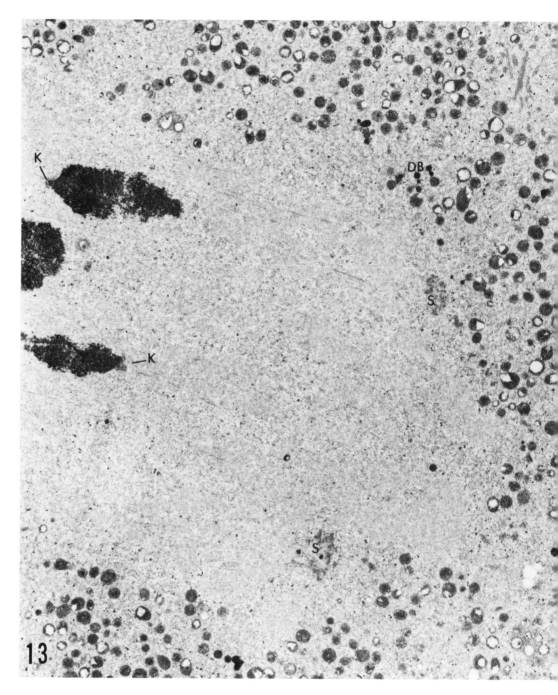

Fig. 4.13 Metaphase I electron micrograph of a section illustrating one half of the spindle with the "satellite" masses (S) at one pole and the chromosomes in the center. Note the mitochondria surrounding the spindle area. K, kinetochore. DB, dark body cluster (X 8,500).

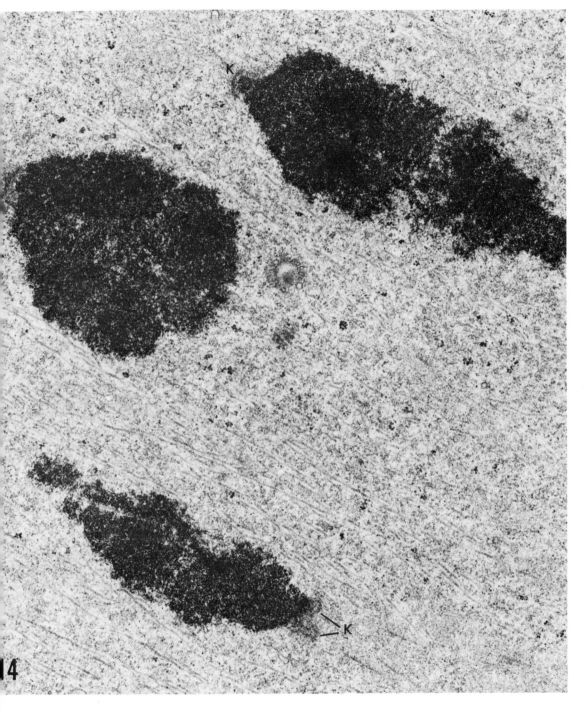

Fig. 4.14 Enlargement of chromosomes in Fig. 4.13. Note the two kinetochores (K) on the chromosome at the bottom of the figure (X 24,100).

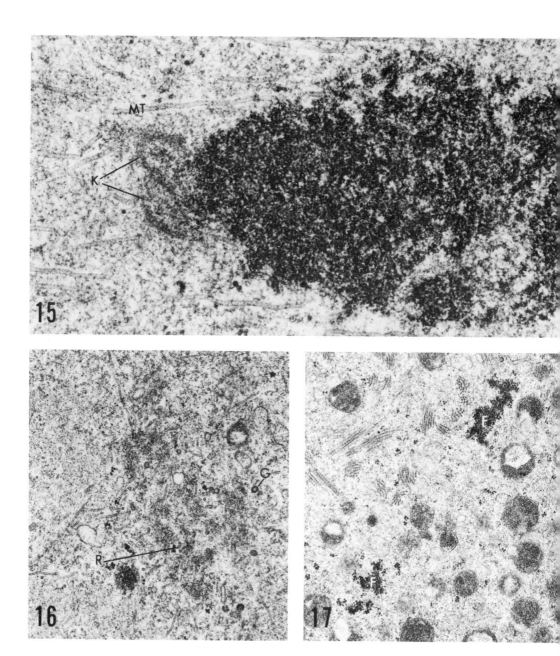

Fig. 4.15 Portion of one metaphase I **chromosome**. Note the two kinetochores (K). MT, microtubule (X 60,750).

Fig. 4.16 "Satellite-like" mass from a metaphase I pole. G, coated vesicles probably derived from the Golgi apparatus. R, ribosomes (X 25,600).

Fig. 4.17 Metaphase I. Fibrous patches (F) surrounded by ribosomes and found in the vicinity of the spindle (X 20,400).

interpreted as being barrel-shaped. Kinetochores appear double, i.e., one per chromatid (Figs. 4.13-4.15) and are located at varying distances from the center of the spindle. This distance from the center of the spindle is determined by the length of the chromosome. The broad poles of the spindle are composed of several aggregates of medium dense "satellite" material (Figs. 4.13, 4.16). These satellite areas in metaphase I oocytes are similar to satellites seen near the nuclear envelope during chromatin condensation. Metaphase I satellites, however, have coated vesicles derived from the Golgi apparatus and ribosomes associated with them (compare Fig. 4.16 to Fig. 4.7). Mitochondria surround the metaphase spindle (Fig. 4.13) and are of the vacuolated type described by Wischnitzer (1967). Clusters of membrane-bound dark bodies (DB) are seen close to the metaphase spindle (Fig. 4.13). These dark bodies are usually observed near the periphery of the egg before germinal vesicle breakdown, near the nucleus during chromatin condensation and in the region of the spindle during meiotic divisions. These dark-body clusters may be analogous to the membrane-bounded osmiophilic bodies (MBOB) reported in mitotic cells (Robbins and Gonatas, 1964). Other structures seen only during maturation and located near or within the mitochondrial sheath surrounding the spindle are electron-dense fibrous patches surrounded by ribosomes (Fig. 4.17).

F. Anaphase I

Concomitant with anaphase I movements of the chromosomes, the spindle migrates from the center to the periphery of the oocyte maturing *in vitro* (Fig. 4.18). One of the spindle poles leads the movement of the spindle to the periphery. Significantly, the mitochondrial sheath is now open at this end and does not migrate with the spindle; thus, most of the mitochondria will stay in the oocyte when polar body abstriction occurs. The underlying mechanism of this spindle migration is not known, but "satellites" do exhibit microtubules oriented away from the spindle (aster-like?) which could function in "moving" the spindle to the periphery. Figure 4.19 illustrates two kinetochores on an anaphase homolog.

G. Telophase I

During telophase I the midbody is formed and the first polar body is pinched off, the midbody remaining with the polar body. Telophase I kinetochores are occasionally seen and appear partly buried in the chromosomes (Figs. 4.20, 4.21). Some kinetochores exhibit circular and ellipsoidal profiles (Fig. 4.20). Short sections of microtubules contact the kinetochore, but most microtubules in the vicinity of the chromosomes are intimately associated with ribosomes (Figs. 4.20, 4.22). The ribosomal-microtubule complex is also seen in the first polar body. Since treatment of the oocyte at telophase I with puromycin blocks the normal scattering of polar body chromosomes and causes oocyte chromosomes to remain in a ball (Donahue, personal communication), one can speculate that a ribosomal-microtubule interaction is necessary for further chromosome movement. This interaction could be related to the formation of new intermicrotubule bridges (Hepler, McIntosh and Cleland, 1970) or to the growth of microtubules for the second spindle. It should be noted that ribosome-microtubule complexes are never found in the first meiotic

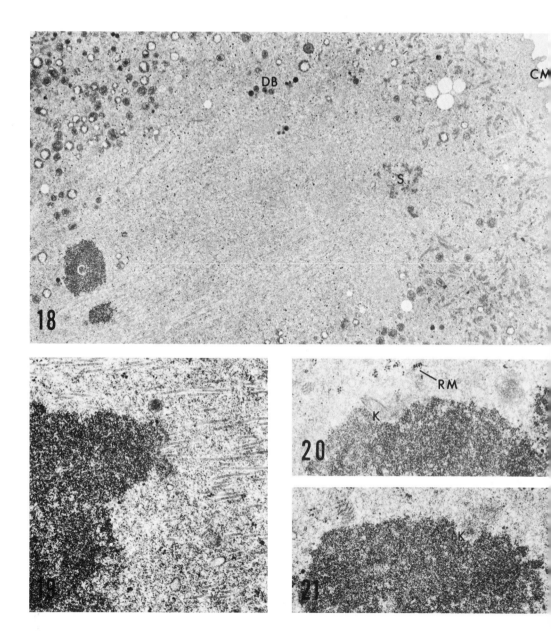

Fig. 4.18   Anaphase I. Note that the mitochondrial sheath is not continuous near the upper right. The entire spindle is closer to the cell margin (CM) than at metaphase I. S, "satellite". DB, dark bodies. C, chromosome   (X 6,850).

Fig. 4.19   Part of an anaphase I chromosome showing two kinetochores   (X 25,500).

Fig. 4.20   Telophase I. Note the ribosome-microtubule association (RM). K, kinetochore   (X 22,350).

Fig. 4.21   Telophase I chromosome with kinetochore (K)   (X 29,300).

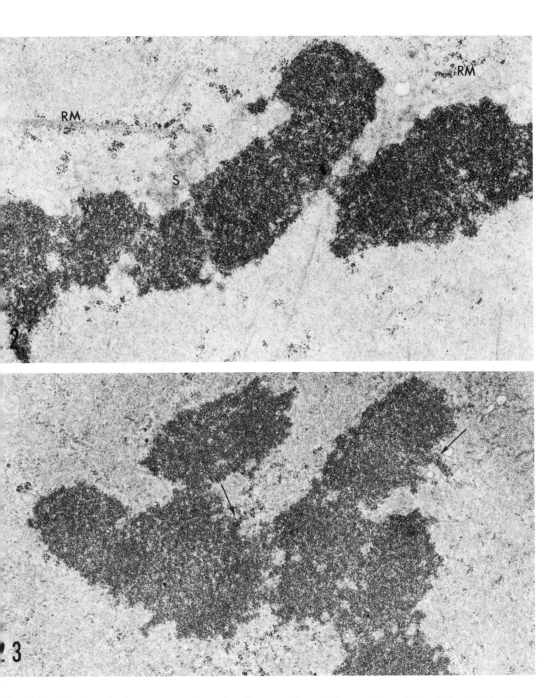

Fig. 4.22 Telophase I chromosomes. Note the ribosome-microtubule complexes (RM). S, "satellite" (X 24,000).

Fig. 4.23 Early chromatin mass stage. Arrows indicate a ribosome-microtubule complex contacting the chromatin (X 20,500).

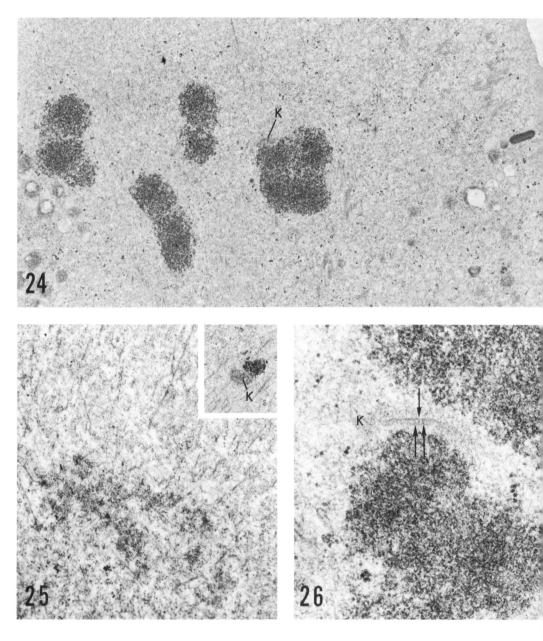

Fig. 4.24   Prometaphase II. The spindle is aligned parallel to the cell surface. K, kinetochore   (X 10,700).

Fig. 4.25   "Satellite" mass from one pole of a metaphase II ovum   (X 23,100). *Insert.* Circular kinetochore (K) and portion of a chromosome from a metaphase II ovum   (X 14,150).

Fig. 4.26   Portion of the metaphase II chromatin from an ovulated and centrifugally stratified egg. Note the two layers to the kinetochore. Arrows indicate cross linkages between these layers   (X 28,900).

spindle. A "satellite-like" mass is occasionally observed quite near the telophase I chromosomes (Fig. 4.22).

H. Chromatin Mass

Telophase I is followed by the chromatin mass stage characterized by a fully formed first polar body and a compact crescent-shaped chromatin group in the oocyte (Donahue, 1968). To date, kinetochores and "satellites" have not been identified in the compact chromatin mass stage. Neither have kinetochores been observed on polar body chromosomes at this stage. However, it is possible that the area of contact of microtubules with the chromosomes corresponds to kinetochore sites (Fig. 4.23). The association of ribosomes with chromosomal microtubules is still evident in early chromatin mass stages, but later in this stage these ribosomal-microtubule complexes are not seen.

I. Prometaphase II

In prometaphase II discrete dyads with kinetochores are seen at varying distances from the poles (Fig. 4.24). The second meiotic spindle is located near the oocyte periphery and the long axis of the spindle is oriented roughly parallel to the cell surface. There is very little difference between prometaphase II and metaphase II except for the alignment of the dyads in the center of the spindle. "Satellite" material of medium electron density makes up the poles of the metaphase II spindle (Fig. 4.25). Kinetochores are occasionally sectioned to reveal circular profiles.

Further evidence on kinetochore fine structure was obtained during the study of centrifugally stratified metaphase II eggs. Kinetochore microtubules are not visible after centrifugation and some substructure of the kinetochore is revealed (Fig. 4.26). Here the kinetochore is composed of two layers approximately 9 nm wide which are separated by a 18 nm space. Some cross linkage is observed between these two layers at 18 nm intervals. Alternatively, this structure could be interpreted as two helically twisted fibrils. Cohelical fibrils have been reported to make up the axial elements of mitotic kinetochores (Brinkley and Stubblefield, 1970).

IV. SUMMARY

Kinetochores are first observed in mouse oocytes maturing *in vitro* during chromatin condensation. They often appear near areas of more condensed chromatin interpreted to be the centromeric heterochromatin. Microtubules already are associated with the kinetochores when they are first observed. The present work suggests that kinetochores are double, i.e., one per chromatid, as soon as they can be recognized as being associated with their specific homologs. It seems probable that they arise during chromatin condensation as one single unit per chromatid.

Following first polar body extrusion, kinetochores are not observed, but ribosomal-microtubule complexes are often seen in association with the chromosomes. Kinetochores are again detected at prometaphase II, at which time the chromosomes are aligning themselves on the second spindle.

The results of this study suggest that the meiotic kinetochore in mice is disc-shaped. Further, the wall of this disc-shaped kinetochore is approximately 30-36 nm thick and appears to be composed of two dense layers separated by a lighter space.

Condensing chromosomes in maturing mouse oocytes exhibit large numbers of chromatin granules. Probably part of each condensing chromosome is associated with the nuclear envelope. Breakdown of the nuclear envelope proceeds with the nuclear envelope becoming progressively more convoluted, resulting in the formation of nuclear envelope doublets.

Clusters of "satellite" material, probably serving as focal origins for microtubules, surround the nucleus during chromatin condensation and later form the two broad poles of the metaphase spindle.

V. ACKNOWLEDGMENTS

This work supported by USPHS Postdoctoral Fellowship grant 1 F02 HD 24170-01 and by USPHS research grant GM 16598 both from the National Institutes of Health.

The author gratefully acknowledges the contributions to this research made by Dr. Daniel Szollosi and Dr. Roger Donahue, the helpful criticisms of this manuscript by Dr. John Luft, and the expert technical assistance of Mr. John Trotter and Mrs. Jean Greenaway.

VI. REFERENCES

Bajer, A. and Mole-Bajer, J. (1969) Formation of spindle fibers, kinetochore organization and behavior of the nuclear envelope during mitosis in endosperm. Fine structural and *in vitro* studies. *Chromosoma.* **27**, 448.

Brinkley, B. R. and Stubblefield, E. (1966) The fine structure of the kinetochore of a mammalian cell *in vitro*. *Chromosoma.* **19**, 28.

Brinkley, B. R. and Stubblefield, E. (1970) Ultrastructure and interaction of the kinetochore and centriole in mitosis and meiosis. *In* "Advances in Cell Biology". (D. M. Prescott, L. Goldstein and E. McConkey, eds.) Vol. 1, p. 119. New York, Appleton- Century-Crofts.

Darlington, C. D. (1937) Recent Advances in Cytology. 2nd ed. Philadelphia, P. Blakiston's Sons and Co.

Donahue, R. P. (1968) Maturation of the mouse oocyte *in vitro*. I. Sequence and timing of nuclear progression. *J. Exp. Zool.* **169**, 237.

Hepler, P. K., McIntosh, J. R. and Cleland, S. (1970) Intermicrotubule bridges in mitotic spindle apparatus. *J. Cell Biol.* **45**, 438.

Inoue, S. (1964) Organization and function of the mitotic spindle. *In* "Primitive Motile Systems in Cell Biology". (R. D. Allen and N. Kamiya, eds.) p. 549. New York, Academic Press, Inc.

Jokelainen, P. T. (1967) The ultrastructure and spatial organization of the metaphase kinetochore in mitotic rat cells. *J. Ultrastruct. Res.* **19**, 19.

Journey, L. J. and Whaley, A. (1970) Kinetochore ultrastructure in vincristine-treated mammalian cells. *J. Cell Sci.* **7**, 49.

Krishan, A. (1968) Fine structure of the kinetochores in vinblastine sulfate-treated cells. *J. Ultrastruct. Res.* **23**, 134.

Krishan, A. and Buck, R. C. (1965) Structure of the mitotic spindle in L strain fibroblasts. *J. Cell Biol.* **24**, 433.

Lima-de-Faria, A. (1958) Recent advances in the study of the kinetochore. *In* "International Review of Cytology". (G. H. Bourne and J. E. Danielli, eds.) New York, Academic Press, Inc.

Luft, J. H. (1961) Improvements in epoxy resin embedding methods. *J. Biophys. Biochem. Cytol.* **9**, 409.

Luykx, P. (1965) The structure of the kinetochore in meiosis and mitosis in *Urechis* eggs. *Exp. Cell Res.* **39**, 643.

Nebel, B. R. and Coulon, E. M. (1962) The fine structure of chromosomes in pigeon spermatocytes. *Chromosoma.* **13**, 272.

Pardue, M. L. and Gall, J. G. (1970) Chromosomal localization of mouse satellite DNA. *Science.* **168**, 1356.

Reynolds, E. S. (1963) The use of lead citrate at high pH as an electron-opaque stain in electron microscopy. *J. Cell Biol.* **17**, 208.

Richardson, K. C., Jarett, L. and Finke, E. H. (1960) Embedding in epoxy resins for ultrathin sectioning in electron microscopy. *Stain Tech.* **35**, 313.

Robbins, E. and Gonatas, N. K. (1964) The ultrastructure of a mammalian cell during the mitotic cycle. *J. Cell Biol.* **21**, 429.

Sparvoli, E., Gay, H. and Kaufman, B. P. (1964) Number and pattern of association of chromonemata in chromosomes of *Tradescantia*. *Chromosoma.* **16**, 415.

Szollosi, D. (1964) The structure and function of centrioles and their satellites in the jellyfish *Phialidium gregarium*. *J. Cell Biol.* **21**, 465.

Szollosi, D. and Calarco, P. (1970) Nuclear envelope breakdown and reutilization. *Proc. 7th International Congress of Electron Microscopy*. (Grenoble, France). 275.

Watson, M. (1958) Staining of tissue sections for electron microscopy with heavy metals. *J. Biophys. Biochem. Cytol.* **4**, 475.

Watson, M. L. (1962) Observations on a granule associated with chromatin in the nuclei of cells of rat and mouse. *J. Cell Biol.* **13**, 162.

Wischnitzer, S. (1967) Intramitochondrial transformations during oocyte maturation in the mouse. *J. Morph.* **121**, 29.

Woolam, D. H. M., Miller, J. W. and Ford, E. H. R. (1967) Points of attachment of pachytene chromosomes to the nuclear membrane in mouse spermatocytes. *Nature (London).* **213**, 298.

5

THE LOCALIZATION OF ACID PHOSPHATASE AND THE UPTAKE OF HORSERADISH PEROXIDASE IN THE OOCYTE AND FOLLICLE CELLS OF MAMMALS

Everett Anderson

I.   Introduction
II.   Materials and methods
        A.   General
        B.   Acid phosphatase
        C.   Horseradish peroxidase
III.   Observations
        A.   Organization of the Graafian follicle
        B.   Fine structure of differentiating oocytes
                1.   Golgi complex
                2.   Oolemma
                3.   Cytology of follicle-granulosa cells
        C.   Acid phosphatase
                1.   Differentiating oocytes
                2.   Follicle-granulosa cells
        D.   Horseradish peroxidase
                1.   Differentiating oocytes
                2.   Follicle-granulosa cells
IV.   Discussion
V.   Summary
VI.   Acknowledgments
VII.   References

I. INTRODUCTION

The adult mammalian ovary is a complex, highly compartmentalized, morphogenetic system some of whose functional units, the Graafian follicles, are in varied stages of cytomorphosis. As in other morphogenetic systems, the normal differentiating processes in the ovary are accompanied by regression or degeneration. As suggested by DeDuve and Wattiaux (1966), DeDuve (1969) and Novikoff (1961), certain aspects of the normal economy and autolysis of cells is a function of a group of subcellular, membrane-bounded structures rich in acid hydrolases designated as lysosomes.

Among one of the constituent enzymes of the lysosome that may be identified visually is acid phosphatase (Novikoff, 1961). Much of what we know concerning acid phosphatase in the mammalian ovary, particularly as it relates to oocytes and accompanying follicle cells, comes from the light microscopic studies made by Arvy (1960), Lobel, Rosenbaum and Deane (1961), Banon, Brandes and Frost (1964), Adams and Hertig (1964), and Dott (1969). In view of the fact that the hydrolytic enzymes of lysosomes may be involved in the economy of cells as well as physiological autolysis, it seemed of interest to submicroscopically localize acid phosphatase in developing oocytes (Anderson, 1970) and their companion cells with the hope of yielding a better understanding of certain processes of the adult mammalian ovary.

Previous studies (Anderson, 1967) have shown that mammalian oocytes take up, by the process of pino- or endocytosis, the plant protein horseradish peroxidase when administered exogenously. A second purpose of the present study is to extend the aforementioned observations and thus to provide a more detailed picture of the uptake of peroxidase.

II. MATERIALS AND METHODS

A. General

Animals used in this study were rabbits (Dutch Belted), guinea pigs obtained from a local breeder, and mice (CD strain, 7-week-old) obtained from Charles River, Charles River, Mass. For general electron microscopy, ovaries were removed from Dibutal anesthesized animals and cut into small pieces and prefixed for 1½ hours in 3% glutaraldehyde buffered (pH 7.5) with 0.1 M sodium cacodylate. After fixation the material was washed in buffer and subsequently post-fixed in a 1% osmium tetroxide solution buffered (pH 7.5) with 0.1 M sodium cacodylate. The osmium fixed material was dehydrated in a graded series of ethanol, infiltrated and embedded in Epon (Luft, 1961). Thick sections of the Epon embedded material were made and stained with 1% fast green or according to the recommendations of Ito and Winchester (1963). Thin sections were cut with a Porter Blum MT-2 ultramicrotome stained with uranyl acetate followed with lead (Venable and Coggeshall, 1965) and examined with the Philips 200 and RCA EMU 3H electron microscopes.

B. Acid Phosphatase

For the localization of acid phosphatase ovaries were removed and fixed for one hour in 3% glutaraldehyde buffered (pH 7.4) with 0.1 M sodium cacodylate. After fixation the ovaries were cut approximately 25 μm thick with the TC-2 tissue sectioner. The sections were incubated for 1½ hours during which time they were shaken with a reciprocating shaker, in Gomori's medium as modified by Barka and Anderson (1962). After incubation the sections were post-fixed in osmium buffered with s-collidine, stained *en bloc* according to the recommendation of Karnovsky (1967) and processed for electron microscopy.

Controls were prepared by incubating the sections in Gomori's medium without the substrate β-glycerophosphate.

C. Horseradish Peroxidase

For the localization of horseradish peroxidase the organism used for these experiments were 7-week-old mice. Each of twelve animals was injected, via the tail vein, with 12 mg of horseradish peroxidase dissolved in mammalian Ringers. Three animals were killed 15 minutes subsequent to the injection; three were also killed at 30 min, 1 and 2 hours post-injection. Cotran and Karnovsky (1967) have reported that the injection of peroxidase caused temporary venular leakage in rats. Therefore, in a second series of experiments each of nine animals was injected intraperitoneally with 10 mg of horseradish peroxidase dissolved in mammalian Ringers. Three animals were killed 1 minute after the injection, and 3 five and 3 fifteen minutes subsequent to the injection. All of the ovaries were fixed in 3% glutaraldehyde and sectioned about 25 μm with the TC-2 tissue sectioner and incubated in the substrate 3,3' diaminobenzidine (Graham and Karnovsky, 1966) and processed for electron microscopy as indicated above.

For controls mice were injected with mammalian Ringers solution for the appropriate times, ovaries removed, fixed, sliced with the tissue sectioner and placed in Karnovsky's medium containing the substrate. After appropriate incubation times in the substrate, the tissue was processed for light and electron microscopy as indicated above.

III. OBSERVATIONS

In general the submicroscopic organization of the ovaries of the animals examined in this study are similar and reasonably well known. Minor differences in ovarian structures however do exist and these will be noted where appropriate (Adams and Hertig, 1964; Anderson and Beams, 1960; Odor and Blandau, 1969; Zamboni and Mastroianni, 1966).

To clarify the data obtained with the electron cytochemical and tracer

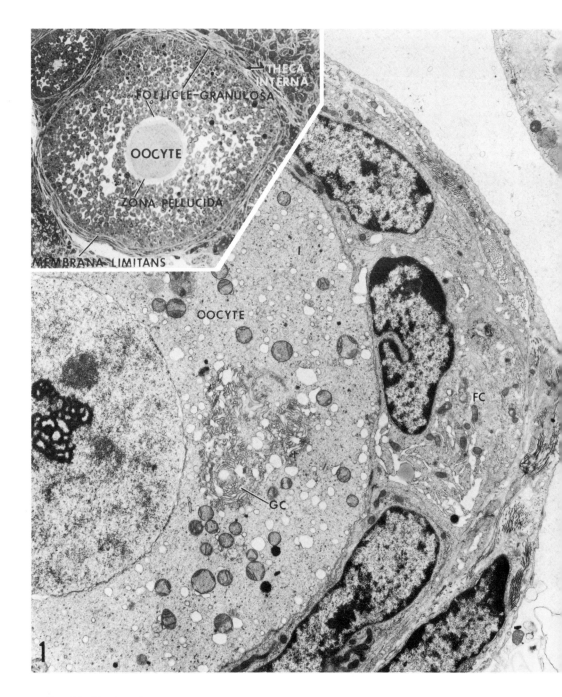

Fig. 5.1 The inset is a photomicrograph of a section through a differentiating Graafian follicle of the mouse illustrating the theca interna, follicle-granulosa cells, oocyte, zona pellucida and membrana limitans (X 210). The electron micrograph shows a young oocyte from the guinea pig. GC, Golgi complex. FC, follicle cells (X 8,300).

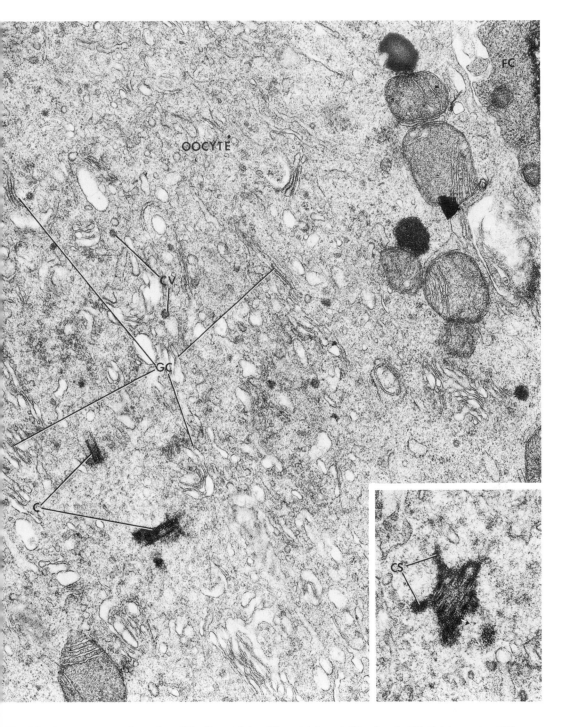

Fig. 5.2 A young oocyte from the rabbit. C, centrioles. CS, centriolar satellites (inset). GC, Golgi complex. CV, coated vesicles (X 5,400).

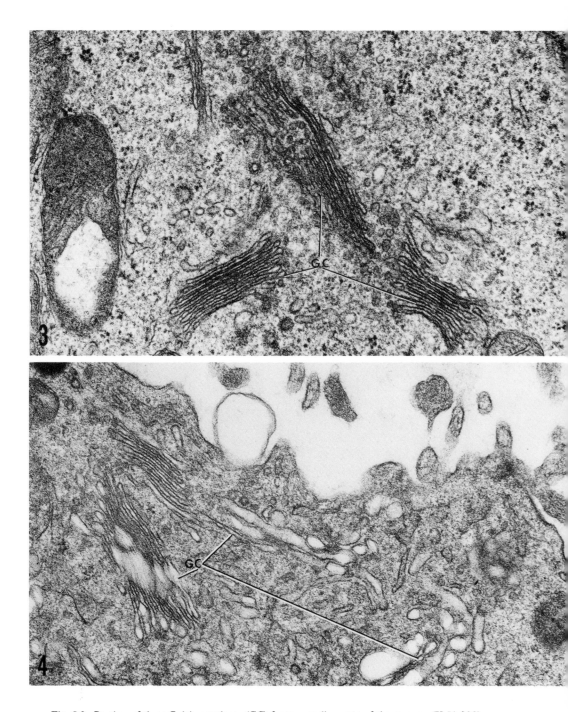

Fig. 5.3   Portion of three Golgi complexes (GC) from a small oocyte of the mouse   (X 51,200).

Fig. 5.4   A small portion of the peripheral ooplasm of the rabbit illustrating a Golgi complex (GC)   (X 51,200).

techniques, we will call attention to (a) the organization of the follicle and (b) the disposition and architecture of specific organelles that are relevant to this exploration during the follicle's differentiation.

A. Organization of the Graafian Follicle

Each Graafian follicle may be divided into the following three physiological areas (Fig. 5.1 inset): (a) the vascularized theca interna, (b) the avascular follicle and granulosa cells and (c) the oocyte. The oocyte is encompassed by a non-cellular layer, the zona pellucida. The zona pellucida may also be considered as a physiological area. The theca is separated from the granulosa cells also by a non-cellular stratum referred to as the glassy membrane or membrana limitans (preferably basement lamina).

B. Fine Structure of Differentiating Oocytes

1. Golgi complex. In young oocytes that are surrounded by a single layer of flattened follicle cells (Figs. 5.1, 5.2, 5.5, 5.10, and 5.14, FC) the Golgi complex is paranuclear. It is a rather large organelle in the guinea pig (Fig. 5.1, GC) as compared to that in young oocytes of mice and rabbits (Fig. 5.2, GC). In all instances the Golgi complex consists of a number of saccules and associated coated vesicles (Fig. 5.2, CV).

As frequently illustrated by classical cytologists (Balbiani, 1893; Henneguy, 1893; van der Stricht, 1923; Wilson, 1927), but infrequently demonstrated by contemporary cell biologists, the large juxtanuclear Golgi complex of young oocytes is usually associated with centrioles (Fig. 5.2, C). The two centrioles are not closely associated with each other; however, they are oriented perpendicular to each other. Associated with the centrioles are satellites (Fig. 5.2, CS, inset).

As differentiation proceeds, there is an increase in the diameter of the oocyte and the number of follicle cells. The enlargement of the oocyte is accompanied by hyperplasia of the Golgi complexes (Fig. 5.3, GC). Each Golgi complex is composed of a variable number of saccules and associated coated vesicles. With further differentiation of the oocyte, each Golgi complex migrates towards the periphery of the oocyte (Figs. 5.4 and 5.5, GC). Therefore the oocyte is now composed of many Golgi complexes, but presumably involved in various metabolic activities (Anderson, 1969b). Some of the Golgi complexes are thought to be involved in the synthesis and concentration of precursors utilized in the construction of cortical granules (Anderson, 1968a and b; Szollosi, 1967). In Figure 5.5 (CG) the dense material within the Golgi-associated vesicles is identified as "immature" cortical granules. Other Golgi complexes are affiliated with multivesicular bodies (Fig. 5.6, MV), whereas still others are intimately associated with cisternae of the rough endoplasmic reticulum. Frequently, a vesicle is fused with the multivesicular body (Fig. 5.6, FV). Figure 5.7 depicts evaginations (arrows) of a cisterna, of the rough endoplasmic reticulum adjacent to the so-called forming face of a Golgi complex

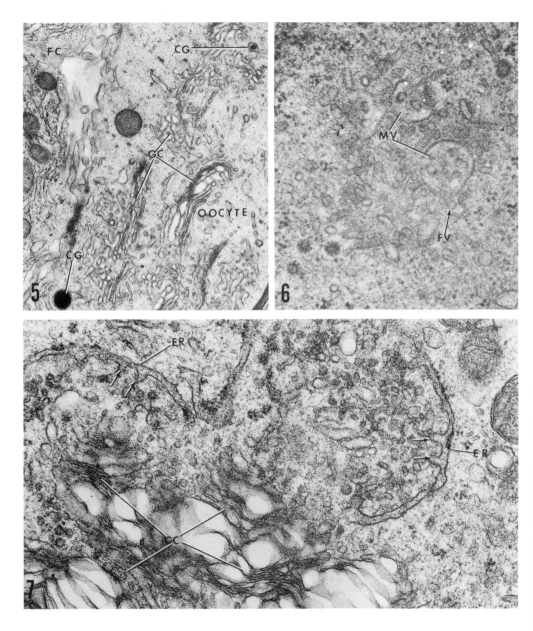

Fig. 5.5 The peripheral ooplasm of an oocyte from the guinea pig. GC, Golgi complex. CG, "immature" cortical granules. FC, follicle cell (X 15,100).

Fig. 5.6 Multivesicular bodies (MV) within the ooplasm of an oocyte from the mouse. Note vesicle (FV) fused with the multivesicular body (X 30,200).

Fig. 5.7 A small portion of the peripheral ooplasm of an oocyte from the rabbit. ER, cisternae of the endoplasmic reticulum, evaginations of the cisternae of the endoplasmic reticulum. GC, Golgi complex (X 45,400).

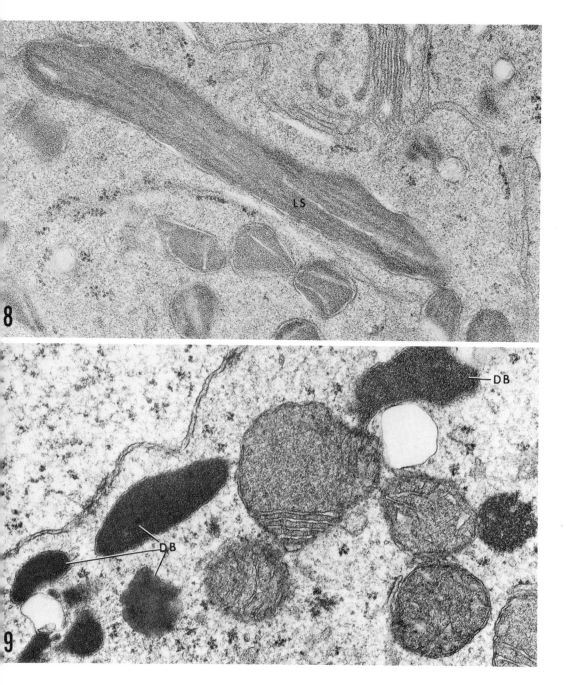

Fig. 5.8 The ooplasm of a young oocyte of a guinea pig showing an elongate lamella structure (LS) (X 49,700).

Fig. 5.9 A small portion of the ooplasm of an oocyte from a rabbit illustrating some irregularly shaped membrane-bounded dense bodies (DB) (X 49,700).

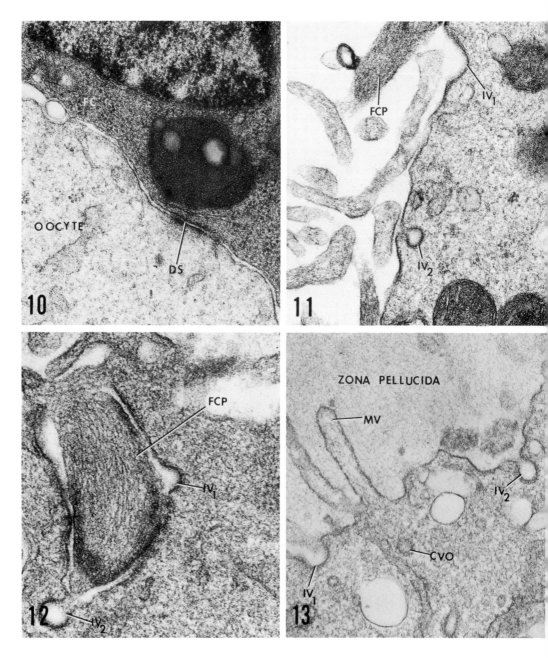

Fig. 5.10   A very young oocyte (OC) of the mouse illustrating a desmosome (DS). FC, follicle cell   (X 48,200).

Figs. 5.11 - 5.13   Small portions of the periphery of oocytes from the rabbit illustrating the zona pellucida, cytoplasmic projections of follicle cells (FCP), microvilli of the oocyte (MV) and varying degrees of invaginations (IV$_1$, IV$_2$) of the oolemma. CVO, coated vesicle within the ooplasm   (Fig. 5.11 X 24,100) (Figs. 5.12 and 5.13 X 56,300).

(Novikoff *et al.,* 1964).

In the guinea pig one frequently sees some elongate structures like that shown in Fig. 5.8 (LS) and which have been identified as modified mitochondria (Anderson and Beams, 1960; Anderson, Condon, and Sharp, 1970).

Scattered within the ooplasm of young oocytes of rabbits are some irregularly shaped dense bodies (Fig. 5.9, DB). The origin of these structures is unknown.

2. Oolemma. The oolemma shows invaginations of varied sizes. The invaginations possess a fuzzy coat on their ooplasmic side (Figs. 5.11-5.13, IV_1 and IV_2). There appear to be more of these invaginations at the surface of larger oocytes than at the surface of small ones. Fuzzy coated vesicles not associated with the Golgi complex (Fig. 5.13, CVO) are also present in the ooplasm. Such coated vesicles are thought to be derived from the invaginations of the oolemma. The polysaccharide-rich zona pellucida (Fig. 5.13) overlies, but does not enter, the invaginations.

3. Cytology of Follicle-Granulosa Cells. It has been reported that there are functional differences between the follicle (Biggers, Whittingham and Donahue, 1967) and granulosa cells (Bjorkman, 1962). This investigation however has revealed no unique submicroscopic morphological cytoplasmic features of these cells. Follicle-granulosa cells are associated with each other by both gap junctions and desmosomes (Anderson *et al.,* 1970) whereas the follicle cells are coupled with the oocyte only by desmosomes (Fig. 5.10, DS) (Anderson and Beams, 1960). The plasmalemma of follicle-granulosa cells also possess invaginations of varied sizes which also possess a fuzzy coat on their cytoplasmic side.

Follicle and granulosa cells contain a cilium (9+0) that is directed towards the oocyte (Figs. 5.14-5.15) (Beams and King, 1938). Pertinent to the present report is the fact that the Golgi complex is associated with the basal body-centriole complex (Figs. 5.14-5.15, BB) or centrosome. Like the cilium, the Golgi complex is also polarized toward the oocyte (Figs. 5.14-5.16, GC). Associated with the saccules of the Golgi complexes are fuzzy coated vesicles that presumably originate from the saccules of the Golgi complex (Fig. 5.16, CVF). Multivesicular bodies and some membrane-bounded structures with dense interiors appear in the vicinity of the Golgi complex in some of the follicle-granulosa cells.

C. Acid Phosphatase

1. Differentiating Oocytes. Fig. 5.17 (GC) is the Golgi complex of a young oocyte from the guinea pig showing the distribution of acid phosphatase. The enzyme appears in certain saccules and vesicles of the organelle. In older oocytes, acid phosphatase is located in a few of the saccules and vesicles of the peripherally situated Golgi complex (Figs. 5.18-5.20, GC). The acid phosphatase positive vesicles are interpreted as primary lysosomes. In addition to the saccules of the Golgi complex and associated vesicles, the multivesicular bodies are acid phosphatase

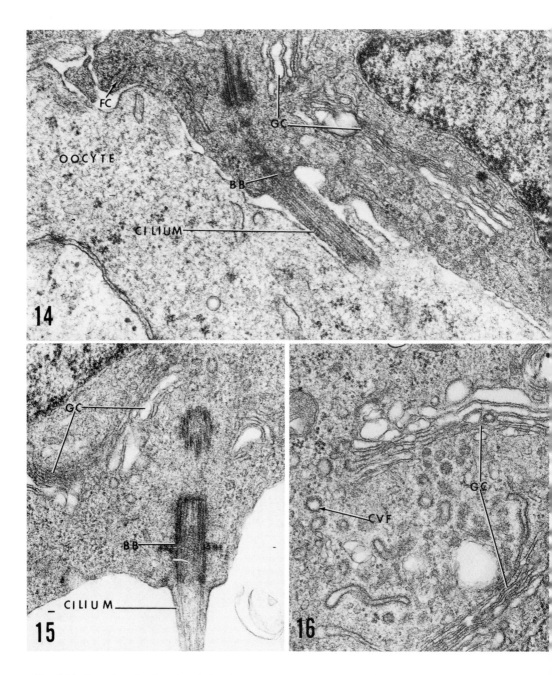

Fig. 5.14 A section showing the relation between a young oocyte and a follicle cell (FC) of the rabbit. Cilium. BB, basal body. GC, Golgi complex (X 9,700).

Fig. 5.15 A section showing a small portion of the apical end of a granulosa cell of the rabbit. Cilium. BB, basal body. GC, Golgi complex (X 9,700).

Fig. 5.16 This image depicts the Golgi complex (GC) of a follicle cell of the rabbit. CVF, coated vesicles (X 48,200).

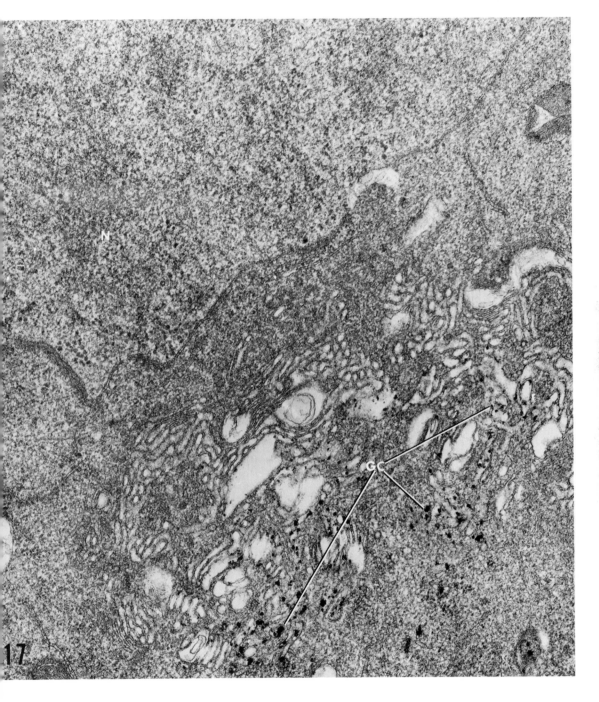

Fig. 5.17 A section through a very young oocyte of the guinea pig illustrating the large paranuclear (N) Golgi complex (GC) containing acid phosphatase (represented by the electron dense material) in some of its saccules and vesicles (X 35,000).

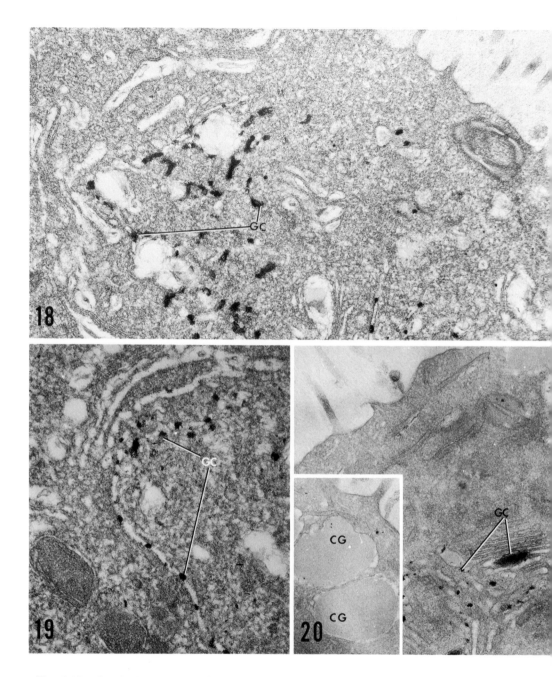

Figs. 5.18 and 5.19 Small portions of the periphery of rather large oocytes of rabbits showing acid phosphatase (electron dense material) associated with the Golgi complex (GC)   (X 48,200).

Fig. 5.20 A section through the periphery of an oocyte of the guinea pig showing acid phosphatase in the Golgi complex (GC). The inset illustrates the acid phosphatase negative cortical granules (CG)   (X 32,200).

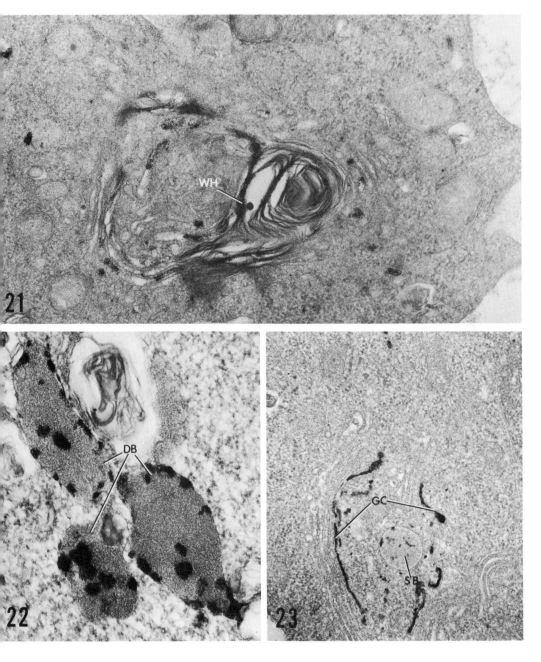

Fig. 5.21 A section through a granulosa cell of the guinea pig showing acid phosphatase associated with membranes (WH) organized in a whorl-like configuration (X 31,200).

Fig. 5.22 Acid phosphatase positive dense bodies (DB) from an oocyte of the rabbit (X 46,800).

Fig. 5.23 Acid phosphatase in the Golgi complex (GC) of a follicle cell from the rabbit. Note the acid phosphatase positive spherical body (SB) (X 31,200).

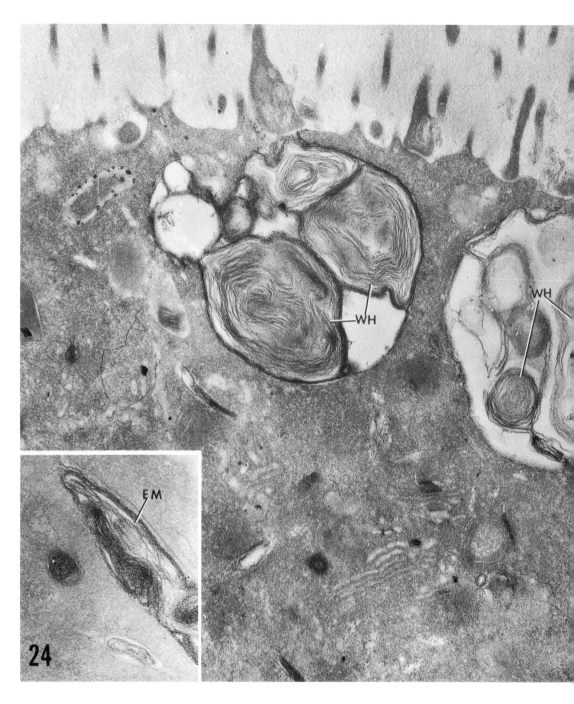

Fig. 5.24 A section through the periphery of a rather large oocyte of the guinea pig. WH, acid phosphatase positive membranes organized in a whorl-like configuration. The inset depicts an elongate acid phosphatase positive membranous (guinea pig) structure (EM)   (Fig. 5.24 X 35,900) (Inset X 62,800).

positive. Cortical granules are acid phosphatase negative (Fig. 5.20, inset, CG).

The dense irregularly shaped structures found in the oocytes of rabbits, and previously pointed out in Figure 5.9 (DB), and the elongate structures encountered in the oocytes of guinea pigs and identified as modified mitochondria (Fig. 5.8, LS) are acid phosphatase postive (Fig. 5.22, DB and Fig. 5.24 inset, EM). These structures are interpreted to be secondary lysosomes.

In many of the oocytes of the animals studied, regardless of their size, are some membranes displaying a whorl-like configuration (Fig. 5.24, WH). These membranes are acid phosphatase positive and are thought to be focal areas of degeneration (Swift and Hruban, 1964), an indication of atresia.

2. Follicle-Granulosa Cells.   The major portion of the acid phosphatase is found associated with the saccules and vesicles of the Golgi complex (Figs. 5.23, 5.25, 5.26, GC) in follicle and granulosa cells. In addition to the acid phosphatase positive dense bodies of follicle and granulosal cells (Fig. 5.23, SB), there are also some membranes displaying a whorl-like configuration that are also acid phosphatase positive (Fig. 5.21, WH). These structures are also thought to be focal areas of degeneration.

D. Horseradish Peroxidase

1. Differentiating Oocytes.   Within two minutes after intraperitoneal injection the horseradish peroxidase travels, unimpeded, to the ovary proper via the space between (IS) the cells of the ovarian epithelium (Fig. 5.27). This is unlike the situation as it exists in the mammalian testis (Fawcett, Leak and Heidger, 1970). Whether the peroxidase is administered intravenously or intraperitoneally, it gains access to the oocyte and follicle-granulosa cells by way of the relatively thick membrana limitans (basement lamina) (Figs. 5.28-5.30, ML), upon which either the follicle or granulosa cells rests.

The information presented below is from animals injected intravenously. Figure 5.28 is a photomicrograph of a group of young oocytes (OC) 30 min after exposure to the protein and Figs. 5.29-5.30 are photomicrographs of older oocytes (OC) after one hour exposure. In older oocytes the reaction product can be seen in the zona pellucida (Figs. 5.29 and 5.30, ZP).

After 15 minutes the reaction product is found at the oocyte-follicle cell interface (Fig. 5.31, IF) of those small oocytes with no microvilli and are completely surrounded by follicle cells. Little or no reaction product is found within the ooplasm of these oocytes. In slightly older oocytes the reaction product is found on the surface of the cytoplasmic projections of follicle cells, on the surface of microvilli of the oocytes, in the zona pellucida, and within the invaginations of the oolemma. We have never observed the peroxidase entering the oocytes via the desmosomes.

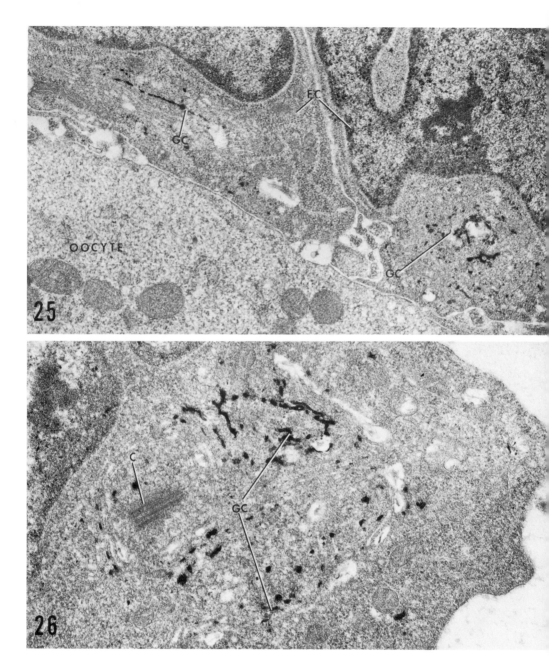

Fig. 5.25 A section showing the relation between two follicle cells (FC) and an oocyte of the rabbit. GC, acid phosphatase positive Golgi complex   (X 24,100).

Fig. 5.26 A granulosa cell from the follicle of a rabbit showing the acid phosphatase positive Golgi complex (GC) organized around a centriole (C)   (X 32,200).

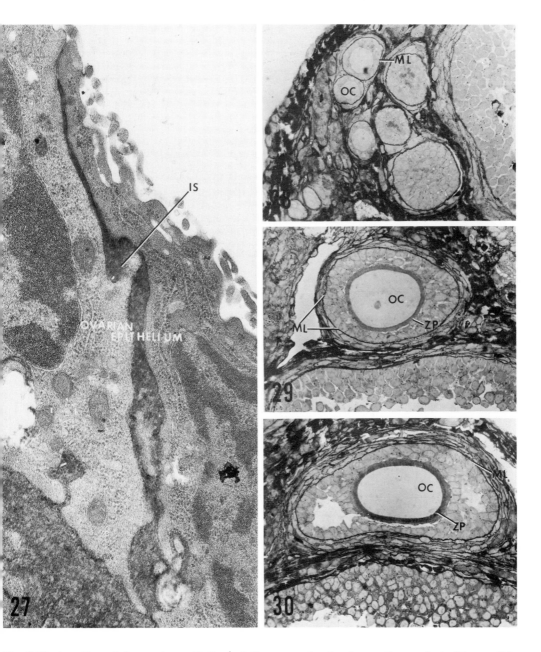

Fig. 5.27 A section of the ovarian epithelium of the mouse showing the reaction product of horseradish peroxidase in the intercellular spaces (IS) (X 31,200).

Figs. 5.28 - 5.30 Photomicrographs of various sized oocytes (OC) from the mouse illustrating the reaction product (the reaction product appears black) in the membrana limitans (ML) and zona pellucida (ZP). Figures 5.28 and 5.29 are from ovaries 30 minutes and Figure 5.30 one hour after injection of horseradish peroxidase (X 195).

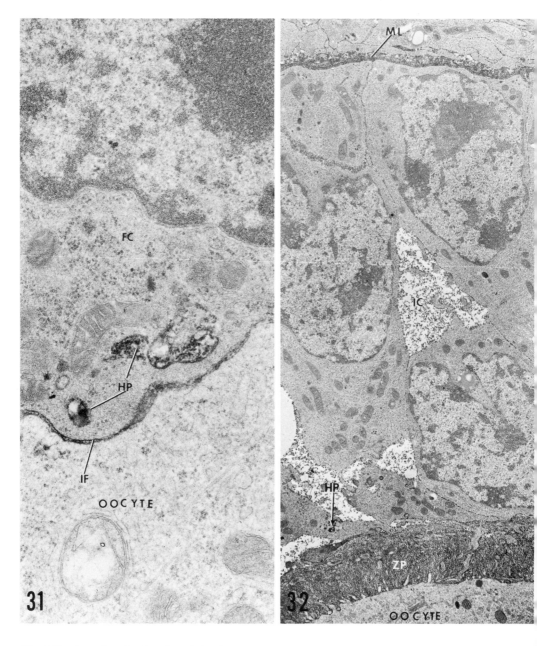

Fig. 5.31 A section showing the relation between a very young oocyte (mouse) and one of its encompassing follicle cells (FC). Thirty minutes after injection of horseradish peroxidase the reaction product is found at the oocyte-follicle cell interface (IF) and within the cytoplasm of follicle cells (HP)   (X 39,000).

Fig. 5.32 A section through a young oocyte of the mouse (one hour after peroxidase injection) showing the distribution of the reaction product in the zona pellucida (ZP) intercellular spaces (IC) within the follicle cells (HP) and membrana limitans (ML)   (X 7,000).

Figure 5.33 is a section taken from an oocyte after 30 minutes exposure to the protein where the material is located in an invagination (IV). Figure 5.34 illustrates a vesicle (HVP) and an invagination (IV) containing the reaction product one hour after exposure. In some oocytes, after one hour exposure to the protein, the reaction product is found within the multivesicular bodies and vesicles associated with some of the Golgi complexes (Fig. 5.35, MBV). This is similar to what is observed in follicle-granulosa cells (see below) where the product appears to concentrate in the region of the Golgi complex. Deep within the ooplasm the reaction product is seen within the vesicles (Fig. 5.37, VHP) and elongate structures (Fig. 5.37, THP). It is believed that the horseradish peroxidase is internalized in coated vesicles. These vesicles appear to fuse with each other producing elongate tubular (THP) structures as in Fig. 5.37. Some vesicles may fuse with multivesicular bodies whereas one may interpret the structure labeled TMV in Figure 5.36 as a tubular structure that fused with a multivesicular body prior to uptake of horseradish peroxidase.

After 1½ - 2 hours no reaction product can be found within the zona pellucida.

2. Follicle-Granulosa Cells.   Fifteen to thirty minutes after exposure to the protein one finds relatively little of the reaction product in follicle cells around young oocytes (Fig. 5.31, HP). This is not the case for follicle-granulosa cells encompassing large oocytes exposed for the same length of time. Figures 5.38 and 5.39, IFC, depict the reaction product between cells, within the invaginations (Fig. 5.39, IVP) and coated vesicles (Fig. 5.39, CVP) fifteen minutes after injection of the tracer. Presumably, as in the case of the oocyte, the horseradish peroxidase gains access to the cytoplasm via the invaginations of the plasmalemma. Once inside of the cytoplasm the coated vesicle loses its coat and fuses with others forming vesicles of varied sizes (Figs. 5.39 and 5.40, VP). These non-coated vesicles and elongate structures (Fig. 5.40, SP and EL) can be found scattered within the cytoplasm; although most congregate within the region of the Golgi complex (GC) and centrioles (C) (Fig. 5.40). Frequently one sees an "empty" appearing vesicle fused (Fig. 5.40, inset EV) with one containing the reaction product (Fig. 5.40, VHP). The "empty" vesicle is thought to be a primary lysosome.

IV. DISCUSSION

One of the observations presented in this paper indicates that in both the oocyte and its accompanying follicle-granulosa cells some saccules and associated coated vesicles of the Golgi complexes are rich in acid phosphatase. The coated vesicles appear to be produced by a pinching off process from the tips of the Golgi saccules. Moreover it was observed that in differentiating oocytes all Golgi complexes are not associated with any one specific structure or organelle. For example, some are affiliated with multivesicular bodies; some with cortical granules; and some with cisternae of the rough endoplasmic reticulum. Multivesicular bodies are also acid phosphatase positive. The cortical granules are acid phosphatase negative. In follicle-granulosa cells, the Golgi complex is polarized in that it is

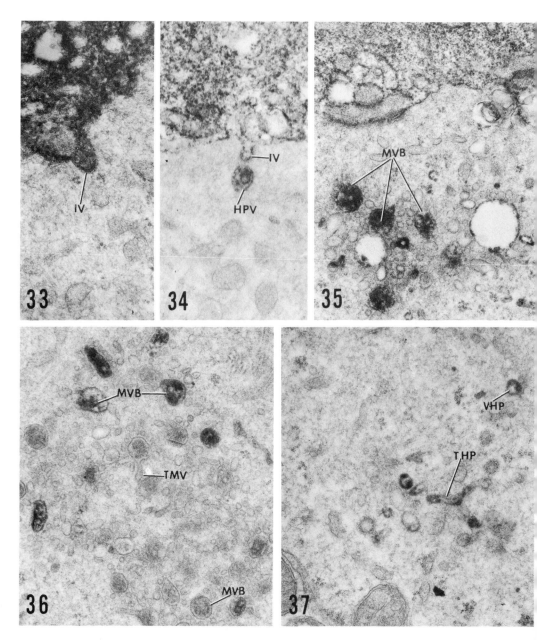

Figs. 5.33 and 5.34 Sections of young oocytes of the mouse injected with horseradish peroxidase 30 minutes and 1 hour respectively prior to fixing. Note the reaction product in the invagination (IV) and within a vesicle (Fig. 5.34, HPV) in the ooplasm   (Fig. 5.33 X 46,800) (Fig. 5.34 X 31,200).

Figs. 5.35 - 5.37 Sections depicting horseradish peroxidase within multivesicular bodies (MVB) and vesicles (VHP) of young mouse oocytes injected with horseradish peroxidase one hour prior to fixation. Note the tubular like structure (Fig. 5.36, TMV) associated with one multivesicular body and tubular structures (THP) containing the reaction product   (X 31,200).

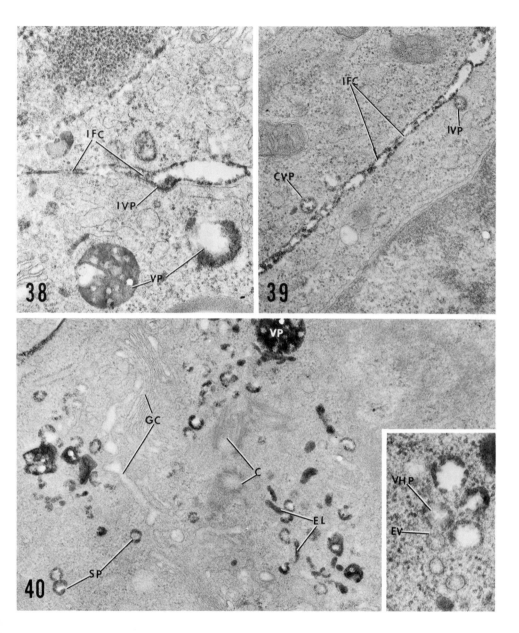

Figs. 5.38 and 5.39 Sections showing the distribution of horseradish peroxidase (mouse) 15 minutes subsequent to the injection of the protein between follicle cells (IFC), in invaginations (IVP), coated vesicles (CVP) and large vacular structures (VP)   (Fig. 5.38 X 30,200) (Fig. 5.39 X 22,700).

Fig. 5.40 This electron micrograph shows the distribution of horseradish peroxidase in the region of the centrosphere in the follicle cell of the mouse 30 minutes subsequent to the injection of protein. The reaction product is found in elongate (EL) large vacular structures (VP) and spherical structures (SP). Note the Golgi complex (GC) and centrioles (C). The inset illustrates an "empty" vesicle (EV) fused with one containing the reaction product (VHP)   (Fig. 5.40 X 30,200) (Inset X 60,500).

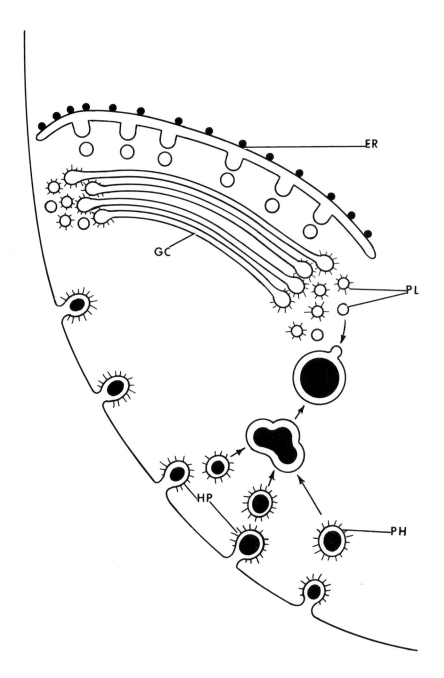

Fig. 5.41 Diagrammatic representation of the fate of the exogeneously administered protein in the cells comprising the Graafian follicle. Enzyme proteins in the lysosome are manufactured by the rough endoplasmic reticulum (ER) and later concentrated by the Golgi complex (GC). The Golgi complex, by a process of budding, produce primary lysosome (PL). The cells of the Graafian follicle take up the horseradish peroxidase by endo- or pinocytosis (HP), a process that produces a membrane-bounded phagosome (PH). The phagosome fuses with a primary lysosome where digestion takes place.

associated with the basal body-centriole complex. In the differentiating mammalian oocyte and the follicle granulosa cell it is possible that the Golgi complexes may play a role in many metabolic activities (Anderson, 1969b) and we envision one of these to be the production of primary lysosomes. The high concentration of enzymes in the lysosome has prompted others to suggest, and we agree, that the acid hydrolases of the organelle are manufactured by the endoplasmic reticulum. The close morphological association between the Golgi complex and endoplasmic reticulum may be viewed as the kind of companionship that facilitates the transfer of the lysosomal enzymes from the endoplasmic reticulum to the Golgi complex (Fig. 5.41). Subsequently, the Golgi saccules, by a process of budding, may produce vesicles containing the hydrolayses and which may be considered primary lysosomes (Novikoff, 1961; Novikoff and Shin, 1964).

What might be the function of primary lysosomes in cells comprising the differentiating Graafian follicle? We have already pointed out that one of the functions of the lysosomes is its involvement in intracellular digestion. In connection with this, some of the data presented in this study reveal that the oocytes and follicle cells take up exogeneously administered horseradish peroxidase by the process of endo- or pinocytosis. The follicle-granulosa cells take up more of the tracer than the oocytes. Large oocytes however take up more of the protein compared to the small oocytes that possess no microvilli and which are completely enringed by a layer of flattened follicle cells. This presumably is neither a function of the availability of the horseradish peroxidase nor a function of the junctional complex between adjacent follicle cells but rather appears to be a function of pinocytotic activity. In any event, once the protein gains access to the ooplasm and the cytoplasm of the follicle-granulosa cells its distribution appears to be somewhat similar in that the protein appears to flow in vesicles directly to the region of the Golgi complex. This situation is much like that of certain blood cells (Cohn and Fedorko, 1969). In the follicle cells we observed a coated vesicle fused with one whose interior contained the reaction product of the protein. Since some of the companion vesicles of the Golgi complex of both the oocyte and follicle-granulosa cells contain acid phosphatase we believe that the coated vesicle contains acid phosphatase and would therefore be considered a primary lysosome adding its content to the ingested material. The picture which emerges from this study is as follows: The oocytes and follicle-granulosa cells take up the horseradish peroxidase by endocytosis, a process that produces a membrane-bounded vacuole or phagosome (Fig. 5.41, PH). The phagosome fuses with a primary lysosome (Fig. 5.41, PL) where digestion takes place uninterrupted (see Friend and Farquhar, 1967; Holtzman and Dominitz, 1968; Straus, 1964a, b, 1967; Ryser, 1968).

When digestion is completed in the oocytes and follicle-granulosa cells the resulting body, as in other cell types, is a residual body. We believe that some of the dense bodies of varying diameters seen within these cells may be residual bodies. If this is the case, the following question arises: What is the fate of residual bodies in cells comprising the differentiating Graafian follicle? The answer to this question as it relates to the Graafian follicle is unknown. It is known however that many cell release residual bodies by the process referred to as exocytosis, the reverse of endocytosis (Maunsbach, 1969). However information obtained by other investigators for other cell types prompts the following discussion: As is well known secretory cells when properly stimulated discharge their secretory material. Moreover in certain cells material can move through the cytoplasm without being digested (diacytosis, Jacques, 1969; cytopensis, Moore and Ruska, 1957). Furthermore, certain substances that are refractive to digestion may be stored, as in lymphatic cells (Leak and Burke, 1968). In such cases the cells become overloaded. About the cell's inability to unload residual bodies, DeDuve and Wattiaux (1966) wrote that, "In certain cells, the system can unload itself in the extracellular medium by exocytosis, a process which will take any character between regurgitation, defecation, and secretion, depending on the relative amounts of original material, undigestible residues, and enzymes discharged. In full "constipated" cells, this traffic leads into a blind alley; the system becomes progressively engorged with residues up to a point when pathological derangements of the system itself and of other cell functions start manifesting themselves. Aging and death may well be the final outcome of these processes."

We have found no morphological evidence in mammalian oocytes for reverse endocytosis. Assuming that the developing oocyte with its restricted metabolic activity (Biggers *et al.,* 1967) lacks the capability to engage in exocytosis, it is feasible that this may signal its autolysis and thereby initiate atresia of the Graafian follicle.

Much has already been written on the atresia of the Graafian follicle of mammalian ovaries and this literature will not be reviewed here. Readers are directed to the reports by Knigge and Leathem (1956) and Ingram (1962).

It might be profitable to make some further comments on the general significance of pinocytosis in differentiating oocytes. In reference to pinocytosis in mammalian oocytes Anderson and Beams (1960) suggested that pinocytosis may serve as a mechanism for internalizing nutritive substances. Since this suggestion was made, it has been found that particular oocytes within the same ovary of a wide variety of organisms remove specific proteins (vitellogenins) from the blood by selective endocytosis and store them as yolk (Anderson, 1964; Telfer, 1960; Roth and Porter, 1964; Raven, 1961). These specific proteins are manufactured by an organ other than the ovary and have been called heterosynthetic by Schechtman (1955). Moreover other eggs synthesize their own yolk, a situation Schechtman (1955) refers to as autosynthetic (Anderson, 1969b; Anderson and Huebner, 1968; Dumont and Anderson, 1967). It is possible that the mechanism of active transport in the form of pinocytosis in differentiating oocytes throughout the animal kingdom, while amplified in some forms such as insects (Anderson, 1964; Anderson and Telfer, 1970; Kunkel and Pan, personal communication), amphibian (Wallace and Dumont, 1968), fish (Anderson, 1968a; Droller and Roth, 1966), aves (Bellairs, 1964) and reptiles (cf. Schjeide *et al.,* 1970), and not exaggerated in forms such as crustacea (Beams and Kessel, 1963) and mammals (Glass, 1966; 1970; Glass and Cons, 1968; Hahn, Church and Gorbman, 1969; Mancini, Villar, Heinrich, Davidson and Alvarez, 1963) may nevertheless be serving the same roles, namely internalizing metabolites for oogenesis and picking and selecting specific macromolecules to be used during embryogenesis or during early postnatal development.

V. SUMMARY

The localization of acid phosphatase and the uptake of horseradish peroxidase by cells comprising the differentiating Graafian follicle of young adult virgin mice, guinea pigs and rabbits was studied by techniques of light and electron microscopy. In developing oocytes and follicle-granulosa cells acid phosphatase activity appears in the Golgi complex, multivesicular bodies and dense bodies of varying sizes and shapes. The cortical granules are acid phosphatase negative.

Horseradish peroxidase enters the oocyte and its encompassing follicle-granulosa cells via coated vesicles formed by endocytosis. These vesicles subsequently lose their coat and some fuse with one another forming rather large smooth vacuoles; some fuse with multivesicular bodies, and others with "empty" appearing vesicles thought to be primary lysosomes derived from the Golgi complex. It is suggested that lysosomes within cells of the Graafian follicle are normally involved in intracellular digestion and may initiate its atresia.

VI. ACKNOWLEDGMENTS

This investigation was supported by a grant (HD-04924) from the National Institute of Child Health and Human Development.

The author wishes to thank Mrs. Gloria S. Lee, Mr. and Mrs. L. Musante and Mr. Richard Letourneau for their technical assistance.

VII. REFERENCES

Adams, E. C., and Hertig, A. T. (1964) Studies on guinea pig oocytes. 1. Electron microscopic observations on the development of cytoplasmic organelles in oocytes of primordial and primary follicles. *J. Cell Biol.* **21**, 397.

Anderson, E. (1964) Oocyte differentiation and vitellogenesis in the roach *Periplaneta americana.* *J. Cell Biol.* **20**, 131.

Anderson, E. (1967) Observations on the uptake of horseradish peroxidase by developing oocytes of the rabbit. *J. Cell Biol.* **35**, (pt. 2), 160A.

Anderson, E. (1968a) Cortical alveoli formation and vitellogenesis during oocyte differentiation in the pipefish, *Syngnathus fuscus,* and the killifish, *Fundulus heteroclitus. J. Morph.* **125**, 23.

Anderson, E. (1968b) Oocyte differentiation in the sea urchin, *Arbacia punctulata* with particula reference to the origin of cortical granules and their participation in the cortical reaction *J. Cell Biol.* **37**, 514.

Anderson, E. (1969) Oogenesis in the cockroach, *Periplaneta americana,* with special reference to the specialization of the oolemma and the fate of coated vesicles. *J. de Microscopie.* **8** 721.

Anderson, E. (1970) The localization of acid phosphatase activity in developing oocytes anc associated follicle cells of mammals. *Anat. Rec.* **166**, 271.

Anderson, E., and Beams, H. W. (1960) Cytological observations on the fine structure of the guinea pig ovary with special reference to the oogonium, primary oocyte and associatec follicle cells. *J. Ultrastruct. Res.* **3**, 432.

Anderson, E., and Huebner, E. (1968) Development of the oocyte and its accessory cells of the polychaete, *Diopatra cuprea* (Bosc). *J. Morph.* **126**, 163.

Anderson, E., Condon, W., and Sharp, D. (1970) A study of oogenesis and early embryogenesis in the rabbit, *Oryctologus cuniculus* with special reference to the structural changes o, mitochondria. *J. Morph.* **130**, 67.

Anderson, L. M., and Telfer, W. H. (1970) Extracellular concentrating of proteins in Cecropi moth follicle. *J. Cell Physiol.* **76**, 37.

Arvy, L. (1960) Contribution a l'histoenzymologie de l'ovaire. *Z. Zellforsch.* **51**, 406.

Balbiani, E. G. (1893) Centrosome et "Dotterkern". *Jour. Anat. et Physiol.* **29**, 145.

Banon, P., Brandes, D., and Frost, J. K. (1964) Lysosomal enzymes in the rat ovary anc endometrium during the estrous cycle. *Acta Cytol.* **8**, 416.

Barka, T. and Anderson, P. J. (1962) Histochemical methods for acid phosphatase using hexazonium pararosanilin as coupler. *J. Histochem. Cytochem.* **10**, 741.

Beams, H. W. and Kessel, R. G. (1963) Electron microscope studies on developing crayfish oocytes with special reference to the origin of yolk. *J. Cell Biol.* **18**, 621.

Beams, H. W. and King, R. L. (1938) A study of the cytoplasmic components and inclusions of the developing guinea pig egg. *Cytologia.* **8**, 353.

Bellairs, R. (1964) Biological aspects of the yolk of the hen's egg. *Advances Morph.* **4**, 217.

Biggers, J. D., Whittingham, D. C., and Donahue, R. P. (1967) The pattern of energy metabolism in the mouse oocyte and zygote. *Proc. Nat. Acad. Sci. U.S.A.* **58**, 560.

Bjorkman, N. (1962) A study of the ultrastructure of the granulosa cells of the rat ovary. *Acta Anat.* **51**, 125.

Cohn, Z. A. and Fedorko, M. E. (1969) The formation and fate of lysosomes. *In* "Lysosomes in Biology and Pathology". (J. T. Dingle and H. B. Fell, eds.) Vol. 1, p. 43. London, North-Holland Publishing Company.

Cotran, R. S. and Karnovsky, M. T. (1967) Vascular leakage induced by horseradish peroxidase in the rat. *Proc. Soc. Exp. Biol. Med.* **126**, 557.

DeDuve, C. (1969) The lysosome in retrospect. *In* "Lysosomes in Biology and Pathology". (J. T. Dingle and H. B. Fell, eds.) Vol. 1, p. 3. London, North-Holland Publishing Company.

DeDuve, C. and Wattiaux, R. (1966) Functions of lysosomes. *Ann. Rev. Physiol.* **28**, 435.

Dott, H. M. (1969) Lysosomes and lysosomal enzymes in the reproductive tract. *In* "Lysosomes in Biology and Pathology". (J. T. Dingle and H. B. Fell, eds.) Vol. 1, p. 330. London, North-Holland Publishing Company.

Droller, M. J. and Roth, T. F. (1966) An electron microscope study of yolk formation during oogenesis in *Labistes reticulatus* Guppyi. *J. Cell Biol.* **28**, 209.

Dumont, J. N. and Anderson, E. (1967) Vitellogenesis in the horseshoe crab, *Limulus polyphemus*. *J. de Microscopie.* **6**, 791.

Fawcett, D. W., Leak, L. V. and Heidger, P. M. (1970) Electron microscopic observations on the structural components of the blood-testes barrier. *J. Reprod. Fert. Suppl.* **10**, 105.

Friend, D. S. and Farquhar, M. G. (1967) Functions of coated vesicles during protein absorption in the rat vas deferens. *J. Cell Biol.* **35**, 357.

Glass, L. E. (1966) Serum antigen transfer in the mouse ovary: Dissimilar localization of bovine albumin and globulin. *Fertil. Steril.* **17**, 226.

Glass, L. (1970) Translocation of macromolecules. *In* "Cell Differentiation". (O. A. Schjeide and J. de Vellis, eds.) p. 42. New York, Van Nostrand-Reinhold.

Glass, L. E. and Cons, J. M. (1968) Stage dependent transfer of systemically injected foreign protein antigen and radiolabel into mouse ovarian follicles. *Anat. Rec.* **162**, 139.

Graham, R. C., Jr. and Karnovsky, M. J. (1966) The early stages of absorption of injected horseradish peroxidase in the proximal tubules of mouse kidney: Ultrastructural cytochemistry by a new technique. *J. Histochem. Cytochem.* **14**, 291.

Hahn, W. E., Church, R. B., and Gorbman, A. (1969) Organ-specific estrogen-induced RNA synthesis resolved by DNA-RNA hybridization in the domestic fowl. *Proc. Nat. Acad. Sci. U.S.A.* **62**, 112.

Henneguy, M. G. (1893) Le corps vitellin de Balbiani dans l'oeuf des vertebres. *Jour. Anat. de Physiol.* **39**, 68.

Holtzman, E. and Dominitz, R. (1968) Cytochemical studies of lysosomes, Golgi apparatus and endoplasmic reticulum in secretion and protein uptake by adrenal medulla cells of the rat. *J. Histochem. Cytochem.* **16**, 320.

Ingram, D. L. (1962) Atresia. *In* "The Ovary". (S. Zuckerman, ed.) Vol. 1, p. 247. New York, Academic Press.

Ito, S. and Winchester, R. J. (1963) The fine structure of the gastric mucosa in the bat. *J. Cell Biol.* **16**, 541.

Jacques, P. J. (1969) Endocytosis. *In* "Lysosomes". (J. T. Dingle and H. B. Fell, eds.) Vol. 2, p. 395. Amsterdam-London, North-Holland Publishing Company.

Karnovsky, M. J. (1967) The ultrastructural basis of capillary permeability studied with peroxidase as a tracer. *J. Cell Biol.* **35**, 213.

Knigge, K. M. and Leathem, J. H. (1956) Growth and atresia of follicles in the ovary of the hamster. *Anat. Rec.* **124**, 679.

Leak, L. V. and Burke, J. F. (1968) Electron microscopic study of lymphatic capillaries in the removal of connective tissue fluids and particulate substances. *Lymphology.* **1**, 39.

Lobel, B. L., Rosenbaum, R. M. and Deane, H. W. (1961) Enzymic correlates of physiological regression of follicles and corpora lutea in ovaries of normal rats. *Endocrinology.* **68**, 232.

Luft, J. H. (1961) Improvements in epoxy resin embedding methods. *J. Biophys. Biochem. Cytol.* **9**, 409.

Mancini, R. E., Villar, O., Heinrich, J. J., Davidson, O. W. and Alvarez, B. (1963) Transference of circulating labeled serum proteins to the follicle of the rat ovary. *J. Histochem. Cytochem.* **11**, 80.

Maunsbach, A. B. (1969) Functions of lysosomes in kidney cells. *In* "Lysosomes". (J. T. Dingle and H. B. Fell, eds.) Vol. 1, Amsterdam-London, North-Holland Publishing Company.

Moore, D. H. and Ruska, H. (1957) The fine structure of capillaries and small arteries. *J. Biophysic. and Biochem. Cytol.* **3**, 457.

Novikoff, A. B. (1961) Lysosomes and related particles. *In* "The Cell". (J. Brachet and A. E. Mirsky, eds.) Vol. 2, p. 424. New York, Academic Press, Inc.

Novikoff, A. B. and Shin, W. Y. (1964) The endoplasmic reticulum in the Golgi zone and its relation to microbodies, Golgi apparatus and autophagic vacuoles in rat liver cells. *J. de Microscopie.* **3**, 187.

Odor, L. D. and Blandau, R. J. (1969) Ultrastructural studies on fetal and early postnatal mouse ovaries. 1. Histogenesis and Organogenesis. *Amer. J. Anat.* **124**, 163.

Raven, C. P. (1961) Oogenesis. The Storage of Developmental Information. New York, Pergamon Press.

Roth, T. F. and Porter, K. R. (1964) Yolk protein uptake in the oocyte of the mosquito *Aedes aegypti* L. *J. Cell Biol.* **20**, 313.

Ryser, H. J. P. (1968) Uptake of protein by mammalian cells: An undeveloped area. *Science.* **159**, 390.

Schechtman, A. M. (1955) Ontogeny of the blood and related antigens and their significance for the theory of differentiation. *In* "Biological Specificity and Growth". (E. G. Butler, ed.) p. 3. Princeton, New Jersey, Princeton University Press.

Schjeide, O. A., Galey, F., Greller, E. A., San Lin, R. I., de Vellis, J., and Mead, J. F. (1970) Macromolecules in oocyte maturation. *Biol. Reprod. Suppl.* **2**, 14.

Straus, W. (1964a) Cytochemical observations on the relationship between lysosomes and phagosomes in kidney and liver by combined staining for acid phosphatase and intravenously injected horseradish peroxidase. *J. Cell Biol.* **20**, 497.

Straus, W. (1964b) Occurrence of phagosomes and phago-lysosomes in different segments of the nephron in relation to the reabsorption, transport, digestion, and extrusion of intravenously injected horseradish peroxidase. *J. Cell Biol.* **21**, 295.

Straus, W. (1967) Methods for the study of small phagosomes and their relationship to lysosomes with horseradish peroxidase as a "marker protein". *J. Histochem. Cytochem.* **15**, 375.

Swift, H. and Hruban, Z. (1964) Focal degradation as a biological process. *Fed. Proc.* **23**, 1026.

Szollosi, D. (1967) Development of cortical granules and the cortical reaction in rat and hamster eggs. *Anat. Rec.* **159**, 434.

Telfer, W. H. (1960) The selective accumulation of blood proteins by the oocytes of saturniid moths. *Biol. Bull.* **118**, 338.

Van der Stricht, O. (1923) Etude comparee des ovules des mammiferes aux differentes periodes de l'ovogenese, d'apres les travaux du Laboratoire d'Histologie et d'Embryologie de l'Universite de Gand. *Arch. Biol.* **33**, 229.

Venable, J. H. and Coggeshall, R. (1965) A simplified lead citrate stain for use in electron microscopy. *J. Cell Biol.* **25**, 407.

Wallace, R. A. and Dumont, J. N. (1968) The induced synthesis and transport of yolk proteins and their accumulation by the oocyte in *Xenopus laevis. J. Cell Physiol. Suppl. 1.* **72**, 73.

Wilson, E. B. (1927) The Cell in Development and Heredity, London, The Macmillan Company.

Zamboni, L. and Mastroianni, L., Jr. (1966) Electron microscopic study on rabbit ova: I. The follicular oocytes. *J. Ultrastruct. Res.* **14**, 95.

NUCLEAR STRUCTURE AND FUNCTION
DURING AMPHIBIAN OOGENESIS

O. L. Miller, Jr.
Barbara R. Beatty
Barbara A. Hamkalo

.   Introduction
I.   Methodology
II.   Nucleolar genes
V.   Lampbrush chromosomes
V.   Nucleoplasm
VI.   Conclusion
VII.   Acknowledgments
VIII.   References

I. INTRODUCTION

Amphibian oocytes possess unique advantages for studies which attempt to correlate structure and function during genetic activity. The very large nucleus present during middle to late oogenesis allows manual isolation and rapid manipulation of contents of individual nuclei for light or electron microscopy. The highly extended chromosomes of these cells exhibit hundreds of active synthetic loci (Gall, 1958; Callan and Lloyd, 1960; Callan, 1963; Hess, 1966). In addition, during early oogenesis there is a large amplification of genes coding for ribosomal RNA (rRNA). These genes are located in the several hundred extrachromosomal nucleoli which appear in each nucleus following pachytene (Brown and Dawid, 1968; Evans and Birnstiel, 1968; Macgregor, 1968; Gall, 1969).

In this paper we review our ultrastructural studies of genetic activity during amphibian oogenesis. Although the main aspects of oogenesis appear to be similar in

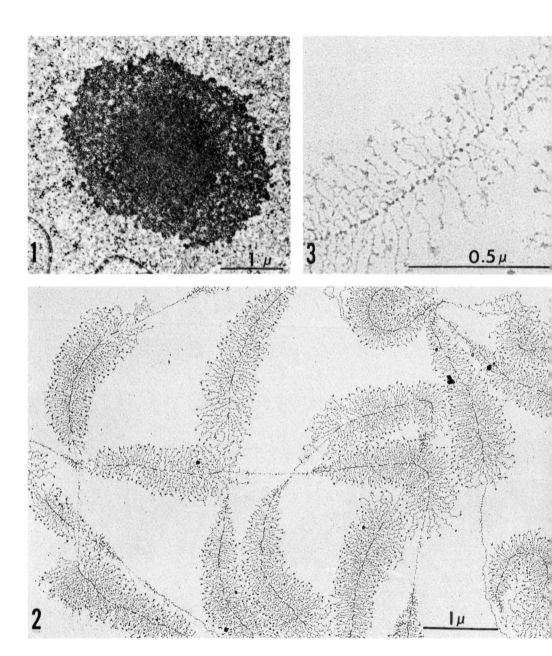

Fig. 6.1 Thin section of extrachromosomal nucleolus, showing compact fibrous core surrounded by a granular cortex. Micrographs for all figures were obtained from preparations of *T. viridescens,* unless otherwise stated.

Fig. 6.2 Portion of dispersed nucleolar core matrix units and matrix-free axis segments.

Fig. 6.3 Portion of a single nucleolar gene showing RNA polymerase molecules located on the DNA axis at the base of each matrix fibril.

most amphibian species, our studies have utilized primarily the South African clawed toad, *Xenopus laevis,* for which extensive biochemical information is available, and the spotted newt, *Triturus viridescens,* which provides superior morphological preparations. This material is more extensively covered by Miller and Beatty (1969a, b, c, d) and Miller, Beatty, Hamkalo and Thomas (1970).

II. METHODOLOGY

Methods of handling amphibian oocyte nuclei for light microscopy have been described by Callan and Lloyd (1960) and by Gall (1966a) and for electron microscopy by Miller (1965a), Gall (1966a), and Miller and Beatty (1969b). Briefly, our current procedures for electron microscope preparations are the following: Nuclear contents, which contain both extrachromosomal nucleoli and lampbrush chromosomes, are isolated in a dispersal medium and then fixed and deposited on carbon-coated grids by low-speed centrifugation (3-4 min, 2350 x) through 0.1 M sucrose-10% formalin (pH 8.5). Grids are rinsed in 0.4% Kodak Photo-flo, dried, stained for 1 min. with 1% phosphotungstic acid (PTA) in 70% ethanol (pH2), rinsed in 95% ethanol, and dried. Silverman and Glick (1969) have shown that PTA staining at low pH is fairly specific for positively charged groups of proteins.

III. NUCLEOLAR GENES

A thin section of an extrachromosomal nucleolus is shown in Fig. 6.1. Such nucleoli are structurally bipartite, with a granular cortex region surrounding a dense core. These two components can be separated and dispersed by the isolation of unfixed nucleoli into very dilute saline. Figure 6.2 is an electron micrograph of a portion of a dispersed nucleolar core. Every core consists of a thin circular axial fiber with repeating matrix units, separated by matrix-free segments. Each matrix unit is made of 80-100 short to long fibrils, and all matrices within a given core exhibit the same polarity along the circular core axis.

Enzymatic treatments of unfixed, dispersed cores give the following results: DNase breaks the core axis, both within and between matrix units; RNase completely removes matrix fibrils; proteases uncover naked DNA. These results, combined with specific protein staining by PTA, allow us to conclude that each core axis is a circular DNA molecule, that the matrix fibrils have RNA axes, and that both nucleic acids are coated with protein.

Available biochemical (Gall, 1966b) and autoradiographic (Macgregor, 1967) evidence indicate that 40S rRNA precursor molecules are synthesized within nucleolar cores. Electron microscopic autoradiography of dispersed nucleolar cores following *in situ* labeling of RNA shows that label is confined to the matrix units of core axes. Since this labeling is correlated in space and time with the synthesis of 40S rRNA precursor molecules, and the length of unstretched matrix units (2.3-2.5μ) is very close to the length of double-helix DNA necessary to code for the precursor molecule ($\sim$2.6μ), we conclude that the DNA within each matrix unit is a gene coding for 40S rRNA precursor molecules. These genes are visible because 80-100 precursor molecules are being synthesized simultaneously on each gene.

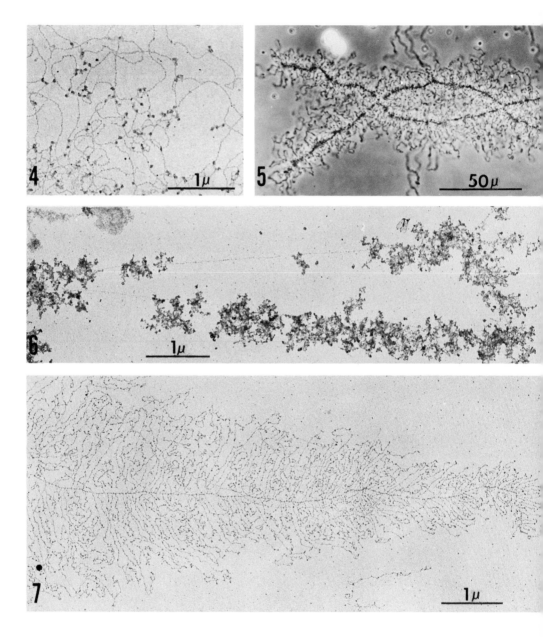

Fig. 6.4   Portion of dispersed nucleolar cortex showing fibro-granular RNP network.

Fig. 6.5   Phase contrast micrograph of isolated lampbrush chromosome (Photograph courtesy of Dr. J. G. Gall, Yale University).

Fig. 6.6   Lampbrush chromosome loop. Part of the RNP matrix was lost during isolation, exposing the loop axis.

Fig. 6.7   Thin insertion end of a lampbrush chromosome loop showing gradients of RNP matrix fibril lengths.

The length of mature matrix fibrils is ~0.5μ. Based on molecular weight determination, a fully extended 40S rRNA precursor molecule should be ~5.2 μ in length. These figures indicate that the length of mature precursor molecules is reduced about tenfold by association with protein.

In preparations with stretched matrices, a granule ~12.5 nm in diameter is seen on the core axis at the base of each matrix fibril (Fig. 6.3). Because of their size and location, these molecules are almost certainly RNA polymerases which were transcribing at the time of isolation.

The number of matrices per core varies widely. For example, cores from *X. laevis* containing from eight to well over one thousand matrix units per core have been observed. The number of redundant rRNA genes per haploid nucleolus organizer locus in *X. laevis* has been variously estimated as ~450 (Brown and Weber, 1968) and 800 (Wallace and Birnstiel, 1966). Although the molecular mechanism of nucleolar gene amplification is unknown at present, it apparently gives rise to nucleoli containing cores both with fewer and with more genes than the chromosomal locus.

Two labile core components, one membranous and one fibrillar, have been observed. No function is known for these structures.

40S rRNA precursor molecules are cleaved to produce 20S molecules which migrate relatively rapidly from the nucleus and give rise to the 18S rRNA of small ribosomal subunits and 30S molecules which remain in the nucleus for some time before appearing in the cytoplasm as the 28S rRNA of large ribosomal subunits (Gall, 1966b).

About the time labeled 30S molecules appear in sucrose gradients, silver grains appear over nucleolar cortices in autoradiographic preparations (Gall, 1966b; Macgregor, 1967). These results suggest that the 30S molecules are localized in the cortices. Consequently, one might expect to find discrete granules or fibrils containing single 30S molecules in cortices. An electron micrograph of isolated cortex (Fig. 6.4), however, shows a network in which all granules are attached to well defined fibrils. The location of 30S molecules within this complex has not been determined.

IV. LAMPBRUSH CHROMOSOMES

Lampbrush chromosomes of amphibian oocytes are highly extended chromosomes in the diplotene stage of meiosis. A phase contrast micrograph of a portion of a lampbrush chromosome from *T. viridescens* is shown in Fig. 6.5, and a schematic diagram of the structure of such chromosomes is depicted in Fig. 6.8. The main axis of each homologue consists of two chromatids arranged as a series of Feulgen-positive chromomeres joined by a thin fiber. At each chromomere, sister

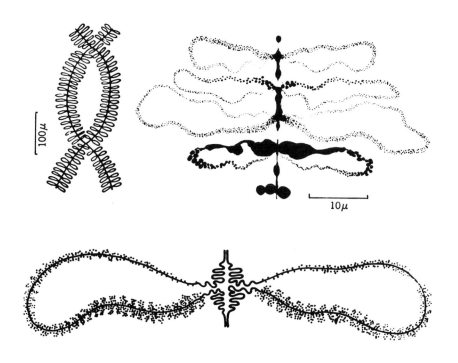

Fig. 6.8 *Upper left:* diagrammatic sketch of a pair of lampbrush chromosomes joined by 2 chiasmata. *Upper right:* drawing of pairs of loops showing differences in morphology and length. *Bottom:* concept of continuity of chromatids in main chromosome axis with those forming axes of the lateral loops. [From Swanson (1957), after Gall (1956)].

chromatids typically separate to form pairs of lateral loops.

On the average, there are approximately 1000 pairs of lateral loops per chromosome and an average loop is ~50μ in length (Gall, 1956). Staining, enzymatic digestion, and autoradiography show that each loop consists of a DNA axis coated with ribonucleoprotein (RNP). The RNP matrix of each loop is asymmetric, forming

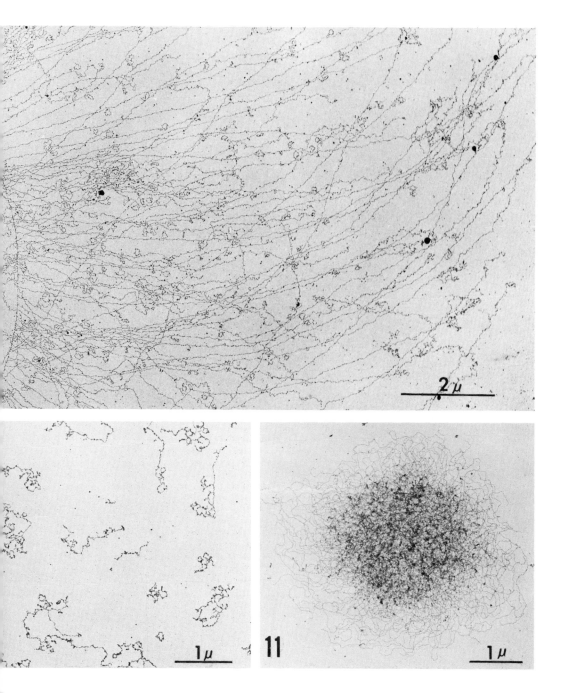

Fig. 6.9   Portion of lampbrush chromosome loop with RNP matrix fibrils over 10 μ long.

Fig. 6.10   RNP fibrils of isolated nucleoplasm.

Fig. 6.11   Fibrous inclusion in the nucleoplasm of *X. laevis*.

a thin and a thick end at the point of loop insertion into the main axis.

Figure 6.7 is an electron micrograph of a portion of a lateral loop whose axis has been partially denuded of matrix during isolation. Limited trypsin digestion reduces the diameter of loop axes (single chromatids) from about 10-3 nm and main axis fibers from about 20-5 nm. These measurements suggest that each chromatid is a single duplex DNA molecule.

The thin insertion end of a typical lampbrush chromosome loop is shown in Fig. 6.8. The RNP matrix is composed of fibrils which are closely spaced and show a gradient of increasing length toward the thick insertion end of the loop. As with nucleolar genes, each fibril is attached to a presumptive RNA polymerase molecule on the loop axis. Again, the fibrils appear to be considerably foreshortened by associated protein.

RNP fibrils measuring well over 10μ in length have been observed at intermediate points along loop axes (Fig. 6.9). Fibrils considerably longer than these would be expected at the thick insertion ends of such loops. If matrix fibrils contain single RNA molecules, it is likely that RNA molecules over 100μ long are synthesized on lampbrush chromosomes.

V. NUCLEOPLASM

The nucleoplasm of amphibian oocytes contains well over half of their nuclear RNA (Gall, 1966b). Biochemical analysis indicates that this RNA consists mainly of 30S and 20S rRNA precursor molecules in the form of RNP particles (Rogers, 1968). Nucleoplasm isolated by our preparative methods consists of RNP fibrils of widely varying lengths (Fig. 6.10).

There is evidence that some of the RNA synthesized on lampbrush chromosomes is stored in the cytoplasm of the oocyte and is utilized for protein synthesis in early embryogenesis (Davidson, Crippa, Kramer and Mirsky, 1966; Crippa, Davidson and Mirsky, 1967). Since almost all of the RNA molecules being synthesized on the lampbrush chromosomes are longer than the longest nucleoplasm fibrils, it is likely that this RNA is processed into shorter lengths prior to migration through the nucleoplasm to the cytoplasm.

In addition to extrachromosomal nucleoli, the nucleoplasm of amphibian oocytes contains several types of spherical inclusions, either fibro-granular or fibrous (Fig. 6.11) in nature. As yet, no function can be ascribed to these bodies.

VI. CONCLUSION

The large nuclear size and high genetic activity of amphibian oocytes have allowed visualization of the ultrastructure of active rRNA genes and lampbrush chromosome loops. Further innovations in preparative techniques should permit the observation of changes in activity of specific genetic loci during the course of amphibian oogenesis and provide information on the mechanisms of gene activation and repression.

VII. ACKNOWLEDGMENTS

Research sponsored by the U. S. Atomic Energy Commission under contract with the Union Carbide Corporation.

VIII. REFERENCES

Brown, D. D. and Dawid, I. B. (1968) Specific gene amplification in oocytes. *Science.* **160**, 272.

Brown, D. D. and Weber, C. S. (1968) Gene linkage by RNA-DNA hybridization. I. Unique DNA sequences homologous to 4S RNA, 5S RNA and ribosomal RNA. *J. Molec. Biol.* **34**, 661.

Callan, H. G. (1963) The nature of lampbrush chromosomes. *In* "Int. Rev. Cytol." (G. H. Bourne and J. F. Danielli, eds.), Vol. 15, 1. New York, Academic Press.

Callan, H. G. and Lloyd, L. (1960) Lampbrush chromosomes of crested newts *Triturus cristatus* (Laurenti). *Proc. Roy. Soc. [Biol.].* **243**, 135.

Crippa, M., Davidson, E. H. and Mirsky, A. E. (1967) Persistence in early amphibian embryos of informational RNA's from the lampbrush chromosome stage of oogenesis. *Proc. Nat. Acad. Sci. U.S.A.* **57**, 885.

Davidson, E. H., Crippa, M., Kramer, F. R. and Mirsky, A. E. (1966) Genomic function during the lampbrush chromosome stage of amphibian oogenesis. *Proc. Nat. Acad. Sci. U.S.A.* **56**, 856.

Evans, D. and Birnstiel, M. L. (1968) Localization of amplified ribosomal DNA in the oocyte of *Xenopus laevis. Biochim. Biophys. Acta.* **166**, 274.

Gall, J. G. (1956) On the submicroscopic structure of chromosomes. *Brookhaven Sympos. Biol.* **8**, 17.

Gall, J. G. (1958) Chromosomal differentiation. *In* "The Chemical Basis of Development". (W. D. McElroy and B. Glass, eds.), p. 103. Baltimore, The Johns Hopkins Press.

Gall, J. G. (1966a) Techniques for the study of lampbrush chromosomes. *In* "Methods in Cell Physiology". (D. M. Prescott, ed.), Vol. 2, p. 37. New York, Academic Press.

Gall, J. G. (1966b) Nuclear RNA of the salamander oocyte. *Nat. Cancer Inst. Monogr.* **23**, 475.

Gall, J. G. (1969) The genes for ribosomal RNA during oogenesis. *Genetics Suppl.* **61**, 121.

Hess, O. (1966) Funktionelle und strukturelle organization der lampenbursten- chromosomen. *In* "Probleme der Biologischen Reduplikation". (P. Sitte, ed.), p. 29. Heidelberg, Springer-Verlag.

Macgregor, H. C. (1967) Pattern of incorporation of (^{3}H) uridine into RNA of amphibian oocyte nucleoli. *J. Cell. Sci.* **2**, 145.

Macgregor, H. C. (1968) Nucleolar DNA in oocytes of *Xenopus laevis. J. Cell. Sci.* **3**, 437.

Miller, O. L., Jr. (1965) Fine structure of lampbrush chromosomes. *Nat. Cancer Inst. Monogr.* **18**, 79.

Miller, O. L., Jr. and Beatty, B. R. (1969a) Nucleolar structure and function. *In* "Handbook of Molecular Cytology". (A. Lima-de-Faria, ed.), p. 605. Amsterdam, North-Holland Publishing Company.

Miller, O. L., Jr. and Beatty, B. R. (1969b) Visualization of nucleolar genes. *Science.* **164**, 955.

Miller, O. L., Jr. and Beatty, B. R. (1969c) Extrachromosomal nucleolar genes in amphibian oocytes. *Genetics Suppl.* **61**, 133.

Miller, O. L., Jr. and Beatty, B. R. (1969d) Portrait of a gene. *J. Cell. Physiol.* **74**, 225.

Miller, O. L., Jr., Beatty, B. R., Hamkalo, B. A. and Thomas, C. A., Jr. (1970) Electron microscopic visualization of transcription. *Cold Spring Harbor Symp. Quant. Biol.* **35**, 505.

Rogers, M. E. (1968) Ribonucleoprotein particles in the amphibian oocyte nucleus. Possible intermediates in ribosome synthesis. *J. Cell Biol.* **36**, 421.

Silverman, L. and Glick, D. (1969) The reactivity and staining of tissue proteins with phosphotungstic acid. *J. Cell Biol.* **40**, 761.

Swanson, C. P. (1957) Cytology and Cytogenetics. Englewood Cliffs, New Jersey, Prentice-Hall Inc.

Wallace, H. and Birnstiel, M. L. (1966) Ribosomal cistrons and the nucleolar organizer. *Biochim. Biophys. Acta.* **114**, 296.

7

UTILIZATION OF GENETIC INFORMATION DURING OOGENESIS

Eric H. Davidson
Barbara R. Hough

I.   General significance of gene action during oogenesis
II.   Comparative evidence on the significance of gene activity during oogenesis
III.   The informational content of oocyte RNA
IV.   Discussion and conclusions
V.   References

I. THE GENERAL SIGNIFICANCE OF GENE ACTION DURING OOGENESIS

For three-quarters of a century biologists have understood that mature eggs contain a stored program of developmental information, on which the events of early embryogenesis to some extent depend. Since these events are part of the heritable characteristics of the organism, the developmental information in the egg must itself be of genetic origin, and must derive from gene read-out during oogenesis. E. B. Wilson applied this line of reasoning to the interpretation of cytoplasmic localization as long ago as 1896. Nonetheless, the molecular events underlying the establishment of cytoplasmic localization patterns remain completely unknown.

Localization, which probably occurs in all metazoan eggs (see review in Davidson, 1968), is partially or wholly responsible for the initial specification of blastomere fate, depending on the organism. That is, the type of differentiation a blastomere and its descendants will undergo is a function of which segment of egg cytoplasm the blastomere will include as cleavage planes divide up the egg. On the other hand, differentiated function in the blastomeres and their descendants ultimately depends on the imposition of certain patterns of transcription in the nuclei of these embryonic cells. A conceptually attractive possibility is that the determinative elements in egg cytoplasm may operate by specifying the patterns of gene activity in the cells of the early embryo. The direct inference would be that egg cytoplasm contains gene regulatory macromolecules, RNA or protein, derived from transcriptive events during oogenesis. Cytoplasmic localization can be interpreted in this way in terms of the gene regulation theory of Britten and Davidson (1969), which proposes that patterns of gene regulation in higher cells are mediated by macromolecules arising from transcription of regulatory DNA sequences. These speculations lead one to regard such putative carriers of genomic regulative information as oocyte RNA with special attention. This view emphasizes the importance of studying gene transcription during oogenesis, as the possible source of the program for early nuclear differentiation in embryogenesis.

Whatever the nature of their function, it is clear that some of the RNA's synthesized during oogenesis serve later as templates for protein synthesis. Stored messenger RNA inherited from oogenesis has been demonstrated in a variety of early embryos (Gross, 1967; Kedes and Gross, 1969), though the diversity of functions attributable to the proteins coded for by the maternal messenger RNA population cannot yet be appreciated. Evidence that informational RNA's synthesized in oogenesis are utilized during development also derives from RNA-DNA hybridization studies with amphibian embryos. These investigations show that a population of informational RNA's transcribed from repetitive DNA sequences during oogenesis are inherited by the embryo and these maternal transcripts do not begin to disappear until well into blastulation (Crippa and Davidson, 1967; Crippa and Gross, 1969). At this point the embryo contains $10^3 - 10^4$ cells. At least in *Xenopus,* the RNA's synthesized by the embryo genome during this stage are not homologous with those inherited from oogenesis (Davidson, Crippa and Mirsky, 1968). The demonstration of maternal messengers in early embryos and the observation that maternal RNA sequences persist well into early development imply independently that the RNA synthesized by the oocyte bears genomic information intended for use during embryogenesis. To investigate the genetic programming of early development, therefore, it is evident that we must turn to the transcriptive processes occurring in the oocyte.

II. COMPARATIVE EVIDENCE ON THE SIGNIFICANCE OF GENE ACTIVITY DURING OOGENESIS

Considering oogenesis as it occurs in a variety of creatures provides perspective on the generality of the assumption that gene products stored from oogenesis are components of the basic developmental process. Particularly striking are patterns which transcend the most ancient division in the metazoan world, that between deuterostomial and protostomial animals. According to most authorities (e.g., see Hyman, 1951) the deuterostomes, which include the chordates, hemichordates and echinoderms, must have diverged from the protostomes, which include the molluscs, arthropods, annelids and lower worms, shortly after the origin of multicellular bilateral forms. Thus it is a remarkable fact that lampbrush chromosomes are found in growing oocytes of both protostomes and deuterostomes. Except for the occurrence of related structures in dipteran spermatocytes they are confined to oocytes throughout the animal world. The phylogenetic distribution of lampbrush chromosomes is summarized in Table 7.1 (for a complete list of references and

Table 7.1 **Occurrence of Lampbrush Chromosomes in Animal Oocytes and the Duration of the Lampbrush Stage**

Species and affiliation of animals in which lampbrush chromosomes have been reported	Estimated duration of lampbrush stage where available
Deuterostome	
Chaetognath	
Arrow worm	
Echinoderm	
Sea urchin	
Chordate	
Cyclostome	Several months in lamprey
Shark	
Teleost	
Amphibian	
Urodele	About seven months in *Triturus*
Anuran	Four to eight months in *Xenopus*
	30-40 days in *Engystomops*
Reptile	Some months in lizards
Bird	Three weeks in chick
Mammal	Perhaps years in man
Protostome	
Mollusk	
Gastropod	
Cephalopod	
Insect	
Orthopteran	Three months in cricket

details relating to the data in Table 7.1 see Davidson, 1968, p. 188). This distribution suggests that lampbrush chromosomes are almost as ancient, evolutionarily, as the processes of oogenesis and the appearance of cellular embryogenesis.

Also presented in Table 7.1 is the available information regarding the duration of the lampbrush phase. This is an interesting and significant point, since it is during the lampbrush period of oogenesis that, in amphibians at least, some of the non-ribosomal, heterogeneous RNA's inherited by the embryo are synthesized (Davidson, Crippa, Kramer and Mirsky, 1966; Crippa *et al.,* 1967; Davidson *et al.,* 1968). In this light the extended length of the lampbrush phase in every animal where it has been studied is suggestive of an accretion process. As Gall and Callan (1962) first demonstrated in *Triturus*, lampbrush chromosomes are found to be active in RNA synthesis wherever studied, in oocytes of the protostome *Locusta migratoria* (Kunz, 1967), for example, as well as in the amphibians. The actual length of the lampbrush chromosome stage in any given organism is notoriously difficult to measure. As the table above indicates, rough estimates exist for several amphibia.

At least one of these estimates of lampbrush stage length is probably fairly exact, however, and that is the period cited for the lampbrush stage of the anuran *Engystomops,* a small neo-tropic amphibian which we have adapted to laboratory culture. *Engystomops* carries out oogenesis synchronously under certain environmental conditions (Davidson and Hough, 1969a). When oogenesis is synchronous all the maturing oocytes in the ovary pass through the lampbrush as well as successive stages of oogenesis in concert. This process can be initiated at will in a "dry-season" *Engystomops* population simply by flooding the culture area and providing breeding ponds. Initiation of the active lampbrush phase follows. About 45 days later the first clutch of newly maturing eggs are shed. Since the post-lampbrush stages occupy only 2 - 3 weeks, an accurate measurement of the lampbrush stage itself, 30 - 40 days, can be obtained. Figure 7.1a is a photograph of this remarkable little toad, and Fig. 7.1b illustrates the appearance of the lampbrush

 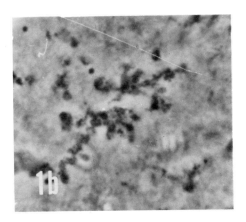

Fig. 7.1a Adult *Engystomops,* laboratory raised. Photograph taken by Dr. William Massover.

Fig. 7.1b Lampbrush chromosomes of stage 4 *Engystomops* oocytes. The loops are not widely extended, giving the chromosomes a bushy appearance.

chromosomes of an *Engystomops* oocyte which has been fixed in epon and sectioned. The loops are very small, compared to the giant loops of urodele lampbrush chromosomes, which may be expected from the relatively small genome of *Engystomops,* 5.5 pg diploid (Hinegardner, personal communication).

A major alternative to the slow form of oocyte maturation observed in organisms bearing lampbrush chromosomes exists in various protostomial groups utilizing nurse cells in the preparation of the oocyte. Nurse cells, which are themselves of germ line origin, typically contain hundreds of times the normal amount of genomic DNA. Bier (1965) has shown that in the dipterans virtually all the newly synthesized RNA of the growing oocyte is made in the nurse cell nuclei and transported into the oocyte via open "ring canals" joining the nurse cell to the oocyte. The oocyte nucleus lacks lampbrush chromosomes and is transcriptionally quiescent. Informative comparisons are thus available among the insects: for example, in the orthopteran, *Acheta,* where the oocyte RNA derives from the 4C lampbrush chromosomes (Kunz, 1967), about 100 days are required to complete the post-synaptic phases of oogenesis, while in the dipteran *Calliphora* about 6 days are required. However, the RNA (and other materials) of the *Calliphora* oocyte are synthesized by the 15 nurse cells, each about 256C in DNA content (Ribbert and Bier, 1969; Kunz and Ribbert, 1969). This comparison suggests that extensive gene activity is needed to create an oocyte, and that the evolution of polytene nurse cells represents (among other things) a means of accumulating the necessary gene products more than an order of magnitude more rapidly than otherwise possible. A further, and clearly provocative correlation in this connection is that those insects utilizing nurse cells are generally characterized by extremely rapid embryological development, rely upon rigid blastoderm localization patterns and extensive imaginal disc systems for larval and adult morphogenesis, and tend to possess genomes which are on the average much smaller than those of insects utilizing lampbrush chromosomes.

III. THE INFORMATION CONTENT OF OOCYTE RNA

RNA-DNA hybridization experiments with newly synthesized lampbrush stage RNA of *Xenopus* have shown that both repetitive and non-repetitive DNA sequences are transcribed in lampbrush chromosomes (Davidson *et al.,* 1966; Davidson and Hough, 1969b). The fraction of a repetitive sequence family hybridizing with an RNA which is utilized in the transcription of the RNA cannot be determined, and therefore we remain ignorant of the actual amount of the repetitive fraction of the genome transcribed. Hybridization experiments indicate that about 6 - 8% of the total repetitive sequence of the *Xenopus* genome is represented in the population of lampbrush transcripts accumulated throughout oogenesis (Davidson *et al.,* 1966; Crippa *et al.,* 1967; Gross and Crippa, 1969), something less than this amount of the total repetitive sequences having been active in the synthesis of this RNA. The complexity of the RNA of course cannot be deduced from these experiments since they involve repetitive sequence transcripts, and the numbers of repetitions per active DNA sequence are unknown. We have recently succeeded in obtaining an experimental estimate of the complexity, or potential genetic information content, of mature *Xenopus* oocyte RNA, by hybridizing the oocyte RNA with the isolated, non-repetitive fraction of the DNA. The mass of RNA hybridized at saturation of the non-repetitive DNA provides a direct measure of the fraction of the total non-repetitive sequence represented in the stored oocyte RNA, and hence of the complexity of this RNA.

About 55% of the *Xenopus* genome behaves as a non-repetitive sequence fraction at a 60° annealing criterion (Davidson and Hough, 1969b, 1971). This fraction was isolated to a high state of purity on hydroxyapatite, using essentially the procedures designed by Britten and Kohne (1966, 1968). Methods and criteria were then developed, as described elsewhere (Davidson and Hough, 1969b, 1970), for the hybridization of oocyte RNA with the purified non-repetitive DNA fraction. In brief, the procedure involves extensive annealing of the DNA with dimethyl-sulfate labelled ^{3}H-oocyte RNA, for up to 10 days at 60°. The annealing mixtures are treated with ribonuclease and the surviving RNA and DNA excluded from a Sephadex G-200 column. Duplexes are next separated on hydroxyapatite, and finally the thermal stability of the hybrids obtained is assayed by melting them from the hydroxyapatite column as the temperature is raised. Figure 7.2 shows such a thermal stability profile, and illustrates the small difference between the thermal stability of the oocyte RNA-DNA hybrids and that of the non-repetitive DNA-DNA duplexes also present in the annealing mixtures. Hybrid formation with the related but non-identical DNA sequences of repetitive sequence families results in significant nucleotide mispairing, and hence lowered thermal stability, while on the other hand,

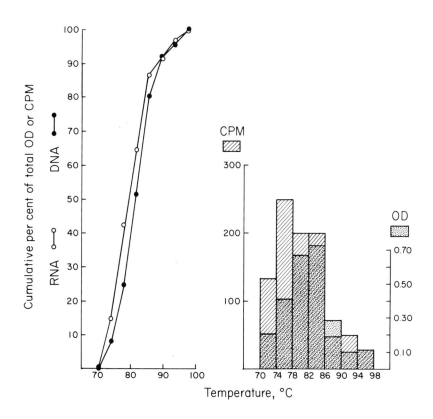

Fig. 7.2 Hydroxyapatite melt of hybrids between H^3-methyl labelled oocyte RNA and the isolated non-repetitive DNA fraction. The input was 2.1 mg RNA, at 15 mg/ml, plus 140 μg DNA, and incubation was for 10 days, to an equivalent Cot of 13,600. After annealing, the hybrids were treated with ribonuclease, excluded from a Sephadex G-200 column, and melted. The data are plotted both integrally (left) and incrementally (right) as a bar graph.

the high thermal stability indicated in Fig. 7.2 indicated the excellent base pairing expected of non-repetitive DNA-RNA hybrids. Hybrid duplexes of comparable thermal stability, relative to DNA-DNA duplexes, have previously been reported only for bacterial systems, which also lack repetitive DNA sequences. To prove that the DNA involved in these hybrids is actually non-repetitive DNA rather than some small repetitive sequence contaminant, we isolated labelled DNA from the hybrids and showed that its rate of reassociation is exactly that expected for the non-repetitive sequence fraction.

Figure 7.3 reproduces a graph describing the series of annealing experiments

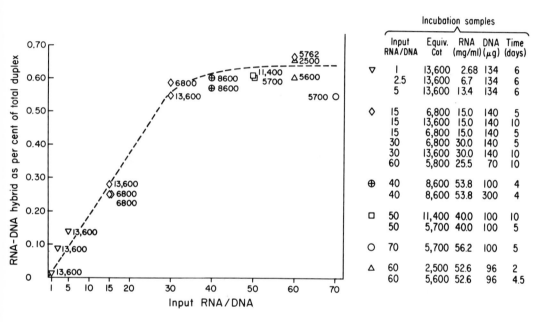

Fig. 7.3 Saturation of non-repetitive DNA with mature oocyte RNA. Characteristics of each incubation sample are given in the table to the right of the saturation curve. The numbers adjacent to each point represent the equivalent DNA Cots to which the samples were incubated. (Cot is expressed as moles nucleotide per liter times time of reaction in seconds, for solutions annealed at 60° in 0.12 M phosphate buffer. The equivalent Cot is the Cot multiplied by the appropriate factor for rate increase in cases in which the incubation was carried out at higher salt concentrations.)

At RNA/DNA inputs 15 and 50 an alteration in annealing procedure was employed, as an additional control: duplicate samples were incubated for 5 days, then exposed to 100° and reincubated for a second 5 day period. This control shows that the second five days of the ten day incubation periods used in other samples at these inputs are as effective as the first 5 days, since no change in the amount of hybridization was observed. During the annealing period the RNA neither hydrolyzes nor suffers any other more subtle forms of degradation detectable by the criterion of hybridization [From Davidson and Hough (1971)].

through which the saturation of the non-repetitive DNA is approached. The representation of non- repetitive DNA sequences in the RNA can thus be estimated, and hence the sequence diversity, or potential information content, of the oocyte RNA. This value is expressed in terms of nucleotide pairs of non-repetitive sequence (at the annealing criterion used, 60°). From the curve plotted in Fig. 7.3 we estimate that the non-repetitive sequence representation in oocyte RNA is something more than 0.6%. On the basis that only one strand is used for transcription this means that at least 1.2% of the total genomic complexity is present in the oocyte RNA. A minimum representation of 1.2% of the non-repetitive sequence in the *Xenopus* genome amounts to some 20×10^6 nucleotide pairs of non-repetitive sequence. This is a surprisingly high value for the complexity of the oocyte RNA, as it suggests that the genetic information content of this RNA is about 4.5 times that of the whole *E. coli* chromosome.

An additional dividend from these experiments is the finding that there may be two orders of magnitude difference in the extent to which non-repetitive sequences and repetitive sequences are presented in the RNA of the mature oocyte. Many of the repetitive sequence transcripts are present at perhaps 50-100 times the concentration of non-repetitive sequence transcripts. As do similar data emerging from other laboratories, this indicates that there is indeed a functional as well as an evolutionary meaning to the distinction between repetitive and non-repetitive sequences deriving from reassociation rate determinations at criteria about 25° below the T_m of native DNA (Britten and Kohne, 1966, 1968).

IV. DISCUSSION AND CONCLUSIONS

A variety of evidence supports the concept that genomic information needed for embryological development is transcribed during oogenesis. Most direct and convincing are the findings that mature oocytes actually contain large stores of heterogeneous RNA, and that this RNA is informationally very rich, evidently representing many thousands of diverse genomic sites. We can at this point merely glimpse the complicated structure of the oocyte RNA stockpile. We know that at least some repetitive sequence transcripts are probably present some 100 times more frequently than most non-repetitive sequence transcripts. We also know that repetitive sequence transcripts from oogenesis survive until well into blastulation, when the first attrition of the set of sequences present initially in the oocyte can be detected (Davidson *et al.,* 1966; Gross and Crippa, 1969). We have no direct information on the fate of the non-repetitive sequence transcripts. On the other hand much evidence indicates that messenger RNA's transcribed in oogenesis are present and functional in early embryos. Since some, if not all, of these messenger RNA's are likely to be non-repetitive sequence transcripts, it is probable that this portion of the oocyte RNA stockpile also survives well into development.

The fundamental problem raised by all these observations is the means by which development is controlled. Most students of early development agree that oocyte cytoplasm contains developmental information, and the deduction that this information must ultimately operate by affecting the pattern of gene activity in the blastomere nuclei is hard to escape. It is the molecular nature of these regulatory pathways which remain totally unknown: are messenger RNA's transcribed during oogenesis translated into regulatory proteins functional in the embryo? Or, are certain oocyte RNA's used directly as regulatory agents at the gene level? Such questions represent a confrontation with one of the basic issues of current biology, the actual mechanism of gene regulation in eukaryotic cells. In the study of genomic information utilization in oogenesis there exists an opportunity to approach these deep issues, as recent theoretical considerations on gene regulation in higher cells have acknowledged (viz., Callan, 1967; Snow and Callan, 1969; Britten and Davidson, 1969, 1971). In any case it seems clear that the phenomena we are dealing with are fundamental in an evolutionary sense, since key elements of the processes of oogenesis are apparently present among the most distantly related animals. Examples are the occurrence of transcriptionally active lampbrush chromosomes in oocytes of both advanced protostomes and advanced deuterostomes, and the invariably lengthy period of oocyte growth, except where the involvement of polytene nurse cells increases the amount of genomic material active in oogenesis by a large factor.

The nature of the unique lampbrush chromosome structures, and their widespread occurrence in female germ-line cells, has been ingeniously interpreted by Callan (1967) and Snow and Callan (1969). Their conception has led to the "master-slave" theory of genomic structure. According to this idea the genome contains linearly repeating gene replicates. Sequence divergence occurring among the replicates is corrected in each generation at the lampbrush chromosome stage of oogenesis, when in each lampbrush loop the replicates (slave genes) are moved across the master gene copy. This theory would provide a neat escape from the dilemma posed by the apparent absence of large-scale polymorphism in most gene products (proteins) on the one hand, and the presence of sequence repetition in transcriptionally active DNA on the other hand. However, there are several serious arguments against the master-slave concept. For one thing, in amphibian lampbrush chromosomes all portions of the loops are active in RNA synthesis, and yet hybridization experiments show that less than 6 - 8% of the repetitive sequences are transcribed during the lampbrush phase (Davidson *et al.,* 1966; Crippa *et al.,* 1967). According to the master-slave theory all of the sequence repeats would have to be corrected during the lampbrush phase, and hence spun out into chromosome loops, where transcription occurs. Thus a large fraction, perhaps all, of the repetitive sequence should be transcribed during the lampbrush phase rather than the minor fraction of repetitive sequence actually represented in the oocyte RNA. Secondly, the theory proposes that repetitive sequences in the genome are in general arranged in tandem, while annealing experiments on the location of repetitive sequences in DNA fragments of different lengths show that in general repetitive sequences are interspersed with other sequences (Britten and Smith, 1970; Britten, personal communication). Recently Thomas, Hamkalo, Misra and Lee (1970) have provided evidence for tandem repetition in the genomes of *Necturus* and several other vertebrates. However, these results merely suggest a certain amount of repetitive sequence clustering. Finally, the key role of lampbrush chromosomes in the master-slave theory raises a most serious problem in that they are totally absent in several groups of organisms. Insects utilizing nurse cells lack lampbrush chromosomes (Bier, 1965; Bier *et al.*, 1969) as mentioned above. However, these are not the only organisms where the absence of lampbrush chromosomes has been indicated with modern procedures. Recent investigations of Franchi and Mandl (1963) suggest that rodent oocytes may lack lampbrush chromosomes, though they are present in other mammals, including man (Baker and Franchi, 1966).

In summary, then, we believe that lampbrush chromosomes are better interpreted as the sites of transcription of those genes bearing the program for early development. Hess (1969) has proposed that the loops of *Drosophila* spermatocyte chromosomes are the sites of synthesis of informational RNA's to be stored for use later in spermiogenesis. Perhaps a somewhat similar view applies to the lampbrush chromosome loops, which could function as the sites where newly formed RNA's destined for embryogenesis are "packaged" within protective protein coats.

V. REFERENCES

Baker, T. G. and Franchi, L. L. (1966) Fine structure of the nucleus in the primordial oocytes of primates. *J. Anat.* **100**, 697.

Bier, K. (1965) Zur Funktion der Nahrzellen in meroistis chen Insektenovar unter besonderer Berucksichtigung der Oogenese Adephager Coleopteren. *Zool. Jahrb. Physiol.* **71**, 371.

Bier, K., Kunz, W. and Ribbert, D. (1969) *In* "Chromosomes Today". (C. D. Darlington and K. R. Lewis, eds.) Vol. 2, p. 107. Edinburgh, Oliver and Boyd, Ltd.

Britten, R. J. and Davidson, E. H. (1969) Gene regulation for higher cells: a theory. *Science.* **165**, 349.

Britten, R. J. and Davidson, E. H. (1971) Repetitive and non-repetitive DNA sequences and a speculation on the origins of evolutionary novelty. *Quart. Rev. Biol.* (In press).

Britten, R. J. and Kohne, D. E. (1966) Nucleotide sequence repetition in DNA. *Carnegie Inst. of Wash. Year Book.* **65**, 78.

Britten, R. J. and Kohne, D. E. (1968) Repeated sequences in DNA. *Science.* **161**, 529.

Britten, R. J. and Smith, J. (1970) A bovine genome. *Carnegie Inst. of Wash. Year Book.* **68**, 378.

Britten, R. J. and Smith, J. Unpublished data.

Callan, H. G. (1967) The organization of genetic units in chromosomes. *J. Cell Sci.* **2**, 1.

Crippa, M., Davidson, E. H. and Mirsky, A. E. (1967) Persistence in early amphibian embryos of informational RNA's from the lampbrush stage of oogenesis. *Proc. Nat. Acad. Sci. U.S.A.* **57**, 885.

Crippa, M. and Gross, P. R. (1969) Maternal and embryonic contributions to the functional messenger RNA of early development. *Proc. Nat. Acad. Sci. U.S.A.* **62**, 121.

Davidson, E. H. (1968) *In* "Gene Activity in Early Development". ch. 2. New York, Academic Press.

Davidson, E. H., Crippa, M., Kramer, F. R. and Mirsky, A. E. (1966) Genomic function during the lampbrush chromosome stage of amphibian oogenesis. *Proc. Nat. Acad. Sci. U.S.A.* **56**, 856.

Davidson, E. H., Crippa, M. and Mirsky, A. E. (1968) Evidence for the appearance of novel gene products during amphibian blastulation. *Proc. Nat. Acad. Sci. U.S.A.* **60**, 152.

Davidson, E. H. and Hough, B. R. (1969a) Synchronous oogenesis in *Engystomops pustulosus*, a neotropic anuran suitable for laboratory studies: localization in the embryo of RNA synthesized at the lampbrush stage. *J. Exp. Zool.* **172**, 25.

Davidson, E. H. and Hough, B. R. (1969b) High sequence diversity in the RNA synthesized at the lampbrush stage of oogenesis. *Proc. Nat. Acad. Sci. U.S.A.* **63**, 342.

Davidson, E. H. and Hough, B. R. (1971) Genetic information in oocyte RNA. *J. Molec. Biol.* **56**, (In press).

Franchi, L. L. and Mandl, A. M. (1963) The ultrastructure of oogonia and oocytes in the foetal and the neonatal rat. *Proc. Roy. Soc. B.* **157**, 99.

Gall, J. G. and Callan, H. G. (1962) ^{3}H-uridine incorporation in lampbrush chromosomes. *Proc. Nat. Acad. Sci. U.S.A.* **48**, 562.

Gross, P. R. (1967) The control of protein synthesis in embryonic development and differentiation. *Current Topics in Develop. Biol.* **2**, 1.

Hess, O. (1967) Complementation of genetic activity in translocated fragments of the Y chromosome in *Drosophila hydei. Genetics.* **56**, 283.

Hyman, L. H. (1951) *In* "The Invertebrates". Vol. II, p. 3, New York, McGraw-Hill.

Kedes, L. and Gross, P. R. (1969) Synthesis and function of messenger RNA during early embryonic development. *J. Molec. Biol.* **42**, 559.

Kunz, W. (1967) Funktionsstrukturen Im Oocytenkern von Locusta Migratoria. *Chromosoma.* **20**, 332.

Ribbert, D. and Bier, K. (1969) Multiple nucleoli and enhanced nucleolar activity in the nurse cells of the insect ovary. *Chromosoma. (Berl.)* **27**, 178.

Snow, M. H. L. and Callan, H. G. (1969) 1. Evidence for a polarized movement of the lateral loops of newt lampbrush chromosomes during oogenesis. *J. Cell Sci.* **5**, 1.

Thomas, C. A. Jr., Hamkalo, B. A., Misra, D. N. and Lee, C. S. (1970) Cyclization of eucaryotic deoxyribonucleic acid fragments. *J. Molec. Biol.* **51**, 621.

Wilson, E. B. (1896) On cleavage and mosaic work. *Arch. Entwicklangs-mechanik Organ.* **3**, 19.

8

CHROMOSOMAL PROTEINS

Ru Chih C. Huang
William Cieplinski

I.   Introduction
II.   Basic proteins
        A.   Histones
                1.   Histone characterization
                2.   Description of histone fractions
                3.   Histone synthesis
                4.   Histone degradation
                5.   Histone changes in development
                6.   Chemical modifications and microheterogeneity
                7.   Phosphorylation
                8.   Interactions between histones and DNA
        B.   Protamines
III.   Acidic proteins
        A.   Structural proteins and enzymes
        B.   Proteins associated with chromosomal RNA
IV.   Role of chromatin components in controlling transcription
V.   Concluding remarks
VI.   Acknowledgments
VII.   References

I. INTRODUCTION

Advances in the understanding of gene regulation in prokaryotes have prompted the isolation of chromatin from higher organisms so that its role in cellular differentiation may be studied. Differentiation does not appear to involve loss of genetic material in somatic cells since (a) the nucleus of a totally differentiated cell from the intestine, when implanted in an anucleate ovum, can direct development of a new organism (Gurdon and Uehlinger, 1966) and (b) each cell has an exact complement of chromosomes. These results suggest that differentiation involves a process of selective specific repression of the genome rather than a process of selective DNA synthesis.

The first successful attempt to isolate chromatin was made by Zubay and Doty (1959), later modified by Bonner, Chalkley, Dahmus, Fambrough, Fujimura, Huang, Huberman, Jensen, Marushige, Ohlenbusch, Olivera, and Widholm (1968). Recently a chromatin isolation procedure involving separation of intact nuclei has succeeded in facilitating the purification of nucleoprotein complexes, containing over 90% of the cell DNA along with chromosomal RNAs and associated proteins (Panyim and Chalkley, 1969a, b; Shaw and Huang, 1970). These proteins, complexed with the nucleic acids in chromatin, are the subject of this paper. This is not, however, a comprehensive review of our present knowledge on chromatin structure and function. For more detailed information we refer the reader to the reviews by Stellwagen and Cole (1969), Fambrough (1969), Elgin, Froehner, Smart and Bonner (1970) and Kleiman and Huang (1970).

Two types of classification for chromosomal proteins are possible. They can be classified either on their chemical properties into basic and acidic proteins, or by their biological function into three groups: (a) enzymes; (b) structural proteins; (c) regulatory proteins. Since detailed information is lacking on the biological activities of most of these proteins, classification will be by their chemical properties (Table 8.1).

II. BASIC PROTEINS

A. Histones

Stedman and Stedman (1950) first suggested that histones might be involved in the repression of genes during differentiation. This possibility has generated a great interest in histones, and as a result they have been the most thoroughly studied of the nuclear proteins.

Histones are chemically characterized as very basic proteins with a relatively high content of lysine and arginine. Amino acid analysis has also shown an absence of tryptophan. These compounds are associated with DNA in a complex with a weight ratio of approximately 1:1, and they can be obtained from either nuclei or chromatin using dilute acid or concentrated salt solutions.

Histones have been grouped into two major classes: lysine-rich and arginine-rich, each class having several protein species (Table 8.1). All tissues examined so far have been found to contain both classes. In genetically repressed tissues, such as avian nucleated erythrocytes, a new type of lysine-rich histone, histone V, was first described by Neelin and Butler (1961). It is present in the chromatin accompanying the other histones. Further studies have shown that

Table 8.1 **Classification of Chromatin Proteins**

I.	Basic proteins		
	1. Histones (lysine rich)	Histone I	$[F_1]$
		Histones IIb_1, IIb_2	[F2b and $F2b_2$]
		Histone V	[F2c]
	2. Histones (arginine rich)	Histone III	[F3]
		Histone IV	[F2a]
	3. Protamines		
II.	Acidic proteins		
	1. Enzymes		
	2. Structural proteins		
	3. Regulating proteins		

Figures in brackets give alternative histone nomenclature (Johns, 1964).

histone V is also present in the nucleated erythrocytes of some amphibians and fish. Paoletti and Huang (1969) demonstrated the presence of histone V in another type of totally repressed tissue, sea urchin sperm. These cells are also characterized by their lack of histone I.

There are two species of arginine-rich histones found both in plants and in animals. Arginine-rich histone III is the only histone found which contains cysteine. A much more basic arginine-rich histone is the histone IV. This histone molecule has a lys-arg ratio of 0.7 and a basic amino acid to acidic amino acid ratio of 3.9. A general summary of the unique features of each histone is presented in Figure 8.1 and their electrophoretic pattern in acrylamide gel system recorded in Fig. 8.2.

1. Histone characterization. Histones have been found in practically all tissues of a variety of higher organisms, both plants and animals, and this suggests their possible role as repressors. Because of this fact, a search was made for histone heterogeneity, in both tissues and species. However, this search has not been fruitful for the most part, as different tissues from the same animal, and organisms as far apart phylogenetically as pea and calf, show strikingly similar histone complements. More careful studies like the one by Crampton, Stein and Moore (1957) on moderately lysine-rich histones have shown that the heterogeneities described in early papers

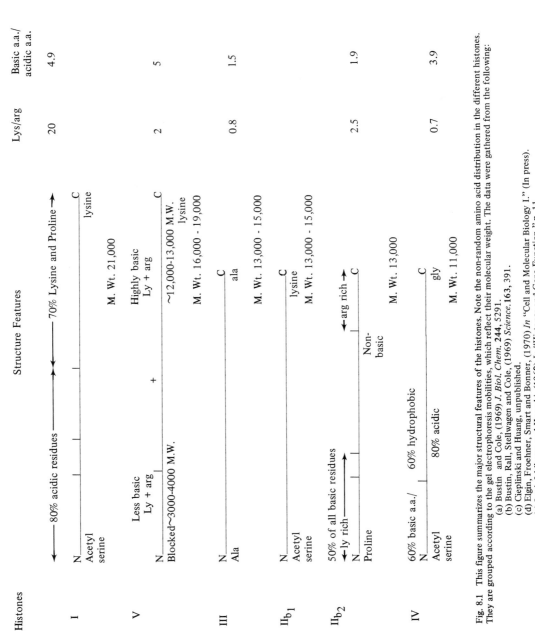

Histones	Structure Features	Lys/arg	Basic a.a./acidic a.a.
I	80% acidic residues — 70% Lysine and Proline. N Acetyl serine ... C lysine. M. Wt. 21,000	20	4.9
V	Less basic Ly + arg. Highly basic Ly + arg. N Blocked ~3000-4000 M.W. + ... ~12,000-13,000 M.W. ... C lysine. M. Wt. 16,000 - 19,000	2	5
III	N Ala ... C ala. M. Wt. 13,000 - 15,000	0.8	1.5
IIb₁	N Acetyl serine ... C lysine. M. Wt. 13,000 - 15,000		
IIb₂	50% of all basic residues. ly rich — arg rich. N Proline ... Non-basic ... C. M. Wt. 13,000	2.5	1.9
IV	60% basic a.a./ 60% hydrophobic. N Acetyl serine 80% acidic ... gly C. M. Wt. 11,000	0.7	3.9

Fig. 8.1 This figure summarizes the major structural features of the histones. Note the non-random amino acid distribution in the different histones. They are grouped according to the gel electrophoresis mobilities, which reflect their molecular weight. The data were gathered from the following:

(a) Bustin and Cole, (1969) J. Biol. Chem. **244**, 5291.
(b) Bustin, Rall, Stellwagen and Cole, (1969) Science. **163**, 391.
(c) Cieplinski and Huang, unpublished.
(d) Elgin, Froehner, Smart and Bonner, (1970) In "Cell and Molecular Biology I." (In press).
(e) Iwai, Ishikawa and Hayashi, (1969) In "Histones and Gene Function." p. 11.
(f) De Lange, Fambrough, Smith and Bonner, (1969) J. Biol. Chem. **244**, 319.

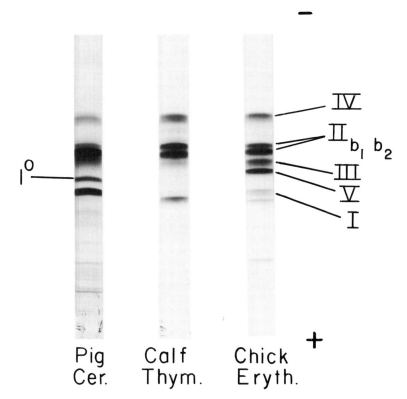

Fig. 8.2 Gel electrophoresis pattern comparison of total histones from pig cerebellum, chicken erythrocyte and calf thymus. Note that there is no equivalent in the pig cerebellum and calf thymus histones for histone V, the specific histone of chicken erythrocyte, while all the other histone classes are represented in the three gels. Electrophoresis was performed using the method of Panyim and Chalkley (1969a, b). Pig cerebellar histones pattern from Shaw and Huang (1970).

were artifacts due to proteolysis, degradation, aggregation and, possibly, contamination with other nuclear or even cytoplasmic proteins. Therefore, to obtain chromosomal proteins with the lowest degree of contamination it is best to start from purified chromatin, as acid extraction of nuclei can result in contamination by other acid soluble proteins. When HCl is used the histone chlorides obtained are not totally precipitable from acid ethanol so dilute sulfuric acid (0.4N) extraction is recommended for total histones (Fambrough and Bonner, 1968).

To obtain individual histones, selective histone fractionation is desired and the method of Johns (1964) allows the stepwise fractionation to obtain the different histone species. It has to be noted that this method does not give pure fractions and that its effectiveness varies from tissue to tissue and therefore should only be used as an initial step preceding other purification methods.

Acid extracted histone can be further purified by chromatography on ion exchange column IRC-A50 Amberlite using a guanidine-HCl gradient (Luck,

Rasmussen, Satake and Tsvetikov, 1958; Rasmussen, 1962). The fractions are eluted in order of increasing arginine content with the histone I coming first followed by slightly lysine-rich histones IIb_1 and IIb_2, then histone V and finally the arginine-rich histones III and IV. These fractions may still not be completely pure and rechromatography using a shallower gradient sometimes is necessary for the obtention of homogeneous histone samples. Exclusion chromatography has largely failed as a method of fractionation, probably due to aggregation of the histones. A successful instance is the isolation of histone IV by Fambrough and Bonner (1969) which has a molecular weight of 11,000, the smallest among all histones.

As an analytical tool, the polyacrylamide gel electrophoresis method (Bonner *et al.,* 1968) has many advantages, especially with the improved resolution given by the modifications of Panyim and Chalkley (1969a, b). This method requires very little material and it gives clear and reproducible bands with wide separation.

With the above methods the characterization of the histones has been done and it is now believed that there are only a few classes of histones, and that they are the same classes for all tissues and species of distant phylogenetical origins. Of course, this does not exclude microheterogeneity, like the conservative substitution of one amino acid for another in going from one species to another species. But even this is remarkably constant as evidenced in the amino acid sequence of histone IV from pea and calf thymus where there are only two amino acid substitutions, both of the conservative type, involving the substitution of one amino acid by another of the same chemical nature. In contrast, 30% of the residues of cytochrome *c* are substituted in wheat and calf.

Microheterogeneity can also exist as a result of chemical modification of the histones, by methylation, acetylation and phosphorylation. This will be discussed later.

2. Description of histone fractions. The very lysine-rich group consists only of histone I, but the fraction is not homogeneous. Molecular weight has been calculated to be around 21,000. It has a C-terminal lysine and the N-terminal is acetylated serine. Species differences were described by Kinkade (1969) and microheterogeneities have been shown to exist in the histone I of rabbit thymus, where up to nine fractions were described by Bustin and Cole (1968). These last authors specifically ruled out degradation showing that all the fractions possess the lysyl-lysine C-terminal and blocked N-terminal characteristic of histone I. In spite of their high basicity they move the slowest in polyacrylamide gels which is to be expected if we remember that in gels of above 7.5% in concentration, molecular sieve effects are the main determinants.

Bustin and Cole (1969) and Rall, Stellwagen and Cole (1969) have studied the amino acid distribution in the molecule of histone I by dividing it into 4 fragments of different sizes. The carboxy-terminal fragment constitutes slightly more than half of the molecule and it contains approximately 70% of the lysine residues but only

20% of the acidic amino acids. It also contains 70% of the proline residues. The amino-terminal fragment constitutes one-fourth of the total molecule and it contains only 15% of the lysines of the molecule. It contains, on the other hand, the rest of the proline residues of the molecule. This means that the two middle fragments, that together constitute the other one-fourth of the molecule, have no proline at all. It is also interesting to note that the amino-terminal fragment, together with the two middle fragments, account for 80% of the acidic residues and most of the hydrophobic residues. This observation suggests that the carboxy-terminal half is the one complexed with the phosphate backbone of the DNA while the amino-terminal conceivably reacts with other components of the complex. The physical accomodation of histone I when complexed with the DNA has been studied in different ways. It is known that histone I is the most sensitive fraction to proteolytic degradation by the neutral protease associated with the chromatin. It is also known that, in spite of the high density of positive charges, histone I is the first one removed by salt gradient dissociation. This fact suggests that histone I is highly accessible and exposed to its chemical environment.

Model building suggests that histone I can fit into the major groove of the DNA. The most acceptable model requires that histone I neutralizes approximately 50% of the phosphates on the DNA backbone, a prediction that has been verified experimentally by *in vitro* dye-binding studies (Miura and Ohba, 1967). Other work that supports the correctness of this model has been done by Olins (1969), who found that histone I when complexed with DNA does not inhibit actinomycin D binding which occurs in the minor groove. However, the complex does interfere with glucosylation of non-glucosylated T2 DNA which occurs in the major groove.

The moderately lysine-rich histones are a group characterized by a lys/arg ratio of approximately 2.5. These substances are also less basic with a basic amino acid/acidic amino acid ratio of 1.9.

The two components of group II, histones IIb_1 and IIb_2, account for approximately 40% of the total histones of most tissues. In calf thymus they have the same C-terminal lysine, and the N-terminal of IIb_1 is acetylated while the N-terminal of IIb_2 is proline. Histone IIb_2 was recently sequenced in its entirety by Iwai, Ishikawa and Hayashi (1970), and here, as in histone I, we can observe a dissymmetric distribution of basic and acidic amino acids. In this case it is the N-terminal side that is rich in lysine with the C-terminal side enriched in arginine. The region in between contains an accumulation of acidic and aromatic residues. The molecule has a total of 125 amino acids.

Histone V was first described in mature erythrocytes of the chicken by Neelin and Butler (1961), Neelin, Callahan, Lamb and Murray (1964), and Hnilica (1964). It has a lys/arg ratio of approximately 2, and a basic amino acid/acidic amino acid ratio of 5. It is probably the most basic histone of all, as almost 40% of its residues are basic. Many procedures have been used for its purification (Neelin *et al.,* 1964; Hnilica, 1964; Murray, Vidali and Neelin, 1968; Johns and Forrester, 1969), and it has been purified to homogeneity (Cieplinski and Huang, in preparation) as determined by gel electrophoresis, ultracentrifugation, and terminal group determinations. Histone V has a molecular weight of 16,000-19,000 as determined by the sodium dodecyl sulphate (SDS) gel method described by Weber and Osborn (1969).

The N-terminal of histone V is blocked and the C-terminal is lysine. Histone V runs in gel electrophoresis between monomer histone III and histone I, and no equivalent of it can be seen when compared with coelectrophoresed calf thymus histones (Table 8.1). The molecule has only one methionine residue, and on cleavage with cyanogen bromide, gives rise to two fragments that can be separated by Sephadex chromatography and acrylamide gel electrophoresis. Amino acid analysis suggests that basicity exists throughout the molecule with changes between the fragments consisting merely of differences in the lysine/arginine ratio. However, the NH_2-terminal fragment has a lower content of basic residues. These results have been confirmed and extended in a recent paper by Greenaway and Murray (1971) who separated histone V into two subcomponents V_a and V_b that apparently differ from one another by residue 15 which is glutamine in V_a and arginine in V_b. They also sequenced the small cyanogen bromide fragment which they showed to contain 30 residues of which only 5 are basic amino acids.

The arginine-rich histones are histone III and histone IV. Histone III is the only histone fraction that contains cysteine, and this fact accounted for the early reports of heterogeneity as it forms dimers and trimers giving a great number of bands in electrophoresis. However, work by Fambrough and Bonner (1968) has shown that treatment with β-mercaptoethanol will reduce it to the monomer state, giving only one band in electrophoresis. Histone III has a molecular weight of approximately 13,000 - 15,000, has a lys/arg ratio of 0.8 and a basic/acidic amino acid ratio of 1.5. Both its N- and C-terminals are alanine residues and in the case of pea bud it has only one cysteine while calf thymus histone III has two.

Histone IV is the smallest histone, and therefore moves the fastest in electrophoresis. It has a molecular weight of around 11,000 and a lys/arg ratio of 0.7 with a basic/acidic amino acid ratio of 3.9. This histone has also been completely sequenced (De Lange, Fambrough, Smith and Bonner, 1969a, b), and as noted earlier it shows a very high degree of conservation, the difference between peas and calf thymus being only two amino acids (Table 8.2.) It is unlikely that the genes coding for histone IV are more resistant to random mutation. Thus it is much more possible that the reason for the absence of evolutionary change in the amino acid sequence of histone IV is that most mutations would be lethal, possibly due to their hampering the association of the histone with the phosphate groups of DNA. The amino acid sequence of this histone also shows dissymmetry. The N-terminal section contains over 60% of the basic residues and the C-terminal region contains most of the hydrophobic amino acids. It is interesting to note the frequency of occurrence, in those molecules whose sequence is known, of two consecutive basic residues and of a basic residue every five residues, a periodicity that might also be significant in

Table 8.2 **The Complete Sequences of Calf and Pea Histone IV**

Ac-Ser-Gly-Arg-Gly-Lys-Gly-Gly-Lys-Gly-Leu-Gly-Lys-Gly-Gly-Ala-

Lys(Ac)-Arg-His-Arg-Lys(Me)-Val-Leu-Arg-Asp-Asn-Ile-Gln-Gly-Ile-Thr-Lys-
Lys

Pro-Ala-Ile-Arg-Arg-Leu-Ala-Arg-Arg-Gly-Gly-Val-Lys-Arg-Ile-Ser-Gly-Leu-

Ile-Tyr-Glu-Glu-Thr-Arg-Gly-Val-Leu-Lys-Val-Phe-Leu-Glu-Asn-Val-Ile-Arg-
Ile

Asp-Ala-Val-Thr-Tyr-Thr-Glu-His-Ala-Lys-Arg-Lys-Thr-Val-Thr-Ala-Met-Asp-
Arg

Val-Val-Tyr-Ala-Leu-Lys-Arg-Gln-Gly-Arg-Thr-Leu-Tyr-Gly-Phe-Gly-Gly-COOH.

A comparison of the amino acid sequences of calf thymus Histone IV with pea bud Histone IV. Note that the two amino acid substitutions at residues 60 and 77 are of the conservative type, that is, an amino acid of the same chemical nature substitutes the one existing in the other sequence. The third difference involves the specific methylation of the lysine in position 20 in calf thymus, but not in the pea. [From De Lange, Fambrough, Smith, and Bonner, 1969a, b].

binding to the phosphate backbone of DNA.

3. Histone synthesis.   Another approach in studying the importance of histones in regulation is to study the site and timing of their synthesis and degradation. It has been shown that certain types of nuclei possess the apparatus for protein synthesis, and the one best known is the thymocyte. In this system, Allfrey, Littau and Mirsky (1964) showed incorporation of precursor amino acids into histones, and that this synthesis was inhibited by puromycin. Reid and Cole (1964) record a similar observation and cite the lack of sensitivity to ribonuclease and the dependency on sodium ions as previously described by Allfrey *et al.* as characteristic of intranuclear synthesis in the thymocyte system.

On the other hand, there is considerable evidence to suggest that synthesis of histones in other systems occurs in the cytoplasm. An isolation procedure involving separation from the cytoplasm of small polyribosomes, with nascent polypeptides containing high lysine and low tryptophan, has been described in HeLa cells by Borun, Scharff and Robbins (1967). These polyribosomes appear only during the S phase of the cell cycle when the synthesis of histones is known to occur.

The *in vitro* synthesis of histones using cytoplasmic microsomes from HeLa cells has been achieved by Gallwitz and Mueller (1969a, b), and the products co-electrophorese with native histones.

There is no conclusive evidence therefore in favor of a special site of histone synthesis. It might occur in the nucleus in some species and in the cytoplasm in others. We cannot rule out either the possibility of synthesis in both parts of the cell, be it simultaneously or at different stages of the cell development.

The timing of synthesis is an important factor to be studied. If the role of histones is only structural, one would then expect the synthesis to occur almost simultaneously with DNA synthesis. The evidence from most systems studied suggests that synthesis occurs mostly, if not only, during the S phase of the cell cycle, coinciding or slightly preceding DNA synthesis (Robbins and Borun, 1967; Takei, Borun, Muchmore and Lieberman, 1968; Prescott, 1966). However, Spalding, Kajiwara and Mueller (1966) discovered independent metabolic patterns for the different histone fractions in HeLa cells with most of the synthesis coinciding with the DNA synthesis. Marzluff and McCarty (1970) studied insulin-induced development of mouse mammary gland, and also concluded that histones are synthesized concurrently with DNA and are stable. Both the lysine-rich histones and arginine-rich histones are synthesized at the same rate.

Sadgopal and Kabat (1969), however, observed synthesis of arginine-rich histones without concomitant DNA synthesis. The conclusion was based on the result of synthesis studies with radioactive precursors and the histone fractions separated by means of Amberlite chromatography. The radioactive labeled histone chromatographed with arginine-rich histones. Since chick erythrocyte specific histone (Fraction V) was found to have eluted with arginine-rich histones on an Amberlite column under the conditions used by Sadgopal and Kabat (Gershey, Haslett, Vidali and Allfrey, 1969; Cieplinski and Huang, 1971), it is not possible to conclude which histone is actually synthesized. An acrylamide gel electrophoresis assay system probably should also be employed to rule out the possibility that acidic protein that aggregates with histones is in fact the one synthesized.

4. Histone degradation. Very little is known about the degradation processes affecting histones. Most evidence indicates a very low turnover in the normal cell. A neutral protease, apparently associated *in vivo* with the chromatin and degraded histones, has been described by Phillips and Johns (1959) and Panyim, Jensen and Chalkley (1968). This enzyme, however, might be normally inhibited *in vivo*, as Furlau and Jericijo (1967) showed an increase in its activity following gel filtration. Another protease with an acid pH optimum has been described by Dounce and Umana (1962), Dounce and Hilgartner (1964), Dounce, Seaman and McKay (1966), and Dounce and Ickowicz (1969), but there is a possibility that it is actually a cytoplasmic contaminant. In a recent paper, Bartley and Chalkley (1970) reported that the susceptibility of the five major calf thymus histone-fractions to proteolysis is critically dependent upon whether the histones form a complex with DNA. In the intact nucleohistone, three groups of histones are almost totally resistant to proteolytic attack while the lysine-rich histone I and arginine-rich histone III are rapidly degraded. If histones are freed from DNA only the lysine-rich histone is resistant to protease; all other fractions are rapidly degraded. The proteolytic activity is reversibly inhibited by either the presence of sodium bisulfate or by storing the nucleohistone at low ionic strength (5×10^{-4}). The authors felt that the enhanced resistance of these histone fractions to the proteolytic enzyme probably lies in a protective effect of their interaction with DNA. Histones III and IV are likely to be in a much more intimate association with DNA than are histones I and II (IIb_1 and IIb_2).

5. Histone changes in development. Considerable attention has been given to the role of histones in development because it was expected that where a multipotential fertilized zygote divides and differentiates, some changes in the histone complement would be seen. Also, changes had been observed during the process of spermatogenesis and sperm maturation. Bloch and Hew (1960) studied spermatogenesis and early fertilization in the snail with the use of cytochemical techniques. They showed that sperm maturation involved the substitution of the histones present in early spermatids first by an "arginine-rich histone" and finally by protamines. In early stages of development (after fertilization), the protamines disappeared and gave way to a new histone that did not stain with fast green and with the promatine specific stain, but gave a weakly positive reaction with bromphenol blue. These "cleavage" histones are found in the male and female pronuclei, the early polar body chromosomes and the nuclei of the cleaving egg and morula stages. By the time gastrulation occurs the histone complement is similar to the adult somatic type, at least qualitatively. This work, however, relies only on staining techniques, and is subject to error if the permeability of the fertilized egg to the specific stains is different from the cell membrane permeability at later stages of development. Similar observations have been made also in several arthropods, molluscs and vertebrates, and the results are summarized by Bloch (1969). Later stages of development were studied by Dingman and Sporn (1964) and Kischer and Hnilica (1967) who studied differences between adult and chick embryos. The former found that the chromatins of both have the same total histone/DNA ratio while the latter showed that the chick embryo and adult histones coelectrophoresed and had similar amino acid composition. However, quantitative differences in the histone proportions can be seen in different organs (Lindsay, 1964), and the nucleated erythrocyte (Champagne and Mazen, 1969). Little work has been done on the early stages but Marushige and Ozaki (1967) did gel electrophoresis of histones from blastula stages of sea urchin embryos and found almost no difference compared to later stages.

As an alternative approach, Kedes, Gross, Cognetti and Hunter (1969) studied the early stages of sea urchin development by looking for the appearance of newly synthesized basic proteins in the nucleus. They used as their criteria the incorporation of labeled amino acids into proteins, defining as histones those proteins with a high lysine/tryptophan ratio. They also found by autoradiography that in the early stages of differentiation, 40-60% of the proteins synthesized accumulate in the nuclei and are associated with the chromosomes. They confirmed the observation of Borun *et al.* (1967) that the synthesis of the proteins of high lysine/tryptophan ratio is associated with small polyribosomes that sediment slowly. They were able to block most of the histone-like protein synthesis by the use of Actinomycin-D, an antibiotic that blocks transcription. The absence of an absolute block on histone synthesis suggests that even though most of the nuclear proteins are synthesized after fertilization, some histones present in the early stages of the embryo originate from long-lived templates stored in the cell during oogenesis.

6. Chemical modifications and microheterogeneity. A basic requirement for histones to be specific repressors is heterogeneity, in order to enable them to interact specifically with certain genes and not others. As we have seen before, major heterogeneities seem to have been ruled out as evidenced by the presence of the same classes of histones with similar amino acid composition in the different tissues and species which have been studied. This has focused attention on the possibilities of microheterogeneity. Microheterogeneity is defined as chemical modification of specific amino acid residues of the protein during or after its synthesis. Chemical modification of histones is a frequent observation in the different systems observed. The modifications seen are methylation, acetylation and phosphorylation. The donor of the methyl group is S-adenosyl methionine and the apparent receptor is the ϵ-amino group of lysine. Arginine can be methylated also; Paik and Kim (1967, 1968) have described enzymes associated with the chromatin that methylate specifically lysine and arginine. The significance of methylation is not known as studies on regenerating liver by Tidwell, Allfrey and Mirsky (1968) failed to show a temporal relationship between the burst of DNA, protein synthesis and histone methylation, the latter being a relatively late event.

In contrast, acetylation has shown more correlation with increased RNA polymerase activity and RNA synthesis (Pogo, Pogo, Allfrey and Mirsky, 1968; Pogo, Pogo and Allfrey, 1969), but obviously this does not mean that a causal relationship exists. Acetylation can occur at the N-terminal or at different parts of the molecule. Acetylation occurs after the chain synthesis is finished and seems to be very specific, involving different residues in different organisms. Some enzymes that have been isolated seem to have preference for arginine-rich histones (Allfrey, Littau and Mirsky, 1964; Vidali, Gershey and Allfrey, 1968).

Marzluff and McCarty (1970) have reported that there are two classes of histone acetylation in developing mouse mammary gland, one at the NH_2-terminal serine of histone I, histone IIb_2 and histone IV, another on the ϵ-NH_2 group of the internal lysine residues of histone III and histone IV. The NH_2-terminal acetylation is metabolically stable and is inhibited by cycloheximide. The internal ϵ-NH_2 group acetylation of histones III and IV, on the other hand, is quite reversible. Marzluff and McCarty suggest that amino-terminal acetylation probably represents a mechanism of polypeptide chain initiation. In contrast, the internal acetylation could function in control of RNA transcription.

7. Phosphorylation. Phosphorylation affects principally the serine residues in all types of histones. Occasionally a small amount of phosphothreonine can be detected. The phosphorylation process is energy dependent and a specific kinase has been described that will phosphorylate all histones (but not non-histone protein), with an apparent preference for histones I, IIb_1 and IIb_2. Phosphatase activity has also been described with the same specificities (Langan, 1968; Meisler and Langan, 1967, 1969).

Phosphorylation has also been involved in gene regulation, because it tends to decrease with increasing mitotic activity (Gutierrez and Hnilica, 1967). Phosphorylated F_1 is not as effective inhibiting an *in vitro* RNA polymerase system as non-phosphorylated F_1. Phytohemagluttinin increases RNA synthesis in equine lymphocytes within minutes and this is correlated, at least temporarily, with an increase in phosphorylation and acetylation of histones. However, inhibitors of RNA synthesis do not inhibit histone phosphorylation, indicating that RNA synthesis is not an absolute requirement for phosphorylation (Kleinsmith, Allfrey and Mirsky, 1966; Pogo, Pogo, Allfrey and Mirsky, 1968; Pogo, Pogo and Allfrey, 1969). Histone phosphorylation is stimulated by cyclic AMP suggesting that the process is under hormonal control.

The time course of DNA synthesis in regenerating rat liver coincides with increased F_1 phosphorylation (Stevely and Stocken, 1968a, b). Phosphorylation here might be playing a role in transporting the histones from the cytoplasm to the nucleus rather than in genetic control *per se*. All this evidence is indirect, but it suggests nevertheless that chemical modifications might play a role in gene expression and repression. It is not clear yet what that role might be, and whether it is specific or just a non-specific event during derepression.

8. Interaction between histones and DNA. Huang and Bonner (1962) showed a marked increase in synthesis of RNA primed by DNA when histones were added to the reaction mixture. Later they demonstrated that native chromatin was a poor primer compared to pure DNA in supporting RNA synthesis (Bonner and Huang, 1963). Since then, researchers have tried to study the mechanisms of this interaction in more detail.

One approach has been to study the physical associations that occur when histones or model systems like the synthetic polypeptides, polylysine and polyarginine are mixed with DNA. Leng and Felsenfeld (1966) using this approach showed that polylysine has a preference for A.T base pair rich regions, while polyarginine preferred G.C base pair rich regions. The effect was more than just charge neutralization as shown by Olins, Olins, and von Hippel (1968) who observed that tetralysine does not preferentially stabilize A.T base pair regions, but polymers bigger than 8 residues do. However, Leng and Felsenfeld did their work using 1.0 M NaCl, and when using 0.1 M NaCl the preferential stabilization effect disappeared. Another objection to considering homopolymer - DNA complexes as good model systems comes from optical rotary dispersion studies by Shapiro, Leng and Felsenfeld (1969), whose results suggest a large change of the DNA secondary structure when complexed with polylysine. However, DNA-histone I complexes show almost no change from the spectra obtained with DNA alone (Olins, 1969). This last result is contradicted by Fasman, Schaffhausen, Goldsmith and Adler (1970) who did find a large change in the circular dichroism of DNA complexed with histone I.

The method of reassociation is also of importance as results obtained when association is measured by thermal stabilization will vary with the method used. Chromatin characteristics and specificities seemed to be retained when reassociation was carried out by mixing the DNA and the protein in high ionic strength buffer and then slowly dialysing the salt by gradient dialysis in the presence of 5M urea (Huang, 1968; Huang and Huang, 1969; Bekhor, Kung, and Bonner, 1969). Further physical studies on such reconstituted chromatin seem to be one of the most important tasks for the future.

B. Protamines

Protamines are very basic molecules of low molecular weight, around 5000. Their basicity is due principally to their high arginine content. They are found in the sperm of certain fish, birds, and many invertebrates. The best studied class of protamines are the clupeines obtained from herring sperm. The clupeines have been resolved into three components, each one having approximately 30-31 residues of which 20 are arginine (Ando and Suzuki, 1966). The appearance of protamines in sperm is associated with cessation of RNA synthesis so it has been thought that they act as non-specific repressors. Protamines can also be phosphorylated (Dixon, Inglis, Jergil, Ling and Marushige, 1969), and this occurs in the cytoplasm of the sperm cell after the synthesis of the polypeptide chain is finished. Phosphorylation of protamines might also be a step in their transport to the nucleus.

III. ACIDIC PROTEINS

A. Structural proteins and enzymes

By definition, protein other than histones and protamines which associate with DNA to form chromatin constitute acidic proteins. It is a heterogeneous group, including enzymes like DNA polymerase, RNA polymerase, proteases, the enzymes associated with chemical modification of histones, some other assorted enzymes, and a group of proteins without enzymatic activity but possibly involved in structural and functional aspects of chromatin.

The problem of determining what constitutes a non-basic chromosomal protein is major, since there is no specific test to differentiate between contaminants and authentic chromosomal proteins. The seriousness of this problem is exemplified by the fact that the amount of non-histone proteins in chromatin varies, depending on the method used to isolate the chromatin and the extraction procedure used to obtain its associated proteins.

Johns and Forrester (1969) have shown, however, that some contamination by non-chromosomal acid- and dilute salt-soluble proteins can be eliminated when chromatin is isolated from calf thymus in dilute saline (0.14 M NaCl). Reextraction with 0.3 M NaCl will further remove two-thirds of the adsorbed non-histone protein. Chromatin preparations such as that of Bonner *et al.* (1968) still contain some of the supposed contaminant in that extraction with 0.3 M NaCl still removes 10% of the non-histone protein (Smart and Bonner, 1970). Kleiman and Huang (1970) remove contaminating proteins by passing the sheared chromatin through a Bio-Gel A50 column.

Conventional methods of isolation of non-histone proteins described in the literature (Marushige, Brutlag and Bonner, 1968; Benjamin and Gellhorn, 1968) use acid to remove the histones. The residual proteins are then solubilized by using sodium dodecyl sulfate (Marushige, Brutlag and Bonner, 1968), or cesium chloride (Benjamin and Gellhorn, 1968). Another method of extracting acidic proteins has been described by Wang (1967), who extracted calf thymus chromatin with 1 M NaCl and then diluted the extract to 0.14 M NaCl, thereby obtaining precipitation of nucleohistone. The supernatant contained acidic protein and had DNA polymerase activity.

Shaw and Huang (1970) and Kleiman and Huang (1970) have developed a method of extraction of all chromosomal proteins using 3M NaCl in Bio-Gel A50 chromatography and separated histones from non-histones by gel electrophoresis, at pH 2.7. At this pH, both basic and non-basic chromosomal proteins migrate toward the cathode to give up to 20 electrophoretic bands. SDS gel molecular weight determinations of these proteins by the technique of Weber and Osborn (1969) gives an average value of about 50,000. Amino acid analysis of the proteins recovered from these bands gives an acidic/basic amino acid ratio of 1.4 confirming their acidic nature. So far very little is known about the tissue and species differences of the non-histone proteins. The patterns obtained seem similar for the different tissues studied and no major absences or band additions have been noted, the differences so far being quantitative. Improved resolution in electrophoresis is needed before a firm conclusion can be reached concerning tissue specificity of acidic proteins.

Non-basic proteins seem to be involved in both structure and function of chromatin. Reagents that break sulfhydryl bonds disrupt chromosomal gels made by complexing DNA and residual non-histone proteins (Hilgartner, 1968). Their involvement as derepressors was suspected by Dingman and Sporn (1964) when they found that the amount of non-histone protein correlates with the type and physiological state of the chicken tissues they used. Frenster (1965) studied calf thymus lymphocytes where up to 80% of the chromatin is in the form of heterochromatin. By isolating separately the heterochromatin and the euchromatin he found a lower T_M for euchromatin. (When heated, the two DNA strands separate, a phenomenon that is accompanied by an increase in the absorbance at 260 mμ. The temperature at which 50% of this increase in absorbance is reached is called the T_M.)

This result is particularly interesting in that no qualitative or quantitative (relative to DNA) differences between the two types of chromatin were found as far as histones are concerned. Euchromatin, however, contained twice as much non-histone proteins as heterochromatin. We will discuss more evidence for the role of non-histone proteins later. Components other than the structural proteins are the nuclear enzymes. Wang (1968b) has partially purified DNA polymerase from calf thymus. The enzyme needs all four deoxyribonucleoside phosphates for optimal activity, but it can synthesize poly (dT) from dTTP, with an efficiency of 30%. The enzyme prefers heat denatured DNA as a template, and divalent cations are also required. Moreover, the enzyme is active with a variety of templates but its highest activity is observed with an homologous template (Howk and Wang, 1969). Later experiments showed that the preparation had some terminal transferase activity.

DNAase activity giving a $3'$ hydroxyl-terminal product has been found associated with chromatin prepared by precipitation from 0.2 M sodium phosphate buffer (Paul and Gilmour, 1966), but chromatin prepared by the method of Zubay and Doty (1959) is free of this enzyme.

RNA polymerase was first shown in chromatin by Huang, Maheshwari and Bonner (1960) and Weiss (1960). It is not clear whether all the enzyme is associated *in vivo* with the chromatin, and participating in RNA synthesis, or whether part is in solution depending on the physiological state of the nucleus. Considerable attention has been given to the study of nuclear protease activity (Reid and Cole, 1964; Furlan and Jericijo, 1967).

Some enzymes involved in histone modification have been isolated. Protein phosphokinase isolated by Langan (1966) is found in the acidic protein group (Kleinsmith and Allfrey, 1969). Comb, Sarkar and Pinzino (1966) associate a methylase with the chromosomal proteins. Finally Wang (1967) describes in his acidic protein extract a number of assorted enzymatic activities including glutamate dehydrogenase, glutamate oxalacetate, transaminase, lactate dehydrogenase and ATPase. The biological role of these enzymes in the nucleus is not known.

B. Proteins associated with chromosomal RNA

RNAs present in isolated chromatin can be generally grouped into two classes. One class consists of RNAs larger than 4S, the other class consists of those RNAs which are 4S and smaller. The two groups can be separated from each other by Bio-Gel column chromatography. When chromatin is dissociated by high salt (3 M NaCl), the class of small RNAs (3S) can be easily separated from DNA and other RNA by Bio-Gel A50 (Shaw and Huang, 1970). This small size RNA is eluted with the chromosomal proteins and can be separated from them by means of CsCl equilibrium centrifugation. The isolated RNA, after further purification by disc electrophoresis, is unique in its covalent linkage to amino acids, peptides or a polypeptide. In the chick embryo system, the polypeptide (approximately 3000 molecular weight) is linked to N^4 of the dihydropyrimidine base of the RNA through a peptide bond. Analysis indicated that the amino acid composition of the linked polypeptide is acidic in nature. Similar types of chromosomal RNAs were also found in the chromatin of rat Novikoff ascites cells (Dahmus and McConnell, 1969), leukaemic leukocytes (Getz and Saunders, 1970), pea embryos (Huang and Bonner, 1965), calf thymus (Shih and Bonner, 1970), HeLa cells (DeFillippes, 1970), and recently rat liver (Mascia, 1970). In the case of chick embryo chromsomal RNA, the structure is visualized as a continuous RNA molecule of 40 to 50 nucleotides containing several dihydropyrimidine bases which are linked through an amide bond to an amino acid, peptide or polypeptides. If this structure of protein-bound RNA can be generalized to the RNA found in the other system, we will indeed have a candidate for the regulatory molecule in interphase chromosomes. There is evidence suggesting the RNA sequence can base-pair with DNA, and the attached polypeptide may prove to be essential for specific interaction with other chromosomal proteins or hormones (Huang and Smith, 1971). Future experiments examining these theories should be very rewarding.

IV. ROLE OF CHROMATIN COMPONENTS IN CONTROLLING TRANSCRIPTION

Chromatin has only a fraction of the capacity to prime RNA synthesis, as compared with deproteinized DNA. Histones have been found to be responsible for the repression of RNA transcription. Evidence for the repression being tissue specific comes from the work of Paul and Gilmour (1968) and Smith, Church and McCarthy (1969), who demonstrated by hybridization techniques that the RNA produced by chromatin template *in vitro* is indistinguishable from the one produced by the same tissue *in vivo*.

Chromatin appears to be, then, an active complex in which each one of its components plays a vital role in tissue specific gene transcription. The dynamic role of all the components is shown when we observe that the lack of one component disrupts the system. When histones are removed but non-histone proteins remain, the complex has the template activity of DNA (Marushige and Bonner, 1966). This is true also when native DNA is complexed with non-histone proteins (Marushige, Brutlag and Bonner, 1968). DNA histone complexes have no template activity. The repression can be overcome by adding a non-specific acidic protein, like bovine serum albumin; but for the reconstitution to be specific, the specific non-histone chromosomal acidic fraction and slow dialysis with urea is needed (Gilmour and Paul, 1969). Whether the specificity resides on all the acidic proteins, or few proteins and/or the associated chromosomal RNA in this fraction, remains to be answered.

The importance of chromosomal RNA for specific reconstitution has also been implicated in two systems (Huang, 1968; Huang and Huang, 1969; Bekhor, Kung and Bonner, 1969). The precise mechanism for specific reconstitution is currently being investigated.

V. CONCLUDING REMARKS

This paper briefly discusses the present state of our knowledge on the structure of chromosomal components and their possible role in controlling transcriptional processes. The information suggests a delicate balance between the different constituents, whose precise interaction give the cellular genome the specificity which is the basis of cell differentiation.

The challenges for future research in this area are several. As the mechanisms for protein synthesis in bacteria are being described a great interest has been generated in the study of their counterpart in higher organisms. Ideally, the function of chromatin should be examined through its transcriptional properties in direct specific protein synthesis. Assay systems of cell free protein synthesis as established by Lockard and Lingrel (1970) and Stavnezer and Huang (1971) seem to be best suited for this study. When certain tissues are actively synthesizing one particular protein in their development, is the genomic chromatin turned on for transcription of the messenger RNA for this protein? If so, the way the chromatin components exert their effects upon transcription can be extensively studied.

Another intriguing problem is the one concerning chromosome replication. Its study will require not only the characterization of the enzymes and cofactors involved but also the optimal conditions needed. In the enzymatic aspect some progress has already been achieved as a DNA polymerase has been described and characterized in several eukaryotes (Mantsavinos, 1964; Bellair, 1968; Wang, 1968; Chang and Hodes, 1968; Loeb, 1969; Howk and Wang, 1969; De Recondo and Fichot, 1970). It is conceivable to expect that the enzyme involved would transcribe reconstituted chromosomal nucleoprotein at least as efficiently as DNA. As more studies are being made on the functional aspects, it will become possible to study the molecular basis of chromosome replication.

VI. ACKNOWLEDGMENTS

This work was supported in part by Research Grant GM 13723, National Institutes of Health, and the American Cancer Society ACS MD 69-70. W.C. is a recipient of a Belding Fellowship from the Association for the Aid of Crippled Children, New York.

VII. REFERENCES

Allfrey, V. G., Littau, V. C. and Mirsky, A. E. (1964) Methods for the purification of thymus nuclei and their application to studies of nuclear protein synthesis. *J. Cell Biol.* 21, 213.

Allfrey, V. G., Faulkner, R. and Mirsky, A. E. (1964) Acetylation and methylation of histones and their possible role in the regulation of RNA synthesis. *Proc. Nat. Acad. Sci. U.S.A.* **51**, 786.

Bartley, J. and Chalkley, R. (1970) Further studies of a thymus nucleohistone-associated protease. *J. Biol. Chem.* **245**, 4286.

Bekhor, I., Kung, Grace M., and Bonner, James (1969) Sequence-specific interaction of DNA and chromosomal protein. *J. Molec. Biol.* **39**, 351.

Bellair, J. T. (1968) The rapid isolation of a regenerating rat liver DNA nucleotidyltransferase fraction which shows a preference for native DNA as primer. *Biochim. Biophys. Acta.* **161**, 119.

Benjamin, W. B. and Gellhorn, A. (1968) Acidic proteins of mammalian nuclei: Isolation and Characterization. *Proc. Nat. Acad. Sci. U.S.A.* **59**, 262.

Bloch, D. P. and Hew, H. Y. (1960) Schedule of spermatogenesis in the pulmonate snail *Helix aspersa,* with special reference to histone transition. *J. Biophys. Biochem. Cytol.* **7**, 515.

Bloch, D. P. (1969) A catalog of sperm histones. *In* "International symposium on nuclear physiology and differentiation," Belo Horizonte, Brazil, 1968. (R. P. Wagner, ed.) *Genetics.* (Suppl.) **61**, 93.

Bonner, J., Chalkley, R. G., Dahmus, M., Fambrough, D., Fujimura, F., Huang, R. C., Huberman, J., Jensen, R., Marushige, K., Ohlenbusch, H., Olivera, B. and Widholm, S. (1968) Isolation and characterization of chromosomal nucleoproteins. *In* "Methods in Enzymology". (L. Grossman and K. Moldave, eds.) Vol. XII, p. 3, New York, Academic Press.

Bonner, J. and Huang, R. C. (1963) Properties of chromosomal nucleohistones. *J. Molec. Biol.* **6**, 169.

Borun, T. W., Scharff, M. D. and Robbins, E. (1967) Rapidly labeled, polyribosome associated RNA having the properties of histone messenger. *Proc. Nat. Acad. Sci. U.S.A.* **58**, 1977.

Bustin, M. and Cole, R. D. (1968) Species and organ specificity in very lysine-rich histones. *J. Biol. Chem.* **243**, 4500.

Bustin, M. and Cole, R. D. (1969) Bisection of a lysine-rich histone by N-bromosuccinimide. *J. Biol. Chem.* **244**, 5291.

Bustin, M., Rall, S. C., Stellwagen, R. H. and Cole, R. D. (1969) Histone structure: Asymmetric distribution of lysine residues in lysine-rich histone. *Science.* **163**, 391.

Champagne, M. and Mazen, A. (1969) Variation quantitative de la fraction d'histone specifique aux erythrocytes, au cours du developpement du poulet. 6th F.E.B.S. Meeting. Madrid. 720 (Abstracts).

Chang, L. M. S. and Hodes, M. E. (1968) The deoxyribonucleic acid nucleotidyltransferase activities of the Shope fibroma. *J. Biol. Chem.* **243**, 5337.

Crampton, C. F., Stein, W. H. and Moore, S. (1957) Comparative studies on chromatographically purified histones. *J. Biol. Chem.* **225**, 363.

Comb, D. G., Sarkar, N. and Pinzino, C. J. (1966) The methylation of lysine residues in protein. *J. Biol. Chem.* **241**, 1857.

Dahmus, M. and McConnell, D. J. (1969) Chromosomal ribonucleic acid of rat ascites cells. *Biochemistry.* **8**, 1524.

De Filippes, F. M. (1970) Simple purification of HeLa chromatin associated RNA. *Biochim. Biophys. Acta.* **199**, 562.

De Lange, R. J., Fambrough, D. M., Smith, E. L. and Bonner, J. (1969a) The complete amino-acid sequence of calf thymus histone IV. *J. Biol. Chem.* **244**, 319.

De Lange, R. J., Fambrough, D. M., Smith, E. L. and Bonner, J. (1969b) Calf and pea histone IV. Complete amino-acid sequence of pea seedling histone IV; comparison with the homologous calf thymus histone. *J. Biol. Chem.* **244**, 5669.

De Recondo, A. M. and Fichot, O. (1969) Purification partielle et proprietes de la desoxynucleotidyltransferase replicative de foie de rat. *Biochim. Biophys. Acta.* **186**, 340.

Dingman, W. C. and Sporn, M. D. (1964) Studies on chromatin: I. Isolation and characterization of nuclear complexes of DNA, RNA and protein from embryonic and adult tissues of the chicken. *J. Biol. Chem.* **239**, 3483.

Dixon, G. H., Inglis, C. J., Jergil, B., Ling, V. and Marushige, K. (1969) Protein transformations during differentiation of trout testis. *In* "Canadian Cancer Conference", p. 76. Toronto, Pergamon Press.

Dounce, A. L. and Umana, R. (1962) The proteases of isolated cell nuclei. *Biochemistry.* **1**, 811.

Dounce, A. L. and Hilgartner, C. A. (1964) DNA nucleoprotein gels and residual protein. *Exp. Cell Res.* **36**, 228.

Dounce, A. L., Seaman, F. and McKay, M. (1966) Effect of pH of isolation on the quantities and properties of histones extractable from calf thymus and rat liver nuclei. *Arch. Biochem.* **117**, 550.

Dounce, A. L. and Ickowicz, R. (1969) A study of optimal conditions for separating protein fractions from isolated cell nuclei. *Arch. Biochem.* **131**, 210.

Elgin, S. C. R., Froehner, S. C., Smart, J. E. and Bonner, J. (1970) *In* "Advances in Cell and Molecular Biology", (E. J. DuPraw, ed.) New York, Academic Press. (In press).

Fambrough, D. M. (1968) Studies on plant and animal histones. Ph.D. Thesis. California Institute of Technology.

Fambrough, D. M. and Bonner, J. (1968) Selective dissociation of pea-bud nucleohistone. *Biochim. Biophys. Acta.* **154**, 601.

Fambrough, D. M. (1969) Nuclear protein fractions. *In* "Handbook of Molecular Cytology." (A. Lima de Faria, ed.) Amsterdam, London, North Holland Publishing Co.

Fasman, G. D., Schaffhausen, B., Goldsmith, L. and Adler, A. (1970) Conformational changes associated with F1 histone-DNA complexes. Circular dichroism studies. *Biochemistry.* **9**, 2814.

Frenster, J. H., Allfrey, V. G. and Mirsky, A. E. (1963) Repressed and active chromatin isolated from interphase lymphocytes. *Proc. Nat. Acad. Sci. U.S.A.* **50**, 1026.

Frenster, J. H. (1965) Nuclear polyanions as de-repressors of synthesis of RNA. *Nature (London).* **206**, 680.

Furlan, M. and Jericijo, M. (1967) Protein catabolism in thymus nuclei. I. Hydrolysis of nucleoproteins by proteases present in calf thymus nuclei. II. Binding of histone-splitting nuclear proteases to DNA. *Biochim. Biophys. Acta.* **147**, 135 and 145.

Gallwitz, D. and Mueller, G. C. (1969a) Histone synthesis *in vitro* by cytoplasmic microsomes from HeLa cells. *Science.* **163**, 1351.

Gallwitz, D. and Mueller, G. C. (1969b) Histone synthesis *in vitro* on HeLa cell microsomes. The nature of the coupling to DNA synthesis. *J. Biol. Chem.* **244**, 5947.

Gershey, E. L., Haslett, G. W., Vidali, G. and Allfrey, V. G. (1969) Chemical studies of histone methylation. *J. Biol. Chem.* **244**, 4871.

Getz, M. J. and Saunders, G. F. (1970) Chromosomal RNA of human leukemic leukocytes. *Fed. Proc.* **29**, 671.

Gilmour, R. S. and Paul, J. (1969) RNA transcribed from reconstituted nucleoprotein is similar to natural RNA. *J. Molec. Biol.* **40**, 137.

Greenaway, P. J. and Murray, K. (1971) Heterogeneity and polymorphism in chicken erythrocyte histone fraction V. *Nature (London).* **229**, 233.

Gurdon, J. B. and Uehlinger, V. (1966) "Fertile" intestine nuclei. *Nature (London).* **210**, 1240.

Gutierrez, R. M. and Hnilica, L. S. (1967) Tissue specificity of histone phosphorylation. *Science.* **157**, 1324.

Hilgartner, C. A. (1968) The binding of DNA to residual protein in mammalian nuclei. *Exp. Cell Res.* **49**, 520.

Hnilica, L. (1964) The specificity of histones in chicken erythrocytes. *Experientia.* **20**, 13.

Howk, R. and Wang, T. Y. (1969) DNA polymerase from rat liver chromosomal proteins. I. Partial purification and general characteristics. *Arch. Biochem.* **133**, 238.

Huang, R. C., Maheshwari, N. and Bonner, J. (1960) Enzymatic synthesis of RNA. *Biochem. Biophys. Res. Commun.* **3**, 689.

Huang, R. C. and Bonner, J. (1962) Histone, a suppressor of chromosomal RNA synthesis. *Proc. Nat. Acad. Sci. U.S.A.* **48**, 1216.

Huang, R. C. and Bonner, J. (1965) Histone-bound RNA, a component of native nucleohistone. *Proc. Nat. Acad. Sci. U.S.A.* **54**, 960.

Huang, R. C. (1968) The control of transcription in higher organisms. *In* "Recent Development of Biochemistry". Taiwan, Academia Sinica.

Huang, R. C. and Huang, P. C. (1969) Effect of protein-bound RNA associated with chick embryo chromatin on template specificity of the chromatin. *J. Molec. Biol.* **39**, 365.

Huang, R. C. and Smith, M. M. (1971) Nucleic acid hybridization concerning chromosomal protein bound RNA. *In* "Results and Problems in Cell Differentiation". (H. Ursprung, ed.) Vol. III Nucleic acid hybridization in the study of cell differentiation. Berlin-Heidelberg-New York, Springer Verlag. (In press).

Inoue, S. and Ando, T. (1966) Interaction of clupeine and DNA. *Biochim. Biophys. Acta.* **129**, 649.

Iwai, K., Ishikawa, K. and Hayashi, H. (1970) Amino acid sequence of slightly lysine-rich histone. *Nature (London).* **226**, 1056.

Johns, E. W. (1964) Studies on histones. Preparative methods for histone fractions from calf thymus. *Biochem. J.* **92**, 55.

Johns, E. W. and Diggle, J. H. (1969) A method for the large scale preparation of the avian erythrocyte specific histone F_{2c}. *Europ. J. Biochem.* **11**, 495.

Johns, E. W. and Forrester, S. (1969) Studies on nuclear proteins. The binding of extra acidic proteins to deoxyribonucleoprotein during the preparation of nuclear proteins. *Europ. J. Biochem.* **8**, 547.

Kedes, L. H., Gross, P. R., Cognetti, G. and Hunter, A. L. (1969) Synthesis of nuclear and chromosomal proteins on light polyribosomes during cleavage in the sea-urchin embryo. *J. Molec. Biol.* **45**, 337.

Kinkade, J. M. (1969) Qualitative species differences and quantitative tissue differences in the distribution of lysine-rich histones. *J. Biol. Chem.* **244**, 3375.

Kischer, C. N. and Hnilica, L. S. (1967) Analysis of histones during organogenesis. *Exp. Cell Res.* **48**, 424.

Kleiman, L. and Huang, R. C. (1970) Specificities in the structure and function of interphase chromosomes, *In* "Growth and Differentiation." Cambridge University Press. (In press).

Kleinsmith, L. J., Allfrey, V. G. and Mirsky, A. E. (1966) Phosphoprotein metabolism in isolated lymphocyte nuclei. *Proc. Nat. Acad. Sci. U.S.A.* **55**, 1182.

Kleinsmith, L. J. and Allfrey, V. G. (1969) Nuclear phosphoproteins I. Isolation and characterization of a phosphoprotein fraction from calf thymus nuclei. *Biochim. Biophys. Acta.* **175**, 123.

Langan, T. A. (1966) A phosphoprotein preparation from liver nuclei and its effect on the inhibition of RNA synthesis by histones. *In* "Regulation of Nucleic Acid and Protein Biosynthesis." Proc. Internat. Symp. Lunteren, Netherlands. (V. V. Konigsberger and L. Bosch, eds.) p. 233. Amsterdam, Elsevier.

Langan, T. A. (1968) Phosphorylation of proteins of the cell nucleus. *In* "Regulatory Mechanisms for Protein Synthesis in Mammalian Cells." (A. San Pietro, M. Lamborg and F. T. Kenney, eds.) p. 101. New York, Academic Press.

Leng, M. and Felsenfeld, G. (1966) The preferential interactions of polylysine polyarginine with specific base sequences in DNA. *Proc. Nat. Acad. Sci. U.S.A.* **56**, 1325.

Lindsay, D. T. (1964) Histones from developing tissues of the chicken. Heterogeneity. *Science.* **144**, 420.

Lockard, R. and Lingrel, J. B. (1970) Identification of mouse hemoglobin messenger RNA. *Fed. Proc.* **29**, (2), Abst. No. 2394.

Loeb, L. A. (1969) Purification and properties of DNA polymerase from nuclei of sea urchin embryos. *J. Biol. Chem.* **244**, 1672.

Luck, J. M., Rasmussen, P. S., Satake, K. and Tsvetikov, A. N. (1958) Further studies on the fractionation of calf thymus histone. *J. Biol. Chem.* **233**, 1407.

Mantsavinos, R. (1964) Studies on the synthesis of deoxyribonucleic acid by mammalian enzymes. I. Incorporation of deoxyribonucleoside-5-triphosphates into DNA by a partially purified enzyme from regenerating rat liver. *J. Biol. Chem.* **239**, 3431.

Marushige, K. and Bonner, J. (1966) Template properties of liver chromatin. *J. Molec. Biol.* **15**, 160.

Marushige, K. and Ozaki, H. (1967) Properties of isolated chromatin from sea urchin embryo. *Develop. Biol.* **16**, 474.

Marushige, K., Brutlag, D. and Bonner, J. (1968) Properties of chromosomal nonhistone protein of rat liver. *Biochemistry.* **7**, 3149.

Marzluff, William and McCarty, K. (1970) Two classes of histone acetylation in developing mouse mammary gland. *Biochemistry.* (In press).

Mascia, M. (1970) (Personal communication). See Kleiman, L. and Huang, R. C. Specificities in the structure and function of interphase chromosomes. *In* "Growth and Differentiation." Cambridge University Press. (In press.).

Meisler, M. H. and Langan, T.A. (1967) Enzymatic phosphorylation and dephosphorylation of histones and protamines. *J. Cell Biol.* **35**, 91A.

Meisler, M. H. and Langan, T. A. (1969) Characterization of a phosphatase specific for phosphorylated histones and protamines. *J. Biol. Chem.* **244**, 4961.

Miura, A. and Ohba, Y. (1967) Structure of nucleohistone. III. Interaction with toluidine blue. *Biochim. Biophys. Acta.* **145**, 436.

Murray, K., Vidali, G. and Neelin, J. M. (1968) The stepwise removal of histones from chicken erythrocyte nucleoprotein. *Biochem. J.* **107**, 207.

Neelin, J. M. and Butler, G. C. (1961) A comparison of histones from chicken tissues by zone electrophoresis in starch gel. *Can. J. Biochem. Physiol.* **39**, 485.

Neelin, J. M., Callahan, P. X., Lamb, D. C. and Murray, K. (1964) The histones of chicken erythrocyte nuclei. *Can. J. Biochem.* **42**, 1743.

Olins, D. E., Olins, A. L. and von Hippel, P. H. (1968) On the structure and stability of DNA-protamine and DNA polypeptide complexes. *J. Molec. Biol.* **33**, 265.

Olins, D. E. (1969) Interaction of lysine rich histones and DNA. *J. Molec. Biol.* **43**, 439.

Paik, W. J. and Kim, S. (1967) Enzymatic methylation of protein fractions from calf thymus nuclei. *Biochem. Biophys. Res. Commun.* **29**, 14.

Paik, W. J. and Kim, S. (1968) Protein methylase I. Purification and properties of the enzyme. *J. Biol. Chem.* **243**, 2108.

Panyim, S., Jensen, R. H. and Chalkley, R. (1968) Proteolytic contamination of calf thymus nucleohistone and its inhibition. *Biochim. Biophys. Acta.* **160**, 252.

Panyim, S. and Chalkley, R. (1969a) The heterogeneity of histones. I. A quantitative analysis of calf histones in very long polyacrylamide gels. *Biochemistry.* **8**, 3972.

Panyim, S. and Chalkley, R. (1969b) A new histone found only in mammalian tissues with little cell division. *Biochem. Biophys. Res. Commun.* **37**, 1042.

Paoletti, R. and Huang, R. C. (1969) Characterization of sea urchin sperm chromatin and its basic proteins. *Biochemistry.* **8**, 1615.

Paul, J. and Gilmour, R. S. (1966) Template activity of DNA is restricted in chromatin. *J. Molec. Biol.* **16**, 242.

Paul, J. and Gilmour, R. S. (1968) Organ specific restriction of transcription in mammalian chromatin. *J. Molec. Biol.* **34**, 305.

Paul, J. and Gilmour, R. S. (1969) RNA transcribed from reconstituted nucleoprotein is similar to natural RNA. *J. Molec. Biol.* **40**, 137.

Phillips, D. M. P. and Johns, E. W. (1959) A study of the proteinase content and the chromatography of thymus histones. *Biochem. J.* **72**, 538.

Pogo, B. G. T., Pogo, A. O., Allfrey, V. G. and Mirsky, A. E. (1968) Changing patterns of histone acetylation and RNA synthesis in regeneration of the liver. *Proc. Nat. Acad. Sci. U.S.A.* **59**, 1337.

Pogo, B. G. T., Pogo, A. O. and Allfrey, V. G. (1969) Histone acetylation and RNA synthesis in rat liver regeneration. *Genetics.* (Suppl.) **61**, 373.

Prescott, D. M. (1966) The synthesis of total macronuclear protein histone and DNA during the cell cycle in *Euplotes eurystomus. J. Cell Biol.* **31**, 1.

Rasmussen, P. S., Murray, K. and Luck, J. M. (1962) On the complexity of calf thymus histone. *Biochemistry.* **1**, 79.

Reid, B. R. and Cole, R. D. (1964) Biosynthesis of a lysine-rich histone in isolated calf thymus nuclei. *Proc. Nat. Acad. Sci. U.S.A.* **51**, 1044.

Robbins, E. and Borun, T. W. (1967) The cytoplasmic synthesis of histone in HeLa cells and its temporal relationship to DNA replication. *Proc. Nat. Acad. Sci. U.S.A.* **57**, 409.

Sadgopal, A. and Kabat, D. (1969) Synthesis of chromosomal proteins during the maturation of chicken erythrocytes. *Biochim. Biophys. Acta.* **190**, 486.

Shapiro, K., Leng, M. and Felsenfeld, G. (1969) DNA - polylysine complexes. Structure and nucleotide specificity. *Biochemistry.* **8**, 3219.

Shaw, L. and Huang, R. C. (1970) A description of two procedures which avoid the use of extreme pH conditions for the resolution of components isolated from chromatins prepared from pig cerebellar and pituitary nuclei. *Biochemistry.* **9**, 4530.

Shih, T. Y. and Bonner, J. (1970a) Thermal denaturation and template properties of DNA complexes with purified histone fractions. *J. Molec. Biol.* **48**, 469.

Shih, T. Y. and Bonner, J. (1970b) Template properties of DNA polypeptide complexes. *J. Molec. Biol.* **50**, 333.

Smart, J. E. and Bonner, J. (1970) *In* "Advances in Cell and Molecular Biology". (E. J. DuPraw, ed.) New York, Academic Press. (In press).

Smith, K. D., Church, R. B. and McCarthy, B. J. (1969) Template specificity of isolated chromatin. *Biochemistry.* **8**, 4271.

Spalding, J., Kajiwara, K. and Mueller, G. C. (1966) The metabolism of basic proteins in HeLa cells. *Proc. Nat. Acad. Sci. U.S.A.* **56**, 1535.

Stavnezer, J. and Huang, R. C. (1971) *In vitro* synthesis of mouse immunoglobulin light chain in rabbit reticulocyte cell free system. *New Biology.* **230**, 172.

Stedman, E. and Stedman, E. (1950) Cell specificity of histones. *Nature.* **166**, 780.

Stellwagen, R. H. and Cole, R. D. (1969) Chromosomal proteins. *In* "Ann. Rev. Biochem." (E. E. Snell, ed.) Vol. 38, p. 951. Palo Alto, California, Annual Reviews, Inc.

Stevely, W. S. and Stocken, L. A. (1968a) Histone phosphorylation and cell division. *Biochem. J.* **109**, 24.

Stevely, W. S. and Stocken, L. A. (1968b) Variations in the phosphate content of histone F_1 in normal and irradiated tissues. *Biochem. J.* **110**, 187.

Takai, S., Borun, T. W., Muchmore, J. and Lieberman, I. (1968) Concurrent synthesis of histone and DNA in liver after partial hepatectomy. *Nature (London).* **219**, 860.

Tidwell, T., Allfrey, V. G. and Mirsky, A. E. (1968) The methylation of histones during regeneration of the liver. *J. Biol. Chem.* **243**, 707.

Vidali, G., Gershey, E. L. and Allfrey, V. G. (1968) Chemical studies of histone acetylation. *J. Biol. Chem.* **243**, 6361.

Wang, T. Y. (1967) The isolation, properties and possible functions of chromatin acidic proteins. *J. Biol. Chem.* **242**, 1220.

Wang, T. Y. (1968a) Restoration of histone-inhibited DNA-dependent RNA synthesis by acidic chromatin proteins. *Exp. Cell Res.* **53**, 288.

Wang, T. Y. (1968b) Partial purification of chromosomal DNA polymerase from rat Walker Tumor. *Proc. Soc. Exp. Biol. Med.* **129**, 469.

Weber, K. and Osborn, M. (1969) The reliability of molecular weight determinations by dodecyl sulfate polyacrylamide gel electrophoresis. *J. Biol. Chem.* **244**, 4406.

Weiss, S. B. (1960) Enzymatic incorporation of ribonucleoside triphosphates into the interpolynucleotide linkages of RNA. *Proc. Nat. Acad. Sci. U.S.A.* **46**, 1020.

Zubay, G. and Doty, P. (1959) The isolation and properties of deoxyribonucleoprotein particles containing single nuclei and molecules. *J. Molec. Biol.* **1**, 1.

9

RIBONUCLEIC ACID SYNTHESIS DURING OOGENESIS IN *XENOPUS LAEVIS*

Peter J. Ford

I.   Introduction
    A.   Stages of oogenesis
    B.   Sources of DNA in *Xenopus* ovary
        1.   Nuclear DNA
        2.   Cytoplasmic DNA
II.   Types of RNA synthesis
    A.   Non-nuclear RNA
        1.   Cytoplasmic RNA
        2.   Follicle cell RNA
    B.   Nuclear RNA
        1.   DNA-like RNA
        2.   4S RNA
        3.   18S and 28S rRNA
        4.   5S RNA
III.   Nature of ribonucleic acids in the immature ovary
    A.   Non-coordinate synthesis of 5S RNA
    B.   Cytoplasmic location of the 5S RNA
    C.   A ribonucleoprotein particle containing 5S RNA
    D.   Physical and chemical properties of the 42S material
    E.   Is the 42S material a homogenization artifact?
IV.   Possible explanations for excess synthesis of 5S RNA by immature oocytes
    of *Xenopus*
        Control of ribosomal RNA synthesis in *Xenopus laevis*
V.   Acknowledgments
VI.   References

I. INTRODUCTION

There are many reasons why knowledge of synthesis of the different types of RNA during oogenesis is important. Information on the synthesis of these different RNAs is central for understanding: (1) transcription and information storage, (2) mechanisms controlling RNA synthesis, and (3) the role of the RNAs in the process of oocyte differentiation. The mature oocyte contains information for considerable protein synthesis (Davidson, Crippa, Kramer and Mirsky, 1966), yet much of it is expressed only after fertilization in sea urchins (Stavy and Gross, 1967; 1969a, b) and during ovulation in *Rana pipiens* (Smith, Ecker and Subtelny, 1966; Smith and Ecker, 1969). The information is expressed even after physical enucleation of newly fertilized eggs (Ecker and Smith, 1968), suggesting that it is pre-existing mRNA that is being translated.

The oocyte contains a prodigious number of ribosomes for a single cell (Brown and Littna, 1964a, b), and these are largely inactive in protein synthesis (Ford, 1966; Cox, Ford and Pratt, 1970). The stored ribosomes, however, are sufficient to allow development of all major tissues in the absence of new ribosome synthesis (Brown and Gurdon, 1964). Large amounts of 4S RNA accumulate in the oocyte; however, the ratio of 4S RNA to the number of ribosomes is very low (Brown and Littna, 1966a). Finally the enormous replication of mitochondrial DNA (Dawid, 1966), which takes place largely at a specific time in oogenesis, is probably correlated with an increased synthesis of mitochondria specific RNA, since it seems certain that mitochondrial DNA is transcribed to give mitochondrial RNA (Zylber and Penman, 1969; Zylber, Vesco and Penman, 1969; Vesco and Penman, 1969).

A. Stages of Oogenesis

In *Xenopus* the oocyte passes through a series of stages characterized by morphological and biochemical means, and correlated with size (Table 9.1). The mature oocyte takes roughly 18 months to produce, and measures 1.3 - 1.4 mm in diameter. It has been found convenient for certain biochemical studies to use immature ovaries in which the oldest oocytes are all roughly synchronized at one stage of growth. However, since the rate of development depends on many external factors, such as season of the year, temperature, crowding and food supply, it is extremely variable and therefore time is not a good basis for measuring oocyte growth and differentiation.

B. Sources of DNA in *Xenopus* Ovary

Since RNA is synthesized in animal cells probably from DNA templates only, it is important to consider all sources of DNA in the oocyte. We may classify DNAs in two groups according to whether they are cytoplasmic or nuclear.

1. Nuclear DNA. The oocyte nucleus (germinal vesicle) should contain the 4C amount of DNA, which is 12 pg in *Xenopus* (Dawid, 1965). However, Izawa, Allfrey

Table 9.1 **Stages of Oocyte Growth in *Xenopus laevis***

Stage				Diameter microns	Morphological criteria	Biochemical criteria
Premeiotic				5 - 10	Rapid cell division, special histochemical reactions (Blackler, 1958)	-
Meiotic	Preamplification			10 - 15	Enlarged cell and nucleus, two prominent nucleoli	-
	Amplification			15 - 20	Feulgen positive cap appears, nucleoli are absent (Gall, 1968)	rDNA synthesis (Gall, 1968)
	Postamplification	1		20 - 100	Diplotene, cap disintegrates, nucleoli appear, nuclear membrane regular	4S RNA and 5S RNA in excess of 18 and 28S rRNA; DNA-like RNA synthesis (Davidson, Crippa, Kramer and Mirsky, 1966)
		2		100 - 300	Cap completely disappeared, many small nucleoli with feulgen positive granules associated	18 and 28S rRNA begins but still below 4S and 5S rRNA, dRNA synthesis continues
		3		300 - 500	Many nucleoli, nuclear membrane irregular, yolk nucleus present, mitochondria increase	18 and 28S rRNA synthesis at maximum rates, 4S, 5S and dRNA continues, presumably mitochondrial DNA & RNA synthesis
		4		500 - 1000	Yolk nucleus disappears, pigment appears, mitochondria move peripherally, yolk accumulation begins	18 and 28S rRNA synthesis maintained, protein synthesis declines
		5		1000 - 1300	Nuclear membrane becomes regular, nucleoli enlarge, yolk accumulates	18 and 28S rRNA synthesis declines, protein synthesis very low
		6		1300 - 1400	Ovulation, nuclear membrane disintegrates, 1st and 2nd meiotic divisions and polar body formation	Small amount of dRNA synthesis (Brown and Littna, 1964a) protein synthesis begins (Smith and Ecker, 1969), DNA synthesis begins (Gurdon, 1968)

and Mirsky (1963b) found about four times the 4C amount in *Triturus viridescens,* while in *Xenopus* Brown and Dawid (1968) found between five and six times the amount of low density 'chromosomal' DNA in preparations of isolated germinal vesicles. They explain this excess as probably due to contamination by mitochondria, since mitochondrial DNA is present at 200 times the 4C value and is virtually indistinguishable from chromosomal DNA by buoyant density (Dawid, 1966). Other amphibians examined by Brown and Dawid (1968) show a very much closer agreement of germinal vesicle 'chromosomal' DNA to the predicted 4C value. It may be concluded that the germinal vesicle probably contains the 4C amount of chromosomal DNA, but that really good agreement between measured and predicted values has not been obtained yet.

Brown and Dawid (1968), Gall (1968), and Perkowska, MacGregor and Birnstiel (1968) have shown that there is an excessive amount of extra-chromosomal DNA in *Xenopus* germinal vesicles. The extra DNA is associated with nucleoli (Evans and Birnstiel, 1968), is unmethylated in contrast to somatic cell rDNA (Dawid, Brown and Reeder, 1970), and is complementary to 18S and 28S ribosomal RNA (Brown and Dawid, 1968). It has been calculated (Perkowska, MacGregor and Birnstiel, 1968) that there are 4×10^6 18S and 28S rRNA cistrons per oocyte which represents a $2.5 - 5.0 \times 10^3$ increase over the normal diploid value.

In view of this apparently selective amplification of one type of gene it is relevant to ask whether amplification of other genes also takes place. In particular, Brown and Weber (1968) and Brown and Dawid (1968) have looked at the ratio of genes for 5S ribosomal RNA and 4S RNA to 18S and 28S genes in preparations of DNA from various somatic tissues as well as egg DNA. They found that the ratio is constant for all somatic tissues but that egg DNA shows a greatly reduced amount of DNA complementary to 4S and 5S RNA relative to that complementary to 18S and 28S RNA. They conclude that detectable amplification of the 4S RNA and 5S RNA genes does not occur in *Xenopus* oocytes.

2. Cytoplasmic DNA. A review of this subject appears in chapter 11 of this book. Cytoplasmic DNA occurs in mitochondria (Dawid, 1966) and in yolk (Hanocq-Quertier, Baltus, Ficq and Brachet, 1968). The function of yolk DNA is unknown, while the mitochondrial DNA is probably transcribed to give mitochondria specific RNA.

II. TYPES OF RNA SYNTHESIS

A. Non-Nuclear RNA

1. Cytoplasmic RNA. Recent reports by Craig (1970) and Selvig, Gross and Hunter (1970) suggest that RNA may be synthesized in the cytoplasm of developing sea urchin embryos. DNA-RNA hybridization suggests that the RNA made hybridizes in part (32%) to mitochondrial DNA, while none hybridizes to sperm DNA (Craig, 1970). In view of the enormous amount of mitochondrial DNA in developing oocytes it seems reasonable to suppose that similar synthesis of mitochondrial RNA takes place during oogenesis. Since mitochondrial replication seems to occur at a specific stage of oocyte growth it may be that mitochondrial RNA is also synthesized during a limited period of oogenesis.

2. Follicle cell RNA. Each oocyte is surrounded by a number of follicle cells active in both protein and RNA synthesis. It is important in studies on RNA synthesis during oogenesis to obtain preparations free of follicle cells; these preparations may be achieved by microdissection or differential homogenization. The nature of the RNA and protein made by follicle cells is unknown, though the RNA seems to be rather unstable (Ford, 1967) since a seventeen hour chase after a three hour label with [^{3}H] uridine removes all the label from follicle cells, although after a 20-hour labeling nearly 100% of follicle cells are labeled.

B. Nuclear RNA

Gall and Callan (1962) have shown that the loops of lampbrush chromosomes are sites of active RNA synthesis, as are many of the nucleoli in *Triturus cristatus.* The loops contain DNA (Callan and MacGregor, 1958) and the RNA synthesis on them may be inhibited by Actinomycin D (10 μg/ml) (Izawa, Allfrey and Mirsky, 1963a) which is accompanied by loop retraction. Edstrom and Gall (1963) drew attention to the fact that the RNA of nucleoli in *Triturus* oocytes was more closely related, on the basis of nucleotide composition, to the RNA of the cytoplasm than to the RNA of the nuclear sap or chromosomes.

1. DNA-like RNA. The term DNA-like RNA was introduced to describe RNA which resembled the bulk DNA in its base composition. Unfortunately it is probably too broad a term, since at least two types of DNA-like RNA are known. The first is nuclear heterogeneous RNA, described by Houssais and Attardi (1966) in Hela cells and by Attardi, Parnas, **Huang** and Attardi (1966) and **Scherrer**, Marcaud, Zagdela, London and Gros (1966) in immature duck erythrocytes. It is characterized by large sedimentation rate, rapid labeling, DNA-like base composition and completely nuclear localization. No simple precursor product relationship between nuclear heterogeneous RNA and so-called messenger RNA isolated from Hela cell polyribosomes could be demonstrated, and indeed there were consistent differences in the base compositions determined for these two types of RNA (Houssais and Attardi, 1966). Contrary to this observation Brown and Gurdon (1966), studying the size distribution and stability of DNA-like RNA synthesized during the development of anucleolate *Xenopus* embryos concluded that there may be a precursor product relationship between "heavy" and "light" dRNA, and that the stability of the RNA was inversely related to its sedimentation rate.

The second type of DNA-like RNA is messenger RNA (mRNA). Messenger RNA isolation has recently been claimed from three different animal cell systems: the rabbit reticulocyte which synthesizes haemoglobin (Laycock and Hunt, 1969; Labrie, 1969), the immune rabbit lymph-node which synthesizes γ globulin

(Kuechler and Rich, 1969), and the sea urchin embryo which synthesizes a great deal of histone protein (Kedes and Gross, 1969). These RNA molecules have been found associated with polyribosomes, and they are not rapidly labeled nor are they unstable. In each case mRNA has been isolated from a cell specialized for the production of one or a few particular types of protein; however, the growing oocyte is not specialized in such a way. At present no oocyte mRNA has been isolated or characterized.

Approximately 1% of mature oocyte RNA, which has been transcribed from 3.0% of the genome, belongs to this class and is retained throughout oogenesis but disappears gradually during early development (Davidson, Crippa, Kramer and Mirsky, 1966; Crippa, Davidson and Mirsky, 1967; Davidson and Hough, 1969).

2. 4S RNA. The mature oocyte contains 50 ng 4S RNA (Brown and Littna, 1966a). Although this is a very large amount for a single cell it represents only one or two 4S RNA molecules per ribosome. In later development there are 15-20 4S RNA molecules per ribosome.

3. 18S and 28S rRNA. It is well established that a single precursor RNA molecule gives rise to both the 18S and 28S ribosomal RNA (rRNA) in animal cells (for review see Darnell, 1968). The nature of the processing of the precursor is not fully understood; however, a chart of the relationship between the various intermediate molecules has been established for Hela cells (Weinberg and Penman, 1970) and includes an estimate that one half of the precursor is discarded during processing. The size of the precursor and the amount discarded seems to vary among species (Greenberg, 1969). For instance, Loening, Jones and Birnstiel (1969) have shown that the precursor in mammals has a molecular weight of 4.4×10^6 daltons while in *Xenopus* it is $2.5 - 2.6 \times 10^6$ daltons. The amount of precursor discarded in mammals is about 2.0×10^6 daltons, while in *Xenopus* it is $0.3 - 0.4 \times 10^6$ daltons.

Ribosomal precursor RNAs are located in nucleoli, where they are in association with protein (Liau and Perry, 1969; Miller and Beatty, 1969; Narayan and Birnstiel, 1969). Similar particles have been isolated from germinal vesicle preparations of *Triturus* and *Amblystoma* oocytes (Rogers, 1968). It would appear that 18S and 28S rRNA synthesis proceeds in oocytes in the same way it does in tissue culture cells (Gall, 1966; Rogers, 1968).

Ninety-five per cent of the RNA in a mature oocyte is 18S and 28S rRNA. Figure 9.1 shows the amount of RNA in oocytes of different sizes. It is apparent that about 80% of the RNA is made in oocytes greater than 0.3 mm in diameter and that there is a linear increase of RNA content with oocyte diameter up to 1.0 mm (Davidson, Crippa, Kramer and Mirsky, 1966; Brown and Littna, 1964a, b). It takes approximately 9 - 12 months for a developing ovary to produce oocytes of 0.3 mm diameter, and only 6 - 9 months after this there will be mature oocytes present. The rate of RNA synthesis, during oocyte growth from 0.3 - 1.0 mm diameter, averages five times greater than during its growth to 0.3 mm diameter. Brown and Littna

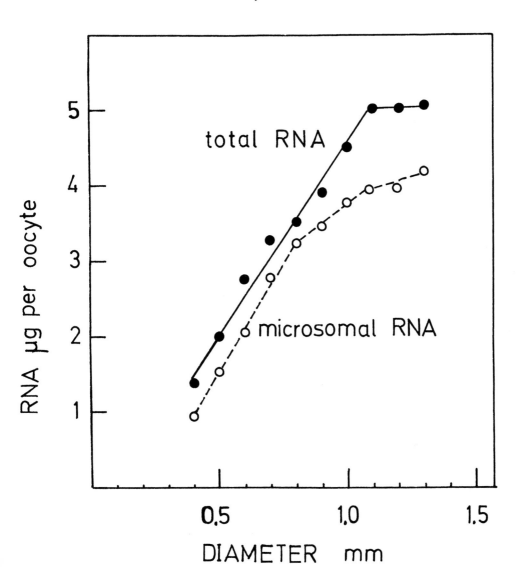

Fig. 9.1   RNA was estimated (Fleck and Munro, 1962) on the whole homogenate and microsome fractions of different size classes of oocytes isolated by hand. The microsome fraction was prepared by centrifugation of the homogenate at 15,000 rpm for 15 minutes followed by removal of the supernatant which was centrifuged at 40,000 rpm for 2 hrs at 4° in an MSE 10 x 10 rotor giving a microsome pellet.

(1964a, b) have shown that there is very little or no rRNA synthesis in mature oocytes.

It appears that the increase in RNA synthesis during the vitellogenic stages of oogenesis is due entirely to the activation of the amplified ribosomal cistrons, by increased recruitment of genes, by increased loading of the genes with polymerase, or by decreased transcription times. It is not possible at present to decide between these possibilities. However, it is apparent that amplification of the ribosomal RNA cistrons is insufficient in itself to account for the increased rates of RNA synthesis, since it has taken place many months previously (Gall, 1968). Moreover, cytological and electron microscope studies (Jacob and Birnstiel, personal communication) show that the amplified DNA is only gradually unfolded from the tightly coiled, inactive rDNA of the post amplification, cap cell. The process is reminiscent of the reactivation of the chicken erythrocyte nucleus after its introduction by cell fusion into Hela cell cytoplasm (Harris, 1967), or of *Xenopus* brain and gastrula nuclei after microinjection into growing oocytes (Gurdon, 1968). In both these cases dispersion of chromatin precedes or is correlated with activation of RNA synthesis.

Brown and Littna (1966b) show that the *Xenopus* oocyte synthesizes some DNA-like RNA during hormone induced ovulation, but it cannot be decided whether this is a feature of ovulation or an effect of hormone on the oocyte nucleus. Ford (1967) has shown that human chorionic gonadotrophin will increase the incorporation in tissue culture of [5-^{3}H] uridine into *Xenopus* ovary RNA by about 50%, and that the incorporation is increased in all classes of RNA. The major defect of this experiment is that the RNA included follicle cell RNA; thus it is impossible to be certain that the hormone effects were on the oocytes themselves. However, possible effects of hormones on the rate of RNA synthesis during oogenesis should not be neglected.

Finally, it seems likely that the processing of rRNA and the assembly time for the ribosome subunits may be quite long in amphibian oogenesis. Gall (1968) has demonstrated that germinal vesicles contain concentrations of rRNA precursor molecules which remain labeled for many hours after administration of isotope. Similarly Rogers (1968) shows that there is very little labeled 18S or 28S rRNA in nuclei of immature *Triturus* oocytes even after two days labeling, although there is a great deal of labeled precursor present. Ford (1967) has estimated that in *Xenopus* ovary incubated in tissue culture the mean synthesis time of the small ribosome subunit is 2 hours and of the large subunit 4 hours. These times are six to eight times longer than the mean synthesis times for ribosome subunits in Hela cells (Joklik and Becker, 1965a, b; Latham and Darnell, 1965a, b).

4. 5S RNA. It is well known that 5S RNA occurs attached to the large subunit in all ribosomes (Rosset, Monier and Julien, 1964; Brown and Littna, 1966a; Galibert, Larsen, Lelong and Boiron, 1965; Knight and Darnell, 1967). 5S RNA can be removed only at the cost of considerable alteration in conformation of the large subunit and it is not easily reassociated (Sarkar and Comb, 1969). Although it has not been possible to determine a function for 5S RNA in the ribosome, it must be important in both the assembly and functioning of ribosomes. For instance, the nucleotide sequence of 5S RNA shows evolutionary conservatism (Williamson and Brownlee, 1969; Forget and Weissman, 1969) implying strong selection pressures acting to preserve its structure, presumably for some functional role. It has been determined (Knight and Darnell, 1967) that a pool of 5S RNA exists in the Hela cell nucleus and that a 5S RNA molecule is attached to the precursor ribosome particles n the nucleolus very early in ribosome assembly (Warner and Soeiro, 1967). It is conceivable therefore that 5S RNA may be important in the regulation of ribosome assembly, as well as in ribosome directed protein synthesis.

It has been suggested that in early development of *Xenopus* embryos all rRNAs are synthesized coordinately (Brown and Weber, 1968; Brown and Dawid, 1968) because they appear to accumulate in equimolar amounts. 5S RNA genes are not linked closely to the 18S and 28S rRNA genes (Brown and Weber, 1968) and they are present in the ONu mutant which has lost at least 99% of its ribosomal cistrons. It is clear, therefore, that in order to achieve coordinated synthesis of 5S rRNA with 18S and 28S rRNA a special process must be invoked. This control may formally act in two ways, either at the level of the genes by equalizing the rates of transcription, or at the level of the RNA by destroying excess molecules. It is probably that the latter process may not be important in the *Xenopus* oocyte since all rRNAs seem to be equally stable (Brown and Littna, 1966a; Davidson, Allfrey and Mirsky, 1964). The simplest mechanism for equalizing rates of transcription is for 5S RNA to act as a cofactor in the release of the 18S and 28S precursor molecule. Details of the coordinate control are not known; however, it can be uncoupled in two ways. The first is artificial and involves the use of Actinomycin D at levels which inhibit 18S and 28S rRNA synthesis but not 5S RNA synthesis (Perry and Kelley, 1968; Watson and Ralph, 1966). 5S RNA made under these conditions is more labile than normal and is not readily utilized if ribosome synthesis is allowed to resume, indicating that it is probably abnormal. The second is achieved by the developing oocyte of *Xenopus laevis* and therefore occurs *in vivo*. The experiments to be described will show that 5S RNA is accumulated in early oogenesis to a great molar excess over 18S and 28S rRNA and further that 5S RNA is stored in the cytoplasm in association with protein.

Table 9.2 **Incorporation of 5H^3 uridine into the RNA of Three Fractions of the Immature Ovaries of *Xenopus laevis***

Fraction	Experiment 1		Experiment 2	
	Total cpm x 10^{-4}	% total cpm	Total cpm x 10^{-4}	% total cpm
'top'	9.6	40	6.31	67
42S	3.1	11	1.45	15
80S	11.5	48	1.74	18

Ovaries from immature frogs (60 weeks post-metamorphosis in experiment 1 and 20 weeks post-metamorphosis in experiment 2) were incubated with 5H^3 uridine (100 μCi/ml, 31 Ci/mM) for four days. The 480,000 g supernatant was separated on sucrose gradients (Fig. 9.4) into three fractions. RNA was extracted and the total radioactivity estimated after precipitation with 5% TCA, filtering and washing.

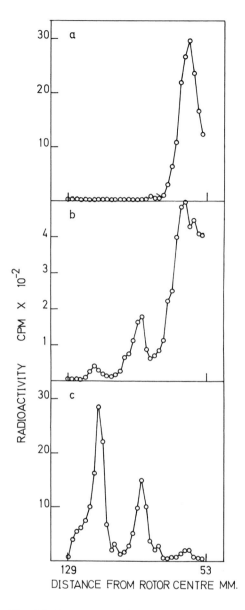

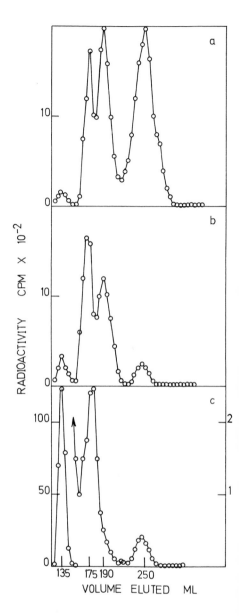

Fig. 9.2 Immature ovaries were incubated with [5-^{3}H] uridine (100 μCi/ml; 24 Ci/mM) at 20° for four days. RNA was prepared from the ribonucleoprotein fractions isolated on sucrose gradients. The RNA was centrifuged in 30 ml gradients at 63,600 g for 17 hours in a Spinco SW 25.1 rotor at 5°. Fractions were collected and precipitated onto filters with TCA adding 0.5 mg bovine serum albumin as carrier. (a) 'Top' fraction.   (b) 42S fraction.   (c) 80S fraction. Ninety-five to 103% of the radioactivity was recovered from the gradients.

Fig. 9.3 The RNA preparations used in the experiment shown in Fig. 9.2 were run through Sephadex G100 columns at room temperature. The fractions were precipitated with TCA and collected on filters. The recovery of radioactivity was 70 – 90%. (a) 'Top' fraction. (b) 42S fraction. (c) 80S fraction.

III. NATURE OF RIBONUCLEIC ACIDS IN
THE IMMATURE OVARY

A series of experiments was undertaken to elucidate the nature of the RNAs present in the immature ovary of *Xenopus.* Immature ovaries (20 and 60 weeks post metamorphosis) were labeled in culture for four days with [5-³H] uridine, and the RNA from different regions of a sucrose gradient separation of the 480,000 g supernatant, extracted in the presence of carrier 80S ribosomes (see Fig. 9.4). The total incorporation into different fractions (Table 9.2) shows that at 20 weeks post metamorphosis the greatest incorporation is into the RNA of the 'top' fraction, while at 60 weeks post metamorphosis there is approximately equal incorporation into the RNA of the 'top' and 80S fractions. The RNA preparations were fractionated by sucrose gradient (Fig. 9.2) and by chromatography on Sephadex G100 (Fig. 9.3). The results show that the radioactivity in the 80S ribosome fraction is in 18S and 28S rRNA almost entirely (Fig. 9.2c), while the incorporation into RNA of the 'top' fraction is into three species of low molecular weight RNA (Fig. 9.3a). One of these RNAs (Kd 0.13) has been identified with 5S RNA by its coelution from Sephadex G100 columns with 5S RNA isolated from 80S ribosomes, by base composition, and by 'fingerprinting' (Williamson and Brownlee, 1969; Williamson, personal communication). A second RNA is 4S RNA (Kd 0.41), which agrees with previous results obtained by chromatography of *Xenopus* 4S RNA on Sephadex G100 (Brown and Littna, 1966a). The third intermediate RNA (Kd 0.2) has not been identified. The general conclusion from these results is that the immature ovary synthesizes a disproportionately large amount of low molecular weight RNA of which a substantial proportion is 5S RNA.

A. Non-coordinate Synthesis of 5S RNA

Immature ovaries at the yolk accumulation stage were labeled with [¹⁴C] uridine in tissue culture for six days. The 80S ribosomes were prepared and total RNA extracted from them. The [¹⁴C] RNA was checked by chromatography on Sephadex G100 to see that it contained amounts of label in each RNA species according to that predicted from their molecular weights and mole per cent UMP content. Assuming that equimolar quantities of 5S, 18S and 28S RNAs are present, it is predicted that 2.0% of the total radioactivity is in 5S RNA. The observed value was 2.2% γ 0.3 (5) of the total radioactivity. The [¹⁴C] preparation was now used as a reference for the test of coordinate synthesis of 5S RNA with 18S and 28S RNA in samples of [³H] uridine-labeled RNA from immature ovaries labeled in culture for one day. The ³H/¹⁴C ratios for high molecular weight RNA (28S or 18S + 28S) were given the value of unity in each case. The normalized ratios are a measure of the molar abundance of [³H] labeled 4S or 5S RNA relative to the 28S RNA or to the void volume (19S + 28S rRNA) RNA. If coordinate synthesis is occurring the normalized values will all be equal to one. Any value in excess of one indicates non-coordinate synthesis of 4S or 5S in molar excess. Any value less than one indicates non-coordinate synthesis of 4S or 5S in molar deficit. The results of this experiment (Table 9.3) show that 18S rRNA is synthesized coordinately with

Table 9.3 **Normalized Ratios of the Test for Coordinate Synthesis of Ribosomal RNAs by Immature** *Xenopus* **Ovaries**

Sample	Ovary age in weeks post-metamorphosis	Normalized ratios					
		Sucrose gradients				G100 columns	
		28S	18S	4+5S	Void	5S	4S
1	2	1.00	0.90	12.05	1.00	16.4	38.2
2	8	1.00	1.12	15.60	1.00	21.95	51.8
3	25	1.00	0.96	5.41	1.00	4.55	22.3
4	25	1.00	0.91	4.60	1.00	5.04	15.3
5	32	1.00	0.84	2.19	1.00	4.44	11.4
6	32	1.00	0.96	2.58	1.00	3.19	13.8

For explanation see text.

28S rRNA at all stages of ovary development. However, 5S RNA is synthesized in 1 - 20 fold molar excess in ovaries of animals 2 - 8 weeks post metamorphosis and this drops to 3 - 5 fold excess by 25 - 32 weeks after metamorphosis. If all this 5S RNA is used in ribosome assembly in later oogenesis, it may follow that there is a period of oogenesis when 5S RNA is synthesized in molar deficit compared to 18S and 28S rRNA.

B. Cytoplasmic Location of the 5S RNA

Histochemically the RNA of the small oocyte can be shown to be present almost entirely in the cytoplasm. Greater than 80% of the RNA of the small oocyte (50 - 100 microns diameter) is in low molecular weight RNA. The existence of 5S RNA in the cytoplasm of a cell other than in association with the 80S ribosome or large ribosome subunit is unusual since previous work on the distribution of 5S RNA in Hela cells shows no such pool (Knight and Darnell, 1967). However, a pool of up to 20% of the total cellular 5S RNA has been found in Hela cell nuclei (Knight and Darnell, 1967), which is in accordance with the observation that 5S RNA attaches very early to the ribosome precursor particles (Warner and Soeiro, 1967). If the 5S RNA in the *Xenopus* oocyte eventually enters the 80S ribosome it may do so either by re-entering the nucleus and assembling with the ribosome precursor in the manner shown for Hela cells, or it may attach to a 5S RNA deficient ribosome or large subunit in the cytoplasm. The latter process would be a novel one, but in either case the situation might be exploited in experiments to determine the role of 5S RNA in the ribosome.

C. A Ribonucleoprotein Particle Containing 5S RNA

I have shown that in early oogenesis 5S RNA is synthesized in 15 - 20 molar excess compared with 18S and 28S rRNA, and probably accumulates in the cytoplasm.

One may inquire whether the 5S RNA is accumulated in association with protein in a cytoplasmic particle fraction. A ribonucleoprotein particle may be demonstrated in homogenates of immature ovaries of *Xenopus* by sucrose gradient centrifugation (Fig. 9.4). Two points are made in this figure; first, a significant proportion of the total optical density of immature oocytes sediments at about 42S, and second, the relative amounts of the 42S and 80S peaks changes from about 4:1 at two weeks after metamorphosis to 0.3:1 at 50 weeks post metamorphosis as the ovary matures. Indeed, in mature oocytes it is not possible to demonstrate any 42S material by optical methods. The RNA from this 42S RNP material has been shown Fig. 9.3b) to contain over 50% 5S rRNA.

In order to examine the possibility that the synthesis of the RNA and protein components in the 42S and 80S fractions may be correlated, ovaries were taken from frogs at different ages after metamorphosis and incubated with [2-^{14}C] uridine and [4,5-^{3}H] lysine for four days. The percentage of total optical density and radioactivity incorporated into each fraction was measured. The results (Fig. 9.5) clearly show that the synthesis of the RNA and protein components of the 42S fraction are not correlated with the synthesis of 80S ribosome components. Since the 42S particle contains over 50% 5S RNA it is concluded that the synthesis of 5S RNA is not correlated with the synthesis of 18S and 28S rRNA. As the ovary grows an increasing proportion of the synthetic effort of the ovary is devoted to 80S ribosome synthesis.

D. Physical and Chemical Properties of the 42S Material

Chemical analysis has shown that the 42S and 80S particles contain very little of any DNA, and have protein to RNA ratios of 3.0 and 1.5 respectively. These values have been confirmed and extended by measurement of buoyant density in cesium chloride after formalin fixation. The results show that the 42S ribonucleoprotein material is resolved into two major and one minor components, while the 80S material has one major and one minor component. The idea that there may be two or more particles with different buoyant densities but similar sedimentation coefficients in the 80S material is supported by the fact that three distinct species of RNA may be isolated from it (Figs. 9.2b and 9.3b). The 18S RNA will be in native ribosome subunits which accounts for the minor component at 1.532 gcm^{-3}, while it may be that the two low molecular weight RNAs are also in particles of different buoyant density, accounting for the major components at .680 and 1.456 gcm^{-3}.

The results may be interpreted alternatively as indicating the dissociation of 42S ribonucleoprotein material, in the presence of high concentrations of salt, into

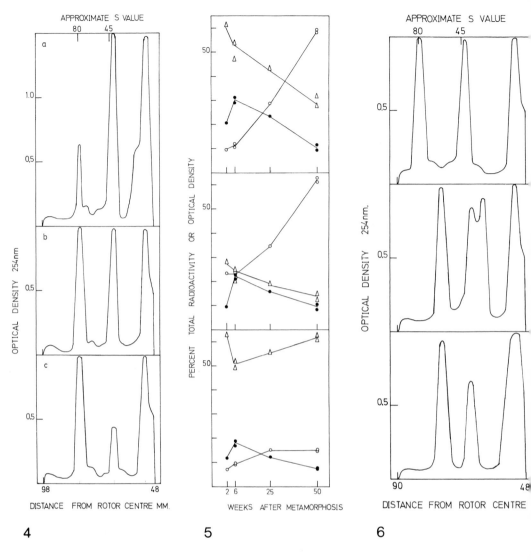

4   5   6

Fig. 9.4 The 480,000 g supernatants of *Xenopus* ovaries were centrifuged into 5 ml sucrose gradients a 100,000 g for 4 hours at 10° in a Spinco SW 39 rotor. The optical density was monitored continuously at 254 nm using an Isco density gradient fractionator. Ovaries from frogs (a) 8 weeks post metamorphosis, (b) 32 week post metamorphosis, and (c) 60 weeks post metamorphosis.

Fig. 9.5 Ovaries of frogs at different ages were incubated with [2-^{14}C] uridine (1μCi/ml, 60.7 mCi/mM) an 4.5 L-[^{3}H] lysine (20μCi/ml, 197 mCi/mM) for 4 days at 20°. The 480,000 supernatants were prepared and centrifuged into sucrose gradients. The gradients were fractionated and the optical density continuously monitored at 260 nm. The fractions were precipitated with TCA at 0°, collected on filters and prepared for liquid scintillation counting. The percentage of (a) total optical density, (b) total [^{14}C] uridine radioactivity and (c) total [^{3}H]lysine radioactivity in three fractions (Fig. 9.4) 'top' △−−−−−△, 42S ●−−−−−●, and 80S ○−−−−−○ are shown plotted against age of frog in weeks post metamorphosis.

Fig. 9.6 The 480,000 g supernatant of immature *Xenopus* ovary was centrifuged into 5 ml sucrose gradients a 110,000 g av for 4.5 hours at 10° in a Spinco SW 39 rotor. The optical density was monitored continuously a 254 nm using an Isco density gradient fractionator. The gradients were linear 10-45% sucrose in 35mM Tris HC pH 7.6, 50mM KCl, and 1.5mM MgCl$_2$ (a), or 35mM Tris HCl pH 7.6, 50mM KCl, and 1mM EDTA (b), o 35mM Tris HCl pH 7.6, 0.5M KCl, and 1.5mM MgCl$_2$ (c).

free RNA and protein. In favor of this idea are the facts that quite low concentrations of monovalent cations do cause complete dissociation of RNA and protein in the 42S material, and that formaldehyde cross linkages are known to be reversible with time. However, since these experiments were done in the presence of 4% formaldehyde, it is considered unlikely that reversal of fixation can have had an effect on the result. Complete dissociation to RNA and protein can be ruled out since they would not have the required buoyant densities.

The 42S ribonucleoprotein material does not cosediment with the small ribosome subunits derived from 80S ribosomes by EDTA or 0.5M KCl treatment (Fig. 9.6). Indeed, the sedimentation coefficient of the 42S material in the presence of 0.5M KCl is 4.6S as measured in the analytical ultracentrifuge. The results of a more detailed study of the effect of KCl concentration on the 42S material revealed that it is converted into 5S material via an intermediate sedimenting at 27S. The same effect may be produced by increased magnesium concentration, by EDTA, and by pronase treatment. In all cases an intermediate sedimenting at 27S is observed before complete reduction of sedimentation constant. Electron microscopy of negatively stained preparations of 42S material (Fig. 9.7) shows that it is heterogeneous with respect to size and that the effect of KCl is to increase the proportion of the smallest particles.

It is apparent that the interaction between protein and RNA in the 42S material is due to ionic bonds almost entirely, since it can be shown by chromatography on Sephadex G100 in the presence of 0.5M KCl that there is complete dissociation between RNA and protein. Interaction may also be due in part to magnesium bridges, since EDTA causes some unfolding or dissociation of the 42S material.

2. Is the 42S Material a Homogenization Artifact?

It may be that the 42S material is a homogenization artifact, since it is well known that exogenous RNA will bind to protein in homogenates and sediment at increased rates (Baltimore and Huang, 1970). However, there are several arguments against this interpretation for the 42S particle. Experiments on the protein components clearly show (Fig. 9.8) that it is not a random sample of cell sap proteins, since there are two major components by acrylamide gel elctrophoresis in acid urea gels, neither of which can be identified with any of the many cell sap proteins. Isoelectric focusing in acrylamide gel slabs (Awdeh, Williamson and Askonas, 1968) reveals one major component with a pI of about 8.5. Chromatography on Sephadex G100 columns has shown that the labeled protein is homogeneous, with a molecular weight of less than 100,000 daltons since it is included by the gel, whereas the bulk of the labeled cell sap proteins are eluted in the void volume. Particles of the same size and structural characteristics as those observed in electron micrographs of negatively stained preparations of 42S material (Fig. 9.7) may be observed in electron micrographs of sections of immature oocytes

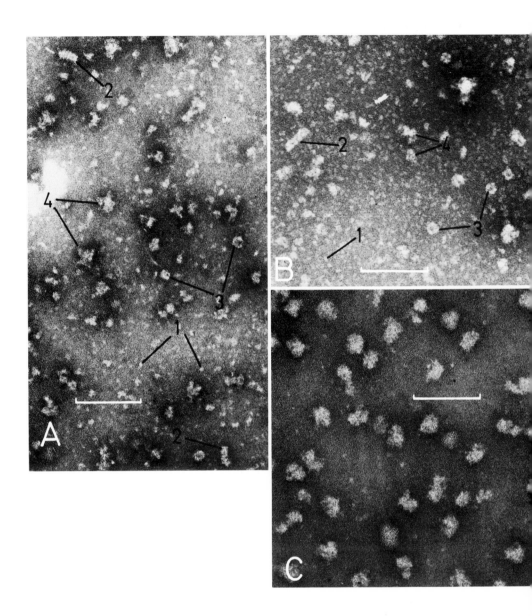

Fig. 9.7 Electron micrographs of negatively stained preparations of ribonucleoprotein particles from immatur
Xenopus ovaries. (a) 42S material, (b) 42S material treated with 0.3M KCl, and (c) 80S material. The scales ar
100 nm.

(Fig. 9.9) (Thomas, 1967). Taken together these results suggest that specifi
interaction between 5S rRNA and protein does occur in the 42S material *in vivo* an
that it is therefore probably not a homogenization artifact.

IV. POSSIBLE EXPLANATIONS FOR EXCESS SYNTHESIS
OF 5S RNA BY IMMATURE OOCYTES OF *XENOPUS*

In the normal diploid *Xenopus* cell there is a 25 - 30 fold excess of 5S RNA genes over 18S and 28S rRNA genes (Brown and Weber, 1968). In the oocyte, however, the number of 18S and 28S rRNA genes is amplified some 5,000 times (Birnstiel, Wallace, Sirlin and Fischberg, 1966), but the number of 5S RNA genes remains constant (Brown and Dawid, 1968). There thus turns out to be a 40 fold excess of 18S and 28S rRNA genes over 5S RNA genes in the oocyte. There is evidence (Miller and Beatty, 1969) that the 18S and 28S genes are maximally packed with polymerase; if one assumes the same is true for the 5S RNA genes and that the duration of the nucleotide step is the same for both sets of genes, then the oocyte is capable of producing 40 fold more 18S and 28S rRNA molecules than 5S RNA molecules. Such a situation would make 5S RNA availability rate limiting for ribosome assembly. I would propose therefore that after rDNA amplification the oocyte produces a store of 5S RNA molecules to enable the amplified 18S and 28S rRNA genes to be transcribed at maximum rates when the time comes.

Control of Ribosomal RNA Synthesis in *Xenopus laevis*

There are five facts that have to be explained by any system for the control of ribosomal RNA synthesis in *Xenopus:*
1. 5S RNA and 18S and 28S rRNA are coordinately synthesized during embryonic development (Brown and Littna, 1966a).
2. 5S RNA is synthesized non-coordinately with 18S and 28S rRNA during early oogenesis.
3. 5S RNA synthesis is not detectable in ONu mutants, although 5S RNA genes are present in normal amount (Brown and Weber, 1968).
4. rRNA synthesis is variable within the growth and differentiation of the oocyte, and also during the early stages of embryogenesis. Different adult tissues synthesize different amounts of rRNA per cell per unit time (Brown, 1967).
5. rRNA synthesis may be inhibited in nuclei already synthesizing rRNA by injection into eggs (Gurdon and Brown, 1965) or activated in nuclei not previously synthesizing RNA by injection into immature oocytes (Gurdon, 1968).

The scheme shown in Figure 9.10 has been designed to meet these requirements. 5S RNA participation in the release of 40S precursor RNA is the simplest way of accounting for coordinate synthesis of 5S with 18S and 28S rRNA. It could just as easily participate in the initiation of the 40S synthesis were it not for the evidence which suggests that the 18S part of the 40S molecule is made first, which would presumably mean that the 5S binding site does not appear until near the end of the transcription unit. Control of 5S RNA synthesis by feedback of unused molecules, a possible interpretation of the work with the ONu mutant, is ruled. out by the situation in the oocyte where large amounts of 5S RNA

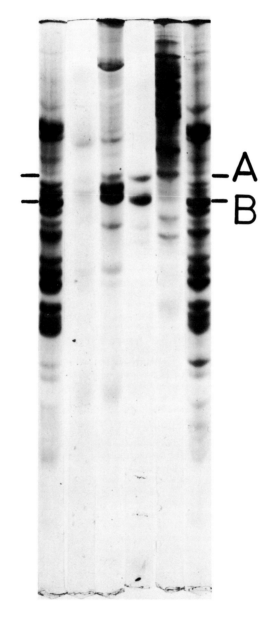

Fig. 9.8 Protein from various fractions of immature *Xenopus* oocytes was electrophoresced in 10% acrylamide gels. Gel buffer - 50mM sodium acetate/acetic acid pH 4.5; electrode buffer - 0.066 M 2-alanine-acetic acid pH 4.5. Both buffers were 6M in urea. Gels were run at 2.5 mA per gel for 16 hours at 2°. Gels 1 and 6 are proteins from 80S ribosomes prepared from mature ovary, 0.2mg; Gel 2. is protein from 80S fraction of immature ovary, 0.05 mg; Gel 3 is protein from the fraction intermediate between 42S and 80S material of immature ovary, 0.2 mg; Gel 3. is protein from the 42S fraction of immature ovary, 0.1 mg; Gel 4. is protein from the 'top' fraction of immature ovary, 0.2 mg.

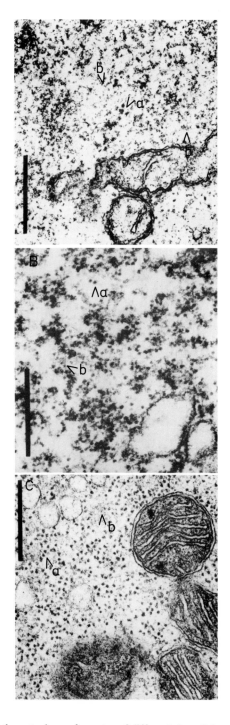

Fig. 9.9 Electron micrographs of the cytoplasm of oocytes of different sizes. It is possible to see the appearance of ribosomes (arrows A) and disappearance of the diffuse material (arrows B) as the oocyte grows. (a) oocyte diameter 100 μ; (b) oocyte diameter 300 μ; (c) oocyte diameter 500 μ. The scales are 500 nm.

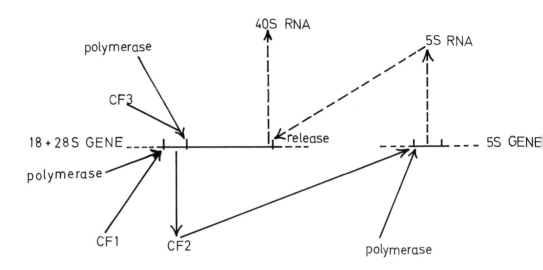

CONTROL OF RIBOSOMAL RNA SYNTHESIS

Fig. 9.10

accumulate. The alternative explanation, adopted here, is that a controlling factor (CF2) is produced by some part of the nucleolar DNA which is essential for 5S gene activity; this factor may be a polymerase or it may interact with polymerase in some way. Production of CF2 by the nucleolar DNA is probably itself controlled by a factor (CF1) polymerase interaction, which may not be the same as the factor (CF3) controlling transcription of the 18S and 28S rRNA gene since it would then be difficult to account for the situation in the oocyte where 5S RNA accumulates in the presence of much lower 18S and 28S synthesis. Variation in rate of RNA synthesis in different tissues and at different stages of embryogenesis is difficult to rationalize at present and is presumably determined by the stoichiometry of the various controlling reactions. The presence of a cytoplasmic inhibitor of rRNA synthesis in *Xenopus* eggs has been suggested (Gurdon and Brown, 1965). It seems that this inhibitor is released from the oocyte nucleus during maturation when the germinal vesicle breaks down (Gurdon, 1968). An inhibitory protein has been isolated from mature eggs of *Xenopus* and partially characterized (Crippa, 1970 and this volume, chapter 10). In view of these observations it could be argued that initiation of rRNA synthesis is accomplished by the removal or neutralization of this inhibitor by the hypothetical CF3 of Figure 9.10. Relevant to this question would be information concerning the specificity of the inhibitor for 18S and 28S rRNA synthesis compared to 5S RNA.

Predictions from the model are that:
    1. 5S RNA binds to nascent ribosome precursor particles and facilitates

their release.

2. The 5S genes are transcribed by a polymerase different from the one transcribing the 18S and 28S genes.

3. There is a region of the nucleolar DNA responsible for the production of a factor essential for 5S RNA synthesis, and that this region is controlled independently of the 18S and 28S rRNA genes.

4. The ONu mutant should contain the nucleolar specific RNA polymerase. If it does not it would suggest that the nucleolar specific polymerase was itself a product of the nucleolar DNA.

V. ACKNOWLEDGMENTS

The author is grateful for support by the Medical Research Council. The work described in this paper was done while at the National Institute for Medical Research, Mill Hill, London, N.W.7. Drs. Alan Williamson, Robert Dourmashkin, John Armstrong, Ken Kanagalingam, and Robert Williamson have each contributed to this work. Dr. Max Birnstiel and Dr. R. A. Cox provided helpful criticism and encouragement.

VI. REFERENCES

Attardi, G., Parnas, H., Huang, M. I. H. and Attardi, B. (1966) Giant size rapidly labelled nuclear RNA and cytoplasmic mRNA in immature duck erythrocytes. *J. Molec. Biol.* **20**, 145.

Awdeh, Z., Williamson, A. R. and Askonas, B. A. (1968) Isoelectric focusing in polyacrylamide gel and its application to immunoglobulins. *Nature (London).* **219**, 66.

Baltimore, D. and Huang, A. S. (1970) Interaction of Hela cell proteins with RNA. *J. Molec. Biol.* **47**, 263.

Birnstiel, M. L., Wallace, H., Sirlin, J. L. and Fischberg, M. (1966) Localization of the ribosomal DNA complements in the nucleolar organizer region of *Xenopus laevis. Nat. Cancer Inst. Monogr.* **23**, 431.

Brown, D. D. (1967) The genes for ribosomal RNA and their transcription during amphibian development. *Current Topics in Develop. Biol.* **2**, 48.

Brown, D. D. and Dawid, I. (1968) Specific gene amplification in oocytes. *Science.* **160**, 272.

Brown, D. D. and Gurdon, J. B. (1964) Absence of ribosomal RNA synthesis in anucleolate mutant of *Xenopus laevis. Proc. Nat. Acad. Sci., U. S. A.* **51**, 139.

Brown, D. D. and Gurdon, J. B. (1966) Size distribution and stability of DNA-like RNA synthesized during development of anucleolate embryos of *Xenopus laevis. J. Molec. Biol.* **19**, 399.

Brown, D. D. and Littna, E. (1964a) RNA synthesis during development of *Xenopus laevis,* the South African clawed toad. *J. Molec. Biol.* **8**, 669.

Brown, D. D. and Littna, E. (1964b) Variations in the synthesis of stable RNAs during oogenesis and development of *Xenopus laevis. J. Molec. Biol.* **8**, 688.

Brown, D. D. and Littna, E. (1966a) Synthesis and accumulation of low molecular weight RNA during embryogenesis of *Xenopus laevis. J. Molec. Biol.* **20**, 95.

Brown, D. D. and Littna, E. (1966b) Synthesis and accumulation of DNA-like RNA during embryogenesis of *Xenopus laevis. J. Molec. Biol.* **20**, 81.

Brown, D. D. and Weber, C. S. (1968) Gene linkage by RNA-DNA hybridization. I: On unique DNA sequences homologous to 4S RNA and ribosomal RNA. *J. Molec. Biol.* **34**, 661.

Callan, H. G. and Macgregor, H. C. (1958) Action of deoxyribonuclease on lampbrush chromosomes. *Nature (London).* **181**, 1479.

Cox, R. A., Ford, P. J. and Pratt, H. (1970) Ribosomes from *Xenopus laevis* ovaries and the polyuridylic acid-directed biosynthesis of polyphenylalanine. *Biochem. J.* **119**, 161.

Craig, S. P. (1970) Synthesis of RNA in non-nucleate fragments of sea urchin eggs. *J. Molec. Biol.* **47**, 615.

Crippa, M. (1970) Regulatory factor for the transcription of the ribosomal genes in amphibian oocytes. *Nature (London).* **227**, 1138.

Crippa, M., Davidson, E. H. and Mirsky, A. E. (1967) Persistence in early amphibian embryos of informational RNAs from the lampbrush chromosome stage of oogenesis. *Proc. Nat. Acad. Sci., U.S. A.* **57**, 885.

Darnell, J. E. (1968) Ribonucleic acids from animal cells. *Bact. Rev.* **32**, 262.

Davidson, E. H., Allfrey, V. G. and Mirsky, A. E. (1964) On the RNAs synthesized during the lampbrush phase of amphibian oogenesis. *Proc. Nat. Acad. Sci., U.S.A.* **52**, 501.

Davidson, E. H., Crippa, M., Kramer, F. R. and Mirsky, A. E. (1966) Genomic function during the lampbrush chromosome stage of amphibian oogenesis. *Proc. Nat. Acad. Sci., U.S.A.* **56**, 856.

Davidson, E. H. and Hough, B. R. (1969) High sequence diversity in the RNA synthesized at the lampbrush stage of oogenesis. *Proc. Nat. Acad. Sci., U.S.A.* **63**, 342.

Dawid, I. B. (1965) Deoxyribonucleic acid in amphibian eggs. *J. Molec. Biol.* **12**, 581.

Dawid, I. B. (1966) Evidence for the mitochondrial origin of frog egg cytoplasmic DNA. *Proc. Nat. Acad. Sci. U.S.A.* **56**, 269.

Dawid, I. B., Brown, D. D. and Reeder, R. H. (1970) Composition and structure of chromosomal and amplified ribosomal DNAs of *Xenopus laevis. J. Molec. Biol.* **51**, 341.

Ecker, R. E. and Smith, L. D. (1968) Kinetics of protein synthesis in enucleate frog eggs. *Science.* **160**, 1115.

Edstrom, J. E. and Gall, J. G. (1963) The base composition of RNA in lampbrush chromosomes, nucleoli, nuclear sap and cytoplasm of *Triturus* oocytes. *J. Cell. Biol.* **19**, 279.

Evans, D. and Birnstiel, M. L. (1968) Localization of amplified ribosomal DNA in the oocyte of *Xenopus laevis. Biochim. Biophys. Acta.* **166**, 274.

Fleck, A. and Munro, H. N. (1962) The precision of ultraviolet absorption measurements in the Schmidt-Thannhauser procedure for nucleic acid estimation. *Biochim. Biophys. Acta.* **55**, 571.

Ford, P. J. (1966) A comparative study *in vivo* and *in vitro* of the ability of ribosomes from *Xenopus* liver and ovary to incorporate L-[U[^{14}C] leucine. *Biochem. J.* **101**, 369.

Ford, P. J. (1967) Macromolecular synthesis during oogenesis in *Xenopus laevis.* Ph.D. Thesis, University of Oxford.

Forget, B. G. and Weissman, S. M. (1969) The nucleotide sequence of ribosomal 5S ribonucleic acid from KB cells. *J. Biol. Chem.* **244**, 3148.

Galibert, F., Larsen, C. L., Lelong, J. C. and Boiron, M. (1965) RNA of low molecular weight in ribosomes of mammalian cells. *Nature (London).* **207**, 1039.

Gall, J. G. (1966) Nuclear RNA of the salamander oocyte. *Nat. Cancer Inst. Monogr.* **23**, 475.

Gall, J. G. (1968) Differential synthesis of the genes for ribosomal RNA during amphibian oogenesis. *Proc. Nat. Acad. Sci., U.S.A.* **60**, 553.

Gall, J. G. and Callan, H. G. (1962) [^{3}H] uridine incorporation in lampbrush chromosomes. *Proc. Nat. Acad. Sci., U.S.A.* **48**, 562.

Greenberg, J. R. (1969) Synthesis and properties of ribosomal RNA in Drosophila. *J. Molec. Biol.* **46**, 85.

Gurdon, J. B. (1968) Changes in somatic cell nuclei inserted into growing and maturing amphibian oocytes. *J. Embryol. Exp. Morph.* **20**, 401.

Gurdon, J. B. and Brown, D. D. (1965) Cytoplasmic regulation of RNA synthesis and nucleolus formation in developing embryos of *Xenopus laevis. J. Molec. Biol.* **12**, 27.

Hanocq-Quertier, J., Baltus, E., Ficq, A. and Brachet, J. (1968) Studies on the DNA of *Xenopus laevis* oocytes. *J. Embryol. Exp. Morph.* **19**, 273.

Harris, H. (1967) The reactivation of the red cell nucleus. *J. Cell Sci.* **2**, 23.

Houssais, J. F. and Attardi, G. (1966) High molecular weight non-ribosomal type nuclear RNA and cytoplasmic messenger RNA in Hela cells. *Proc. Nat. Acad. Sci., U.S.A.* **56**, 616.

Izawa, M., Allfrey, V. G. and Mirsky, A. E. (1963a) The relationship between RNA synthesis and loop structure in lampbrush chromosomes. *Proc. Nat. Acad. Sci., U.S.A.* **49**, 544.

Izawa, M., Allfrey, V. G. and Mirsky, A. E. (1963b) Composition of the nucleus and chromosomes in the lampbrush stage of the newt oocyte. *Proc. Nat. Acad. Sci., U.S.A.* **50**, 811.

Joklik, W. K. and Becker, Y. (1965a) Studies on the genesis of polyribosomes. I: Origin and significance of the subribosomal particles. *J. Molec. Biol.* **13**, 496.

Joklik, W. K. and Becker, Y. (1965b) Studies on the genesis of polyribosomes. II: The association of Nascent messenger RNA with the 40S subribosomal particle. *J. Molec. Biol.* **13**, 511.

Kedes, L. H. and Gross, P. R. (1969) Identification in cleaving embryos of three RNA species serving as templates for the synthesis of nuclear proteins. *Nature (London).* **223**, 1335.

Knight, E. and Darnell, J. E. (1967) Distribution of 5S RNA in Hela cells. *J. Molec. Biol.* **28**, 491.

Kuechler, E. and Rich, A. (1969) Sequential synthesis of messenger RNA and antibodies in rabbit lymph nodes. *Nature (London).* **222**, 544.

Labrie, F. (1969) Isolation of an RNA with the properties of haemoglobin messenger. *Nature (London).* **221**, 1217.

Latham, H. and Darnell, J. E. (1965a) Distribution of mRNA in the cytoplasmic polyribosomes of the Hela cell. *J. Molec. Biol.* **14**, 1.

Latham, H. and Darnell, J. E. (1965b) Entrance of mRNA into Hela cell cytoplasm in puromycin treated cells. *J. Molec. Biol.* **14**, 13.

Laycock, D. G. and Hunt, J. A. (1969) Synthesis of rabbit globin by a bacterial cell free system. *Nature (London).* **221**, 1118.

Liau, M. C. and Perry, R. P. (1969) Ribosome precursor particles in nucleoli. *J. Cell Biol.* **42**, 272.

Loening, U. E., Jones, K. and Birnstiel, M. L. (1969) Properties of the ribosomal RNA precursor in *Xenopus laevis;* comparison to the precursor in mammals and plants. *J. Molec. Biol.* **45**, 353.

Miller, O. L., Jr. and Beatty, B. R. (1969) Visualization of nucleolar genes. *Science.* **164**, 955.

Narayan, S. K. and Birnstiel, M. L. (1969) Biochemical and ultrastructural characteristics of ribonucleoprotein particles isolated from rat liver cell nucleoli. *Biochim. Biophys. Acta.* **190**, 470.

Perkowska, E., Macgregor, H. C. and Birnstiel, M. L. (1968) Gene amplification in the oocyte nucleus of mutant and wild-type *Xenopus laevis. Nature (London).* **217**, 649.

Perry, R. P. and Kelley, D. E. (1968) Persistent synthesis of 5S RNA when production of 28S and 18S rRNA is inhibited by low doses of actinomycin D. *J. Cell Physiol.* **72**, 235.

Rogers, M. E. (1968) Ribonucleoprotein particles in the amphibian oocyte nucleus. *J. Cell Biol.* **36**, 421.

Rosset, R., Monier, R. and Julien, J. (1964) Les ribosomes d'*Escherichia coli*. I. Mise en évidence d'un RNA ribosomique de faible poids moléculaire. *Bull. Soc. Chim. Biol.* **46**, 87.

Sarkar, N. and Comb, D. G. (1969) Studies on the attachment and release of 5S ribosomal RNA from the large ribosomal subunit. *J. Molec. Biol.* **39**, 31.

Scherrer, K., Marcaud, L., Zajdela, J., London, I. and Gros, F. (1966) Patterns of RNA metabolism in a differentiated cell: a rapidly labelled, unstable 60S RNA with messenger properties in duck erythroblasts. *Proc. Nat. Acad. Sci., U.S.A.* **56**, 1571.

Selvig, S. E., Gross, P. R. and Hunter, A. L. (1970) Cytoplasmic synthesis of RNA in the sea urchin embryo. *Develop. Biol.* **22**, 343.

Smith, L. D. and Ecker, R. E. (1969) Role of the oocyte nucleus in physiological maturation in *Rana pipiens*. *Develop. Biol.* **19**, 281.

Smith, L. D., Ecker, R. E. and Subtelny, S. (1966) The initiation of protein synthesis in eggs in *Rana pipiens*. *Proc. Nat. Acad. Sci., U.S.A.* **56**, 1724.

Stavy, L. and Gross, P. R. (1967) The protein synthetic lesion in unfertilized eggs. *Proc. Nat. Acad. Sci., U.S.A.* **57**, 735.

Stavy, L. and Gross, P. R. (1969a) Protein synthesis *in vitro* with fractions of sea urchin eggs and embryos. *Biochim. Biophys. Acta.* **182**, 193.

Stavy, L. and Gross, P. R. (1969b) Availability of mRNA for translation during normal and transcription-blocked development. *Biochim. Biophys. Acta.* **182**, 203.

Thomas, C. (1967) Evolution des structures ribosomales au cours de l'Oogenese chez *Xenopus laevis*. *Arch. Biol. (Liege)*. **78**, 347.

Vesco, C. and Penman, S. (1969) The cytoplasmic RNA of Hela cells: New discrete species associated with mitochondria. *Proc. Nat. Acad. Sci., U.S.A.* **62**, 218.

Warner, J. R. and Soeiro, H. C. (1967) Nascent ribosomes from Hela cells. *Proc. Nat. Acad. Sci., U.S.A.* **58**, 1984.

Watson, J. D. and Ralph, R. K. (1966) The nature of 7S RNA in mammalian cells. *J. Molec. Biol.* **22**, 67.

Weinberg, R. A. and Penman, S. (1970) Processing of 45S nucleolar RNA. *J. Molec. Biol.* **47**, 169.

Williamson, R. and Brownlee, G. G. (1969) The sequence of 5S ribosomal RNA from two mouse cell lines. *FEBS Letters.* **3**, 306.

Zylber, E. and Penman, S. (1969) Mitochondrial associated 4S RNA synthesis inhibited by ethidium bromide. *J. Molec. Biol.* **46**, 201.

Zylber, E., Vesco, C. and Penman, S. (1969) Selective inhibition of the synthesis of mitochondria-associated RNA by ethidium bromide. *J. Molec. Biol.* **44**, 195.

10

REGULATION OF RIBOSOMAL RNA SYNTHESIS DURING OOGENESIS OF *XENOPUS LAEVIS*

Marco Crippa
Glauco P. Tocchini-Valentini
Francesca Andronico

I.   Introduction
II.   RNA polymerase from *Xenopus laevis* ovaries
III.   Injection of bacterial RNA polymerase into *Xenopus* oocytes
IV.   Inhibitory factor in mature oocytes
V.   Characterization of the inhibitory factor
VI.   Effects of the inhibitory factor on the *in vitro* transcription of the ribosomal genes
VII.   Acknowledgments
VIII.   References

I. INTRODUCTION

The amphibian oocyte provides a unique system for the study of the possible factors which are involved in the control of ribosomal RNA synthesis. Indeed the oocyte can be considered a cell which became highly specialized in producing ribosomal RNA. As Gall (1968) put it "the oocyte faces a unique problem: its single nucleus must supply RNA to a mass of cytoplasm which in other tissues would contain several thousand nuclei." It has been shown that this goal is achieved at least partially through the selective amplification of the ribosomal genes during the pachytene stage of the meiotic prophase (Perkowska, Macgregor and Birnstiel, 1968; Brown and Dawid, 1968; Evans and Birnstiel, 1968; Gall, 1968). In this way the number of ribosomal cistrons present in the oocyte nucleus increases more than a thousand times over the expected tetraploid value. As a result of this process, more than 97% of the RNA which is synthesized by the oocyte during the time of its maximal RNA synthesis (lampbrush stage) is ribosomal.

The lampbrush stage oocyte (stage 4) has a large nucleus which can be seen to contain over one thousand nucleoli. Autoradiographic studies (Davidson and Mirsky, 1965) have shown that these nucleoli are extremely active in RNA synthesis; [^{3}H] uridine incorporation can also be detected in correspondence of the chromosomal loops. Later in oogenesis the chromosomal loops contract and no longer synthesize any RNA whereas the RNA synthesis at the nucleolar level still continues even if at a lower rate. Finally, in the mature oocyte ready for ovulation (stage 6) no ribosomal RNA synthesis occurs even though all the extrachromosomal copies of the ribosomal genes are still present in the germinal vesicle.

From the picture so far outlined it is clear that the amphibian oocyte has special characteristics which turn out to be very useful for studying regulation of transcription:

(1) A very well defined class of genes, the ribosomal genes (rDNA) represent a conspicuous part of the total oocyte nuclear DNA and can be isolated fairly easily. Good amounts of ribosomal genes can also be obtained from somatic cells of the same animal. According to the extensive studies of Dawid, Brown and Reeder (1970) and of Reeder and Brown (1970), the oocyte and somatic rDNAs, even if they have different densities, behave identically in an *in vitro* transcription assay. Moreover, the *in vivo* transcription product of the ribosomal genes is an RNA molecule with very well defined characteristics which therefore can be easily identified and studied.

(2) The internal structure of the ribosomal genes is now being clarified (Loening, Jones and Birnstiel, 1969; Dawid *et al.,* 1970).

(3) During different stages of oogenesis the ribosomal genes are at one time very actively transcribed and at another time completely inactive in RNA synthesis. A consideration is however necessary at this point: the special characteristics of the oocyte which are so useful probably also place limits on this experimental model that we chose to investigate. Brown and Dawid (1968) already discussed the fact that the oocyte synthesizes ribosomes essentially as a storage product to be used months later during embryogenesis. Such a large accumulation of ribosomes inside of the cytoplasm creates a condition which in bacteria and in somatic cells would prevent any further ribosomal RNA synthesis whereas no effect is seen on the ribosomal RNA synthesis of the oocyte. It is therefore possible that some mechanisms which seem to regulate the transcription of the ribosomal genes in the oocytes only apply to these cells without having any role in the control of ribosomal RNA synthesis in other biological systems.

Recent observations strongly suggest that regulation of RNA synthesis in bacteria involves regulatory modifications of a single "core" RNA polymerase (Geiduschek and Haselkorn, 1969; Burgess, Travers, Dunn and Bautz, 1969; Di Mauro, Snyder, Marino, Lamberti, Coppo and Tocchini-Valentini, 1969). Sigma (σ)-like factors appear to be a general feature in the regulation of gene expression in bacteria and phages, each factor determining the regions of DNA templates that can be transcribed by the "core" RNA polymerase (Travers, 1969; Summers and Spiegel, 1969). We have been interested in the hypothesis that hierarchies of σ-like factors might play a role in the regulation of gene expression in eukaryotic cells and we have decided to use *Xenopus·laevis* because of the great advantages offered by this organism. Our first step was to try to establish if a special RNA polymerase was responsible for the massive synthesis of ribosomal RNA which occurs in the lampbrush stage oocytes. We will report here data on the isolation and purification of two different RNA polymerases from *Xenopus* ovaries; only one of them is localized in the nucleolus. On the basis of localization, binding to ribosomal DNA and response to inhibitors, we conclude that the nucleolar RNA polymerase is responsible for ribosomal RNA synthesis *in vivo*.

II. RNA POLYMERASES FROM *XENOPUS LAEVIS* OVARIES

A procedure has been worked out to isolate RNA polymerases from ovaries; the tissues were homogenized in buffer containing 1 mM Tris pH 7.9 and 1.4 mM β-mercaptoethanol, and the homogenate was centrifuged at low speed after filtration through cheese-cloth. The chromatin was precipitated by increasing the ionic strength to 0.1 M KCl and the precipitate was collected by centrifugation at high

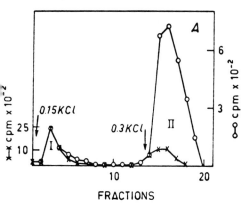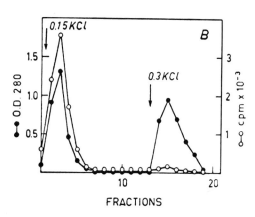

Fig. 10.1 DEAE-cellulose chromatography of *Xenopus laevis* RNA polymerase. DEAE cellulose was equilibrated with 1 mM Tris pH 7.9 and 1.4 mM β-MSH. The chromatin eluate, prepared as described in the text, was applied to the column and after washing with standard Tris-MSH buffer the ionic strength was raised to 0.15 M KCl. The subsequent step elution was at 0.3 M KCl. The fractions were assayed for incorporation activity (-o-o-o-) and the assay mixture (0.3 ml) contained, per ml, 100 μmoles Tris HCl pH 7.9, 3 μmoles MnCl$_2$, 1.8 μmoles ATP, GTP and CTP, 2.5 μC [^{14}C]UTP (395 mC/mole), 0.526 mg phosphoenol pyruvate, 8.8 μg pyruvate kinase, 30 μmoles (NH$_4$)$_2$SO$_4$. The incubation was for 30 min at 30°C. The binding capacity (-x-x-x-) was determined by measuring the retention of the filter of [^{3}H]rDNA. The binding assay incubation mixture (0.3 ml) contained 0.5 μg of rDNA, 0.1 ml of the fraction to be assayed, 0.01 M Tris, 0.01 M MgCl$_2$, 0.01 M KCl. The incubation was 10 min at 30°C and the reaction was stopped by dilution with 0.01 M Tris, 0.01 M MgCl$_2$ and 0.01 M KCl (binding buffer). The content of the tube was filtered through Millipore filters and the filters subsequently washed with binding buffer. (A) DEAE profile of enzymes extracted from ovaries. (B) DEAE profile of enzymes extracted from purified nucleoli.

speed. RNA polymerase activity was solubilized by sonication in 0.75 M KCl and the subsequent steps involved reduction of the ionic strength and chromatin removal by centrifugation. The chromatin eluate was chromatographed on DEAE cellulose and Fig. 10.1A shows that two peaks of RNA polymerase activity can be eluted respectively with 0.15 M (peak I) and with 0.3 M KCl (peak II).

The two peaks were also assayed for their capacity of retaining [^{3}H] rDNA on millipore filters (Jones and Berg, 1966). Figure 10.1A sshows that the rDNA binding capacity and the polymerizing activity are eluted together from the chromatograph.

It has recently been reported that a-amanitin (Fiume and Stirpe, 1969; Seifart, 1969; Chambon, Ramuz, Mandel and Doty, 1968; Jacob, Sajdel and Munro, 1970), a toxin from *Amanita phalloides,* is a potent inhibitor of the calf thymus RNA polymerase activity, eluting at high ionic strength. Table 10.1 shows that a-amanitin completely inhibits the oocyte RNA polymerase in peak II and is completely inactive against the activity eluting at 0.15 M KCl.

Table 10.1 **Effect of a-amanitin on *Xenopus laevis* Polymerases**

Enzyme source		a-amanitin	cpm
	peak I	—	285
		+	290
total ovary			
	peak II	—	766
		+	10
purified nucleoli	peak I	—	344
		+	350

The incorporation assay was done in 0.3 ml of a mixture which contained, per ml, 100 μmoles Tris HCl pH 7.9, 3 μmoles MnCl$_2$, 1.8 μmoles ATP, GTP and CTP, 2.5 μC ^{14}C UTP (395 mC/mmole), 0.526 mg phosphoenol pyruvate, 8.8 μg pyruvate kinase, 30 μmoles (NH$_4$)$_2$SO$_4$. The incubation was for 30 min at 30°C. When indicated the samples contained 10 μg/ml of a-amanitin.

The significance of the two peaks of RNA polymerase could be clarified if it were possible to show that the two enzymes have different specificities for different DNA regions.

Roeder and Rutter (1970) and Jacob *et al.* (1970) reported that one of the activities common to all animal systems so far analyzed is preferentially localized in the nucleolus.*Xenopus* young oocytes are a good source of clean nucleoli since in each cell there are more than 1000 extrachromosomal copies of this organelle which can be isolated with little nucleoplasm contamination. Nucleoli were isolated from ovaries manually dissected from young tadpoles; RNA polymerase was extracted as described above. Figure 10.1B shows that the only RNA polymerase activity extractable from nucleoli is the one eluting from DEAE at low ionic strength (peak I). This activity proved to be completely insensitive to α-amanitin (Table 10.1).

The nucleolar localization of RNA polymerase activity suggested to Roeder and Rutter (1970) and to Jacob *et al.* (1970) that this polymerase is involved in ribosomal RNA synthesis; because of the availability of pure rDNA we were able to further substantiate this hypothesis. We found that *Xenopus* polymerases, like bacterial polymerases, are capable of retaining DNA on millipore filters. The material in peak I and peak II were challenged with different kinds of DNA: ribosomal DNA (68% GC) (Reeder and Brown, 1969), *Xenopus* bulk DNA (42% GC) (Dawid, 1966) free of ribosomal cistrons. The molecular weight of these DNAs was around 1.2×10^7 daltons. Figure 10.2 shows that the material eluting as peak I

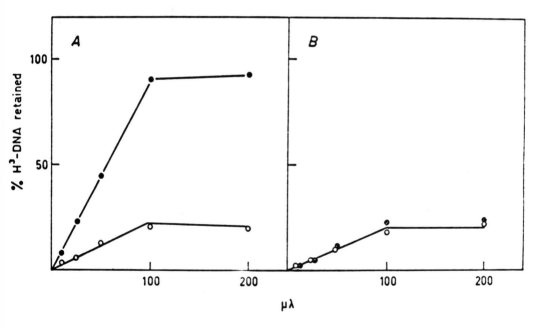

Fig. 10.2 Retention to Millipore filters of *Xenopus* ribosomal DNA and *Xenopus* bulk DNA free of ribosomal cistrons by peak I and peak II. The incubation mixture (0.4 ml) contained 0.5 µg of [^{3}H]DNA, increasing amount of enzyme, 0.01 M Tris, 0.01 M KCl and 0.01 M MgCl$_2$. The purified cistrons banded as an homogeneous peak at a density of 1.724 and when hybridized with purified *Xenopus* ribosomal RNA showed a satruation plateau of 23% (Birnstiel *et al.* 1968). The *Xenopus* bulk DNA was free of ribosomal cistrons as detected by hybridization with high specific activity *Xenopus* ribosomal RNA. -o-o-o- Retention of rDNA; -o-o-o- Retention of *Xenopus* bulk DNA. (A) Retention of [^{3}H]DNA by peak I (tube 31). (B) Retention of [^{3}H]DNA by peak II (tube 42).

is capable of retaining 90% of rDNA but only 20% of bulk DNA *(Xenopus)* free of ribosomal cistrons. The material in peak II can retain only 20% of the DNAs tested. When the fraction of ribosomal DNA not retained by the material in peak II was challenged with peak I material, again 90% of the input was retained. On the contrary no further retention was obtained when the same experiment was performed using peak II material in the second assay. These results demonstrate that in purified ribosomal DNA there are sites that are recognized only by peak I material and, provided that this DNA binding capacity is indeed due to RNA polymerase, the data also suggest that the nucleolar enzyme is involved in rRNA synthesis *in vivo.*

Table 10.2   **Effect of α-amanitin on Stage 4 Oocytes RNA Synthesis**

α-amanitin μg/cell	cpm x 50 oocytes
—	7980
5	8123
10	8000
20	8029

Stage 4 oocytes were injected with approximately 0.1 λ of Gurdon's saline buffer containing 0.15 mC/ml of 5 H^3-uridine and different amounts of α-amanitin to give the reported concentration per cell. After injection the oocytes were incubated for 8 h, collected and homogenized in 0.1 M acetate pH 5.2. The homogenate was TCA precipitated, filtered on "Millipore" and counted.

If the nucleolar RNA polymerase is only responsible for ribosomal RNA synthesis one would expect the latter to be insensitive to α-amanitin *in vivo.* We tested this prediction by micro-injecting α-amanitin into *Xenopus* stage 4 oocytes which are known to synthesize predominantly ribosomal RNA. Table 10.2 shows that α-amanitin does not inhibit incorporation of [^{3}H] uridine into TCA precipitable material when injected into stage 4 oocytes. When the RNA synthesized in the presence of α-amanitin is extracted and analyzed on a sucrose density gradient, the classical profile of ribosomal RNA is observed (Fig. 10.3).

III. INJECTION OF BACTERIAL RNA POLYMERASE INTO *XENOPUS* OOCYTES

There are two kinds of experiments which are generally performed in studies of genetic transcription: one is to take apart the components of the system (DNA,

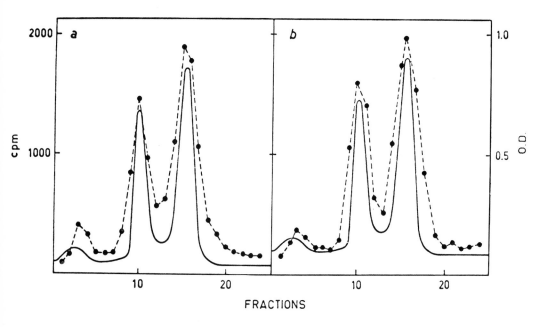

Fig. 10.3   Sucrose density analysis of RNA synthesized by stage 4 oocytes in the absence and in the presence of
α-amanitin. Stage 4 oocytes were injected as described under Table 10.2; after injection the oocytes were
incubated 8 hrs and the RNA was extracted according to the procedure of Brown and Littna (1964). The RNA
was precipitated with alcohol, dialyzed against 0.1 M acetate pH 5.0 and displayed on 15-30% sucrose gradient.
The centrifugation was carried out for 18 hrs at 22,500 r.p.m. in a SW 25.1 Spinco rotor. The gradients were
collected with a continuous recording apparatus, the RNA was TCA precipitated and counted.
(a) Control.   (b)   α-amanitin was injected to reach a concentration of 10 μg/ml inside of the oocyte.

RNA polymerase and factors) and put them together again; the other is to perturb
the intact cell and study the resulting effects. We propose a kind of hybrid
experiment: i.e. perturbation of the cell by injecting RNA polymerase and
associated factors. We have injected *E. coli* RNA polymerase into stage 4 oocytes;
this enzyme was purified by a procedure (Di Mauro *et al.* (1969) involving
streptomycin precipitation of nucleic acids, ammonium sulphate fractionation and
DEAE cellulose chromatography. A mixture containing RNA polymerase labeled
precursors and, when indicated, the drug to be tested, has been used for injection.

Table 10.3 shows that injected *E. coli* RNA polymerase stimulates [5-³H]
uridine incorporation into acid-insoluble material; this material is alkali and RNAse
sensitive. The RNA synthesis stimulated by the bacterial preparation is
DNA-dependent because 100% of the stimulation is abolished by actinomycin D.
The stimulation obtained by injecting the bacterial preparation is certainly due to
RNA polymerase because of its sensitivity to the antibiotic rifampicin. This drug, on
the contrary, does not affect the incorporation due to oocyte polymerase. When the
injected polymerase is derived from an *E. coli* strain resistant to rifampicin, the
stimulation is resistant to the drug.

Table 10.3 **Effect of injected** *E. coli* **RNA polymerase**

Sample injected	cpm	Stimulation ratio
−RNA polymerase	725	1
+RNA polymerase	2750	3.8
−RNA polymerase + rifampicin	720	1
+RNA polymerase + rifampicin	720	2

Stage 4 oocytes were microinjected with 0.1 μl of a mixture containing:

	50 μl of DEAE cellulose-RNA polymerase (5.0 μg protein)
or	50 μl of a buffer containing 0.01 M Tris pH 7.9 0.1 M $MgCl_2$
	0.0001 M EDTA

50 μl of Gurdon's buffer
10 μl of water
or   10 μl of a 5 mg/ml solution of rifampicin
10 μl of 1.3 mC/ml of [5-^{3}H] uridine in Gurdon's buffer

After the injection the oocytes were kept for 6 hr at 22°C in Barth's solution X, collected and homogenized in 0.1 M Na acetate pH 5.2. The homogenate was TCA precipitated, filtered on "Millipore" and counted.

RNA polymerase can be fractionated by phosphocellulose chromatography (Burgess *et al.* 1969) into two components: the core enzyme (PC), eluting at 0.4 M KCl, and the σ-factor recovered in the flow-through. We investigated which component of the polymerase was responsible for the stimulation of RNA synthesis in the oocyte by assaying two regions of the phosphocellulose profile by micro-injection. The fractions where σ-activity was present produced stimulation when injected into the oocyte, but no stimulation was obtained when the fraction containing the PC activity was injected (Fig. 10.4). When run in a glycerol gradient, the oocyte-stimulating activity cosediments with the σ-activity at approximately 5 S (Fig. 10.5); moreover, both activities are inactivated at the same temperature.

Since it has been shown that the sensitivity to rifampicin of bacterial RNA polymerase resides in the PC component (Di Mauro *et al.* 1969), it could then be expected that when σ alone is injected, the stimulation would be insensitive to the antibiotic. Table 10.4 shows that this prediction has been verified. We believe that when a mixture of RNA polymerase, (PC + σ) and rifampicin are injected together, no stimulation occurs because, due to the antibiotic, the σ-factor remains in the inactive complex and is not made available. The σ-factor is unable to produce RNA *in vitro* by itself but requires the interaction of the PC component. Which component participates in RNA synthesis with the bacterial σ inside the oocyte?We suggest that in animal cells, as in bacteria, a mechanism for positive control exists through factors and core polymerases, and that σ is interacting with one or more PC-like components of the oocyte polymerases.

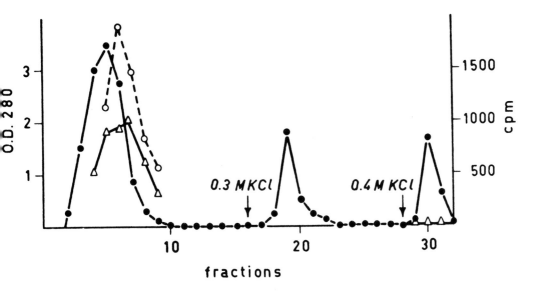

Fig. 10.4   Components of *E. coli* RNA polymerase. Phosphocellulose column profile. Phosphocellulose column was run according to Burgess *et al.* (1969). The PC component elutes at 0.4 M KCl; 10 μl of tube 30, when assayed as previously described (Di Mauro *et al* 1969), catalyzed the incorporation of 450 cpm of [^{14}C] ATP using calf thymus DNA as a template, the cpm incorporated were 120 using bacteriophase α-DNA. The addition of 20 μl of tube 6 to 20 μl of tube 30 stimulated incorporation on α-DNA by a factor of 20. The δ activity peak (-o-o-o-) was identified by assaying for its ability to stimulate RNA synthesis by the PC component on α-DNA (Burgess *et al.* 1969; Di Mauro *et al.* 1969). The oocyte stimulating activity (-△-△-△-) was measured by injecting into each oocyte 0.1 μl of a mixture containing: 50 μl of the phosphocellulose fraction, 50 μl of Gurdon's buffer, 50 μl of water and 10 μl of 1.3 mC/ml of [5-^{3}H]uridine in Gurdon's buffer.

Since in stage 4 oocytes RNA synthesis is completely insensitive to α-amanitin one could suspect that only α-amanitin insensitive activity, i.e. activity in peak I, is present in these cells. Our RNA polymerase is always prepared from oocytes of different stages and therefore the fact that we found two activities does not prove that this situation also refers to stage 4 oocytes. However, by micro-injecting bacterial σ-factor into stage 4 oocytes we were able to show that ~50% of the RNA synthesis due to σ stimulation is α-amanitin sensitive (Table 10.5). Since probably the bacterial σ exerts its role by interacting with a cellular PC-like component, it seems likely that nucleolar polymerase (activity I) and activity II are both present. From the data presented so far we can conclude that in stage 4 oocytes the nucleolar RNA polymerase is responsible for the great amount of ribosomal RNA synthesis which characterizes these cells. A second RNA polymerase activity is however also present. The finding that the injection of *E. coli* σ-factor stimulates both polymerases suggests that the endogenous initiation factor(s) could be limiting. This fact seems to be rather suggestive evidence for positive control in higher organisms.

Table 10.4 **Effect of Injected δ Factor**

Sample injected	cpm	Stimulation ratio
Control	600	1
6	2,172	3.6
6 + rifampicin	2,300	3.8
5 heated 5 min. at 50°	738	1.2

Stage 4 oocytes were injected as described under Table 5.3. The injection mixture contained 50 μl of fractions of the phosphocellulose column presented in Fig. 10.4 instead of the DEAE cellulose RNA polymerase. The δ factor completely loses the ability to stimulate RNA synthesis by the PC component on a-DNA, when heated 5 min at 50°C.

Table 10.5 **Effect of a-amanitin on Stage 4 Oocytes Stimulated RNA Synthesis**

δ factor	a-amanitin μg/cell	cpm x 50 oocytes
—	—	7,125
+	—	27,600
+	5	15,427
+	10	16,061
+	20	15,743

The oocytes were injected and processed as described under the legends of Tables 10.2 and 10.3.

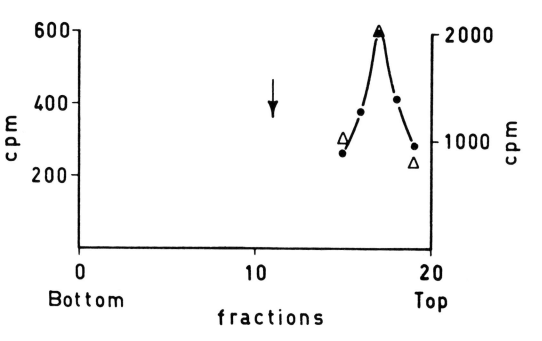

Fig. 10.5 Analysis of oocyte stimulating activity by zone centrifugation. A sample containing the oocyte stimulating activity, prepared by phosphocellulose chromatography, was layered on top of a linear 10-30% glycerol gradient containing 0.01 M Tris HCl buffer pH 7.9, 0.01 M $MgCl_2$, 0.15 M KCl, 0.1 mM EDTA and 0.1 mM dithiothreitol. The gradient was centrifuged in an SW 25 rotor at 4°C for 21 hrs at 23,000 r.p.m. The δ activity (-o-o-o-) was assayed for its ability to stimulate RNA synthesis by the PC component on α-DNA. The oocyte stimulating activity (-$\triangle$-$\triangle$-$\triangle$-) was measured by injecting into each oocyte 0.1 μl of a mixture containing: 50 μl of the gradient fraction, 50 μl of Gurdon buffer, 50 μl of water and 10 μl of 1.3 mC/ml of [5-^{3}H]uridine in Gurdon's buffer. *Escherichia coli* β- galactosidase was used as a molecular weight marker. The arrow indicates the position at which "core" polymerase bands when run in a parallel gradient.

IV. INHIBITORY FACTOR IN MATURE OOCYTES

As previously described in stage 6 oocytes, the amplified rDNA is still present but no transcription activity takes place. The first question asked was whether the nucleolar RNA polymerase is present in these cells. The data obtained with a-amanitin inhibition showed that in the mature oocytes both RNA polymerase activities, as identified by their a-amanitin sensitivity, are present. Even if this finding does not rule out the possibility that an initiation factor is missing, we decided to search for other mechanisms which could be responsible for the lack of ribosomal RNA synthesis in stage 6 oocytes. We now have evidence for a protein factor in mature oocytes which specifically inhibits the transcription of ribosomal genes.

Female *Xenopus laevis* were injected with 1 mC of a [^{14}C]amino-acid mixture lacking arginine and lysine 45 days after ovulation had been hormonally stimulated. After 3 days mature stage 6 oocytes were taken from the ovary and their germinal vesicles isolated. The germinal vesicles were homogenized at low ionic strength (0.07 M) and the homogenate centrifuged at a low speed and precipitated with ammonium sulphate. The fraction precipitated, which was between 35 and 45% saturated and contained 53% of the radioactivity incorporated into protein in the total homogenate, was redissolved in Alberts' buffer. When micro-injected into stage 4 oocytes, this fraction had a strong inhibitory effect on the synthesis of all RNA species, possibly because of the presence of toxic contaminants.

The fraction was then applied to a DNA-cellulose column prepared with *Xenopus* erythrocyte DNA enriched in ribosomal cistrons by polylysine precipitation of A-T rich sequences, according to Dawid, Brown and Reeder (1970). Only a small part of the inhibitory activity in the ammonium sulphate precipitate bound to the column while the remaining activity was recovered in the eluate. Seventy-five per cent of the inhibitory activity bound to the column was eluted at 0.6 M NaCl. When injected into stage 4 oocytes this fraction inhibited RNA synthesis (measured as incorporation of [^{3}H] uridine into trichloracetic acid (TCA) precipitable material) by approximately 30%. Hardly any specificity of inhibition of RNA species was detected when the RNA was extracted and analyzed on sucrose gradients.

The 0.6 M fraction from the DNA-cellulose column was adsorbed to DEAE-Sephadex and stepwise eluted by increasing the ionic strength to 0.15, 0.3, 0.6 and 1.0 M KCl. When injected into stage 4 oocytes the 0.3 M fraction seemed preferentially to inhibit synthesis of ribosomal RNA. The 0.3 M fraction was then dialyzed against 0.1 M KCl and re-chromatographed on DEAE-Sephadex with a 0.1-0.3 M KCl gradient. Because it was practically impossible to assay each tube for effects on RNA synthesis, the tubes corresponding to optical density or radioactivity peaks were pooled and labeled as indicated in Fig. 10.6.

Each of the peaks in Fig. 10.6 was checked for its effect on RNA synthesis by micro-injection into young stage 4 oocytes. After injection oocytes were allowed to incorporate [^{3}H]uridine for 2 hr and then the RNA was extracted and analyzed on sucrose gradients. To avoid complications due to possible effects on the processing mechanisms of the precursor molecule, only [^{3}H]uridine incorporation into the 40 S ribosomal precursor was taken as a measure of ribosomal RNA synthesis. It is clear from Fig. 10.7 that the amount of incorporation into 28 S and 18 S regions closely follows that into 40 S molecules. Of the other RNA species, only 4 S RNA synthesis was followed, both because it is the only RNA species, apart from ribosomal RNA, that can be measured with reasonable accuracy in sucrose gradients, and because it is known that 4 S RNA synthesis is regulated independently of ribosomal RNA. No attempt has been made to estimate incorporation into DNA-like RNAs. Figure 10.7 shows that when peak R material is injected into stage 4 oocytes, it strongly inhibits the incorporation of [^{3}H]uridine into the 40 S ribosomal precursor, while synthesis of 4 S RNA is only slightly affected.

Table 10.6 summarizes the results obtained when the other peaks eluted from DEAE-Sephadex were used for injection. Peak R clearly contains a factor which specifically inhibits ribosomal RNA synthesis. We think that the 15 per cent inhibition of the 4 S RNA synthesis obtained with peak R material, as well as the slight inhibitory effects of the other peaks on both 40 S and 4 S RNAs, are probably a consequence of unspecific damage to the cells.

V. CHARACTERIZATION OF THE INHIBITORY FACTOR

The peak R material has a UV-spectrum characteristic of a protein. Its inhibitory effect disappears after trypsin digestion and heat denaturation and is

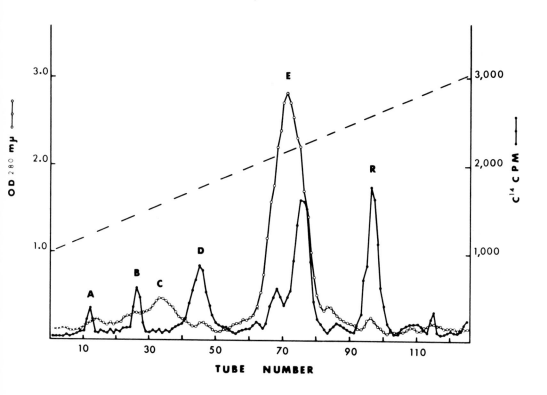

Fig. 10.6 **DEAE-Sephadex chromatography** of a fraction of the DNA-binding proteins from stage **6 oocytes.** The 0.6 M fraction from the DNA-cellulose column was adsorbed to DEAE-Sephadex and stepwise eluted as described in the text. The resulting of 0.3 M fraction was dialyzed against 0.1 M KCl, 10 mM Tris (pH 7.4), 10 mM Mg^{2+}, 6 mM mercaptoethanol, 0.1 mM EDTA, applied to a second DEAE-Sephadex column and eluted with a linear KCl gradient (0.1-0.3 M KCl). The loading and the elution were done at 4°C at a flow rate of 3 ml/hr. The peaks were arbitrarily identified in the following way: (A) tubes 10-14; (B) tubes 24-29; (C) tubes 31-38; (D) tubes 41-45; (E) tubes 62-83; (R) tubes 91-101.

Table 10.6   **Effects of DEAE-"Sephadex" Peaks on RNA Synthesis**

| | 40 S RNA | | 4 S RNA | |
	c.p.m.	Per cent inhibition	c.p.m.	Per cent inhibition
Control	6,215	—	3,950	—
Peak A	5,519	11.2	3,389	14.2
Peak B	5,683	8.6	3,637	8.0
Peak C	5,625	9.5	3,411	13.7
Peak D	5,477	11.9	3,206	18.9
Peak E	5,289	14.9	3,271	17.2
Peak R	1,390	71.7	3,435	13.1

The tubes corresponding to the peaks indicated in Fig. 10.6 were pooled and their protein concentration was determined by Lowry's method. The fractions were then diluted to the same protein concentration and injected as described in the legend of Fig. 10.7. Oocytes were allowed to incorporate H^3-uridine for 2 h, and RNA was extracted and analyzed on sucrose gradients. The radioactivity incorporated into the 40 S and the 4 S regions was precipitated with TCA and measured in a liquid scintillation spectrometer.

resistant to ribonuclease. Although the phloroglucinol reaction for RNA gave negative results a few ribonucleotides could still be bound to the material. By acrylamide gel electrophoresis the material is resolved into three main bands all moving to the anode but only one of them radioactive. By preparative acrylamide gel electrophoresis, we were able to obtain the radioactive component 70% pure. This component which will be referred to as the R protein, has a molecular weight of approximately 65,000. The *in vivo* micro-injection test showed that the R protein possesses more than 80% of the total inhibitory activity present in the peak R region of the DEAE-Sephadex column. The results of control experiments rule out the possibility that the inhibitory effect is due to inhibition of the [3 H] uridine phosphorylation process or to a nuclease activity specific for ribosomal RNA.

When extraction and fractionation procedures were repeated with germinal vesicles isolated from stage 4 oocytes active in RNA synthesis, the DEAE-Sephadex elution profile was different but no peak showed an inhibitory effect higher than 10-15%.

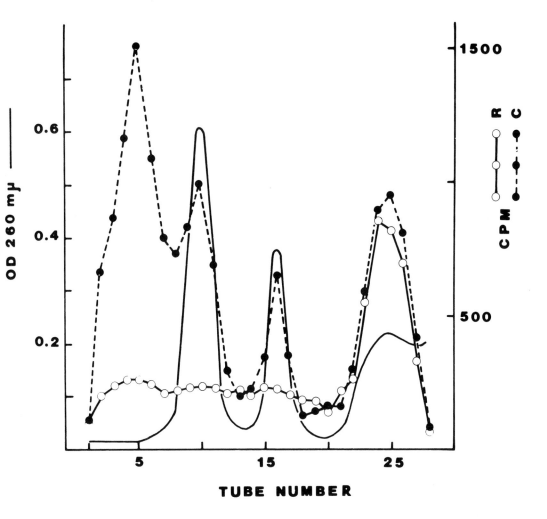

Fig. 10.7 Sedimentation profiles of radioactive RNA extracted from control (•---•) and treated (0—0) stage 4 oocytes. The control oocytes were injected with approximately 1/10 of their volume of Gurdon's saline buffer (88 mM NaCl; 1.0 mM KCl; 0.82 mM $MgSO_4$; 2.4 mM $NaHCO_3$; 0.41 mM $CaCl_2$; 0.33 mM Ca $(NO_3)_2$; 2.0 mM Tris HCl, pH 7.6) containing [^{3}H]uridine (0.06 μC per oocyte). The treated oocytes were injected with an equal volume of saline buffer containing, in addition to [^{3}H]uridine 1 ng of peak R protein. Seventy-five oocytes were used for each gradient. After injection the oocytes were kept in Gurdon's saline buffer. At the end of the incorporation period (2 hrs) the oocytes were homogenized in a Dounce homogenizer. The RNA was extracted in 0.1 M acetate, pH 5.0 containing 0.5% sodium dodecyl sulphate, precipitated with alcohol and layered over a 28 ml 15-30% sucrose gradient. The gradients were centrifuged for 18 hrs in a Spinco SW 25.1 rotor and collected. Absorbance was recorded by Gilford continuous recorder. All fractions were precipitated with TCA and counted in a liquid scintillation spectrometer (Nuclear Chicago). The radioactivity profiles of two gradients are superimposed; the first profile (•---•) was obtained with 75 oocytes which did not receive peak R material; the second profile (0—0) was obtained with 75 oocytes injected with 1 ng each of peak R protein. Using actinomycin I established that at least 85% of counts in the 4 S region were a result of real synthesis and not terminal turnover.

   The data presented here seem to indicate that the mature stage 6 oocytes contain a protein, that we named R protein, with regulatory properties specific for the transcription of the ribosomal cistrons. To substantiate further the specificity of the R protein, its binding properties were investigated using purified ribosomal cistrons (Birnstiel, 1967; Dawid, Brown and Reeder, 1970) and DNA virtually free of ribosomal cistrons as detected by hybridization with high specific activity *Xenopus* ribosomal RNA. Figure 10.8 shows that the radioactive component of peak R material specifically binds to purified ribosomal cistrons. If the DNAs were previously denatured, only 20% of the original binding is observed and the specificity shown in Fig. 10.8 is greatly reduced.

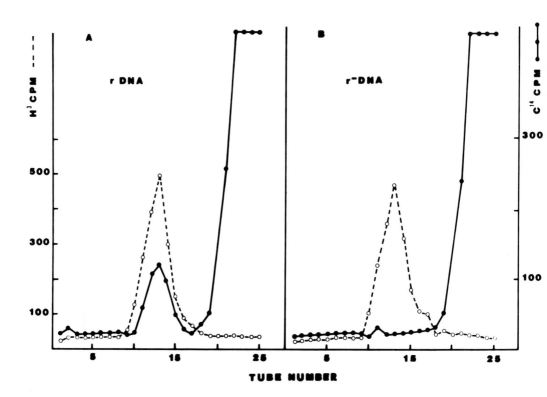

Fig. 10.8 Specific binding of peak R protein to purified ribosomal cistrons. Purified ribosomal cistrons (A) and DNA free of ribosomal cistrons (B) were prepared as described in the text. Both DNAs were tritiated. The [^{14}C] R protein was concentrated and mixed with rDNA or r̄DNA. The incubation mixture contained, per ml: 10 µg of DNA, 10 µg of protein, 50 µmoles of KCl and 1 µmole of EDTA. After 10 min of incubation at 37°C, 0.2 ml of the mixture was layered on a 4.5 ml, 5-30% sucrose gradient. After centrifugation for 4 hrs at 39,000 r.p.m. in a SW 39 Spinco rotor fractions were collected after puncturing the bottom of the tube. Each fraction was then precipitated with TCA and counted.

   Using [^{14}C] R protein of known specific activity, a DNA saturation experiment, of the kind used by Riggs, Suzuki and Bourgeois (1970) for the lac repressor was performed. Given the molecular weight of the protein (6.5×10^4) and of the DNA (1.2×10^7) it was possible to express the data in terms of protein/DNA molar ratios. In Fig. 10.9 are shown the results which demonstrate that the DNA saturation is reached at a molar ratio of 1.2 as if one molecule of DNA had only one binding site for the R protein.

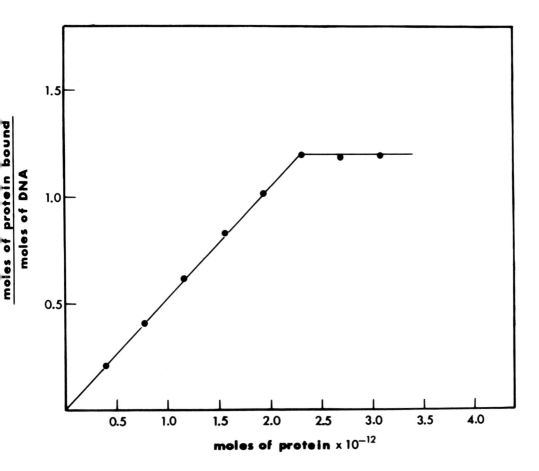

Fig. 10.9 *Xenopus* rDNA saturation with $[^{14}C]R$ protein. The binding mixture contained 0.01 M $MgCl_2$, 0.01 M KCl, 0.1 mM EDTA, 0.1 mM dithiothreitol, 0.01 M Tris HCl pH 7.4, 2 μg of *Xenopus* ribosomal DNA and increasing amounts of R protein. The binding was for 10 min at 30°C. The amount of R protein bound to DNA was determined by centrifugation of the mixture in a glycerol gradient.

Table 10.7 **Effect of "R" Protein on the *In vitro* RNA Synthesis Directed by Different Templates**

DNA	c.p.m. incorporated without "R" protein	c.p.m. incorporated with "R" protein
a	21,670	22,445
T4	17,437	16,811
Xenopus ribosomal	1,970	1,340
Xenopus ribosomal free	859	871

The assay mixture (0.25 ml) contained, per ml, 100 μmoles of Tris pH 7.5, 1 μmole of unlabeled GTP, ATP and UTP, 5 μc of H^3CTP (50 mC/mg), 10 μmoles $MgCl_2$, 4 μmoles of phosphoenol pyruvate, 10 μg of pyruvate kinase, 4 μg of DNA and when indicated 0.4 μg of "R" protein.

Table 10.8 **Binding of *E. Coli* RNA Polymerase and Oocyte "R" Protein to Ribosomal DNA**

H^3-RNA polymerase	^{14}C-"R" protein	H^3-c.p.m. bound	^{14}C-c.p.m. bound
+	—	2450	—
—	+	—	320
+	+	2275	309

The binding conditions were as described in the legend of Figure 10.9. The polymerase or "R" protein bound were monitored using the filter assay of Jones and Berg (1966).

VI. EFFECTS OF THE INHIBITORY FACTOR
ON THE *IN VITRO* TRANSCRIPTION
OF THE RIBOSOMAL GENES

Reeder and Brown (1970) have shown that *E. coli* RNA polymerase can transcribe purified ribosomal RNA genes with a considerable degree of fidelity. Moreover, they showed that most polymerase molecules initiate transcription at the beginning of the 18 S DNA sequences. Because only small amounts of the nucleolar RNA polymerase responsible for ribosomal RNA synthesis *in vivo* was obtained from *Xenopus* ovaries, the system of Reeder and Brown was used to test the effect of the R protein on *in vitro* transcription of the ribosomal genes. Table 10.7 shows the results obtained using *E. coli* RNA polymerase with *Xenopus* rDNA and with other DNAs as controls. A clear template specificity is present. It was not possible, however, to obtain any inhibitory effect on RNA synthesis higher than 30% even by increasing the amount of R protein to values which from the DNA saturation curve of Fig. 10.9 were certainly in excess.

RNA polymerase starts transcription also from "wrong" sites along the DNA sequences. If it is assumed that one molecule of rDNA of 1.2×10^7 molecular weight essentially represents one ribosomal gene with one 28 S, one 18 S and one space region, one could expect to have only one polymerase starting site per DNA molecule. Figure 10.10 shows a DNA saturation curve obtained with radioactive *E. coli* RNA polymerase purified on glycerol gradient. The DNA saturation is reached at a polymerase/DNA molar ratio of approximately 4. This means that under binding conditions (low ionic strength) each DNA molecule has an average of 4 polymerase binding sites. The possibility exists, however, that some of the polymerase binding monitored could be due to aggregation between one molecule of polymerase bound to the DNA and other polymerase molecules.

Are the polymerase binding sites the same as the R protein binding sites? Table 10.8 shows the results obtained in a binding competition experiment. The presence of polymerase does not prevent the R protein from binding and *vice versa*. It seems reasonable to conclude that the R protein and the *E. coli* RNA polymerase have different binding sites. Nevertheless, the possibility exists that these results are biased by an interaction between the two protein molecules (polymerase and R protein) once one of them is bound to the DNA.

No definite conclusion can be drawn about the mode of action of the R protein. There are hopes, however, that better knowledge of the actual mechanisms involved will be obtained by exploiting the advantages offered by the *Xenopus* system. Indeed, at least two other very well known RNA molecules can be usefully studied: the 5 S RNA which is synthesized in a coordinate fashion with ribosomal RNA (Brown, 1967) and the 4 S RNA whose transcription is independent from ribosomal RNA synthesis (Brown and Littna, 1966). One way to achieve functional coordination among different genes physically located in different regions of the genome is to have in all these genes similar or identical regulatory regions. In this situation one would expect any regulatory factor to be able to recognize common

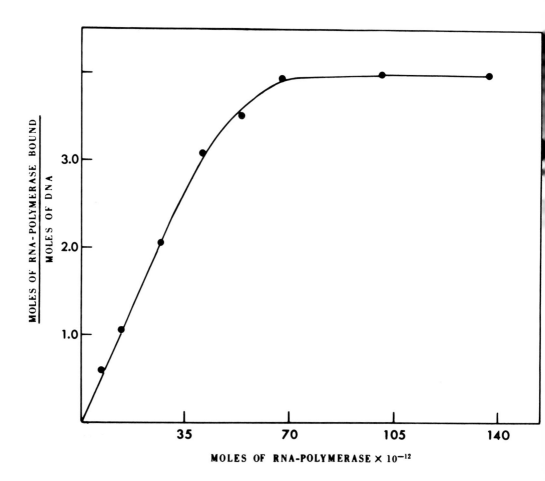

Fig. 10.10 *Xenopus* rDNA saturation with *E. coli* [³H]RNA polymerase. The binding assay was done under the same conditions as explained in the legend of Fig. 10.9.

sequences only on genes which are coordinately transcribed. The isolation and the purification of the 5 S and 4 S genes would be a very useful step in improving our understanding of the mechanisms which control transcription in this system.

VII. ACKNOWLEDGMENTS

We are grateful to Professors Fiume, Lancini, Silvestri and Wieland for the generous gift of inhibitors of RNA synthesis and thank B. Esposito, A. Lamberti and G. Locorotondo for expert technical help.

VIII. REFERENCES

Birnstiel, M. (1967) Some experiments relating to the homogeneity and arrangement of the rRNA genes of *Xenopus laevis. In* "Cell Differentiation", CIBA Found. Symp. (A. V. S. deReuck and J. Knight, eds.), p. 178. Boston, Little Brown.

Birnstiel, M., Speir, J., Purdom, I., Jones, K. and Loening, V. E. (1968) Properties and composition of the isolated ribosomal DNA satellite of *Xenopus laevis. Nature (London.).* **219**, 454.

Brown, D. D. (1967) The genes for ribosomal RNA and their transcription during amphibian development. *In* "Current Topics in Developmental Biology". (A. Monroy and A. Moscona, eds.), Vol. 2, p. 47. New York, Academic Press.

Brown, D. D. and Dawid, I. B. (1968) Specific gene amplification in oocytes. *Science.* **160**, 272.

Brown, D. D. and Littna, E. (1964) Variations in the synthesis of stable RNAs during oogenesis and development of *Xenopus laevis. J. Molec. Biol.* **8**, 688.

Brown, D. D. and Littna, E. (1966) Synthesis and accumulation of low molecular weight RNA during embryogenesis of *Xenopus laevis. J. Molec. Biol.* **20**, 95.

Burgess, R. R., Travers, A. A., Dunn, J. J. and Bautz, E. K. F. (1969) A factor stimulating transcription by RNA polymerase. *Nature (London.).* **221**, 43.

Chambon, P., Ramuz, M., Mandel, P. and Doty, J. (1968) The influence of ionic strength and a polyanion on transcription *in vitro.* I. Stimulation of the aggregate RNA polymerase from rat liver nuclei. *Biochim. Biophys. Acta.* **157**, 504.

Crippa, M. and Tocchini-Valentini, G. P. (1970) Performance of a bacterial RNA polymerase factor in an amphibian oocyte. *Nature (London.).* **226**, 1243.

Davidson, E. H. and Mirsky, A. E. (1965) Gene activity in oogenesis. *Brookhaven Symp. Biol.* **18**, 77.

Dawid, I. (1966) Evidence for the mitochondrial origin of frog egg cytoplasmic DNA. *Proc. Nat. Acad. Sci. U.S.A.* **56**, 269.

Dawid, I., Brown, D. D. and Reeder, R. (1970) Composition and structure of chromosomal and amplified ribosomal DNAs of *Xenopus laevis. J. Molec. Biol.* **51**, 341.

Di Mauro, E., Snyder, L., Marino, P., Lamberti, A., Coppo, A. and Tocchini-Valentini, G. P. (1969) Rifampicin sensitivity of the components of DNA-dependent RNA polymerase. *Nature (London.).* **222**, 533.

Evans, D. and Birnstiel, M. L. (1968) Localization of amplified ribosomal DNA in the oocyte of *Xenopus laevis. Biochim. Biophys. Acta.* **166**, 274.

Fiume, L. and Stirpe, F. (1969) Decreased RNA content in mouse liver nuclei after intoxication with α-amanitin. *Biochim. Biophys. Acta.* **123**, 643.

Gall, J. G. (1968) Differential synthesis of the genes for ribosomal RNA during amphibian oogenesis. *Proc. Nat. Acad. Sci. U.S.A.* **60**, 553.

Geiduschek, E. P. and Haselkorn, R. (1969) Messenger RNA. *Ann. Rev. Biochem.* **38**, 647.

Jacob, S. T., Sajdel, E. M. and Munro, H. N. (1970) Different responses of soluble whole nuclear RNA polymerases and soluble nucleolar RNA polymerase to divalent actions and to inhibition by α-amanitin. *Biochem. Biophys. Res. Commun.* **38**, 765.

Jones, O. W. and Berg, P. (1966) Studies on the binding of RNA polymerase to polynucleotides. *J. Molec. Biol.* **22**, 199.

Perkowska, E., Macgregor, H. C. and Birnstiel, M. L. (1968) Gene amplification in the oocyte nucleus of mutant and wild-type *Xenopus laevis. Nature (London.).* **217**, 649.

Reeder, R. and Brown, D. D. (1969) The fidelity of transcription of the ribosomal RNA genes from a toad by bacterial RNA polymerase. Lepetit Colloquium on RNA polymerase. Florence, North Holland Publ., p. 233.

Reeder, R. H. and Brown, D. D. (1970) Transcription of the ribosomal RNA genes of an amphibian by the RNA polymerase of a bacterium. *J. Molec. Biol.* **51**, 361.

Riggs, A. D., Suzuki, H. and Bourgeois, S. (1970) Lac repressor-operator interaction. I. Equilibrium studies. *J. Molec. Biol.* **48**, 67.

Roeder, R. G. and Rutter, W. J. (1970) Specific nucleolar and nucleoplasmic RNA polymerases. *Proc. Nat. Acad. Sci. U.S.A.* **65**, 675.

Seifart, H. K. (1969) Some characteristics of mammalian RNA polymerase and factors influencing its activity. Lepetit Colloquium on RNA polymerase. Florence, North Holland Publ., p. 233.

Summers, W. C. and Spiegel, R. B. (1969) Control of template specificity of *E. coli.* RNA polymerase by a phage-coded protein. *Nature (London.).* **223**, 1111.

Travers, A. A. (1969) Bacteriophage sigma factor for RNA polymerase. *Nature (London.).* **223**, 1107.

11

CYTOPLASMIC DNA

Igor B. Dawid

I.   Introduction
II.   Cytoplasmic DNA in eggs
        A.   Enrichment of mitochondrial DNA in eggs
        B.   Is there DNA in yolk platelets?
        C.   Why does the oocyte accumulate so many mitochondria?
III.   The function of mitochondrial DNA
IV.   Mitochondrial genetics: the continuity of the mitochondrial genome
V.   References

I. INTRODUCTION

This article summarizes present knowledge on two aspects of the subject "Cytoplasmic DNA." We shall consider the situation in eggs, where relatively large amounts of DNA are commonly found outside the nucleus and where the term cytoplasmic DNA has been most often used. Then we shall relate this phenomenon to the situation in eucaryotic cells in general. In animal eggs and other cells cytoplasmic DNA is localized in the mitochondria, i.e., it is mitochondrial DNA. We shall not discuss another type of cytoplasmic DNA, chloroplast DNA, whose presence is restricted to green plants, nor more specialized DNAs like kinetoplast DNA. Mitochondria are universally present in eucaryotic cells and all mitochondria studied so far contain DNA. We shall summarize the available evidence regarding the function of mitochondrial DNA in eucaryotic cells in general and in eggs and embryos in particular.

II. CYTOPLASMIC DNA IN EGGS

A. The Enrichment of Mitochondrial DNA in Eggs

In somatic cells most of the DNA is localized in the nucleus and is organized in chromosomes. The DNA content of the chromosome set of any species is a stable characteristic of that species; consequently, the DNA content of the different somatic cells in all animals of any species is related to the cell's ploidy. If only diploid cells are considered one finds that the DNA content per cell is constant (Vendrely and Vendrely, 1948). The presence of mitochondrial DNA does not invalidate this constancy rule in somatic cells since mitochondrial DNA amounts to less than one percent of the total DNA present, which is well within the limit of experimental error of DNA determinations. However, eggs are very large cells having a low nucleo-cytoplasmic ratio. Each egg contains thousands of times as many mitochondria as, for example, a liver cell. Therefore, mitochondrial DNA is much enriched in eggs relative to the nuclear DNA. The ratio of nuclear to mitochondrial DNA, which is about 100:1 in somatic cells, changes to between 1:1 and 1:100 in eggs.

Although it is likely that all eggs have a relatively high content of mitochondrial DNA this question has been studied in only a few species so far. Part of the reason for this fact is the technical difficulty involved in the study of egg DNA. The enrichment of mitochondrial DNA in eggs is a relative one, caused by the very low concentration of nuclear DNA in this material, rather than by a particularly high concentration of mitochondrial DNA. In fact, the concentration of mitochondrial DNA in eggs is usually lower by about a factor of ten than in actively metabolizing tissues, mainly because of the accumulation of large quantities of yolk. The presently available information on egg DNA is summarized in Table 11.1. The eggs of all organisms studied contain considerably more DNA than is accounted for by the chromosomal contribution; most of this excess is mitochondrial DNA. The work summarized in Table 11.1 shows that most, and perhaps all the cytoplasmic DNA in animal eggs is mitochondrial DNA. In plants, proplastid DNA has been reported in addition to mitochondrial DNA (Bell and Muhlethaler, 1964).

B. Is There DNA in Yolk Platelets?

This question has repeatedly been answered in the affirmative by Brachet and his collaborators. Baltus and Brachet (1962) and Hanocq-Quertier, Baltus, Ficq, and Brachet (1968) measured the deoxyribose content of amphibian eggs by a fluorimetric method and concluded that these eggs contained much more "DNA" than could be accounted for by the mitochondrial component and that this excess was localized in the yolk fraction. The low concentration of DNA in eggs and the presence of interfering substances of unknown chemistry make the interpretation of such assays difficult (see discussion in Dawid, 1965; 1968a). Furthermore, even if the deoxyribose measurements were correct they would not constitute proof that the material measured is actually DNA. Varying properties have been ascribed to this "cytoplasmic DNA." It has been reported to be soluble in cold acid and therefore of low molecular weight (Brachet and Ficq, 1965; Brachet and Quertier, 1963). Hanocq-Quertier *et al.* (1968) stated that treatment of preparations with N HCl at 60° for up to one hour did not remove the "cytoplasmic DNA." Either property, cold acid solubility or hot acid insolubility, should exclude any compound from being regarded as DNA.

Recently Baltus, Hanocq-Quertier and Brachet (1968) reported the isolation of DNA from oocytes of *Xenopus laevis* and termed this material "yolk DNA." It cannot be related to the previously described substances since it is DNA of moderately high molecular weight which undoubtedly is insoluble in cold acid and hydrolyzable by hot acid. However, it is debatable whether it is yolk DNA or follicle cell nuclear DNA. The experiments were carried out with ovarian oocytes from which, the authors state, all follicle cells had been removed. However, there are very large numbers of such cells around each oocyte and the concentration of DNA in oocytes is much lower (by several orders of magnitude) than in follicle cells. Before any DNA isolated from ovarian material can be considered as a component of the oocyte the absence of follicle cells from the sample must be rigorously established.

I therefore suggest that the material isolated by Hanocq-Quertier *et al.* (1968) is follicle cell nuclear DNA. This suggestion is strengthened by the following facts. First, the physical properties of the DNA obtained suggested that it is nuclear DNA. Second, I did not find "yolk DNA" in ovulated eggs which are naturally freed of their follicle cells. In addition to earlier work using extraction with or without pronase (Dawid, 1965) I carried out experiments using the method of Kirby and Cook (1967), the same method employed by Hanocq-Quertier *et al.* (1968). The results obtained with all methods were essentially the same (Table 11.1). No evidence could be obtained for either the presence of DNA in yolk nor for an unidentified major DNA component other than mitochondrial DNA in whole eggs.

Comparative and theoretical considerations also argue against the presence of DNA in yolk platelets. In sea urchin eggs DNA found in association with the yolk fraction had the physical properties of mitochondrial DNA (Piko, Tyler and Vinograd, 1967); this material is now considered to be derived from mitochondria which contaminated the yolk fraction (Berger, 1968). In our study of egg DNA in *Urechis caupo* (Dawid and Brown, 1970) we found no evidence for a major DNA component in yolk; a minor component could have been missed. Finally, no suggestion has been advanced for a function of DNA in yolk platelets. It would be difficult to account for any such function in the light of what we know about yolk accumulation and use. The bulk of the yolk components are synthesized in the liver and transported to the oocyte (Wallace and Jared, 1969). No genetic continuity has been suggested for yolk platelets, as exists in mitochondria, plastids and kinetoplasts (see below). There exists a type of organelle which is surrounded by a double membrane, shows continuity through the life cycle of the cell and organism in which it is present, and carries some autonomous genetic information in the form of organelle DNA. Yolk platelets do not have these characteristics and they do not contain any DNA.

C. Why Does the Oocyte Accumulate So Many Mitochondria?

The mature oocyte has low metabolic activity, and it does not appear to

Table 11.1 **Content and Distribution of DNA in Eggs and Oocytes**

Species	DNA content pg	DNA content Haploid equivalents	Nuclear	Recovered mitochondria	Distribution of DNA, per cent Mitochondrial Estimated from physical properties	Ref.
Xenopus laevis	3,100	1,000	about 1[a]	78	about 100	Dawid (1965, 66, 68a,b)
Rana pipiens	4,500	600	about 1[a]	70	about 100	Piko *et al.* (1967, 1968)
Lytechinus pictus	8.3	9.5	about 12[b]	55[c]	80 to 90	
Strongylocentrotus purpuratus	3.3	4.3	—	—		Schwartz, (1969)
Urechis caupo	10	10	40[d]	60	60	Dawid and Brown (1970)

[a] Ovulated frog eggs are post-meiotic and contain one haploid equivalent of chromosomal DNA. The amplified copies of ribosomal DNA are located in the cytoplasm at this point, but are derived from the germinal vesicle (Brown and Dawid, 1968). I therefore consider them as "nuclear DNA" and included them under this heading.

[b] These eggs are post-meiotic and haploid; 12% of the total DNA corresponds to about one haploid complement.

[c] About 20% of the egg's DNA was found associated with yolk platelets; this DNA had the physical properties of mitochondrial DNA. More recently it was concluded (Berger, 1968) that the DNA appearing in the yolk fraction was due to contaminating mitochondria; furthermore, Piko *et al.* (1968) state that 80 to 90% of the total DNA in *L. pictus* eggs is mitochondrial.

[d] In *Urechis* analysis was made on pre-meiotic oocytes; the nucleus thus contains 4 haploid equivalents of DNA.

"need" a large number of mitochondria itself. It would appear that the purpose of the observed accumulation of these particles is their storage for later use by the embryo. In this sense mitochondria are a storage product - however, one may suggest that their fate differs in an important aspect from the fate of such a typical storage material as yolk. Whereas yolk is used as raw material which is broken down in the embryo and converted into different substances, it appears likely that mitochondria are conserved as such. In this respect their developmental fate would resemble that of ribosomes stored in the oocyte, which are stable during embryogenesis and support protein synthesis during this period (Brown, 1965). If this view of the function and fate of egg mitochondria is correct we can make several predictions. We expect the egg's mitochondria to be preserved during embryogenesis, where they support the respiratory metabolism of the embryo. Because of the large reserve of inherited mitochondria the embryo has no need to synthesize new ones until later in development. The total number or mass of mitochondria is therefore expected to be constant in early development; before an increase in this mass becomes apparent we expect to see synthesis of mitochondrial RNA molecules which, as we shall see below, are synthesized under the control of mitochondrial DNA and function in the biogenesis of mitochondria.

In the frog *X. laevis* these predictions have been borne out in the work of Chase (1970, and unpublished). In agreement with an earlier report of Boell and Weber (1955) he could show that the total mass of mitochondria and the activity of cytochrome oxidase do not change for over two days of development (to stage 38) at which time the embryo has reached a swimming tadpole stage and all major organ systems are laid down. Chase further showed that the quantity of mitochondrial DNA per embryo does not change for an even longer period, up to stage 41. The synthesis of mitochondrial RNA is not detectable during cleavage and begins at or shortly after gastrulation, well before the accumulation of new mitochondrial protein. Particularly interesting in this system is the regulation of mitochondrial RNA synthesis. It proceeds from an undetectable level to strong activity while the template for this synthetic activity, i.e., the mitochondrial DNA, does not replicate and does not change in concentration.

Whereas a certain degree of genetic autonomy is ascribed to the mitochondria, their activity is coordinated with the activity of the rest of the cell. The drastic, qualitative changes in the synthesis of nuclear RNA during oogenesis and embryogenesis (Brown and Littna, 1964a and b; Brown, 1967), and the equally drastic changes in mitochondrial RNA synthesis during development make the frog embryo a very suitable system for the study of interrelations between these two activities.

III. THE FUNCTION OF MITOCHONDRIAL DNA

The mitochondrial DNA of all metazoans studied so far is a circular molecule with a contour length of 5 to 6 μm, corresponding to a molecular weight of 10 to 12 million daltons, or about 16,000 base pairs (Borst and Kroon, 1969; Wolstenholme,

Koike and Renger, 1970). This is a very small amount of genetic information. The size constancy of mitochondrial DNA is remarkable; when measured under the same conditions the DNA circles from the marine worm *U. caupo* and from *X. laevis* are indistinguishable (Dawid and Brown, 1970). However, some differences in the size of mitochondrial DNA occur between certain species (Wolstenholme and Dawid, 1968). Its small size limits the functions that animal mitochondrial DNA can carry out, but facilitates the analysis of these functions. In yeast, mitochondrial DNA is larger, likely being present as circles 25 μm long (Hollenberg, Borst, Thuring and Van Bruggen, 1970). The sequence complexity of *Neurospora* mitochondrial DNA has been reported at 66 x 10⁶ daltons (Wood and Luck, 1969), but the physical structure of this DNA is not known.

Biochemical studies in fungi and in animals have shown that mitochondrial DNA codes for several RNA species which are part of the protein synthesizing machinery of the mitochondrion. Mitochondria contain machinery which is distinct from that of the cytoplasm (see Roodyn and Wilkie, 1968; Ashwell and Work, 1970). Part of the mitochondrial proteins are synthesized inside the mitochondrion using this special machinery; however, it is not known exactly what fraction of the mitochondrial proteins is made inside (estimates range from 5 to 20%: see Ashwell and Work, 1970; Hawley and Greenawalt, 1970), nor which proteins are synthesized there. More is known about the machinery itself. Special ribosomes have been isolated from the mitochondria of *Neurospora* (Kuntzel and Noll, 1967; Rifkin, Wood and Luck, 1967) and *X. laevis* (Swanson and Dawid, 1970). These ribosomes are different from the cytoplasmic ribosomes of the same organism. In *Neurospora* the sedimentation coefficients of mitochondrial and cytoplasmic ribosomes are rather similar, whereas in *X. laevis* the mitochondrial ribosome is smaller. Both mitochondrial ribosomes so far described are clearly distinct from bacterial ribosomes, and the 2 major RNA molecules they contain are distinct from bacterial ribosomal RNA.

Mitochondrial ribosomal RNA of *X. laevis* has been studied in some detail (Dawid, 1968b, 1970 and unpublished). The two RNA molecules display somewhat unusual physical properties as do the mitochondrial RNAs of *Aspergillus niger* (Edelman, Verma and Littauer, 1970). Relative to *Escherichia coli* ribosomal RNA markers the mitochondrial RNAs of *X. laevis* sediment at 17S and 13S, but in polyacrylamide gel electrophoresis they travel as expected for 21S and 13S RNA. Since these molecules were first detected by gel electrophoresis and are best studied by this technique, and since the 17S position is preempted by the smaller cytoplasmic ribosomal RNA, we shall refer to the mitochondrial components as "21S" and 13S RNA. While their molecular weights are not exactly known they are clearly smaller than either bacterial or cytoplasmic ribosomal RNAs. They hybridize well with mitochondrial DNA and show no detectable sequence homology with cytoplasmic ribosomal RNA. It appears likely that each circle of mitochondrial DNA carries one gene each for "21S" and 13S RNA. Close to 15 percent of the genetic information available in mitochondrial DNA is used to code for these two ribosomal RNAs.

In addition to unique ribosomal RNAs mitochondria contain unique transfer RNAs as well (Barnett and Brown, 1967; Epler, 1969; Buck and Nass, 1969). At least some of these RNA molecules hybridize with mitochondrial DNA and have no sequence homologies with cytoplasmic transfer RNA (Nass and Buck, 1969). However, it appears that mitochondrial DNA codes only for an incomplete set of perhaps 10 to 15 tRNA molecules; whether additional tRNAs are imported is not known at present. Approximately 6% of the information in mitochondrial DNA is used in coding for these tRNA molecules (Dawid, 1970).

Thus, about one fourth of the information in mitochondrial DNA is known to code for the RNA components of a special protein synthesizing system. What about the rest? While minor mitochondrial RNA species have been detected, nothing is known about their nature or function. The remaining sequences in mitochondrial DNA could code for mitochondrial proteins. Indeed, it has often been suggested that the function of mitochondrial DNA is to specify the structure of certain membrane proteins. However, the evidence on this question is rather fragmentary and largely based on the implied assumption that proteins synthesized inside the mitochondria should be coded for by mitochondrial DNA. This is not necessarily so; one can envision mechanisms in which messengers of nuclear origin enter the mitochondria and there direct the synthesis of proteins on mitochondrial ribosomes. This possibility appears quite attractive in view of the paucity of sequence information in animal mitochondrial DNA and the relatively large number of special mitochondrial proteins which cannot all be coded for by this small DNA molecule. If the mitochondria import any messenger RNA molecules at all they might perhaps import all messenger molecules used in mitochondrial protein synthesis. I therefore proposed the hypothesis (Dawid, 1970) that animal mitochondrial DNA codes only for the mitochondrial protein synthesizing machinery. We already know about mitochondrial ribosomal and transfer RNAs. If the mitochondrial ribosomal proteins were also coded for by mitochondrial DNA this would exhaust about all the available information. Alternately, the mitochondrial ribosomal proteins might be made under nuclear control, leaving only the already described mitochondrial RNA molecules as products of mitochondrial DNA. The remaining mitochondrial sequences would then be silent, i.e., "spacer DNA," in analogy to the situation found in the nuclear genes for cytoplasmic ribosomal RNA (Dawid, Brown, and Reeder, 1970).

IV. MITOCHONDRIAL GENETICS: THE CONTINUITY OF THE MITOCHONDRIAL GENOME

This section reviews evidence that some of the information utilized in the biogenesis of mitochondria is inherited cytoplasmically. The most extensive work on this subject has been done in yeast and *Neurospora*. In these organisms mutant strains with impaired respiratory capacity are known in which the mutation is inherited cytoplasmically (see Roodyn and Wilkie, 1968). The unit carrying the cytoplasmic determinant has been identified as the mitochondrion itself, since in "petite" mutants of yeast the mitochondrial DNA is physically altered (Monoulou,

Jacob and Slonimski, 1966; Bernardi, Carnevali, Nicolaieff, Piperno and Tecce, 1968). The usefulness of mutations leading to respiratory deficiency is limited in that no detailed genetic analysis can be carried out with them. Therefore, the discovery of cytoplasmically inherited antibiotic resistance mutants in yeast is of great importance (Thomas and Wilkie, 1968; Coen, Deutsch, Netter, Petrochilo and Slonimski, 1970). In these mutants the protein synthesizing machinery of the mitochondria is altered in various ways, thereby providing genetic evidence that parts of this machinery are produced under the control of mitochondrial DNA. Furthermore, these mutants allow the elaboration of detailed mitochondrial genetics.

The genetic evidence in the fungi shows that some of the information used for the biogenesis of mitochondria is contained in these particles. In animals, no genetic evidence is available on this point. However, the biochemical work summarized above demonstrates that in animals as in fungi mitochondrial DNA codes for mitochondrial ribosomal and transfer RNAs which almost certainly play a part in the biogenesis of mitochondria. Against this background one may ask whether mitochondrial DNA is continuous during the life cycle of the organism and during evolution, and whether it is inherited as an independent cytoplasmic factor in all cases. In fungi this question has been answered some time ago. The existence of cytoplasmically inherited mutations shows that no nuclear "master copy" of the mitochondrial genome exists, at least not in a form that could be put to use - and that is operationally equivalent to non-existence. Cytoplasmic continuity of mitochondrial DNA has been demonstrated explicitly by Reich and Luck (1966) who showed that it was inherited maternally in an interspecific cross in *Neurospora.*

The absence of known mitochondrial genetic markers in animals makes it impossible to follow the inheritance of mitochondrial genes by standard genetic methods. However, it has become possible to investigate the cytoplasmic continuity of mitochondrial DNA in an interspecific cross in amphibians. If we make the reasonable assumption that looking at mitochondrial DNA is equivalent to looking at mitochondrial genes we are now in a position to do, in a limited way, mitochondrial genetics in these animals. In brief, the experiment goes as follows. Two related species of frogs, *X. laevis* and *Xenopus mulleri,* can be induced to mate and produce viable offspring (Blackler, unpublished). We then asked whether the mitochondrial DNAs of the two *Xenopus* species could be distinguished, and an answer to this question was obtained by molecular hybridization experiments. RNA was synthesized on both DNAs *in vitro* using *E. coli* RNA polymerase, and the complementary RNAs (cRNAs) were hybridized to both DNAs. Under the conditions of the experiment each cRNA hybridized five to tenfold better with the template from which it had been transcribed than with the mitochondrial DNA of the other species. This very different hybridization efficiency gives us a simple and sensitive way to distinguish the mitochondrial DNAs of *X. laevis* and *X. mulleri.* It was thus easy to determine the nature of the mitochondrial DNAs in several crosses between the two species. The result was unambiguous: in each case the hybrid frog carried the mitochondrial DNA of the female parent. This experiment then shows that, at least in *Xenopus,* the egg's mitochondria populate the soma of the offspring. It supports the idea, now universally held, that mitochondria multiply by division and that mitochondrial DNA replicates inside these organelles rather than being provided to them by the nucleus. It also shows that the embryo's mitochondria are derived from the egg rather than the sperm; this conclusion is expected because of the very large number of mitochondria in the egg and because of cytological observations showing, in some animals, that the sperm mitochondria do not enter the egg. The experiments on *X. laevis - X. mulleris* crosses do not, so far, prove cytoplasmic continuity for mitochondrial DNA. To do this the oocytes of the hybrid frogs, will be examined.

V. REFERENCES

Ashwell, M. and Work, W. S. (1970) The biogenesis of mitochondria. *Ann. Rev. Biochem.* **39**, 251.

Baltus, E. and Brachet, J. (1962) Le dosage de l'acide desoxiribonucleique dans les oeufs de bectraciens. *Biochem. Biophys. Acta.* **61**, 157.

Baltus, E., Hanocq-Quertier, J. and Brachet, J. (1968) Isolation of deoxyribonucleic acid from yolk platelet of *Xenopus laevis* oocyte. *Proc. Nat. Acad. Sci. U.S.A.* **61**, 469.

Barnett, W. E. and Brown, D. H. (1967) Mitochondrial transfer ribonucleic acids. *Proc. Nat. Acad. Sci. U.S.A.* **57**, 452.

Bell, P. R. and Muhlethaler, K. (1964) Evidence for the presence of deoxyribonucleic acid in the organelles of the egg cells of *Pteridium aquilinum. J. Molec. Biol.* **8**, 853.

Berger, E. R. (1968) A quantitative study of sea urchin egg mitochondria in relation to their DNA content. *J. Cell Biol.* **39**, 12a.

Bernardi, G., Carnevali, F., Nicolaieff, A., Piperno, G. and Tecce, G. (1968) Separation and characterization of a satellite DNA from a yeast cytoplasmic "petite" mutant. *J. Molec. Biol.* **37**, 493.

Boell, E. J. and Weber, R. (1955) Cytochrome oxidase activity in mitochondria during amphibian development. *Exp. Cell Res.* **9**, 559.

Borst, P. and Kroon, A. M. (1969) Mitochondrial DNA: Physicochemical properties, replication, and genetic function. *Int. Rev. Cytol.* **26**, 107.

Brachet, J. and Quertier, J. (1963) Cytochemical detection of cytoplasmic DNA in amphibian ovocytes. *Exp. Cell Res.* **32**, 410.

Brachet, J. and Ficq, A. (1965) Binding sites of 14 C-actinomycine in amphibian ovocytes and an autoradiography technique for the detection of cytoplasmic DNA. *Exp. Cell Res.* **38**, 153.

Brown, D. D. and Littna, E. (1964a) RNA synthesis during the development of *Xenopus laevis,* the South African Clawed Toad. *J. Molec. Biol.* **8**, 669.

Brown, D. D. and Littna, E. (1964b) Variations in the synthesis of stable RNA's during oogenesis and development of *Xenopus laevis. J. Molec. Biol.* **8**, 688.

Brown, D. D. (1965) RNA synthesis during early development. *In* "Developmental and Metabolic Control Mechanisms and Neoplasia." p. 219. Baltimore, Williams & Wilkins Co.

Brown, D. D. (1967) The genes for ribosomal RNA and their transcription during amphibian development. *In* "Current Topics in Developmental Biology." (A. A. Moscona and A. Monroy, eds.), p. 47. New York, Academic Press.

Brown, D. D. and Dawid, I. B. (1968) Specific gene amplification in oocytes. *Science.* **160**, 272.

Buck, C. A. and Nass, M. M. K. (1969) Studies on mitochondrial RNA from animal cells. I. *J. Molec. Biol.* **41**, 67.

Chase, J. W. (1970) Studies on mitochondrial RNA in frog eggs and embryos. *Carnegie Inst. Year Book 68.* 517.

Coen, D., Deutsch, J., Netter, P., Petrochilo, E. and Slonimski, P. P. (1970) Mitochondrial genetics I. methodology and phenomenology. *In* "The Control of Organelle Development, Symposium 24 of the Soc. for Exp. Biology." (P. L. Miller, ed.), p. 449. Cambridge University Press.

Dawid, I. B. (1965) Deoxyribonucleic acid in amphibian eggs. *J. Molec. Biol.* **12**, 581.

Dawid, I. B. (1966) Evidence for the mitochondrian origin of frog egg cytoplasmic DNA. *Proc. Nat. Acad. Sci. U.S.A.* **56**, 269.

Dawid, I. B. (1968a) The cytoplasmic genetic material. *J. Anim. Sci.* **27**, Suppl. I, 61.

Dawid, I. B. (1968b) RNA in frog oocyte mitochondria. *Carnegie Inst. Year Book 67.* 418.

Dawid, I. B. (1970) The nature of mitochondrial RNA in oocytes of *Xenopus laevis* and its relation to mitochondrial DNA. *In* "The Control of Organelle Development, Symposium 24 of the Soc. for Exp. Biology." (P. L. Miller, ed.), p. 227. Cambridge University Press.

Dawid, I. B. and Brown, D. D. (1970) The mitochondrial and ribosomal DNA components of oocytes of *Urechis caupo. Develop. Biol.* **22**, 1.

Dawid, I. B., Brown, D. D. and Reeder, R. H. (1970) Composition and structure of chromosomal and amplified ribosomal DNA's of *Xenopus laevis. J. Molec. Biol.* **51**, 341.

Edelman, M., Verma, I. M., and Littauer, U. Z. (1970) Mitochondrial ribosomal RNA from *Aspergillus nidulans:* characterization of a novel molecular species. *J. Molec. Biol.* **49**, 67.

Epler, J. L. (1969) The mitochondrial and cytoplasmic transfer ribonucleic acids of *Neurospora crassa. Biochemistry.* **8**, 2285.

Hanocq-Quertier, J., Baltus, E., Ficq, A. and Brachet, J. (1968) Studies on the DNA of *Xenopus laevis* oocytes. *J. Embryol. Exp. Morph.* **19**, 273.

Hawley, E. S. and Greenawalt, J. W. (1970) An assessment of *in vivo* mitochondrial prote in synthesis in *Neurospora crassa. J. Biol. Chem.* **245**, 3574.

Hollenberg, C. P., Borst, P., Thuring, R. W. J. and Van Bruggen, E. F. J. (1969) Size, structure and genetic complexity of yeast mitochondrial DNA. *Biochem. Biophys. Acta.* **186**, 417.

Kirby, K. S. and Cook, E. A. (1967) Isolation of deoxyribonucleic acid from mammalian tissues. *Biochem. J.* **104**, 254.

Kuntzel, H. and Noll, H. (1967) Mitochondrial and cytoplasmic polysomes from *Neurospora crassa. Nature (London).* **215**, 1340.

Mounolou, J. C., Jacob, H. and Slonimski, P. P. (1966) Mitochondrial DNA from yeast "petite" mutants: specific changes of buoyant density corresponding to different cytoplasmic mutations. *Biochem. Biophys. Res. Commun.* **24**, 218.

Nass, M. M. K. and Buck, C. A. (1969) Comparative hybridization of mitochondrial and cytoplasmic amnioacyl transfer RNA with mitochondrial DNA from rat liver. *Proc. Nat. Acad. Sci. U.S.A.* **62**, 506.

Piko, L., Tyler, A. and Vinograd, J. (1967) Amount, location, priming capacity, circularity and other properties of cytoplasmic DNA in sea urchin eggs. *Biol. Bull.* **132**, 68.

Piko, L., Blair, D. G., Tyler, A. and Vinograd, J. (1968) Cytoplasmic DNA in the unfertilized sea urchin egg: physical properties of circular mitochondrial DNA and the occurrence of catenated forms. *Proc. Nat. Acad. Sci. U.S.A.* **59**, 838.

Reich, E. and Luck, D. J. L. (1966) Replication and inheritance of mitochondrial DNA. *Proc. Nat. Acad. Sci. U.S.A.* **55**, 1600.

Rifkin, M. R., Wood, D. D. and Luck, D. J. L. (1967) Ribosomal RNA and ribosomes from mitochondria of *Neurospora crassa. Proc. Nat. Acad. Sci. U.S.A.* **58**, 1025.

Roodyn, D. B. and Wilkie, D. (1968) The Biogenesis of Mitochondria. London, Methuen & Co.

Schwartz, M. (1969) Nucleic acid metabolism in eggs and embryos of the Echiuroid *Urechis caupo.* Ph.D. thesis, The Johns Hopkins University.

Swanson, R. F. and Dawid, I. B. (1970) The mitochondrial ribosome of *Xenopus laevis. Proc. Nat. Acad. Sci. U.S.A.* **66**, 117.

Thomas, D. Y. and Wilkie, D. (1968) Inhibition of mitochondrial synthesis in yeast by erythromycin: cytoplasmic and nuclear factors controlling resistance. *Genet. Res.* **11**, 33.

Vendrely, R. and Vendrely, C. (1948) La teneur du noyau cellulaire en acide desoxyribonucleique a travers les organs, les individus, et les especes animales. *Experientia.* **4**, 434.

Wallace, R. A. and Jared, D. W. (1969) Studies on amphibian yolk. VIII. The estrogen- induced hepatic synthesis of a serum lipoprotein and its selective uptake by the ovary and transformation into yolk platelet proteins in *Xenopus laevis. Develop. Biol.* **19**, 498.

Wolstenholme, D. R. and Dawid, I. B. (1968) A size difference between mitochondrial DNA molecules of urodele and anuran amphibia. *J. Cell Biol.* **39**, 222.

Wolstenholme, D. R., Koike, K. and Renger, H. C. (1970) Form and structure of mitochondrial DNA. *Proc. Xth Int. Cancer Congress, Houston, Texas.* (In press).

Wood, D. D. and Luck, D. J. L. (1969) Hybridization of mitochondrial ribosomal RNA. *J. Molec. Biol.* **41**, 211.

12

PROTEIN SYNTHESIS DURING OOCYTE MATURATION

L. Dennis Smith

I.   Introduction
II.   Induction of maturation
III.   Cytoplasmic regulation of protein syntheses during maturation
IV.   Localization of cytoplasmic proteins during maturation
V.   Discussion
VI.   Acknowledgments
VII.   References

I. INTRODUCTION

In amphibians (and most animals), the major part of oocyte growth occurs with the nucleus (germinal vesicle) in an extended first meiotic prophase. Oogenesis is characterized by the acquisition of many kinds of storage materials used later in development, and is generally considered to be a period of intense metabolic activity. Certainly, there is good evidence for the synthesis and sequestration of RNA, particularly during the lampbrush chromosome stage of oocyte growth. Most if not all of the ribosomal RNA needed for early development is synthesized at this time on multiple nucleoli in the germinal vesicle (Brown, 1967; Gall, 1969). Transfer RNA is also synthesized during oogenesis and conserved (Brown and Littna, 1966). Finally, significant amounts of "template active" RNA are synthesized and apparently conserved for later use (Davidson, 1968; Davidson and Hough, 1969). The developing oocyte also acquires large quantities of protein, but for technical reasons it has been difficult to distinguish between synthesis within the oocyte versus accumulation of exogenously synthesized material. In fact, there is good evidence that the bulk of oocyte protein (yolk) is synthesized in the liver and transplanted to the oocyte (Wallace and Jared, 1969). Thus, it remains to be determined just how active growing oocytes are in the synthesis of proteins.

There is general agreement that, by the end of the growth period, metabolic activity in the oocytes has been drastically curtailed if not shut off. These "dormant" oocytes can be maintained for weeks or months, depending on environmental conditions, awaiting the proper stimulus to continue development. The stimulus in amphibians (and vertebrates in general) is hormonal. Soon after the appropriate hormone stimulation, the large germinal vesicle breaks down, releasing all of its contents into the cytoplasm, and meiosis progresses to the second meiotic metaphase, before stopping again.

It has been recognized for years that the fertile union of gametes only takes place when both have attained a certain state of physiological maturity (see Wilson, 1925). This is not necessarily the same as the completion of the maturation (meiotic) divisions. In invertebrates, for example, eggs of the mollusk, *Spisula solidissima* represent one extreme in that they are shed with intact germinal vesicles and fertilization initiates completion of the meiotic divisions. At the other extreme, sea urchin eggs have completed both meiotic divisions prior to shedding and fertilization. In amphibians, eggs are first physiologically mature (ripe) when they reach the second meiotic metaphase, and fertilization or parthenogenetic activation results in the completion of the meiotic divisions. While strictly speaking it may not be justified, the term maturation in reference to studies on amphibians has been used to refer to the attainment of physiological maturity.

The period of maturation is fundamental to further development, not only because it includes the terminal events of meiosis but also because of the dispersal of morphogenetic substances localized in the germinal vesicle (see Wilson, 1925; Briggs and King, 1959; Briggs and Justus, 1968; Smith and Ecker, 1969a). Beyond this, however, there has been the notion that developmental information laid down during oogenesis is activated only after fertilization or artificial activation. This has derived from studies on invertebrates, primarily sea urchins, which showed that incorporation of labeled amino acids into protein is either greatly increased in rate or switched on at fertilization (reviews by Gross, 1967; Monroy and Tyler, 1967). A similar observation had been reported in amphibian eggs (Brachet, 1964).

Several years ago, R. E. Ecker and I began studies on protein synthesis in fertilized or artificially activated *Rana pipiens* eggs (Smith and Ecker, 1965). By studying amino acid pool kinetics, we were able to measure in a precise way the rates of protein synthesis in these eggs (Ecker and Smith, 1966). However, kinetics of protein synthesis in eggs artificially activated by an injection of labeled amino acid were not consistent with the hypothesis that artificial activation "switched on" protein synthesis, and we subsequently showed that such synthesis occurs during the period of hormone-induced maturation (Smith, Ecker and Subtelny, 1966). The present paper summarizes experiments on this protein synthesis.

II. INDUCTION OF MATURATION

Before discussing experimental data on protein synthesis in maturation, it is necessary to describe briefly the methods used to induce maturation in *Rana pipiens* oocytes. Maturation, along with ovulation, is experimentally induced *in vivo* by injecting a mature female with a homoplastic pituitary suspension. The germinal vesicle begins to undergo dissolution while oocytes still are within the ovary, and ovulated eggs continue meiosis to the second meiotic metaphase during passage through the body cavity, down the oviducts, and into the uteri. By the time they are available in reasonable quantities, uterine eggs are almost always in the second meiotic metaphase. Since this represents the first point at which eggs are capable of being fertilized, or artificially activated, uterine eggs are usually mature eggs. Pituitary homogenates can also induce maturation and ovulation *in vitro* in ovarian fragments containing small numbers of oocytes. In this case, however, certain features of the system make it difficult to perform experimental manipulations on oocytes prior to germinal vesicle dissolution. The ovarian fragments must be maintained in the continuous presence of pituitary suspensions until just prior to germinal vesicle breakdown, or maturation will not occur (Dettlaff, Nikitina and Stroeva, 1964). Likewise, dissection of oocytes from their ovarian follicles during this hormone dependent period considerably reduces the percentage of maturing oocytes, even when they are still maintained in the presence of pituitary hormones (Masui, 1967; Schuetz, 1967; Smith, Ecker and Subtelny, 1968).

Recently, we (Smith *et al.*, 1968), along with others (Masui, 1967; Schuetz, 1967), demonstrated that maturation can be induced *in vitro* in oocytes dissected from their ovarian follicles prior to hormone exposure. This system requires only that the oocytes be exposed briefly to steroid hormones such as progesterone, followed by incubation in a suitable culture medium such as amphibian Ringer's solution. Oocytes exposed to progesterone for only a few minutes undergo maturation on essentially the same time scale as similar oocytes induced to mature *in vivo*. In addition, these eggs are capable of perfectly normal cleavage and subsequent development (Smith *et al.*, 1968; Smith and Ecker, 1969a). Thus, the system in every way duplicates the *in vivo* one. In fact, studies on the mechanism of steroid action in the induction of maturation suggest that this is the normal *in vivo* process; pituitary hormones acting indirectly on oocytes through the mediation of follicular tissue (Smith and Ecker, 1969a; 1970c; unpublished). Finally, various kinds of experimental manipulations on full grown oocytes prior to hormone stimulation *in vitro* do not interfere subsequently with the induction of maturation. These results have made possible studies on the role of the germinal vesicle in the maturational processes (Smith and Ecker, 1969a, b; 1970a, c).

III. CYTOPLASMIC REGULATION OF PROTEIN SYNTHESIS DURING MATURATION

The mature amphibian egg is impermeable to almost all exogenous materials, and this restriction extends to much of the period of maturation. In such a closed system, the synthesis of new products necessarily must occur at the expense of endogenous storage materials. The amino acid pool in *Rana pipiens* eggs appears to maintain a steady state, being fed by the breakdown of storage proteins and depleted at the same rate by polypeptide assembly. Thus, measurements of protein synthesis, utilizing radioactive precursors, require a means of introducing labeled amino acids into the endogenous pools. This can be done most effectively by injecting them directly into individual eggs. When a small accurately measured volume of radioactive amino acid is injected into the egg, the endogenous pool becomes labeled almost instantaneously, and the amino acid dilutes out of the pool and is incorporated into new proteins with very predictable kinetics. Incorporation kinetics of this kind, coupled with knowledge of the pool size of the particular amino acid being utilized, allows calculation of the protein synthetic rate (Ecker and Smith, 1966; Ecker and Smith, 1968).

Rates of protein synthesis have been measured in oocytes at the full grown ovarian oocyte stage through the period of hormone-induced maturation, and for several hours beyond the time of fertilization or artificial activation. In addition, protein synthesis has been measured in enucleated oocytes during these times. The experimental results consistently showed the same pattern, regardless of whether or not the germinal vesicle was removed. Protein synthesis occurs at a very low level in full grown oocytes not exposed to hormones, and this level does not increase appreciably in the absence of hormone treatment. After the induction of maturation, the rate of protein synthesis remains low for several hours and then, rather abruptly, undergoes a rapid increase. The exact time after hormonal stimulation at which this increase occurs may vary, depending on such factors as temperature and season of the year. In general, however, the rate of protein synthesis reaches a maximum level within the time required for nucleated oocytes to extrude the first polar body, and this level is maintained for about a day beyond the time required for nucleated oocytes to reach the second meiotic metaphase. Our best estimate of the actual rate of synthesis, once this maximal level is attained, is about 0.5 μg/hour per oocyte, and there is no indication that it is not constant thereafter. Thus, the hormone-stimulated rate increase is brought about many hours before artificial activation or fertilization are even possible and, in fact, these events as such have no discernible effect. The results also show clearly that neither the presence of the germinal vesicle nor its dispersed contents have any quantitative effect on protein synthesis (Ecker, Smith and Subtelny, 1968; Smith and Ecker, 1969a, 1970c).

Comparisons of the kinds of proteins synthesized in the presence and absence of the germinal vesicle can be made by electrophoretic analysis of the newly synthesized (radioactive) proteins. To be meaningful, however, it is desirable to establish conditions in which most, if not all, of the newly synthesized proteins can be examined. This has been possible by incorporating detergents such as sodium dodecyl sulfate (SDS) into our homogenization and electrophoretic procedures. When oocytes are injected with [3H] leucine and, after a sufficient incorporation time, are homogenized in tris buffer containing SDS, essentially all of the radioactivity can be accounted for in the initial supernatant fraction. Aliquots of this are then subjected to electrophoresis on SDS polyacrylamide gels, and radioactivity in the separated proteins is determined in transverse slices of the supporting medium (Smith and Ecker, 1970c; Ecker and Smith, 1971).

Examination of the patterns of radioactivity from these gels reveals significant differences during the period of maturation. An example of these differences is shown in Fig. 12.1, which compares proteins synthesized at the first meiotic metaphase and at the second meiotic metaphase (A1 and A2 respectively). Also

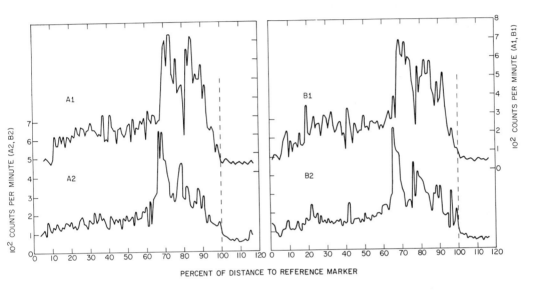

PERCENT OF DISTANCE TO REFERENCE MARKER

Fig. 12.1 Comparison of electrophoretic patterns of proteins synthesized by nucleated and enucleated oocytes during maturation. Oocytes were injected in groups of 20 with 32 nl of 1-leucine-4,5-^{3}H (1 mCi/ml) each and allowed to incorporate for three hours. Each group was homogenized in 0.05 M tris-Cl buffer (pH 7.1) containing 1% sodium dodecyl sulfate (SDS). After centrifugation, the extract was dialyzed overnight against 0.01 M buffer with 0.1% SDS and electrophoresced in 4.5% polyacrylamide gels (0.5 x 10 cm) made up in 0.05M buffer containing 0.1% SDS. Each pattern represents counts from 0.8 mm slices of a single gel. The patterns are as follows: A1, A2. Nucleated oocytes injected 24 and 48 hours respectively after progesterone exposure *in vitro*. B1, B2. Enucleated (germinal vesicle removed) oocytes injected at the same times respectively (see Smith and Ecker, 1970c; Ecker and Smith, 1971 for further details).

shown in the figure are patterns obtained from enucleated oocytes, exposed to progesterone at the same time as the nucleated controls and injected with [^{3}H] leucine at the same times (B1 and B2). These patterns are not significantly different. Whether or not the changing patterns represent the appearance of new protein species or simply altered rates of synthesis of old ones cannot be determined from these experiments. The results do show, however, that the germinal vesicle exerts no obvious control over either the amounts or kinds of protein synthesized during hormone-induced maturation.

IV. LOCALIZATION OF CYTOPLASMIC PROTEINS SYNTHESIZED
DURING MATURATION

Recent autoradiographic evidence provided by Arms (1968) and Merriam (1969) has shown that proteins, synthesized both during cleavage and during gonadotropin-induced maturation in *Xenopus,* can enter transplanted nuclei. In both cases, this phenomenon was related to a specific cytoplasmic effect on the transplanted nuclei, namely the induction of DNA synthesis. We have presented several experiments which show that nuclear localization is a usual fate of many of the proteins synthesized during maturation and early cleavage (Smith and Ecker, 1970a,c; Ecker and Smith, 1971).

When non-stimulated oocytes are injected with [^{3}H] leucine and fixed at progressively later times for autoradiographic analysis, grains can be found over both the cytoplasm and the germinal vesicle. The latter appear to concentrate against a gradient, however, so that by 20 hours after the initial injection, the relative nuclear concentration is about 6 times greater than that in the cytoplasm (Smith and Ecker, 1970c; Ecker and Smith, 1971). Similar oocytes induced to undergo maturation prior to the injection of [^{3}H] leucine show exactly the same overall pattern (increasing nuclear concentration) up to the point of germinal vesicle dissolution. As already pointed out, removal of the germinal vesicle during these times has no effect on protein synthesis. This, coupled with the pattern of nuclear accumulation, leads to the conclusion that most if not all of these nuclear grains are due to the acquisition of proteins synthesized in the cytoplasm. It is unlikely these "nuclear" proteins are involved solely in maturational events, such as dissolution of the nuclear membrane, since they concentrate equally well in the absence of hormone stimulation. Thus, they must be involved in other events (see Smith and Ecker, 1970c; Gurdon, 1968).

Once the germinal vesicle breaks down, it is impossible to discriminate autoradiographically between the labeled "nuclear" proteins and other radioactive cytoplasmic proteins. However, it is possible to follow the subsequent distribution of total radioactivity, particularly during early development. Normal development can be obtained in oocytes induced to mature *in vitro* either by transferring maturing oocytes to foster females and utilizing conventional insemination techniques (see Smith and Ecker, 1969), or by performing nuclear transfers in mature eggs. The results of all such experiments were essentially the same. Regardless of when the radioactive amino acid was injected during maturation, the nuclei in embryos fixed at the blastula stage contained a heavy concentration of grains (Fig. 12.2). While there was considerable variability, grain counts showed that these nuclei contained as much as 40 per cent of the total cellular radioactivity, at a relative concentration as high as 16 times that in the cytoplasm (Smith and Ecker, 1970c; Ecker and Smith, 1971). Essentially, the same results were obtained when eggs, induced to mature *in vivo,* were stripped from an ovulating female and injected with radioactive amino acid after insemination (Smith and Ecker, 1970a,c; Ecker and Smith, 1971).

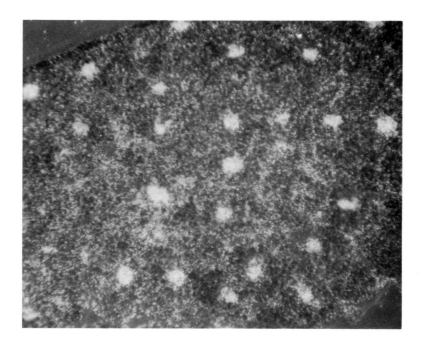

Fig. 12.2 Autoradiograph of a section from a blastula, obtained by transplanting a blastula nucleus into a mature egg which previously had been injected with [^{3}H] leucine at the first meiotic metaphase. The section was stained with hematoxylin-fast green. Photomicrography was performed with incident light using a Leitz ortholux microscope equipped with an ultropak incident light illuminator. Magnification approximately 260X.

Either the nuclear grains observed at the blastula stage resulted totally or in part from the transfer of proteins synthesized in the cytoplasm, or they resulted from intranuclear protein synthesis (see Allfrey, Littau, and Mirsky, 1964; Gallwitz and Mueller, 1969). With the possible exception of the fertilized eggs, there is no doubt that the injected isotope originally was incorporated exclusively into proteins synthesized in the cytoplasm. There is no nucleus, as such, from the time the germinal vesicle breaks down until the time the zygote nucleus forms after fertilization (or until nuclear transfers are performed), a period of at least 20 hours at 18°C. The injected amino acid all would have been incorporated into proteins many hours prior to this (Ecker and Smith, 1966, 1968). Thus, any intranuclear protein synthesis would have to be the result of protein turnover.

It appears extremely unlikely that proteins appear in cleavage nuclei by intranuclear protein synthesis. We have looked repeatedly for turnover, using a variety of techniques, but have never found evidence that it occurs (Ecker and Smith, 1966, 1968; Smith and Ecker, 1970c). In addition, when nuclei are transferred into eggs which already contain labeled proteins but which lack the capacity for additional synthesis because of the presence of puromycin, the nuclei still acquire significant radioactivity (Arms, 1968; Merriam, 1969; Smith and Ecker, 1970a,c; Ecker and Smith, 1971). The only possible source of this nuclear label would have been through accumulation of proteins previously synthesized in the cytoplasm.

V. DISCUSSION

Oocytes from which germinal vesicles were removed prior to hormone exposure still retain the capacity to be artificially activated after subsequent hormone exposure. With the addition of injected germinal vesicle material, these oocytes can also undergo normal cleavage divisions (Smith and Ecker, 1969a,b, 1970c). These observations, together with those just discussed, show that both the morphological and biochemical aspects of maturation are under cytoplasmic control. Several additional lines of evidence suggest that this lack of nuclear control persists through the blastula stage of development. Cleavage can occur in the presence of highly abnormal chromosomal complements (Hennen, 1963; Briggs, Signoret, and Humphrey, 1964), in the absence of functional chromosomes (Briggs, Green and King, 1951), and in chemically enucleated (actinomycin D treated) eggs (Brachet, Denis and de Vitry, 1964). Protein synthesis, known to be essential to cleavage (Brachet *et al.,* 1964; Smith and Ecker, 1970a), is unaffected during this time by treatment of fertilized eggs with actinomycin D or by physical removal of the maternal chromosomes (Smith and Ecker, 1965, 1970a). This evidence all supports the conclusion that the full grown oocyte already contains all the necessary information for maturation as well as early development. Some of this information may exist in the form of stable proteins synthesized during oogenesis (see Briggs and Justus, 1968). We also cannot discount the possibility that cytoplasmic DNA contributes in some way to these events (Dawid, 1966; Baltus, Hanocq-Quertier and Brachet, 1968; Tyler, 1967). However, it seems very likely that most of the stored information exists in the form of stable RNA molecules synthesized during oogenesis (Spirin, 1966; Tyler, 1967; Davidson, 1968; Davidson and Hough, 1969; Crippa and Gross, 1969). Thus, the full grown oocyte would contain a heterogenous population of stable (maternal) templates which code not only for the proteins required during maturation but also for the proteins important to early development.

One of the questions referred to in the Introduction concerns the time at which this stored information is utilized. Does the hormonal induction of maturation start a "program" which includes the so-called "developmental proteins," or are the proteins synthesized during maturation only those required for the immediate events of maturation?

When oocytes are induced to mature *in vitro* and are maintained *in vitro,* they undergo morphogenetic movements which resemble normal gastrulation (Smith and Ecker, 1970b). This process, pseudogastrulation, occurs uniformly in all oocytes and is observed to begin about 24 hours after the second meiotic metaphase is reached.

This length of time corresponds approximately with the period that mature oocytes exhibit elevated rates of protein synthesis when maintained *in vitro*. Moreover, had fertilization occurred just at the time the second meiotic metaphase was reached (the first point at which it is possible), development would have progressed to the early gastrula stage within about 24 hours. While all of these correlations may be simply coincidental, they support the hypothesis that the induction of maturation *in vitro* starts a pattern of protein synthesis which includes proteins associated with early development, and which runs continuously until the stored information has been utilized. From this point of view, fertilization or artificial activation are points superimposed on a continuous biosynthetic process initiated many hours earlier. Development (cleavage), then, would have to begin within a reasonable period of time after oocytes reached the second meiotic metaphase or the synthetic pattern would be expected to progress beyond a point compatible with normal development. This view is consistent with observations (Smith and Ecker, unpublished) which show that normal development in response to transplanted blastula nuclei is obtained only if the transfers are performed within a few hours after attainment of the second meiotic metaphase.

When maturation is induced *in vivo,* however, mature eggs may remain stored in the uteri for several days before they become overripe and undergo cytolysis (see Witschi, 1952). Fertilization and normal development can be obtained at any time during this period. In addition, when mature eggs are stripped from the uteri of ovulated females, they exhibit elevated rates of protein synthesis for about a week; rates comparable to the maximal rate observed in oocytes maintained *in vitro* (Smith and Ecker, 1969b, 1970a, b, c). Since both types of oocytes presumably possess the same quantity of stored information, either the decreased protein synthetic rate which becomes manifest *in vitro* is unrelated to the exhaustion of stored information, or protein synthesis in oocytes maintained *in vivo* is not proceeding uninterrupted at the maximal rate.

Actually, it is impossible to measure protein synthetic rates *in vivo,* using the techniques described in this paper. The best that can be done is to remove oocytes from hormone injected females at various times and measure their rates of protein synthesis *in vitro*. One can then extrapolate to the probable situation *in vivo*. It appears likely that protein synthesis *in vivo* follows the same pattern already described up to about the time ovulated oocytes enter the uteri. Oocytes usually enter the uterus at about the time the second meiotic metaphase is reached. After oocytes enter the uterus, it appears reasonable to suggest that protein synthetic rate is "slowed down" to some level consistent with storage for several days (Smith and Ecker, 1970b). Evidence for a slowdown of protein synthesis is based on the observation that substantial reduction in the level of protein synthesis under anaerobic conditions does not alter subsequent capacity of mature oocytes to undergo artificial activation (Smith and Ecker, 1970b). The resumption of elevated rates of protein synthesis following activation would appear to require a second stimulus. This stimulus may be the environmental change, experienced by mature oocytes expelled from the uteri, and not fertilization or artificial activation as such (Smith and Ecker, 1970b).

Acceptance of this hypothesis does not alter the conclusion that protein synthesis during maturation occurs on cytoplasmic templates, presumably laid down during oogenesis. It is also valid to conclude that a certain level of protein synthesis, stimulated by hormones *in vivo* or *in vitro,* is indispensable to the immediate events of maturation (Smith and Ecker, 1969a,b). However, the possibility exists that much of the protein synthesis observed during maturation *in vitro* is out of sequence compared to its usual time of synthesis *in vivo.* This suggestion poses obvious difficulties if one makes definitive statements concerning whether or not hormones also "activate" the synthesis of protein meaningful to development. On the other hand, if our primary interest concerns the nature of such proteins and the role they play in development, it makes little difference whether they are synthesized precociously *in vitro* as long as it is possible to determine when they are synthesized.

It has been recognized for years that regional differences in the egg cytoplasm specify certain types of cell differentiation later in development (review by Briggs and King, 1959). These cytoplasmic materials probably are synthesized and/or stored in the cytoplasm or germinal vesicle at some time during oogenesis. The nature of this nucleo-cytoplasmic interaction is poorly understood, but presumably involves a form of communication between the heterogeneous cytoplasm and the "identical" cleavage nuclei; a transfer of information which somehow changes nuclear activity (Gurdon, 1969). This communication need not be a one-way process, but the demonstration that cytoplasmic proteins enter the cleavage nuclei suggests a mechanism by which the cytoplasm could exert some control over nuclear function. In this sense, the cytoplasmic localizations could represent regional segregation either of the cytoplasmic templates or their specific protein products. This provides the hope that additional studies on the cytoplasmic proteins, synthesized from these templates during and after maturation, may provide insight into the nature of nucleo-cytoplasmic interactions so critical in early development.

VI. ACKNOWLEDGMENTS

The original research referred to in this paper was conducted at the Argonne National Laboratory, under the auspices of the U. S. Atomic Energy Commission, and at Purdue through support from the U. S. Public Health Service (HD 04229) and the National Science Foundation (GB 8741). The author holds a USPHS research career development award (HD 42549).

VII. REFERENCES

Allfrey, V. G., Littau, V. C. and Mirsky, A. E. (1964) Methods for the purification of thymus nuclei and their application to studies of nuclear protein synthesis. *J. Cell Biol.* **21**, 213.

Arms, K. (1968) Cytonucleoproteins in cleaving eggs of *Xenopus laevis. J. Embryol. Exp. Morph.* **20**, 267.

Baltus, E., Hanocq-Quertier, J. and Brachet, J. (1968) Isolation of deoxyribonucleic acid from the yolk platelets of *Xenopus laevis* oocyte. *Proc. Nat. Acad. Sci. U.S.A.* **61**, 469.

Brachet, J. (1964) Nouvelles observations sur le role des acides nucleiques dans la morphogenese. Symp. Acidi Nucleici e loro Funzione, p. 288, Milan, Fondazione Baselli.

Brachet, J., Denis, H. and de Vitry, F. (1964) Effects of actinomycin D and puromycin on morphogenesis in amphibian eggs and *Acetabularia mediterranea. Develop. Biol.* **9**, 398.

Brachet, J., Hanocq, F. and Van Gansen, P. (1970) A cytochemical and ultrastructural analysis of *in vitro* maturation in amphibian oocytes. *Develop. Biol.* **21**, 157.

Briggs, R. W., Green, E. and King, T. J. (1951) An investigation of the capacity for cleavage and differentiation in *Rana pipiens* eggs lacking "functional" chromosomes. *J. Exp. Zool.* **116**, 455.

Briggs, R. W. and King, T. J. (1959) Nucleocytoplasmic interactions in eggs and embryos. *In* "The Cell". (J. Brachet and A. E. Mirsky, eds.), Vol. 1, p. 537. New York, Academic Press.

Briggs, R. W., Signoret, J. and Humphrey, R. R. (1964) Transplantation of nuclei of various cell types from neurulae of the Mexican axolotl *(Ambystoma mexicanum). Develop. Biol.* **10**, 233.

Briggs, R. W. and Justus, J. T. (1968) Partial characterization of the component from normal eggs which corrects the maternal effect of gene o in the Mexican axolotl *(Ambystoma mexicanum). J. Exp. Zool.* **167**, 105.

Brown, D. D. and Littna, E. (1966) Synthesis and accumulation of low molecular weight RNA during embryogenesis of *Xenopus laevis. J. Molec. Biol.* **20**, 95.

Brown, D. D. (1967) The genes for ribosomal RNA and their transcription during amphibian development. *In* "Current Topics in Developmental Biology". (A. A. Moscona and A. Monroy, eds.), Vol. 2, p. 47. New York, Academic Press.

Crippa, M. and Gross, P. R. (1969) Maternal and embryonic contributions to the functional messenger RNA of early development. *Proc. Nat. Acad. Sci., U.S.A.* **62**, 120.

Davidson, E. H. (1968) Gene Activity in Early Development. New York, Academic Press.

Davidson, E. H. and Hough, B. R. (1969) High sequence diversity in the RNA synthesized at the lampbrush stage of oogenesis. *Proc. Nat. Acad. Sci., U.S.A.* **63**, 342.

Dawid, I. B. (1966) Evidence for the mitochondrial origin of frog egg cytoplasmic DNA. *Proc. Nat. Acad. Sci., U.S.A.* **56**, 269.

Dettlaff, T. A., Nikitina, L. A. and Stroeva, O. G. (1964) The role of the germinal vesicle in oocyte maturation in anurans as revealed by the removal and transplantation of nuclei. *J. Embryol. Exp. Morph.* **12**, 851.

Dettlaff, T. A. (1966) Action of actinomycin and puromycin upon frog oocyte maturation. *J. Embryol. Exp. Morph.* **16**, 183.

Ecker, R. E. and Smith, L. D. (1966) The kinetics of protein synthesis in early amphibian development. *Biochim. Biophys. Acta.* **129**, 186.

Ecker, R. E. and Smith, L. D. (1968) Protein synthesis in amphibian oocytes and early embryos. *Develop. Biol.* **18**, 232.

Ecker, R. E., Smith, L. D. and Subtelny, S. (1968) Kinetics of protein synthesis in enucleate frog oocytes. *Science.* **160**, 1115.

Ecker, R. E. and Smith, L. D. (1971) The nature and fate of *Rana pipiens* proteins synthesized during maturation and early cleavage. *Develop. Biol.* (In press).

Gall, J. G. (1969) The genes for ribosomal RNA during oogenesis. *Genetics.* **61**, 121.

Gallwitz, D. and Mueller, G. C. (1969) Protein synthesis in nuclei isolated from HeLa cells. *Europ. J. Biochem.* **9**, 431.

Gross, P. R. (1967) The control of protein synthesis in embryonic development and differentiation. *In* "Current Topics in Developmental Biology". (A. A. Moscona and A. Monroy, eds.), Vol. 2, p. 1. New York, Academic Press.

Gurdon, J. B. (1968) Changes in somatic cell nuclei inserted into growing and maturing amphibian oocytes. *J. Embryol. Exp. Morph.* **20**, 401.

Gurdon, J. B. (1969) Intracellular communication in early animal development. *Develop. Biol. Suppl.* **3**, 59.

Hennen, S. (1963) Chromosomal and embryological analysis of nuclear changes occurring in embryos derived from transfers of nuclei between *Rana pipiens* and *Rana sylvatica*. *Develop. Biol.* **6**, 133.

Masui, Y. (1967) Relative roles of the pituitary, follicle cells, and progesterone in the induction of oocyte maturation in *Rana pipiens. J. Exp. Zool.* **166**, 365.

Merriam, R. W. (1969) Movement of cytoplasmic proteins into nuclei induced to enlarge and initiate DNA or RNA synthesis. *J. Cell Sci.* **5**, 333.

Monroy, A. and Tyler, A. (1967) The activation of the egg. *In* "Fertilization". (C. B. Metz and A. Monroy, eds.), Vol. 1, p. 369. New York, Academic Press.

Schuetz, A. W. (1967) Action of hormones on germinal vesicle breakdown. *J. Exp. Zool.* **166**, 347.

Smith, L. D. and Ecker, R. E. (1965) Protein synthesis in enucleated eggs of *Rana pipiens. Science.* **150**, 777.

Smith, L. D., Ecker, R. E. and Subtelny, S. (1966) The initiation of protein synthesis in eggs of *Rana pipiens. Proc. Nat. Acad. Sci., U.S.A.* **56**, 1724.

Smith, L. D., Ecker, R. E. and Subtelny, S. (1968) *In vitro* induction of physiological maturation in *Rana pipiens* oocytes removed from ovarian follicles. *Develop. Biol.* **17**, 627.

Smith, L. D. and Ecker, R. E. (1969a) Role of the oocyte nucleus in physiological maturation in *Rana pipiens. Develop. Biol.* **19**, 281.

Smith, L. D. and Ecker, R. E. (1969b) Cytoplasmic regulation in early events of amphibian development. *Proc. Can. Cancer Res. Conf.* **8**, 103.

Smith, L. D. and Ecker, R. E. (1970a) Foundations for the expression of developmental potential. *In* "RNA in Development". (E. W. Hanley, ed.), p. 355. Salt Lake City, Utah, Univ. of Utah Press.

Smith, L. D. and Ecker, R. E. (1970b) Uterine suppression of biochemical and morphogenetic events in *Rana pipiens. Develop. Biol.* **22**, 622.

Smith, L. D. and Ecker, R. E. (1970c) Regulatory processes in the maturation and early cleavage of amphibian eggs. *In* "Current Topics in Developmental Biology". (A. A. Moscona and A. Monroy, eds.), Vol. 5, p. 1. New York, Academic Press.

Spirin, A. S. (1966) On "masked" forms of messenger RNA in early embryogenesis and in other differentiating systems. *In* "Current Topics in Developmental Biology". (A. A. Moscona and A. Monroy, eds.), Vol. 1, p. 1. New York, Academic Press.

Tyler, A. (1967) Masked messenger RNA and cytoplasmic DNA in relation to protein synthesis and processes of fertilization and determination in embryonic development. *Develop. Biol. Suppl.* **1**, 73.

Wallace, R. A. and Jared, D. W. (1969) Studies on amphibian yolk. VIII. The estrogen-induced hepatic synthesis of a serum lipophosphoprotein and its selective uptake by the ovary and transformation into yolk platelet proteins in *Xenopus laevis. Develop. Biol.* **19**, 498.

Wilson, E. B. (1925) The Cell in Development and Heredity. New York, Macmillan.

Witschi, E. (1952) Overripeness of the egg as a cause of twinning and teratogenesis: A review. *Cancer Res.* **12**, 763.

13

METABOLISM OF THE OOCYTE

J. D. Biggers

I.   Introduction
II.   Some nutritional requirements of the mouse oocyte and early embryo *in vitro*
        A.   General techniques
        B.   Oocyte
        C.   Fertilized ovum
        D.   Later stages
        E.   A working hypothesis
III.   Regulation of metabolism
        A.   Permeability changes
        B.   Changes in enzyme activity
IV.   Role of the follicle cells
V.   Conclusion
VI.   Acknowledgments
VII.   References

I. INTRODUCTION

A considerable body of evidence has accumulated which indicates that the fertilized ova of mice and rabbits have unusual metabolic pathways (see Biggers, 1971a; Biggers and Stern, 1972, for recent reviews). Since one function of the ovum is to provide the vehicle which ensures cellular continuity between generations it follows that the metabolic peculiarities of the fertilized ovum are determined during the differentiation of the oocyte. At the present time little is known of the metabolism of the mammalian oocyte. Consequently the main approach in this chapter is to examine the events which occur in early mammalian development, and then use this information to speculate on the changes which occur during mammalian oogenesis. This type of analysis is not new since it is frequently used by cytogeneticists when they speculate on the etiology of aberrant karyotypes.

II. SOME NUTRITIONAL REQUIREMENTS OF THE MOUSE OOCYTE AND EARLY EMBRYO *IN VITRO*

A. General Techniques

A major discovery in the study of mammalian development was made by Whitten (1956) who showed that 8-cell mouse embryos would develop into blastocysts in a simple chemically defined medium based on Krebs-Ringer bicarbonate. It was shown later that oocyte maturation, the first cleavage division, and the development of the 2-cell stage can also occur on similar chemically defined media. The media differed, however, in the substances necessary as energy sources.

Table 13.1 shows the basic medium devoid of substances capable of supplying energy. By the addition of different compounds it has been possible to identify the compounds necessary for the development *in vitro* of the oocyte, zygote, 2-cell and 8-cell stages.

Table 13.1 **A basic chemically defined medium used for the study of the nutrition of mouse oocytes and early embryos.**

The medium is a modified Krebs-Ringer bicarbonate and is supplemented with various energy sources (from Biggers, Whittingham and Donahue, 1967).

Component	Concentration (mM)
NaCl	119.4
KCl	4.78
$CaCl_2$	1.71
KH_2PO_4	1.19
$MgSO_4 \cdot 7H_2O$	1.19
$NaHCO_3$	25.07
Crystalline bovine serum albumin	1 mg/ml
Penicillin G (potassium)	100 U/ml
Streptomycin sulphate	50 μg/ml

B. Oocyte

An adult female mammal has a limited supply of oocytes which are formed during embryonic life. They eventually reach early prophase of the first meiotic reduction division (the diplotene or dictyate stage), where they become arrested. I

1935 Pincus and Enzmann showed that when such arrested rabbit oocytes are placed in culture they resume meiosis and mature to the metaphase II stage. These observations have been confirmed by many investigators in several species (see Biggers, 1971b, for a recent review). In 1967 Biggers, Whittingham and Donahue showed that the mouse oocyte would undergo maturation in the medium shown in Table 13.1 supplemented with either pyruvate or oxaloacetate, but not glucose, lactate or phosphoenolpyruvate (PEP). Subsequently Donahue (1968) and Talbert (1969) showed cytologically and kinetically, respectively, that maturation was essentially normal.

C. Fertilized Ovum

Prior to 1967 it was only possible to culture late 2-cell mouse embryos to blastocysts. Several investigators had noticed that fertilized ova would cleave to the 2-cell stage in a simple chemically defined medium containing lactate and pyruvate, and although the embryos remained alive they developed no further. Whittingham and Biggers (1967) then showed that the first cleavage division was normal, and that under appropriate conditions could give rise to viable embryos. The procedure used to demonstrate the normality of the 2-cell stages was to transfer the arrested embryos into organ cultures of the oviduct. Under these conditions the arrested embryos developed into blastocysts. Some of these blastocysts were then transferred to uterine foster mothers and developed into normal-looking embryos.

Subsequent studies on the ability of fertilized 1-cell mouse embryos to develop to the 2-cell stage in the medium shown in Table 13.1 demonstrated that cleavage only occurred in the presence of pyruvate and oxaloacetate, but not glucose, lactate or PEP. The requirements for the first cleavage division are thus the same as those for maturation of the oocyte.

D. Later Stages

Late 2-cell mouse embryos and 8-cell mouse embryos readily develop in chemically defined media of the type shown in Table 13.1, supplemented with various compounds which can supply energy. The nutritional requirements of the 2-cell stage have been studied by Brinster (1965a) and the 8-cell stage by Brinster and Thomson (1966). The results of this work together with those obtained on the oocyte and 1-cell stage are summarized in Table 13.2. The combined results indicate that as cleavage proceeds the number of substances which can supply energy to the embryos increases.

E. A Working Hypothesis

The results shown in Table 13.2 suggest that the energy metabolic pathways are unusual in the early mouse embryo, and that as development proceeds following fertilization these pathways change and approach the more usual condition. If this

Table 13.2 **Comparison of compounds which support the development** *in vitro*
of mouse oocytes, fertilized ova,
2-cell and 8-cell embryos.

Substrate	Oocyte[1]	1-cell[1]	2-cell[2]	8-cell[3]
lactate	-	-	+	+
pyruvate	+	+	+	+
oxaloacetate	+	+	+	+
phosphoenolpyruvate	-	-	+	+
malate	?	?	-	+
citrate	?	?	-	+
α-oxoglutarate	?	?	-	+
acetate	?	?	-	±
d-glyceraldehyde	?	?	-	-
glucose-6-phosphate	?	?	-	-
fructose-1,6-diphosphate	?	?	-	-
glucose	-	-	-	+
fructose	?	?	-	±

[1] Biggers, Whittingham and Donahue (1967)
[2] Brinster (1965a)
[3] Brinster and Thomson (1966)

interpretation is correct it follows that *during oogenesis the energy metabolic
pathways of the oocyte become modified so that only a few compounds can be used
for metabolic purposes.*

III. REGULATION OF METABOLISM

Theoretically the metabolic peculiarities found in the oocyte could arise in
several ways. Two examples of possible mechanisms are:
(1) Changes in permeability to substrates either of the blastomeres or of
intracellular organelles.
(2) Changes in enzyme activities.

A. Permeability Changes

Some indirect evidence has now accumulated which indicates that
mitochondrial function is changed during oogenesis in the mouse and that during
cleavage the function is restored. There is a considerable literature on the
ultrastructure of fully-grown mammalian oocytes which demonstrates the presence

of atypical mitochondria. These mitochondria have been described in the rat, mouse, guinea pig, hamster, rabbit, cow, monkey and human (see Stern, Biggers and Anderson, 1971, for a review of the literature). The mitochondria are spherical or slightly elongate in shape. The matrix is electron opaque, and nearly all the cristae are concentrically arranged close to the inner mitochondrial membrane. The stage when these mitochondria arise during oogenesis is unknown. The modified mitochondria found in the fully-grown oocyte retain their form after ovulation, and if the ovum is fertilized they persist until the 4- to 8-cell stage. At this period of development the mitochondrial population becomes pleiomorphic. By the morula stage of development the mitochondria are more typical in appearance with many transverse cristae. Morphologically then, a profound change in mitochondrial structure occurs at the 4- to 8-cell stage of development in the mouse.

This change in mitochondrial structure can be correlated with the changes in the types of substances which can support cleavage of the 2- and 8-cell mouse embryos *in vitro* (Table 13.2). Between the 2- and 8-cell stages the embryo acquires the ability to use several compounds of the citric acid cycle - namely malate, citrate and α-oxoglutarate.

The uptake of $[U-^{14}C_1]$ malate by 2-cell and 8-cell mouse embryos was compared by Wales and Biggers (1968), and more recently by Kramen and Biggers (1971). Some results are shown in Table 13.3. The 2-cell embryo accumulates a very small amount of malate, while very much larger amounts are accumulated by the 8-cell stage. Similar studies have been done using $[U-^{14}C]$ α-oxoglutarate and $[1,(5)-^{14}C]$ citrate (Kramen and Biggers, 1971). Some results are also shown in Table 13.3. The 2-cell embryo accumulates very little, if any, α-oxoglutarate and citrate whereas the 8-cell stage accumulates relatively large amounts of both compounds.

It is possible that the change in accumulation of these three compounds between the 2- and 8-cell stages is due to changes in the cell membrane of the blastomeres. However, it is also known that these compounds are transported into the mitochondrial matrix by special transporter carrier molecules located in the inner mitochondrial membrane and the mitochondrial cristae (see Chappell, 1968 and Pressman, 1970 for reviews). The following transporters are recognized:
    (1) dicarboxylic acid transporter (malate, succinate);
    (2) α-oxoglutarate transporter;
    (3) citrate, isocitrate and *cis*-aconitate transporter.
Thus an alternative explanation for the changes in uptake of malate, α-oxoglutarate and citrate by 2- and 8-cell mouse embryos is that an alteration occurs in the transporter substances in the mitochondria. This hypothesis is strongly supported by the fact that extensive changes occur in the internal structure of the mitochondria of the mouse between the 2- and 8-cell stages. Moreover, it is known that some mitochondria lack anion transporter molecules, *e.g.* those in housefly flight muscle

Table 13.3 Uptake *in vitro* of TCA Intermediates by 2- and 8-cell Mouse Embryos

(f-moles ^{14}C substrate / embryo)*
(From Kramen and Biggers, 1971)

Time (min.)	[U-^{14}C] Malic Acid			[U-^{14}C] α-Oxoglutaric Acid			[1,(5)-^{14}C] Citric Acid		
	Number observations	Mean	Confidence limits (P=0.05)	Number observations	Mean	Confidence limits (P=0.05)	Number observations	Mean	Confidence limits (P=0.05)
	2-cell (D.F.=9)			2-cell (D.F.=15)			2-cell (D.F.=9)		
5	10	0.62	0.52 - 0.74	9	0.58	0.42 - 0.81	7	0.83	0.64 - 1.08
15	10	1.42	1.18 - 1.71	11	0.31	0.23 - 0.41	7	0.54	0.42 - 0.71
45	10	1.74	1.45 - 2.07	10	0.53	0.39 - 0.73	5	0.34	0.25 - 0.47
135	6	3.22	2.5 - 4.14	11	0.74	0.55 - 0.99	8	1.05	0.82 - 1.35
	8-cell (D.F.=12)			8-cell (D.F.=12)			8-cell (D.F.=9)		
5	7	17.1	14.7 - 19.9	9	10.7	7.73 - 14.9	7	2.38	1.69 - 3.34
15	8	34.5	30.0 - 39.6	9	21.6	15.6 - 30.0	6	4.98	3.44 - 7.19
45	8	74.2	64.7 - 85.2	9	52.8	38.0 - 73.4	6	11.0	7.64 - 16.0
135	9	168	147 - 192	7	182	124 - 268	7	23.0	16.4 - 32.5

*f-moles = femto moles = 10^{-15} moles.

(Van den Bergh and Slater, 1962; Van den Bergh, 1964). It is possible therefore that during oogenesis the mitochondria are modified by suppression or elimination of their anion transporter mechanisms. Much more work will be needed on the ultrastructure and metabolism of oocytes to confirm this hypothesis.

Table 13.4 **Enzymes whose activity has been measured in several stages of the preimplantation mouse embryo**

Enzyme		Reference
A. Enzymes whose activity increases		
hexokinase	(EC 2.7.1.1)	Brinster (1968)
fructose-1,6-diphosphate aldolase	(EC 4.1.2.13)	Epstein, Wegienka and Smith (1969)
malate dehydrogenase	(EC 1.1.1.37)	Brinster (1966a)
		Epstein *et al.* (1969)
hypoxanthine-quanine phosphoribosyl transferase	(EC 2.4.2.8)	Epstein (1970)
adenine phosphoribosyl transferase	(EC 2.4.2.7)	Epstein (1970)
B. Enzymes whose activity decreases		
lactate dehydrogenase	(EC 1.1.1.27)	Brinster (1965b)
		Epstein *et al.* (1969)
glucose-6-phosphate dehydrogenase	(EC 1.1.1.49)	Brinster (1966b)
		Epstein *et al.* (1969)
glutamic dehydrogenase	(EC 1.4.1.2)	Moore and Brinster (1970)
phosphofructokinase	(EC 2.7.1.11)	Brinster (1971)
C. Enzyme which increases and then decreases		
glycogen synthetase	(EC 2.3.1.11)	Stern (1970)
D. Enzyme which decreases and then increases		
aspartate aminotransferase	(EC 2.6.1.1)	Moore and Brinster (1970)
E. Enzyme which does not change		
isocitrate dehydrogenase (NADP)	(EC 1.1.1.42)	Donahue and Stern (1970)

B. Changes in Enzyme Activity

During the past few years improvements in ultramicrochemical techniques have enabled the measurement of many compounds in early mammalian embryos (see Biggers and Stern, 1972, for a review). Among these are several enzymes (Table 13.4).

The amounts of lactate dehydrogenase and glucose-6-phosphate dehydrogenase is particularly high in the unfertilized mouse ovum, and the amounts decrease as development proceeds. Possibly these enzymes are more important during oogenesis than they are during subsequent development. It is of considerable interest that the lactate dehydrogenase exists in only one isoenzymic form (LDH I) in the mouse (Auerbach and Brinster, 1967; Rapola and Koskimies, 1967; Epstein, Wegienka and Smith, 1969) and in the rat (Cornette, Pharris and Duncan, 1967). This isoenzyme is the adult form associated with aerobic metabolism. Unfortunately the measurement of the absolute or relative amounts of enzymes at different stages tells us little about metabolic function. We need to know whether these enzymes are functional. An enzyme can be present in considerable quantity but if the appropriate co-factors are lacking the enzyme will be inactive. Until more sensitive methods are developed to study the products of enzyme activity in early mammalian embryos and oocytes the role of enzymes in regulating oocyte metabolism will remain obscure.

IV. ROLE OF THE FOLLICLE CELLS

When mouse oocytes are placed in the medium shown in Table 13.1 supplemented with pyruvate or oxaloacetate they undergo meiotic maturation. In contrast, no maturation occurs when only lactate, PEP or glucose is added to the medium. However, Biggers, Whittingham and Donahue (1967) demonstrated that when follicle cells were added to the medium lactate, PEP and glucose supported meiotic maturation. Subsequently Donahue and Stern (1968) showed that cumulus cells *in vitro* can produce pyruvate from these substrates. The results raise the possibility that the follicle cells can supply pyruvate to the oocyte under appropriate stimulation, and induce meiotic maturation. Under natural conditions this stimulus would be the luteinizing hormone (LH).

Recently Hamberger (1968) and Ahren, Hamberger and Rubinstein (1969) have studied the action of LH and the follicle stimulating hormone (FSH) on isolated granulosa and theca cells using a "two compartment" microdiver technique. When isolated granulosa cells were exposed to LH in a medium containing succinate as a substrate a marked increase in respiration occurred. Conversely, when isolated thecal cells were exposed to LH under the same circumstances no increased respiration was observed. When FSH was used no effect was observed on the granulosa cells while a considerable effect was produced on the thecal cells. Thus there is some evidence that the target cell of LH is the granulosa cell, and that its metabolism could be changed so that its products influence the oocyte. Much more quantitative work needs to be done to confirm that such a mechanism operates. (For a more extensive discussion, see Biggers, 1971b).

V. CONCLUSION

The evidence discussed in this paper suggests that the processes concerned with energy metabolism in the mouse oocyte are modified during oogenesis. One possible mechanism involves changes in mitochondrial function. However, since knowledge of the metabolism of the oocyte is so rudimentary it is possible that many other changes also occur. Other results suggest that the follicle cells may regulate meiotic maturation in mammals. However, there is also evidence that the oocyte regulates the differentiation of the granulosa cells into luteinized cells. This phenomenon is discussed in detail by Nalbandov elsewhere in this volume. It seems that the oocyte and follicle cells reciprocally influence each other and form a transitory integrated system, some of whose properties may be lost when we experimentally separate the two components. This complication should be kept in mind in the interpretation of future experiments.

VI. ACKNOWLEDGMENTS

The preparation of this paper has been made possible by grants from the Ford Foundation, the National Institute of Child Health and Human Development and the Population Council.

VII. REFERENCES

Ahren, K., Hamberger, L. A. and Rubinstein, L. (1969) Acute *in vivo* and *in vitro* effects of gonadotrophins on the metabolism of the rat ovary. *In* "The Gonads." (K. W. McKerms, ed.) p. 327. New York, Appleton-Century-Crofts.

Auerbach, S. and Brinster, R. L. (1967) Lactate dehydrogenase isozymes in the early mouse embryo. *Exp. Cell Res.* **46**, 89.

Biggers, J. D. (1971a) Metabolism of mouse embryos. *J. Reprod. Fert.* Suppl. **14**, 41.

Biggers, J. D. (1971b) Oogenesis and ovum maturation. *In* "Regulation of Mammalian Reproduction." (S. J. Segal, R. Crozier and P. A. Corfman, eds.), Springfield, Illinois, Charles C. Thomas. (In press).

Biggers, J. D. and Stern, S. (1972) Metabolism of the preimplantation embryo. *Adv. Reprod. Physiol.* **6**. (In press).

Biggers, J. D., Whittingham, D. G. and Donahue, R. P. (1967) The pattern of energy metabolism in the mouse oocyte and zygote. *Proc. Nat. Acad. Sci, U.S.A.* **58**, 560.

Brinster, R. L. (1965a) Studies on the development of mouse embryos *in vitro*. II. The effect of energy source. *J. exp. Zool.* **158**, 59.

Brinster, R. L. (1965b) Lactate dehydrogenase activity in the preimplanted mouse embryo. *Biochim. Biophys. Acta.* **110**, 439.

Brinster, R. L. (1966a) Malic dehydrogenase activity in the preimplantation mouse embryo. *Exp. Cell Res.* **43**, 131.

Brinster, R. L. (1966b) Glucose-6-phosphate dehydrogenase activity in the preimplantation mouse embryo. *Biochem. J.* **101**, 161.

Brinster, R. L. (1968) Herokinase activity in the preimplantation mouse embryo. *Enzymologia.* **34**, 304.

Brinster, R. L. (1971) Phosphofructokinase activity in the preimplantation mouse embryo. *Wilhelm Roux' Archiv.* **166**, 300.

Brinster, R. L. and Thomson, J. L. (1966) Development of 8-cell mouse embryos *in vitro. Exp. Cell Res.* **42**, 303.

Chappell, J. B. (1968) Systems used for the transport of substrates into mitochondria. *Brit. Med. Bull.* **24**, 150.

Cornette, J. C., Pharris, B. B. and Duncan, G. W. (1967) Lactic dehydrogenase isozymes in the ovum and embryo of the rat. *Physiologist.* **10**, 146.

Donahue, R. P. (1968) Maturation of the mouse oocyte *in vitro.* I. Sequence and timing of nuclear progression. *J. exp. Zool.* **169**, 237.

Donahue, R. P. and Stern, S. (1968) Follicular cell support of oocyte maturation: production of pyruvate *in vitro. J. Reprod. Fert.* **17**, 395.

Donahue, R. P. and Stern, S. (1970) Isocitrate dehydrogenase in mouse embryos: activity and electrophoretic variation. *J. Reprod. Fert.* **22**, 575.

Epstein, C. J. (1970) Phosphoribosyltransferase activity during early mammalian development. *J. biol. Chem.* **245**, 3289.

Epstein, C. J. Wegienka, E. A. and Smith, C. W. (1969) Biochemical development of preimplantation mouse embryos: *In vivo* activities of fructose 1,6-diphosphate aldolase, glucose-6-phosphate dehydrogenase, malate dehydrogenase, and lactate dehydrogenase. *Biochem. Genet.* **3**, 271.

Hamberger, L. A. (1968) Influence of gonadotrophins on the respiration of isolated cells from the prepubertal rat ovary. *Acta physiol. scand.* **74**, 410.

Kramen, M. A. and Biggers, J. D. (1971) Uptake of tricarboxylic acid cycle intermediates by preimplantation mouse embryos *in vitro. Proc. Nat. Acad. Sci., U.S.A.* **68**, 2656.

Moore, R. W. and Brinster, R. L. (1970) Transamination and deamination in mouse preimplantation embryos. *Proc. Soc. Study Reprod.* Columbus, Ohio.

Pincus, G. and Enzmann, E. V. (1935) The comparative behavior of mammalian eggs *in vivo* and *in vitro*. I. The activation of ovarian eggs. *J. exp. Med.* **62**, 665.

Pressman, B. C. (1970) Energy-linked transport in mitochondria. *In* "Membranes of Mitochondria and Chloroplasts." (E. Racker, ed.) p. 213. New York, Van Nostrand.

Rapola, J. and Koskimies, O. (1967) Embryonic enzyme patterns: Characterization of the single lactate dehydrogenase isozyme in preimplanted mouse ova. *Science.* **157**, 1311.

Stern, S. (1970) The activity of glycogen synthetase in the cleaving mouse embryo. *Proc. Soc. Study Reprod.* Columbus, Ohio.

Stern, S., Biggers, J. D. and Anderson, E. (1971) Mitochondria and early development of the mouse. *J. exp. Zool.* **176**, 179.

Talbert, A. J. (1969) Recursive multiple regression with a polychotomous response variable: an application to stagewise biological development data. M.Sc. Thesis: Johns Hopkins University School of Hygiene and Public Health.

Van den Bergh, S. G. and Slater, E. C. (1962) The respiratory activity and permeability of housefly sarcosomes. *Biochem. J.* **82**, 362.

Van den Bergh, S. G. (1964) Pyruvate oxidation and the permeability of housefly sarcosomes. *Biochem. J.* **93**, 128.

Wales, R. G. and Biggers, J. D. (1968) The permeability of two- and eight-cell mouse embryos to L-malic acid. *J. Reprod. Fert.* **15**, 103.

Whitten, W. K. (1956) Culture of tubal mouse ova. *Nature.* **177**, 96.

Whittingham, D. G. and Biggers, J. D. (1967) Fallopian tube and early cleavage in the mouse. *Nature.* **213**, 942.

14

DROSOPHILA OOGENESIS AND
ITS GENETIC CONTROL

Robert C. King

I.   Introduction
II.   The specialized mitosis of germarial cystocytes
III.   Mutations that influence the division and differentiation of cystocytes
IV.   The further differentiation of the oocyte nucleus
V.   Mutations that influence the synapsis of oocyte chromosomes
VI.   Mutations that influence the development of nurse cells
VII.   Yolk synthesis
VIII.   Mutations that influence vitellogenesis
IX.   Acknowledgments
X.   References

I. INTRODUCTION

It is not surprising that the largest and most complex cell produced by insects is the oocyte. Unlike somatic cells, it must contain sufficient nutrient reserves to maintain the potential organism during embryogenesis, because most insect embryos have no means for obtaining exogenous, organic, raw materials. Furthermore, since little transcription of RNA takes place along DNA cistrons during mitosis (Salb and Marcus, 1965; Fan and Penman, 1970), and since a period of rapid mitosis occurs early in embryogenesis (Sonnenblick, 1947), a mechanism must also exist for loading the unfertilized egg with the long lived messenger RNAs, ribosomes, and the transfer RNAs that are required to synthesize the proteins utilized during this early period of development (Lockshin, 1966; Davidson, 1968; Gall, Macgregor, and Kidston, 1969). Many insects have evolved methods for providing (1) highly endopolyploid cells which synthesize the required compounds and (2) a system of canals through which these products can be exported to the oocyte.

Oocytes are also of interest because their chromosomes during meiotic prophase are called upon to undergo a complex series of reactions that chromosomes preparing for somatic mitoses never undertake. Synaptonemal complexes are assembled and meiotic crossing over takes place within the nuclei of oocytes of the fruitfly, *Drosophila melanogaster*. Synaptonemal complexes do not form, and crossing over is suppressed in spermatocytes, presumably to prevent exchange of genes between the dissimilar sex chromosomes of the male.

In *Drosophila* a considerable body of information is now available concerning the details of how reproduction is accomplished and regulated in the female, and many insights are given into the genetic control of oogenesis through studies of the aberrant forms of ovarian development characteristically seen in females possessing certain mutant genes that markedly influence their fertility (King, 1970b). Of special interest are mutations which affect the nuclear behavior of oogonia, oocytes, and their sister nurse cells. In the domesticated silkmoth, *Bombyx mori,* female sterile mutations are also known, and one of these which influences protein uptake by the oocyte during vitellogenesis is under study in Japan.

Females belonging to certain evolutionarily advanced groups of insects (such as the adephagous Coleoptera, the Lepidoptera, the Hymenoptera, and the Diptera) are characterized by ovarioles that contain egg chambers in which the oocyte is one member of a cluster of interconnected cells (Brown and King, 1964; King and Aggarwal, 1965; Cassidy and King, 1969). The other cells of the cluster, the "nurse" cells, grow and simultaneously transfer their cytoplasm to the oocyte via a system of intercellular canals. In *D. melanogaster,* as in most Diptera belonging to the superfamily Muscoidea, each egg chamber consists of an oocyte, its fifteen nurse cells, and a surrounding monolayer of follicle cells (Fig. 14.1b). The egg and its fifteen interconnected nurse cells are fourth generation descendants of a single cell, the germarial "cystoblast" (Fig. 14.2). The interconnected cells formed by the division of a cystoblast are called "cystocytes". It is within the germarial portion of the ovariole that the consecutive divisions occur which produce the sixteen cell cluster, and it is here also that the cluster becomes enveloped by profollicle cells. However, the major growth of the egg chamber occurs in the more posterior part of the ovariole (called the vitellarium) (see Fig. 14.1a). Here, under optimal conditions, it takes only three days for the cytoplasmic volume of the oocyte to increase 90,000 times.

II. THE SPECIALIZED MITOSIS OF GERMARIAL CYSTOCYTES

The finding of clusters of identical mutations among the progeny of individual female *Drosophila* treated as adults with a mutagenic agent constitutes conclusive evidence for the presence of stem-line oogonia in adult germaria (Brown and King, 1962, 1964). Electron microscopic studies of serially sectioned germaria have identified cells representing various developmental stages in the production of egg chambers (Koch and King, 1966; Koch, Smith, and King, 1967; Koch and King,

1969). An apical, mitotically active, germarial region contains in its most anterior portion, one or two single cells, followed posteriorly by clusters of 2, 4, 8, and 16 interconnected cystocytes. The volumes of the cells vary in a characteristic fashion; the single cells are the largest, and the cystocytes from the 16 cell clusters are the smallest. To explain these findings Koch and King (1966) postulated that each germarium contains one (or at most two) stem line oogonia. Each divides forming two daughter cells (see Fig. 14.2). These separate, and the apical one continues to behave as a stem cell, whereas the other functions as a cystoblast. The cystoblast does not double its birth volume before entering mitosis, as does a stem cell; and the daughter cells produced by a cystoblast do not separate, but instead form permanent, interconnecting canals. As the divisions continue the cystocytes become smaller and smaller, and the canal system becomes more complex.

In *Bombyx mori* sister spermatocytes are connected by a canal system similar to the one observed in clusters of ovarian cystocytes (King and Akai, 1971a). The midbody that forms at the site of future canals between sister spermatocytes consists of a disc of electron-dense material in which spindle microtubules are embedded (Fig. 14.2). The micro-tubules extending from one sister cell into the disc terminate at its opposite surface and interdigitate with microtubules extending into the disc in the opposite direction from the other sister cell. Immediately beneath the plasma membrane of the cleavage furrow is an area of dense cytoplasm 30-40 mμ thick. Schroeder (1970) has shown that such contractile rings are composed of an ordered array of filaments (each 4-7 mμ in diameter) that are aligned circumferentially along the equator of the cell. Perhaps the midbodies serve to prevent further constriction of the contractile ring. The canals result once the midbody and its associated microtubules dissolve. If the ring rim is derived from a contractile ring stabilized by interaction with the peripheral tubules of the midbody, then microtubules should be demonstrable in canal rims at some developmental stage. Such microtubules have been observed embedded in the ring canal rims of the ovarian cystocytes of the wasp, *Habrobracon juglandis* (Cassidy and King, 1969).

During the consecutive divisions of *Drosophila* cystocytes the axes of the spindles shift with the result that a branching chain of interconnected cells is formed (see Fig. 14.2). At the end of the fourth division there are two cystocytes with four canals, two with three canals, four with two canals, and eight with one canal. The system of ring canals functions to permit the nurse cells to transfer their cytoplasm to the oocyte as they grow. The canals are wide enough to allow particles as large as mitochondria to move from one cystocyte to another.

What factors cause the cystocytes to stop dividing and to differentiate either as nurse cells or oocytes? As was mentioned earlier, stem cells double their volume before dividing, whereas cystocytes do not, with the result that they become smaller with each division. Fourth generation cystocytes have volumes only 20-25%, the maximum volumes of stem line oogonia. Thus, if one assumes that stem cells and cystocytes divide equally often, then single cells must grow faster between mitoses than do interconnected cells. Perhaps once a cystocyte reaches a critical minimum

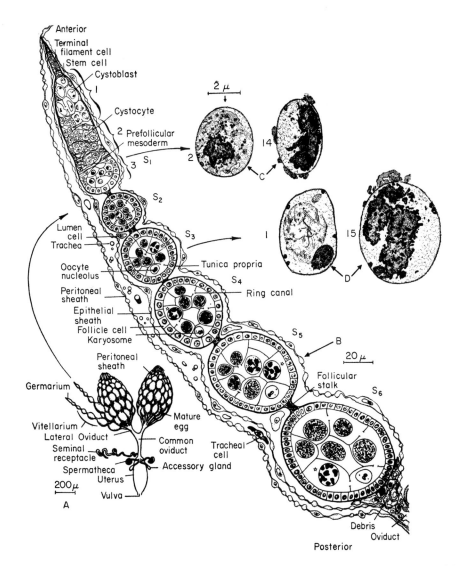

Fig. 14.1 (a) Dorsal view of the internal reproductive system of an adult female *D. melanogaster*. Two ovarioles have been pulled loose from the left ovary. Sperm are stored in the ventral seminal receptacle (which is drawn uncoiled) and in paired spermathecae. The uterus is drawn expanded as it would be when it contains a mature egg. (b) A diagram of a single ovariole and its investing membranes. The nurse cell-oocyte complexes are representative of the first six stages (S1-6) of oogenesis. Within the vitellarium, sectioned egg chambers are drawn so as to show the morphology of the nuclei of the oocyte and 6 of the 15 nurse cells. The distribution of nucleolar material is drawn in the starred nurse cell nucleus, whereas the other five nurse nuclei show the distribution of Feulgen-positive material. The distributions of both DNA and nucleolar RNA are shown for the oocyte and follicle cells. Fragments detach from the oocyte nucleolus during stages 4 through 6. (c) A pro-oocyte nucleus (left) and a nucleus in an adjacent pronurse cell (right). from germarial region 3. Synaptonemal complexes are seen in nuclei of the two pro-oocytes. The nuclei of the 14 pronurse cells lack synaptonemal complexes and contain more nucleolar material (N). Clouds of particulate matter adhere to the surface of these nuclei. (d) The oocyte nucleus (left) and a nucleus from an adjacent nurse cell (right) from a stage-3 egg chamber in the vitellarium. The magnification is the same as in Fig. 14.1c. Nucleolar material is far more abundant in each of the 15 nurse nuclei than in the oocyte nucleus. The synaptonemal complexes have congregated into a central area which under the light microscope (b) appears as a Feulgen-positive mass adjacent to the nucleolus [From Koch, Smith, and King, 1967].

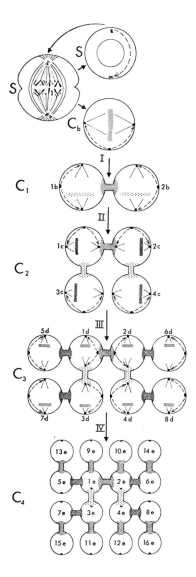

Fig. 14.2 A diagrammatic model showing steps in the production of a cluster of 16 interconnected cystocytes. In this drawing the cells are represented by circles lying in a single plane, and the ring canals have been lengthened for clarity. The area of each circle is proportional to the volume of the cell. The stem cell (S) divides into two daughters, one of which behaves like its parent. The other differentiates into a cystoblast (C_b) which by a series of four divisions (I-IV) produces 16 interconnected cystocytes. C_1, first; C_2, second; C_3, third; and C_4, fourth generation cystocyte. The upper and lower portions of the diagram represent the anterior and posterior portions of germarial region 1, respectively (see Fig. 14.1b). The original stem cell is shown at early anaphase. Each parent-daughter pair of centrioles is attached to the plasma membrane by astral rays. The daughter centriole is drawn slightly smaller than its parent. The daughter stem cell receives one pair or centrioles. One remains in place while the other moves to the opposite pole. This movement is represented by a broken arrow. In the daughter cystoblast and all cystocytes the initial position of the original centriole pair is represented by a solid half circle; whereas their final positions are represented by solid circles. The position of the mid-body is drawn as a strip of defined texture. The canal rim derived from the contractile ring is treated in a similar fashion [From Koch, Smith, and King, 1967].

volume, it can no longer divide and differentiates instead. The concept that cell division may be prevented by reducing the size of the cell below a critical mass has considerable experimental support (see Prescott, 1956; Donachie, 1968).

Brown and King (1964) demonstrated that the *Drosophila* oocyte is invariably one of the two fourth generation cystocytes possessing four ring canals. Subsequently Koch, Smith, and King (1967) reported that as the sixteen cell cluster moved through the germarium both cystocytes possessing four canals (cells 1e and 2e in Fig. 14.2) could be distinguished from the other fourteen cystocytes on the basis of nuclear ultrastructure. The two "pro-oocytes" formed synaptonemal complexes; the fourteen "pro-nurse cells" did not (see Fig. 14.1c and d).

III. MUTATIONS THAT INFLUENCE THE DIVISION AND DIFFERENTIATION OF CYSTOCYTES

Further insights into the factors that control cystocyte division and differentiation are gained through the study of mutations which produce abnormalities in the numbers of cystocyte divisions and in the differentiation of cystocytes into oocytes. The recessive mutations *fused* (*fu*, X-59.5) and *female sterile* (*fes*, 2-5) render homozygous *Drosophila* females sterile because of the production of "ovarian tumors" (King and Burnett, 1957). The tumorous chambers seen in the ovarioles of females homozygous for *fu* and *fes* contain cells which resemble immature germarial cystocytes. Such chambers grow slowly by mitosis, and some eventually come to contain as many as 10,000 cells. The karyotype of the tumor cells is generally normal, and cell divisions continue when tumors are transplanted to new hosts (King and Bodenstein, 1965). However, the tumors do not invade non-ovarian tissues and apparently have little effect upon the viability of the donor or host. Ovaries genetically destined to become tumorous do so even when they are transplanted to the abdomens of normal female adults. Normal ovaries transplanted to the abdomens of females homozygous for such tumor genes do not become tumorous. Therefore there is no evidence for diffusible tumorigenic agents as initiating factors in the formation of the ovarian tumors characteristic of *fu* or *fes*.

Ovaries from newly enclosed *fused* flies contain no tumors. As the flies age, however, tumors appear, and the rate at which they form differs for different *fused* alleles (Smith and King, 1966). In cases where a fusion takes place between a normal egg chamber and a tumorous chamber lying adjacent to it in the ovariole, tumor cells may infiltrate the tissues of the normally developing, egg-nurse cell complex. In this sense *fused* tumor cells are weakly invasive.

The ovaries of *fes* homozygotes differ from those of *fu* homozygotes in that in the former the incidence of ovarian tumors is already 100% in newly enclosed females. Koch and King (1964) reported the results of a reconstruction from serial sections of the three dimensional morphology of a dozen *fes* ovarian tumors. Most tumors were found to be made up of a number of clusters of cystocytes. One tumor, for example, contained 116 tumor cells, and these were distributed among 14

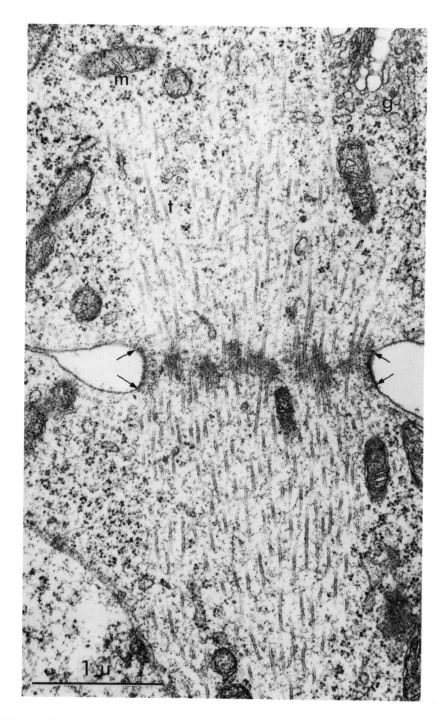

Fig. 14.3 A mid-body forming between two sister spermatocytes in *Bombyx mori*. The midbody contains dense material which is laid down in the region where microtubules extending from each spermatocyte overlap. The arrows show the edges of the contractile ring. g, Golgi material; m, mitochondrion; t, spindle microtubule [From King and Akai, 1971a].

clusters of interconnected cells. The average number of cells per cluster was 8, with a range from 2 - 14. This finding suggests that during the growth of tumors in the vitellarium of the *fes* ovariole, a cell can break loose from a cystocyte cluster and through its subsequent division form an independent, smaller, adjacent cluster.

By appropriate crosses Smith and King (1966) were able to produce females containing both mutant genes. When ovaries from very young females of genotype *fu/fu; fes/fes* were examined only tumorous chambers were seen. Thus *fes* is epistatic to *fu* in terms of ovarian phenotype. The ovaries of females of genotype (*fu/fu; fes/+*) were examined and were found to show a higher incidence of tumors than *fu/fu; +/+* females of the same age. Thus the "recessive" *fes* gene can be detected in the heterozygous condition by its interaction with *fu*.

Under appropriate conditions with respect to age, temperature, and residual genotype, *fes* females will produce chambers in which most cystocytes, instead of continuing to divide, differentiate into cells whose nuclei resemble those of nurse cells. About 2% of the *fes* chambers made up of "nurse" cells also contain an oocyte (King, Koch, and Cassens, 1961). An analysis by Smith and Murphy (cited in King, 1969) of the three-dimensional morphology of *fes* nurse chambers revealed that all oocyte-deficient chambers lacked cystocytes which possessed four ring canals and that all cystocytes with four ring canals had differentiated into oocytes.

The sixteen cystocytes are the mitotic products of a single cell. Mitosis should insure the nuclear equivalence of all sister cystocytes, and the particulate cytoplasmic components should be distributed initially between all sister cells in a random fashion. Yet two specific cells undergo a different type of development from the remaining fourteen. What is the cue that stimulates the nuclei of pro-oocytes to form synaptonemal complexes? Koch and King (1969) have presented evidence that each canal rim marks on a given cystocyte the inner limits of a membranous area of unique age and shape and that the canal rim increases in diameter and thickness and develops along its inner circumference a deposit having different cytochemical properties from the rim itself (Fig. 14.4). This behavior together with its manner of origin suggests that the rim should be thought of as a metabolically active, cortical organelle. Thus a given pro-oocyte is characterized by a plasmalemma that contains a mosaic pattern of cortical structures which is similar to that of the other pro-oocyte, but different from the patterns possessed by any of the remaining pro-nurse cells. Perhaps diffusible compounds secreted at certain cortical areas in pro-oocytes initiate the synthesis of synaptonemal complexes.

Mitoses occur throughout the *fes* germarium instead of being restricted to region 1, and sister cells tend to separate (King and Hewlett, 1970). The germarium becomes packed with single cells and clusters containing 2 or 3 cystocytes. Groups of such cells, enclosed by an envelope of follicle cells, eventually enter the vitellarium. Here the cystocytes continue to divide and so produce the characteristic "ovarian tumor." In the *fes* germarium, all canal rims that have been observed under the electron microscope are immature in the sense that they contain little or no

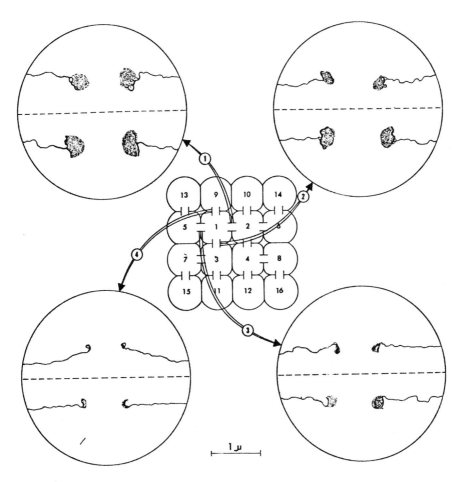

Fig. 14.4 A drawing showing how the canal rims change in morphology with increasing age. Included here are data for two 16-cell clusters, one of which completed the fourth cystocyte division about 24 hours before the other. The appearance of the rims from the younger cluster is illustrated in the upper half of each circle while those of the older cluster are shown in the lower half. The numbers indicate whether the canal was formed at the first, second, third, or fourth division [From Koch and King, 1969].

coating material. Thus, the primary defect in the *fes* ovary appears to be the inability of *fes* cystocytes to form a stable system of ring canals. Since the *fused* gene interacts with *fes*, the products of both genes may play an essential role in the genesis of stable canals.

However, *fes* cystocytes can be induced to stop dividing under special conditions (lowering the environmental temperature is one of these). Since such cells differentiate into nurse cells, and since all wild type cystocytes with less than four ring canals do the same, the obvious conclusion is that cystocytes are programmed to enter the nurse cell developmental pathway automatically once they cease

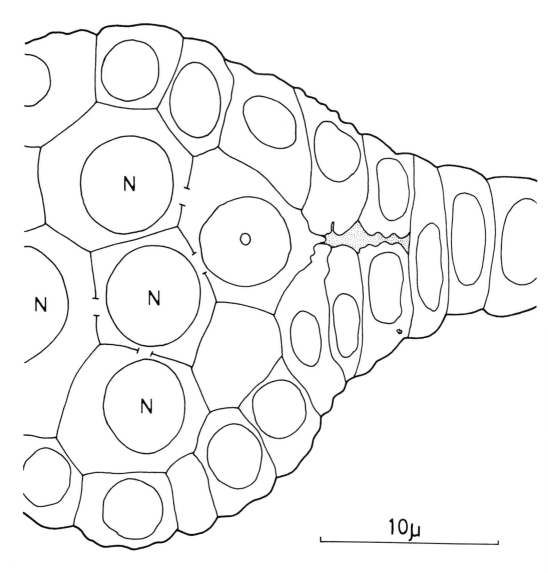

Fig. 14.5 A drawing of a stage 2 chamber and the stalk that connects it to the next chamber in the vitellarium. See Figure 14.1b (S_2 and S_3) for orientation purposes. The canal which connects the oocyte (O) with the most anterior stalk cell is stippled. N, nurse cell nucleus. The remaining nuclei belong to follicle cells.

dividing, and an overriding stimulus is required to make them enter the oocyte developmental pathway. This stimulus is produced only in cystocytes with four ring canals. Since *fes* cystocytes rarely have four canals, oocytes differentiate equally rarely.

Fs(2)D is one of the few dominant, female sterile mutations known in *Drosophila melanogaster.* The mutant resides on the second chromosome, but its precise locus is unknown. Heterozygous males are fertile, whereas heterozygous females are sterile. To obtain females heterozygous for *Fs(2)D*, males of genotype *Fs(2)D/T(1;2)Bld* are mated to virgin females carrying a compound double X chromosome and a normal Y chromosome. The males are kept in a stock with *T(1;2) OR 64/T(1;2)Bld* females. (Consult Lindsley and Grell, 1968, for explanations of the symbols used.) Yarger and King (1971) noted that the production of egg chambers was greatly inhibited in heterozygotes. For example, in 24 hour old females reared at 25°C six or seven egg chambers were found in each wild type ovariole, whereas each *Fs(2)D/+* ovariole held either one chamber or none. Of the relatively few egg chambers produced, between 10 and 20% (depending on the temperature) contained less than 16 cystocytes. The modal number of cystocytes in such abnormal chambers was two. No oocytes were found in the vast majority of chambers with less than 16 cystocytes. This finding is not surprising, since the cystocytes from such small clusters would be expected to have fewer than four ring canals. Since in *Fs(2)D/+* females the ovarian phenotype is mutant, it follows that the division of cystocytes is retarded even though each cell possesses a wild type allele and presumably its product. This conclusion suggests in turn that the mutant allele may code for a mutant protein which has an inhibitory effect on cystocyte mitosis. For example, the incorporation of the mutant protein into some mitotic organelle might impair its functioning even though the normal protein was incorporated simultaneously.

IV. THE FURTHER DIFFERENTIATION
OF THE OOCYTE NUCLEUS

One of the two potential oocytes later loses its synaptonemal complexes and switches to the nurse cell developmental pathway. What is the source of the stimulus that causes the other pro-oocyte to continue developing as an oocyte? Koch and King (1969) have noted that the oolemma and the plasmalemmas of certain follicle cells that lie at the posterior pole of the developing egg chamber interdigitate in a unique fashion, and we speculated that these follicle cells supply the factors which stimulate the continued differentiation of the oocyte. More recently we have studied serial sections through the stalk that connects the first two egg chambers in the vitellarium. As shown in Fig. 14.5 a canal is present which passes from the surface of the oocyte to the most anterior stalk cell. The oocyte is the only one of the 16 sister cystocytes with access to the fluid of the canal.

During an early developmental stage synapsis of the homologous chromosomes is completed and crossing over takes place within the oocyte nucleus. Synaptonemal

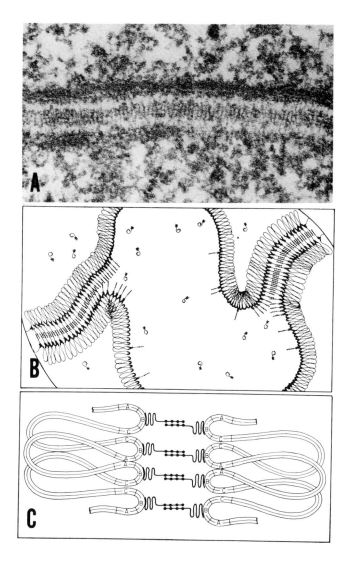

Fig. 14.6 (a) An electron micrograph of an ultrathin section through a synaptonemal complex [King and Akai, 1971b]. (b) The postulated mechanism of the synapsis of homologues to form a meiotic bivalent. The left and right telomeres of a pair of homologous chromosomes attach to a specific area of the inner membrane of the nuclear envelope. The chromosomes shorten and thicken because of the folding that results from the pairing of synaptomeres that are distributed along the chromosomes. Such synaptomeres are represented by segments which are slightly wider than the intervening regions. By the beginning of zygonema the folding is complete. Zygosomes are synthesized in the nucleoplasm, and they attach to synaptomeres and simultaneously uncoil. As a consequence, when each synaptomere possesses a zygosome, a peg extends from the base of each chromosomal fold. Interdigitation of the pegs initially produces short stretches of synaptonemal complex. Eventually, the chromosomes pair throughout their length, and an uninterrupted synaptonemal complex extends from left to right telomeres [From King, 1970a] (c) A small segment of the model synapsed bivalent drawn at higher magnification to elucidate the structure and functioning of synaptomeres and zygosomes. Each synaptomere is composed of three segments, A, B, and C. The lateral elements A and C will pair with the A and C elements, respectively, of any other synaptomere. The B element is the site where the base of a zygosome can attach. Zygosomes projecting from each homolog interdigitate and bind in an overlapping tail-to-tail fashion. Each chromosome is shown to be divided into two chromatids [From King, 1970a].

complexes are assembled during the zygotene and pachytene stages of meiotic prophase, and they function to zip the homologous chromosomes together in preparation for crossing over (see review by King, 1970a). In ultrathin sections a mature synaptonemal complex is seen under the electron microscope (Fig. 14.6a) as a ribbon about 0.2μ wide that is composed of parallel, dense, lateral elements lying to either side of a medial complex. Each lateral element runs the length of one homologue. The majority of the chromosomal material, however, is in the form of loops which are connected to the outer surfaces of the lateral elements (King and Akai, 1971a). The medial complex is made up of an interdigitating system of rods which extend from the inner surfaces of the lateral elements. In *Drosophila melanogaster* meiotic crossing over occurs only in females (Morgan, 1912, 1914), and synaptonemal complexes are not found in the prophase nuclei of primary spermatocytes (Meyer, 1960). In the female, synaptonemal complexes appear in the nuclei of pro-oocytes shortly after their formation at the last cystocyte mitosis (Koch, Smith and King, 1967). Such complexes can often be seen to end at attachment sites on the nuclear envelope.

V. MUTATIONS THAT INFLUENCE THE SYNAPSIS OF OOCYTE CHROMOSOMES

A mutation referred to as the *crossover suppressor of Gowen (c(3)G,* 3-58) was discovered in 1917 by Marie and John Gowen. Subsequently J. Gowen (1933) showed that in females homozygous for *c(3)G* crossing over in the entire chromosomal complement was reduced to a small fraction of the normal value. It was demonstrated subsequently (Hinton, 1966, 1967) that crossing over was lowered to one-half to two-thirds the control value in females heterozygous for the *stubbloid*105 *deficiency (sbd*[105], 3-58.2). Smith and King (1968) reported that the formation of synaptonemal complexes was never initiated in the oocyte nucleus of *c(3)G* homozygotes and that in *sbd*[105]/+ female the period of synthesis of synaptonemal complexes was abbreviated. We concluded that the formation of synaptonemal complexes is an essential prerequisite for meiotic crossing over.

I subsequently proposed (King, 1968) that the synaptonemal complex forms from homologous chromosomes which first attach to the nuclear envelope, next shorten by coiling and finally zip together through the attachment to specific chromosomal segments (synaptomeres) of protein-containing rods (zygosomes) that subsequently interdigitate to form the medial complex (Fig. 14.6b and c). Perhaps the normal allele of the *c(3)G* gene codes for an essential component of the zygosomes. The lateral elements were postulated to represent associations of synaptomeres from the homologous chromosomes. The chromosomal segments between the synaptomeres were visualized as being forced out of the lateral elements to form thousands of loops. The medial complex was postulated to serve to hold the homologues in a parallel orientation.

If homologous chromosomes are not held in parallel, we know from the *c(3)G* data that crossing over does not occur. I therefore concluded that during the period of homologue alignment special enzymes (recombinases) attached to the homologous loops extending from any two non-sister chromatids and moved along them. During such a trip a recombinase could catalyze a crossover event. According to this argument one can estimate the period during which crossing over is occurring by determining the interval during which synaptonemal complexes are present.

In *Drosophila melanogaster* synaptonemal complexes are seen in the nuclei of germarial pro-oocytes and in the oocyte nuclei from the first three chambers in the vitellarium. In a female producing two eggs per ovariole per day, an oocyte will spend about three days in the germarium and one day in the vitellarium in stages during which the synaptonemal complexes are growing (see Koch, Smith and King, 1970, their Fig. 1). Crossing over presumably is taking place during this four-day interval.

In addition to showing that meiotic crossing over was almost completely abolished in female *Drosophila melanogaster* homozygous for *c(3)G*, Gowen (1933) reported that among the offspring of such homozygotes were matroclinous females and patroclinous males, haplo- and triplo-IV males and females, metamales and metafemales, intersexes, and triploids. He concluded that the *c(3)G* females were producing eggs, some of which were diploid for the X, for the fourth, for all autosomes, or for all four chromosomes, and others which lacked the X or the fourth chromosome. Gowen suggested that nondisjunction during the first meiotic division was the phenomenon producing the exceptional offspring because disomic eggs formed by *c(3)G* females always contained chromosomes from both grand-parents. One would expect zygotes containing one or both major autosomes in single or triple dose to be inviable, and such lethal embryos presumably account for the majority of the eggs that fail to hatch (70% in the experiments of Smith and King, 1968). If one concludes that the absence of synaptonemal complexes is the cause of the abolition of meiotic crossing over, it seems reasonable to use the same argument to explain the elevation of nondisjunction.

Darlington (1929) was the first to suggest that homologues must be held together by chiasmata in order to pass to opposite poles at the first meiotic anaphase. He assumed that homologues which were not so attached behaved independently, and since each could go to either pole at anaphase, half of the secondary gametocytes produced would be disomic or nullisomic for a given chromosome. Thus the presence of one or more chiasmata in each metaphase bivalent was necessary to prevent homologues from assorting independently. This line of reasoning obviously leads to the suggestion that synaptonemal complexes promote the normal disjunction of bivalents by allowing crossing over to take place. This in turn insures that each bivalent contains at least one chiasma, since, as Brown and Zohary (1955) have shown, each crossover event generates a chiasma.

The high frequency in man of serious disease associated with chromosomal anomalies arising from nondisjunction has only been appreciated within the past five years (see reviews by Court Brown, 1967; Turpin and Lejeune, 1969). It is now clear that about 0.5 per cent of all newborns have such diseases. Since there are about 120 million births in the world each year, at least about 600,000 babies carrying defects of this type are added to the population annually. The aneuploid children that survive are generally mentally retarded. Monosomic and trisomic children lack a maternally-derived chromosome or possess a supernumary maternal chromosome more often than a paternal one. The observation that the frequency of meiotic nondisjunction rises sharply with advancing maternal age suggests that the probability of homologues assorting independently rises once oocytes pass a critical age. Perhaps chiasmata terminalize slowly in human oocytes, and after thirty or forty years the point is reached where a substantial fraction of all oocytes contain at least one pair of homologues which are no longer attached.

VI. MUTATIONS THAT INFLUENCE THE DEVELOPMENT OF NURSE CELLS

During oogenesis the chromosomes of the nurse cells undergo a series of endomitotic doublings. The first presumably occurs in the germarium (about the time the oocyte doubles its diploid DNA content, prior to entering meiotic prophase), and seven or eight more doublings occur within the vitellarium (Jacob and Sirlin, 1959; Cummings and King, 1969).

According to Ritossa and Spiegelman (1965) there are approximately 130 cistrons for the 18S rRNA and an equal number for the 28S rRNA in each sex chromosome. The provision of the *Drosophila* oocyte with 15 highly polyploid nurse cells serves to multiply the rDNA available for transcription by thousands of times. During the vitellogenic period which lasts about 24 hours the volumes of the nurse cell nucleolus, nucleus, and cytoplasm all double once every 4-5 hours. The nucleolar bodies grow until they form a thick network that is shaped like a shell whose outer boundary lies close to the inner surface of the nuclear envelope. RNA of nucleolar origin continually enters the cytoplasm. The nurse cells empty their cytoplasm into the oocyte which doubles its volume every 2 hours. The 2×10^{10} ribosomes stored in the cytoplasm of the mature oocyte are derived almost exclusively from the nurse cells. About 1000 rRNA molecules are transcribed per rRNA cistron per hour by nurse cells during vitellogenesis (Dapples and King, 1970; Klug, King and Wattiaux, 1970).

Several recessive mutant genes are known in *Drosophila melanogaster* which include among their phenotypic effects the production of cytologically detectable nuclear abnormalities involving the nurse cells. Here I will mention only *morula (mr, 2-106.7)*, *singed* [36a] *(sn[36a], X-21.0)*, *female sterile (2)E, (fs(2)E, 2-57.6)*, and *suppressor of (Hairy wing)[2](su(Hw)[2], 3-54.8)*. In the case of *mr* homozygotes, the compound nurse cell chromosomes fall apart into their component fibrils, condense to metaphase dimensions, and degenerate. The egg chamber breaks down prior to the stage where yolk formation begins. Females homozygous for *fs(2)E* or *sn[36a]* produce egg chambers which proceed further in development. However, in both the

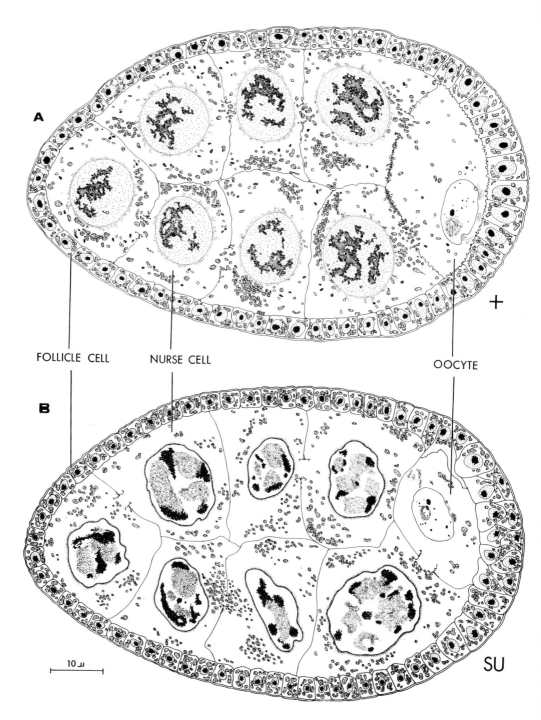

Fig. 14.7 (a) A drawing of a section through a wild type stage 7 egg chamber of *Drosophila melanogaster* (the first 6 stages are illustrated in Fig. 14.1b) as seen with the light microscope. The light stipple in the nuclei of all cells represents sectioned chromosomal fibers, while the denser component in each nucleus is nucleolar ribonucleoprotein. (b) In the equivalent *su(Hw)*[2] chamber, the chromosomal material appears clumped in the majority of the nurse cells, while it is dispersed throughout the nuclei of the wild type nurse cells. The nucleolus in the mutant nurse cell occupies less volume than in the wild type nucleus. The mutant follicle cells form a multiple layered envelope around the oocyte [From Klug, King, and Wattiaux, 1970].

endomitotic replication in the nurse cell nuclei is severely retarded, and vitellogenesis is inhibited (King, 1970b).

In the case of *su(Hw)*[2] the nurse cell chromosomes remain condensed during stages where the normal polytene chromosomes of the wild type nurse cells uncoil and fall apart in preparation for vitellogenesis (Klug, Bodenstein and King, 1968). There is a reduction in the amount of nucleolar material which appears as many small bodies interspersed among the chromatin (Fig. 14.7). The cytoplasm of mutant nurse cells frequently shows a reduction of basophilic staining compared to wild type. Radiotracer studies have been reported by Klug, King and Wattiaux (1970) of the RNA and protein synthesis in *su(Hw)*[2] ovaries. We concluded that the RNA transcribed in *su(Hw)*[2] nurse cell nuclei is not retained in the nucleolus for a normal time interval, and that much of it breaks down. The concentration of cytoplasmic ribosomes is decreased, and cytoplasmic protein synthesis is reduced. We suggested that the normal allele of *su(Hw)*[2] codes for a protein that is an essential component of the ribosomes. The mutant produces defective ribosomes that are unstable. If surviving mutant ribosomes translate with poor fidelity, they might correct some mutant messages and so suppress the phenotypic effects of certain mutant genes. This hypothesis would explain why *su(Hw)*[2] suppresses the expression of specific alleles of many other mutant genes (Lindsley and Grell, 1968).

Sandler (1970) has described still another mutation which may influence ribosome production by the nurse cells. Females homozygous for *abnormal oocyte* *(abo,* 2-38) produce a large excess of female progeny in crosses to homozygous *abo*[+] males that carry an attached - XY chromosome and no free Y. The sex ratio effect is due to lowered viability of the XO male offspring. The viability of male embryos is increased when sex chromatin rDNA is contributed to the zygote by a parental *abo*[+]/*abo*[+] male. Sandler suggests that *abo*[+] is a gene which regulates the synthesis of rRNA and that in mutant homozygotes there is a reduction in rRNA synthesis which is critical only in the case of oogenesis. The deficiency in maternally contributed ribosomes can be made up by new ribosomes synthesized after fertilization, provided the embryo contains sufficient rRNA cistrons in its nuclei.

VII. YOLK SYNTHESIS

In *Drosophila melanogaster* the adult female may live for a month or more, and mature eggs are formed only after the adult emerges. The eggs produced by the adult are made almost exclusively from nutrients ingested during adult life (Sang and King, 1961), and during periods of maximum oviposition, the female ingests daily an amount of yeast almost equal to her body weight and lays a number of eggs amounting to 60% of her weight (King and Wilson, 1955). Ingested material is rapidly converted into cytoplasmic components: for example, 40% of the phosphorus ingested appears the same day in the eggs laid (King and Wilson, 1955), and injected [3H] leucine is incorporated into proteins in the nurse cells in less than 15 seconds (Zalokar, 1960). Ultrastructural investigations suggest that the protein-containing (alpha) yolk spheres originate along the oocyte periphery from membranous sacs derived from coalesced pinosomes. It is assumed that the majority of the protein found in the mature alpha sphere is synthesized by neighboring rough surfaced endoplasmic reticulum (Cummings and King, 1970).

Moths like *Hyalophora cecropia* and *Bombyx mori* do not feed as adults and live only a few days. Vitellogenesis occurs during late pupal development, and during the prolonged time interval that elapses between the end of feeding and the beginning of vitellogenesis, proteins are synthesized and stored in the fat body (Pan, Bell, and Telfer, 1969). These are later transported to the ovary in the blood. One such protein, which is 1000 times more abundant in the blood of female than male cecropia pupae, is taken up preferentially by the ovary and incorporated into protein-rich yolk spheres (Telfer, 1954, 1960, 1961). Subsequent ultrastructural studies (King and Aggarwal, 1965; Stay, 1965) have demonstrated that during the time such yolk spheres are forming the follicle cells surrounding the oocyte separate and thus provide free access of the oocyte surface to protein-rich blood which the oolemma internalizes by both macro- and micropinocytosis.

VIII. MUTATIONS WHICH INFLUENCE VITELLOGENESIS

In *Bombyx mori* Doira (1968) has shown that a female specific protein (FP) also occurs in pupal hemolymph. This protein increases in concentration until it makes up about 17 per cent of the total blood protein. Its concentration then falls, and it is present in only traces in adult females.

Bombyx females homozygous for the *small egg* mutant (*sm*, 3-41.8) produce eggs which lack protein yolk spheres (Otsuki, 1965). Doira has shown (unpublished work) that homyzygous *sm* pupae synthesize FP and that a normal ovary implanted into a *sm* host will undergo normal vitellogenesis. FP is distributed mainly in the yolk fraction of homogenates prepared from normal wild type eggs or from eggs from implanted ovaries obtained from experiments of the type referred to in the previous sentence. The eggs of *sm/sm* females contain only trace amounts of protein, and the FP remains at a high concentration in the blood of late pupae and adults. A similar situation is found with respect to the level of FP in the blood of ovariectomized wild type females. The above results demonstrate that in the absence of the normal allele of *sm*, silkworms are unable to incorporate FP into their oocytes.

In *Drosophila melanogaster* two mutations have been studied in which vitellogenesis is greatly retarded. Females homozygous for *apterous*[4] (*ap*[4], 2-55.4) contain an amount of yolky ooplasm in their ovaries which is no more than 1% that found in wild type females of equivalent age. However, the ovaries from *ap*[4]/*ap*[4] females will produce abundant yolk when implanted into the abdomens of wild type adults (King and Bodenstein, 1965). Obviously, the abdominal environment of *ap*[4] is the cause of sterility, rather than a malfunctioning of the ovary, but what diffusible factors are lacking is unknown.

The growth of ooplasm is also retarded in females homozygous for *tiny (ty,* X-44.5), but mutant ovaries behave autonomously when implanted into the abdomens of wild type adults (King and Bodenstein, 1965). The primary effect of the *tiny* gene seems to be on the follicle cells (Falk and King, 1964). These terminate their migratory phase precociously and enter a secretory state. The result is the production of an abnormally thick follicular epithelium about the nurse cells. Perhaps the flow of nutrients into the nurse cells is impeded, and these compounds serve as precursors for many of the yolk organelles.

Finally, we come to mutations like *deep orange (dor,* X-0.3), *almondex (amx,* X-27.7), *rudimentary (r,* X-54.5), and *fused.* Females homozygous for any of these genes produce offspring, only if they are mated to males bearing the normal allele of the gene in question. All genes are sex linked, and homozygotes produce only daughters (Fig. 14.8). Such maternal effects can be explained by assuming that the

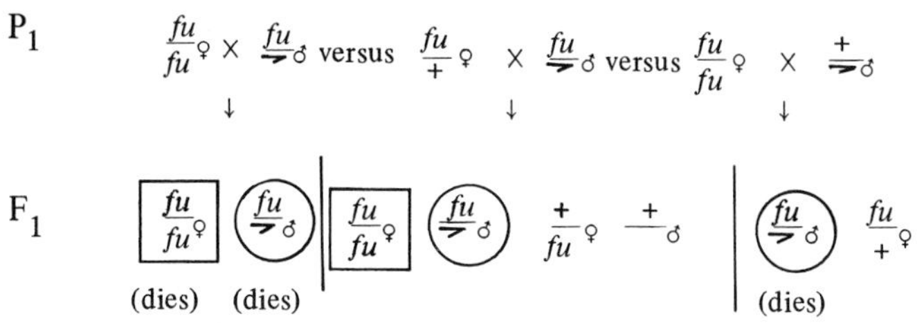

Fig. 14.8   Crosses involving the *fused* mutation which illustrates situations in which individuals of identical genotypes (enclosed) show dissimilar viabilities, presumably because of differences in maternally-contributed cytoplasm.

eggs produced by homozygotes are deficient in some compound essential for normal embryogenesis, but that this substance can be manufactured, subsequent to fertilization, if the normal allele has been contributed by the sperm. However, in heterozygous females the compound is synthesized by the nurse cells and stored in the ooplasm, and as a result homozygous mutant F_1 females and hemizygous mutant F_1 males can develop to adults (King, 1970b).

According to Bahn (1970) homozygous *r* females mated to hemizygous males produce viable offspring of both sexes, if the medium on which they are grown is enriched with cytidine. *Rudimentary* is the first female sterile mutant to have its fertility restored by the addition of a specific compound to the diet.

IX. ACKNOWLEDGMENTS

I acknowledge with gratitude the support of the National Science Foundation (grants GB 7457 and GF 351).

X. REFERENCES

Bahn, E. (1970) Restoration of fertility of the female sterile mutant *rudimentary* on pyrimidine-enriched culture medium. *Drosophila Information Service.* **45**, 99.

Brown, E. H. and King, R. C. (1962) Oogonial and spermatogonial differentiation within a mosaic gonad of *Drosophila melanogaster. Growth.* **26**, 53.

Brown, E. H. and King, R. C. (1964) Studies on the events resulting in the formation of an egg chamber in *Drosophila melanogaster. Growth.* **28**, 41.

Brown, S. W. and Zohary, D. (1955) The relationship of chiasmata and crossing over in *Lilium formosanum. Genetics.* **40**, 850.

Cassidy, J. D. and King, R. C. (1969) The dilatable ring canals of the ovarian cystocytes of *Habrobracon juglandis. Biol. Bull.* **137**, 429.

Court Brown, W. M. (1967) Population Cytogenetics. Amsterdam, North Holland Publishing Co.

Cummings, M. R. and King, R. C. (1969) The cytology of the vitellogenic stages of oogenesis in *Drosophila melanogaster.* I. General staging characteristics. *J. Morph.* **128**, 427.

Cummings, M. R. and King, R. C. (1970) The cytology of the vitellogenic stages of oogenesis in *Drosophila melanogaster.* II. Ultrastructural investigations on the origin of protein yolk spheres. *J. Morph.* **130**, 467.

Dapples, C. C. and King, R. C. (1970) The development of the nucleolus of the ovarian nurse cell of *Drosophila melanogaster. Z. Zellforsch.* **103**, 34.

Darlington, C. D. (1929) Meiosis in polyploids. Part II. Aneuploid hyacinths. *J. Genet.* **21**, 17.

Davidson, E. (1968) Gene Activity in Animal Development. New York, Academic Press.

Doira, H. (1968) Developmental and sexual differences of blood proteins in the silkworm *Bombyx mori. Science Bulletin, Faculty of Agriculture, Kyushu University.* **23**, 205.

Donachie, W. D. (1968) Relationship between cell size and the time of initiation of DNA replication. *Nature (London.).* **219**, 1077.

Falk, G. J. and King, R. C. (1964) Studies on the developmental genetics of the mutant *tiny* of *Drosophila melanogaster. Growth.* **28**, 291.

Fan, H. and Penman, S. (1970) Mitochondrial RNA synthesis during mitosis. *Science.* **168**, 135.

Gall, J., Macgregor, H. and Kidston, M. (1969) Gene amplification in oocytes of dytiscid water beetles. *Chromosoma.* **26**, 169.

Gowen, J. W. (1933) Meiosis as a genetic character in *Drosophila melanogaster. J. Exp. Zool.* **65**, 83.

Hinton, C. W. (1966) Enhancement of recombination associated with the *c(3)G* mutant of *Drosophila melanogaster. Genetics.* **53**, 157.

Hinton, C. W. (1967) Genetic modifiers of recombination in *Drosophila melanogaster. Canad. J. Genet. Cytol.* **9**, 711.

Jacob, J. and Sirlin, J. L. (1959) Cell function in the ovary of *Drosophila.* I. DNA. *Chromosoma.* **10**, 210.

King, R. C. (1968) The synaptomere-zygosome theory of synaptonemal complex formation. *Amer. Zool.* **8**, 822.

King, R. C. (1969) Control of oocyte formation of *female-sterile (fes) Drosophila melanogaster. Natl. Cancer Inst. Monogr.* **31**, 347.

King, R. C. (1970a) The meiotic behavior of the *Drosophila* oocyte. *Int. Rev. Cytol.* **28**, 125.

King, R. C. (1970b) Ovarian Development in *Drosophila melanogaster.* New York, Academic Press.

King, R. C. and Aggarwal, S. K. (1965) Oogenesis in *Hyalophora cecropia. Growth.* **29**, 17.

King, R. C. and Akai, H. (1971a) Spermatogenesis in *Bombyx mori.* I. The canal system joining sister spermatocytes. *J. Morph.* **134**, 47.

King, R. C. and Akai, H. (1971b) Spermatogenesis in *Bombyx mori.* II. The ultrastructure of synapsed bivalents. *J. Morph.* (In press).

King, R. C. and Bodenstein, D. (1965) The transplantation of ovaries between genetically sterile and wild type *Drosophila melanogaster. Z. Naturforsch.* **20b**, 292.

King, R. C. and Burnett, R. G. (1957) Hereditary ovarian tumors in *Drosophila melanogaster. Science.* **126**, 562.

King, R. C. and Hewlett, J. (1970) Studies on the origin of the hereditary ovarian tumors of *fes Drosophila melanogaster. Amer. Zool.* **10**, 526.

King, R. C., Koch, E. A. and Cassens, G. A. (1961) The effect of temperature upon the hereditary ovarian tumors of the *fes* mutant of *Drosophila melanogaster. Growth.* **25**, 45.

King, R. C. and Wilson, L. P. (1955) Studies with radiophosphorus in *Drosophila.* V. The phosphorus balance of adult females. *J. Exp. Zool.* **130**, 71.

Klug, W. S., Bodenstein, D. D. and King, R. C. (1968) Oogenesis in the *suppressor² of Hairy-wing* mutant of *Drosophila melanogaster.* I. Phenotypic characterization and transplantation experiments. *J. Exp. Zool.* **167**, 151.

Klug, W. S., King, R. C. and Wattiaux, J. M. (1970) Oogenesis in the *suppressor² of Hairy-wing* mutant of *Drosophila melanogaster.* II. Nucleolar morphology and *in vitro* studies of RNA and protein synthesis. *J. Exp. Zool.* **174**, 125.

Koch, E. A. and King, R. C. (1964) Studies on the *fes* mutant of *Drosophila melanogaster. Growth.* **28**, 325.

Koch, E. A. and King, R. C. (1966) The origin and early differentiation of the egg chamber of *Drosophila melanogaster. J. Morph.* **119**, 283.

Koch, E. A. and King, R. C. (1969) Further studies on the ring canal system of the ovarian cystocytes of *Drosophila melanogaster. Z. Zellforsch.* **102**, 129.

Koch, E. A., Smith, P. A. and King, R. C. (1970) Variations in the radio-sensitivity of oocyte chromosomes during meiosis in *Drosophila melanogaster. Chromosoma.* **30**, 98.

Lindsley, D. L. and Grell, E. H. (1968) Genetic Variations of *Drosophila melanogaster,* Washington, Carnegie Institution. Publ. No. 627.

Lockshin, R. A. (1966) Insect embryogenesis: macromolecular synthesis during early development. *Science.* **154**, 775.

Meyer, G. F. (1960) The fine structure of the spermatocyte nuclei of *Drosophila melanogaster. Proc. Europ. Regional Conf. Electron Microscopy.* (Delft). **2**, 951.

Morgan, T. H. (1912) Complete linkage in the second chromosome of the male of *Drosophila. Science.* **36**, 719.

Morgan, T. H. (1914) No crossing over in the male of *Drosophila* of genes in the second and third pairs of chromosomes. *Biol. Bull.* **26**, 195.

Otsuki, Y. (1965) Studies on the yolk formation in the silkworm, *Bombyx mori.* I. An analysis of its mechanism revealed by the comparison between normal and abnormally small egg. *Bulletin, Faculty of Industrial and Textile Fibers, Kyoto University.* **4**, 314.

Pan, M. L., Bell, W. J. and Telfer, W. H. (1969) Vitellogenic blood protein synthesis by insect fat body. *Science.* **165**, 393.

Prescott, D. M. (1956) Relation between cell growth and cell division. III. Changes in nuclear volume and growth rate and prevention of cell division in *Amoeba proteus* resulting from cytoplasmic amputation. *Exp. Cell Res.* **11**, 94.

Ritossa, F. M. and Spiegelman, S. (1965) Localization of DNA complementary to ribosomal RNA in the nucleolus organizer of *Drosophila melanogaster. Proc. Nat. Acad. Sci., U.S.A.* **53**, 737.

Salb, J. and Marcus, P. (1965) Translational inhibition in mitotic HeLa cells. *Proc. Nat. Acad. Sci., U.S.A.* **54**, 1353.

Sandler, L. (1970) The regulation of sex chromosome heterochromatic activity by an autosomal gene in *Drosophila melanogaster. Genetics.* **64**, 481.

Sang, J. H. and King, R. C. (1961) Nutritional requirements of axenically cultured *Drosophila melanogaster* adults. *J. Exp. Biol.* **38**, 793.

Schroeder, T. E. (1970) The contractile ring. I. Fine structure of dividing mammalian (HeLa) cells and the effects of cytochalasin B. *Z. Zellforsch.* **109**, 431.

Smith, P. A. and King, R. C. (1966) Studies on *fused,* a mutant producing ovarian tumors in *Drosophila melanogaster. J. Nat. Cancer Inst.* **36**, 445.

Smith, P. A. and King, R. C. (1968) The genetic control of synaptonemal complexes in *Drosophila melanogaster. Genetics.* **60**, 335.

Sonnenblick, B. P. (1947) Synchronous mitoses in *Drosophila,* their intensely rapid rate, and the sudden appearance of the nucleolus. *Rec. Gen. Soc. Am.* **16**, 52.

Stay, B. (1965) Protein uptake in the oocytes of the cecropia moth. *J. Cell Biol.* **26**, 49.

Telfer, W. H. (1954) Immunological studies of insect metamorphosis. II. The role of a sex limited protein in egg formation by the cecropia silkworm. *J. Gen. Physiol.* **37**, 539.

Telfer, W. H. (1960) The selective accumulation of blood proteins by the oocytes of saturiid moths. *Biol. Bull.* **118**, 338.

Telfer, W. H. (1961) The route of entry and localization of blood proteins in the oocytes of saturniid moths. *J. Biophys. Biochem. Cytol.* **9**, 747.

Turpin, R. and Lejeune, J. (1969) Human Afflictions and Chromosomal Aberrations. New York, Pergamon Press.

Yarger, R. J. and King, R. C. (1971) The phenogenetics of a temperature sensitive, autosomal dominant, female sterile gene in *Drosophila melanogaster. Develop. Biol.* (In press).

Zalokar, M. (1960) Sites of ribonucleic acid and protein synthesis in *Drosophila. Exp. Cell Res.* **19**, 184.

15

PARTHENOGENESIS AND HETEROPLOIDY
IN THE MAMMALIAN EGG

R. A. Beatty

I.   Introduction
II.   Earlier reports of parthenogenesis in ovarian eggs
III.   Haploid mouse blastocysts
IV.   Simple haploid parthenogenesis
V.   Gynogenetic haploid embryos
VI.   Diploid parthenogenesis after suppression of the second polar body of the egg
VII.   Parthenogenesis to term in man?
VIII.   Viability of parthenogenetic embryos
IX.   Triploidy after suppression of the second polar body of the egg
X.   Triploidy in man
XI.   Viability of mammalian triploids
XII.   Tetraploidy, pentaploidy, hexaploidy
XIII.   Sex-chromosome aneuploids in relation to meiotic non-disjunction
XIV.   Autosomal aneuploids
XV.   Oocyte mechanisms for the production of chromosomal mosaics
         A.   Equal first meiotic division
         B.   Equal second meiotic division
XVI.   Discussion
XVII.   References

I. INTRODUCTION

In less sophisticated days, genetic variation in vertebrate animals could be accounted for by "billiard-ball" genes strung along chromosomes and inherited by the laws of Mendel according to simple cytogenetic basis whereby a diploid set of chromosomes was reduced to a haploid set at meiosis and diploidy was restored at fertilization. Any genetic oddities could be ascribed to *ad hoc* mutations. More than half a century ago work began on variants from this simple basis, particularly in Amphibia, showing that eggs could develop without fertilization and yield parthenogenetic embryos, or that animals with abnormal numbers of chromosomes could be recognized (heteroploidy).

Parallel studies in mammals lagged behind. Heteroploidy was described in pre-implantation embryos in 1949, in implanted embryos in 1951, and after birth in 1959. The sex-chromatin of cat cells, originally no more than an accessory means of distinguishing between male and female cats, was described in 1949. Subsequently, observation of sex chromatin became of the highest importance as a method for screening large numbers of human beings in order to detect those due for chromosomal analysis, and led directly to the discovery of the first human aneuploids. The number of laboratories throughout the world working on mammalian heteroploidy is now to be numbered in scores. Early parthenogenetic embryos of mammals had been known for a long time. An extensive series of parthenogenetic pre-implantation stages (blastocysts) was described in 1954, implantation was proved in 1970, but born mammalian parthenogenones have yet to be discovered.

The present article has the restricted aim of discussing some mechanisms in the oocyte that give rise to heteroploidy in the zygote or to parthenogenesis in the "zygoid". The spermatocyte receives less attention. I make no attempt to present a balanced account, and many aspects are referenced no more than is necessary to introduce a point of discussion. Some general references are: vertebrates (Beatty, 1957, 1967); mammals (Austin, 1961; Fechheimer, 1968); experimental mammals (Russell, 1962); human (Bartalos and Baramki, 1967; Fechheimer, 1970); sex-chromosomes (Beatty, 1964b; Mittwoch, 1967; Ohno, 1967).

II. EARLIER REPORTS OF PARTHENOGENESIS
IN OVARIAN EGGS

Earlier work on apparent parthenogenetic development of the ovarian eggs of mammals has been summarized by Pincus (1936a). Subsequent claims were made by Saglik (1938) in apes, Krafka (1939) in the human and Dempsey (1939) in the guinea-pig. Most of these claims were interpreted by Thibault (1949, 1952) in terms of degenerative processes. Bacsich and Wyburn (1945) described a guinea-pig ovarian egg with a centrally placed spindle, as if it were about to divide. One is not sure of the chromosome number in these eggs.

III. HAPLOID MOUSE BLASTOCYSTS

In work by the writer and Prof. M. Fischberg from 1949 onwards, summarized in Beatty (1957), several haploid 3½-day mouse embryos were identified, some at least being blastocysts. Virtually all came from matings of a silver strain mother to a non-silver father. Some selection for high yield of heteroploid eggs of all types had been practiced during the breeding of the silver mice. In generations 0 - 3 there was one haploid embryo out of some 800 analyzable embryos, and in generations 4 - 7, six out of 217. This work showed that mouse blastocysts could exist in the haploid state, and the apparent strain difference and apparent response to selection indicated a genetic control of their incidence. All these embryos came from mated mice and there was therefore no formal evidence that they were parthenogenetic rather than androgenetic, except on the comparative grounds that androgenesis (development with the paternal chromosomes only) is the rarer phenomenon in Amphibia.

IV. SIMPLE HAPLOID PARTHENOGENESIS

In simple haploid parthenogenesis, meiosis is normal, but the female gamete with its reduced (haploid) number of chromosomes becomes activated and develops into a haploid embryo (Fig. 15.1).

Direct evidence of spontaneous haploid parthenogenesis in uninseminated ferrets was provided by Chang (1950), who recovered tubal or uterine eggs similar to normal 2 - 6 cell stages, one at least of the eggs having a haploid chromosome complement. Parthenogenesis at the 1-cell stage was reported in rat eggs by Austin (1951) and Austin and Braden (1954b).

Various experimental treatments in several mammals have produced 1- or 2-celled haploid parthenogenetic eggs (Thibault, 1949; Austin and Braden, 1954a, b; Braden and Austin, 1954a). In the rabbit, Pincus (1936b, 1939b) recorded an 8-celled haploid parthenogenetic egg after hypertonic treatment, and Chang (1952) described two 7-celled stages after heat treatment.

Using a technique not previously applied to mammals, Tarkowski, Witkowska and Nowicka (1970) passed an electric current through mouse eggs *in situ*. Four to ten hours afterwards, haploid eggs with a second polar body and one pronucleus were identified. In later development, haploid morulae and blastocysts were identified, as well as haploid/diploid mosaics which presumably arose by chromosome doubling in some of the blastomeres of an initially haploid egg. Some haploids developed better than diploids; this recalls a generalization for mice (Beatty and Fischberg, 1951) that cell number in heteroploids tends to be in inverse proportion to the degree of ploidy. Numerous implantations were observed, the latest stage being an 8-somite embryo at 10 days, but the ploidy of the implanted embryos was unknown. This is the only real evidence of implantation of parthenogenetic embryos in any mammal.

A different approach was made by Graham (1970), who took unfertilized mouse eggs at the metaphase of the second meiotic division, exposed them to hyaluronidase, and cultured them *in vitro*. At 4 hours some had emitted a second polar body, contained a single pronucleus, and must have been haploid. Those cleaving to the 2-cell stage at the normal time were presumably still haploid. Others cleaved late and were presumed to have become diploid by omission of the (normal) first cleavage division of the cytoplasm. Two days later, haploid embryos were

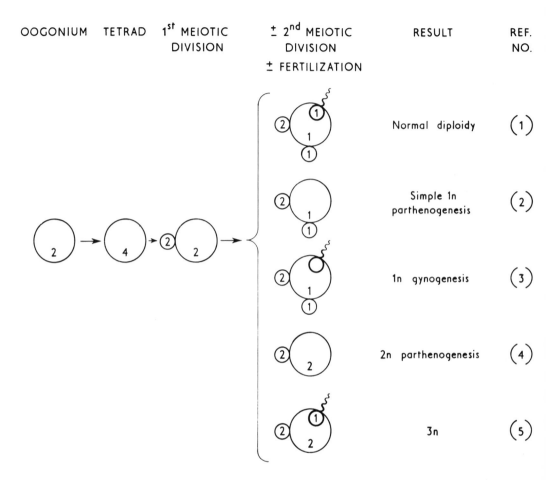

Fig. 15.1 Chromosome distribution in the egg after normal first meiotic division, with normal or suppressed second meiotic division, and with or without fertilization.

The fate of the four chromosome strands of the tetrad is displayed. Numerals represent number of strands. A "suppressed" meiotic division is one in which the cytoplasmic division does not take place but chromosome division proceeds normally. The first meiotic division is shown taking place horizontally on the page, the second meiotic division vertically. The main circles indicate the oogonium or oocyte. Small circles outside the main circles are polar bodies, with 2 chromosome strands in the first polar body and one strand in the second polar body. The spermatozoon is marked by a stylized tail. Possible cleavage of the first polar body is not shown.

identified, though these were not necessarily produced by simple haploid parthenogenesis. A five-day blastocyst transferred to the testis capsule produced a growth from which a tissue-culture was developed. Chromosomes could not be counted in the culture, but the relative light absorbance of the nuclei in Feulgen preparations indicated a haploid DNA content. It is perhaps a little unfortunate that the growth was developed in a testis, which itself contains haploid tissue that might have contributed cells to the growth.

V. GYNOGENETIC HAPLOID EMBRYOS

Normally, a mature egg is in the metaphase or anaphase of the second meiotic division and will degenerate if not entered by a spermatozoon. At normal fertilization the spermatozoon has two functions, the first being to activate the egg to undergo cortical changes in the ooplasm, complete its second meiotic division, and emit a second polar body. The second function is to contribute a haploid chromosome set to the egg. These two functions are separable. Gynogenetic development occurs when the spermatozoon enters and activates the egg but does not contribute effective chromosomes (see review by Beatty, 1964a). The zygoid is therefore haploid, with maternal chromosomes only (Fig. 15.1). Effectively gynogenesis is a kind of parthenogenesis. Gynogenesis can be effected experimentally by "fertilizing" eggs with spermatozoa whose nuclei have been inactivated by ionizing radiation or by use of radiomimetic dyes. Experimental gynogenesis, highly effective in Amphibia, has been applied to mammals. After the pioneer experiment of Amoroso and Parkes (1947) with rabbit eggs "fertilized" by spermatozoa treated with X-rays, Edwards (1954b, 1957a) inseminated mice with spermatozoa treated *in vitro* with various dosage levels of X-rays. With increasing dosage, the number of chromosomes found in the 3½ day blastocysts declined progressively towards the haploid number of 20. In certain mitoses it could be seen that the number of normal and presumably maternal chromosomes was precisely 20, while the few additional and presumably paternal ones were abnormal or fragmented and would be expected to perish. Bruce and Austin (1956) mated female mice to males whose scrotal areas had been irradiated with X-rays and obtained some eggs with a single pronucleus.

In comparable experiments with mouse spermatozoa subjected *in vitro* to ultraviolet light, Edwards (1954b, 1957b) obtained haploid or near-haploid 3½-day embryos, the incidence increasing with dosage. At the optimum dosage, about 60% of all embryos were not diploid, and about one third of these were haploid. There was a tendency towards an all-or-none effect, embryos being either haploid or near-haploid, or else diploid or near-diploid. Radiomimetic dyes and nitrogen mustard, effective in Amphibia, are less so in mammals (Thibault, 1949; Edwards, 1954b, 1958).

VI. DIPLOID PARTHENOGENESIS AFTER
SUPPRESSION OF THE SECOND POLAR BODY
OF THE EGG

With simple haploid parthenogenesis (and haploid gynogenesis) the problem was to activate the egg into emitting its second polar body and continuing with haploid development. With diploid parthenogenesis caused by second polar body suppression, the problem is to activate the egg without emission of the second polar body, so that two maternal sets of chromosomes remain within the developing egg (Fig. 15.1). In a figurative sense, it is as if the oocyte is fertilized by its own second polar body. Techniques for this kind of parthenogenesis have been highly successful in Amphibia.

Parthenogenesis of this kind was demonstrated in the rabbit after chilling unfertilized eggs (Pincus and Shapiro, 1940; Shapiro, 1942; Thibault, 1949). It was amply confirmed by Chang (1952) who chilled unfertilized rabbit eggs *in vitro* and followed their development after transfer to host females. Chang (1954) stored unfertilized eggs at $10°C$ for one day and then transferred them to the Fallopian tubes of host females. At 6 days, about 18% of the transferred ova had developed into blastocysts, many of normal appearance. Chang observed that the second maturation spindle disappeared, that the two sets of chromosomes mingled, that no second polar body was emitted, that the maternal chromosomes transformed into a pronucleus, and that mitosis followed. None of 230 ova transferred came to term, and earlier reports that parthenogenones sometimes come to term were not, therefore, confirmed. The technique of heating unfertilized rabbit eggs gave some early activation (Pincus, 1939a, b) but the parthenogenetic status of a litter of young born from transferred eggs was in doubt, since an incompletely vasectomized male had been used to set the sexual cycle of the host female into operation. A combination of general handling procedure with subjection to abnormal tonicity yielded some cleaved parthenogenetic eggs (Pincus, 1939b; Pincus and Shapiro, 1940). Five young born after such treatments were reported by Pincus (1939a, b). Two of them were not necessarily parthenogenetic (see Beatty, 1957). If we exclude the possibility of some kind of experimental error, then the remaining three must have been parthenogenetic. But there have been no further reports of born parthenogenones in the ensuing thirty years, and the parthenogenetic status of these born young must now be regarded as doubtful.

VII. PARTHENOGENESIS TO TERM IN MAN?

There are several documented case histories of apparent parthenogenetic birth in women. But because of human frailty and because of the powerful social motivations against admitting illegitimacy, no one can feel convinced by these reports. A more scientific approach was made by Balfour-Lynn (1956) who examined 19 women and their allegedly parthenogenetic daughters. Eleven of the claims could easily be eliminated, and seven more because of genetic traits possessed by the daughters but not by their mothers. In the remaining case, the mother and daughter were identical in their blood groups, in ability to taste phenyl thiocarbamide, in secretory activity and in their serum electrophoresis pattern. However, a skin graft from daughter to mother was rejected, and the conservative interpretation of this would be that the daughter contained genes not shared with the mother, and could not therefore be parthenogenetic.

It is sometimes thought that a diploid parthenogenone should be genetically identical with its mother at all loci. This is not true in mammals, where all known mechanisms of parthenogenesis involve meiosis, crossing over, and elimination of some of the maternal genes. Certainly, a diploid parthenogenone must share all its genes with its mother, but the reverse is not true. Absolute genetic identity with the mother can be achieved only if the mother is 100% homozygous, as in a fully inbred strain (in which case ordinary breeding would also produce genetic identity), or else by a mechanism for which there is no known precedent in mammals: ameiotic production of an egg, with persistence of the whole diploid somatic set of the mother.

VIII. VIABILITY OF PARTHENOGENETIC EMBRYOS

Very large numbers of parthenogenetic mammalian embryos have been described, but they do not come to term. It is understandable that haploid parthenogenones should perish. Even in Amphibia, where all kinds of chromosomally abnormal animals can become adult, the haploid is very inviable. But diploid mammalian parthenogenones with their normal chromosome number also fail to be born. The usual explanation is that diploid parthenogenones, whether produced by first or second polar body suppression (and especially if produced by doubling of a haploid chromosome complement in the one-celled zygoid) are highly homozygous and suffer from the equivalent of "inbreeding depression". But reports of born parthenogenones do not come from those who breed inbred strains of mice, in which near-100% homozygosity is evidently no bar to viability. This absence of occasional spontaneous parthenogenesis to birth in inbred strains is, perhaps, surprising. Since the diploid parthenogenone is chromosomally balanced and does not depend on spermatozoan chromosomes for that balance, yet perishes in later embryonic development, it has been suggested that spermatozoa may contribute some cytoplasmic factor, essential for normal later development, but lacking in the diploid parthenogenone (Beatty, 1970a).

In speaking of the identification of parthenogenones, a certain semantic difficulty emerges, for which there is no clear-cut solution, though we can use some common sense. Final proof of parthenogenesis, not yet available, would come from observation of a born parthenogenone. As we move back in development, we have no hesitation in referring to an implanted foetus or to well-developed blastocysts as parthenogenetic. Still further back, in early cleavage stages, normally-developing eggs can be termed "truly" parthenogenetic, but there is not much of a morphological phenotype on which to base this judgment. At the one-celled stage there is very little to go upon, and we begin to speak of "rudimentary" parthenogenesis. Further, at very early stages, we may observe eggs with "blastomeres" of unequal size, with sub-nuclei or no nuclei. We do not always know whether to refer to this as "rudimentary" parthenogenesis or "abortive" parthenogenesis or, more simply as "degeneration".

IX. TRIPLOIDY AFTER SUPPRESSION OF THE SECOND POLAR BODY OF THE EGG

From the point of view of the oocyte, this kind of triploidy arises in exactly the same way as does the diploid parthenogenesis just described; i.e., suppression of the second polar body gives a diploid set of chromosomes to the egg. But a fertilizing spermatozoon adds a third haploid set to give triploidy (Fig. 15.1). This is the dominant form of triploidy in vertebrates other than mammals, and in invertebrates, and the number of known individuals in Amphibia is perhaps to be numbered in thousands.

A series of spontaneously triploid 3½-day mouse blastocysts is described in papers from 1949 onwards by Prof. M. Fischberg and the writer (see summary in Beatty, 1957). In various strains the incidence averaged 0.25%. This percentage increased when one of the parents in a mating came from a *silver* strain, and was particularly high (5.7%) when *silver* females were mated to non-*silver* (CBA) males. Two factors were thought to be operative: (a) a general propensity for the *silver* strain to yield triploid embryos, and (b) a genetic interaction whereby a particular combination of parents of different strains yielded a particularly high incidence. There was no direct proof that these triploids arose by second polar body suppression, but this seemed on *a priori* grounds to be the most likely mode of origin, and was supported by the observation of Braden (1957) that second polar body suppression was typical of *silver* mice. At mid-gestation (9½ days after coitus) several implanted triploids were discovered, somewhat less well developed than accompanying diploids, but with beating hearts, somites and circulating blood. No triploids were observed among several hundred born young from *silver* mice (Beatty, 1964b).

A higher yield of triploid embryos was obtained after experimental treatments. Eleven per cent of 173 analyzable 3½-day mouse embryos were scored as triploid after subjecting eggs near the time of fertilization to a heat shock intended to suppress the second polar body. The method of heat shock had been adapted from Amphibia, where it is a well-known technique, and Braden and Austin (1954b) confirmed in mice that suppression of the second polar body is indeed the main effect of heat shock. A still higher incidence (some 20% at optimum treatment) was recorded by Edwards (1954a) after mating female mice that had been injected with colchicine. Many of his 3½-day triploid embryos seemed to be developing normally. Delayed fertilization also induces triploidy (Vickers, 1969).

In the rat, Piko and Bomsel-Helmreich (1960) injected females with colchicine with the intention of producing second polar body suppression. Triploids were produced by this means, and also by the alternative methods of bringing about dispermic fertilization of a normal haploid female gamete. The triploids implanted, but none were identified after about 12 days of gestation, and the chances of reaching term were thought to be remote. In the rabbit, colcemid treatment of eggs just before fertilization gave an exceptionally high yield of triploid eggs (some 80%), presumed to have arisen by second polar body suppression, but the embryos died at about mid-gestation (Bomsel-Helmreich and Thibault, 1962).

X. TRIPLOIDY IN MAN

Triploidy in man has recently been reviewed by Schindler and Mikamo (1970). The first cases discovered were diploid/triploid mosaics, and triploidy has also been detected in hydatiform moles. The present discussion will be confined to pure triploids. Since one cannot experiment with man, it is difficult to elucidate the cytology of the origin of triploids.

Early triploid embryos are quite common and constitute about one fifth of all spontaneous abortions. The sex-chromosome complements summarized by Schindler and Mikamo for 51 such early triploids were: 17 XXX: 30 XXY: 4 XYY (ratio 1:1.76:0.24). If sigyny is assumed (fertilization of a diploid XX egg by either an X or a Y spermatozoon), the expected proportions are 1 XXX: 1 XXY. If diandry is assumed (fertilization of a haploid X egg by either two spermatozoa, or, possibly, by a diploid spermatozoon) then the expected proportion is 1 XXX: 2 XXY: 1 XYY. The observed figures could best be accounted for by 40% of diandric origin, 60% of digynic origin. However, the scarcity of XXY suggested that they had a decreased viability, and the diandric proportion rises to 80% if the calculation is based only on the XXX and XXY triploids. It seems from this work that suppression of the second polar body, generally taken to be the major cause of triploid embryonic development in animals, plays a relatively smaller role in man. Further, it is only in man that pure triploids have been scored perinatally, three having survived to the stage of a premature birth, while one died just before term. All four were XXY males.

XI. VIABILITY OF MAMMALIAN TRIPLOIDS

It seems now that triploidy is prenatally lethal in mammals (mouse, rat, rabbit, and as a rule in man), with death towards the middle of the gestation period. Up to that point development may be somewhat retarded but otherwise fairly normal. Several reasons for the lack of viability have been suggested. A straightforward possibility is that triploid cells are physiologically inadequate and that embryogenesis cannot proceed normally. One would not have expected this on comparative grounds, since Amphibian adult triploids are sometimes fully viable, but one could say in a vague sense that the later mammalian embryo is at a more complex level of organization than an amphibian and has more to go wrong with it. Stern (1958) suggested a lack of physiological balance between a triploid foetus and a diploid mother, and Bomsel-Helmreich and Thibault (1962) suggest the specific possibility of antigenic incompatibility. It is also possible that triploid embryos, with their retarded development, demand a longer gestation period than that evolved by a species for the purpose of gestating diploid offspring.

Only in man are there records of born triploids. This does not necessarily mean that the human is truly unique in this respect. Every effort is made to keep alive an abnormal human conceptus, and cytogeneticists are used to seeking out abnormal infants as a source of chromosomal abnormalities. But mice are not born in hospitals, and there is no one to assist a parturition, to feed an abnormal offspring too weak to feed itself, or to prevent the mother from eating it. It may be that rare born triploids occur also in other mammals but have not yet reached the hands of cytogeneticists.

One negative point seems very clear. This is, that it is extremely unlikely that there will ever be any economic use of triploid mammals. They are of interest in other connections. They can throw some light on the relative incidence of meiotic abnormalities in the mother and father. They provide gene dosages other than those available from ordinary diploids, and for this reason they have played an invaluable part in formulating the theory of chromosomal sex-determination in mammals, the main lesson being the over-riding importance of the Y chromosome in determining a male gonad. And they have provided a causal explanation for no less than 20% of early spontaneous abortions in man.

XII. TETRAPLOIDY, PENTAPLOIDY AND HEXAPLOIDY

Rare spontaneously tetraploid pre-implantation mouse embryos have been identified (summary in Beatty, 1957). There was a higher incidence after an experimental treatment designed to double chromosome number in a normal diploid egg (Beatty and Fischberg, 1952), and another experimental treatment yielded one tetraploid egg (Vickers, 1969). McFeely (1967) found two spontaneously tetraploid early pig embryos, both XXYY, and Bartalos and Baramki (1967) reference three human abortions that were tetraploid, one of them XXXX, one XXYY. The simplest explanation for the origin of the tetraploidy in all these early embryos is that a suppression of the first cytoplasmic cleavage of normal XX and XY eggs had produced XXXX and XXYY.

Higher ploidies are extremely rare. A pentaploid rabbit blastocyst is known (Shaver and Carr, 1967). Two hexaploid cleavage stages were reported in mice (references in Beatty, 1957).

XIII. SEX CHROMOSOME ANEUPLOIDS IN RELATION
TO MEIOTIC NON-DISJUNCTION

In mammals there is virtually no evidence of partial sex-linkage, nor of cytological crossing-over between X and Y chromosomes. This means that the X and

Y chromosomes should disjoin at the first meiotic division ("pre-reduction"). Pre-reduction has been confirmed cytologically in mammals, though *Apodemus* is an exception (Matthey, 1957). If we use parentheses to indicate cells, a dot to indicate the centromere, and 0 to indicate lack of a sex chromosome, then the meiotic history of sex chromosomes may be illustrated in the normal male as follows. The diploid (X. .Y) constitution of spermatogonium becomes (X.X Y.Y) at the tetrad stage, and the first meiotic division produces (X.X)(Y.Y). Division of the centromeres and completion of the second meiotic division give four haploid cells (X.) (.X) (Y.) (.Y). Non-disjunction means that the centromeres with their attached chromosomal strands are partioned unequally between daughter cells. For instance, non-disjunction at the first meiotic division should yield (0) (X.X Y.Y), and a regular second meiotic division should then produce (0) (0) (X.Y) (X.Y). In female meiosis X replaces Y, and only one of the normal products of meiosis is a functional gamete.

Gametes of the constitutions shown in Fig. 15.2 are therefore possible, the superscripts M and P being used to indicate maternal versus paternal X chromosomes. (In the literature, further possibilities involving "post-reduction" are sometimes included, but these do not apply to mammals in general).

At fertilization, therefore, there are numerous possible sex chromosome constitutions for the zygote, and a surprisingly large number of these is already known. By the use of X-linked genetic markers, such as *tabby* in the mouse, or colour-blindness or Xg in man, it is sometimes possible to allot a paternal or maternal origin to each X chromosome of the zygote. By reference to the gametic types listed in Fig. 15.2, it may then be possible to trace the non- disjunction to the mother or father, and even to infer which meiotic division was involved. This is illustrated in Fig. 15.3 for the mouse. It will be seen that $X^M X^P Y$ and $X^M O$, both common constitutions, necessarily involve non-disjunction in the male, whereas non-disjunction in the female is an unnecessary and unlikely hypothesis. But $X^M X^M Y$, which would have involved non-disjunction in the female, is *not* found, and $X^P O$, which would also have involved the female, has only been found once. In short, it seems to be characteristic for the mouse that these non-disjunctions take place in the male rather than the female.

In man, however, there is a different balance of responsibility for non-disjunction (Fig. 15.4). $X^M X^M Y$, not known in the mouse, is relatively common in man. Either the male or female parent can be implicated in non-disjunction.

In short, the responsibility for non-disjunctional events giving rise to sex-chromosome aneuploids in the zygote is shared between the oocyte and the spermatocyte in man, whereas in the mouse the responsibility lies more on the spermatocyte than on the oocyte.

Although meiotic non-disjunction has predictable and necessary consequences

Female gamete

Normal disjunction	X^M		
Non-disjunction, either meiotic division	O	$X^M X^M$	
Non-disjunction, both meiotic divisions	O	X^M	$X^M X^M X^M X^M$

Male gamete

Normal disjunction	X^P	Y	
Non-disjunction, 1st meiotic division only	O	$X^P Y$	
Non-disjunction, 2nd meiotic division only	O	$X^P X^P$	YY
Non-disjunction, both meiotic divisions	O	X^P	Y	$X^P X^P Y$	$X^P Y Y$	$X^P X^P Y Y$

Fig. 15.2 Possible sex-chromosome constitutions of gametes after non-disjunction in a "pre-reductional" meiosis. The superscripts ^{M} and ^{P} indicate maternal or paternal X chromosomes. The haploid set of autosomes in each gamete is not shown.

Zygote		Presumed gamete constitutions giving rise to zygote		
Incidence	Constitution	Male	Female	Comment on mode of origin
Common	$X^M X^P Y$	$X^P Y$(ND)	X^M	Simplest explanation (1ND)
		$X^P Y$(ND)	or X^M(ND)	Unlikely (3 ND)
Unknown	$X^M X^M Y$	Y	$X^M X^M$(ND)	Simplest explanation (1 ND)
		or Y (ND)	$X^M X^M$(ND)	Unlikely (2 - 3 ND)
Common	$X^M O$	O (ND)	X^M	Simplest explanation (1 ND)
		O (ND)	or X^M(ND)	Unlikely (3 ND)
1 only	$X^P O$	X^P	O (ND)	Simplest explanation
		or X^P(ND)	O (ND)	Unlikely (2 -3 ND)

Fig. 15.3 Origin of some sex-chromosome aneuploids in mice in terms of meiotic non-disjunction (based on Russell, 1962). The presumed gametic constitutions are derived from Fig. 15.2. ND = non-disjunction.

for the sex-chromosome complement of the zygote, there are certain snags when the constitution of the zygote is used to argue backwards towards its mode of origin, though in a brief article I can refer to these complications only cursorily. For instance, non-disjunction in embryonic cells *after* fertilization is a possible mode of origin of some types. However, since an abnormal partition of chromosomes between somatic cells must mean that *both* daughter cells are abnormal, and in different ways, the result is more likely to be identified as a mosaic than as a uniform aneuploid.

XIV. AUTOSOMAL ANEUPLOIDS

Three common anomalies in man are caused by trisomy of one of the autosomes. Polani (1969) has reviewed Patau's syndrome (trisomy of the D_1 chromosome, one of the Nos. 13 - 15 group) and Edwards' syndrome (trisomy of chromosome No. 18). Penrose (1966) has reviewed Down's syndrome, or trisomy of chromosome No. 21. In all three syndromes and especially in Down's syndrome there is evidence of an increase in incidence with increased maternal age. Penrose has set out an hypothesis that the age-dependence of Down's syndrome is connected with a gradual weakening of the kinetochore (centromere) while the chromosomes are in the dictytene stage of early meiosis during the long period lasting from the birth of the mother until ovulation. The chromosome strands of a dyad may fall apart for this reason and be distributed at random at the close of the first meiotic

Zygote constitutions	Parent probably involved in non-disjunction
$X^M X^P Y$ 40%	♂
$X^M X^M Y$ 60%	♀
$X^M O$ 74%	♂
$X^P O$ 26%	♀
$X^M X^M X^M X^M$	♀
$X^M X^M X^M Y$	♀
$X^M X^M X^M X^M Y$	♀
$X^? X^P Y Y$	♂

Fig. 15.4 Some sex-chromosome aneuploids in man interpreted in terms of meiotic non-disjunction in one parent (based on Race and Sanger, 1969).

division, leading sometimes to a female gamete with two chromosome strands instead of one. This is not quite the same as the non-disjunction *sensu strictu* already described for sex chromosomes, where centromere structure is taken to be normal throughout. But the end result, production of a gamete with an incorrect number of chromosomes, is the same. It is perhaps worth underlining the fact that it is becoming customary to explain sex chromosome aneuploidy by non-disjunction *sensu strictu*, and chromosome No. 21 aneuploidy by centromere deterioration, though either chromosomal anomaly could, theoretically, be ascribed to either mode of origin.

XV. OOCYTE MECHANISMS FOR THE PRODUCTION OF CHROMOSOMAL MOSAICS

So far, we have considered parthenogenones or heteroploids with a uniform ploidy in all their cells. But animals mosaic for whole sets of chromosomes are well known. Because of recent trends in the literature, I would like to expand the possibilities of a particular oocyte mechanism by which they might arise, though it is not the only mechanism. Normally, a meiotic division in the egg is cytoplasmically unequal, producing a small non-functional polar body and a larger functional oocyte. There is increasing evidence that this division is sometimes equal, so that two functional oocytes are produced, attached to one another and potentially capable of further development as a single zygote or zygoid.

A. Equal First Meiotic Division

When the first polar body and the oocyte are equal in size we may speak of an "equal first meiotic division". Several instances have been observed in the eggs of mammals and non-mammals (summary by Austin, 1961). A theoretical scheme for the further development of such two-celled oocytes in given in Fig. 15.5. The possibilities depend on whether one, both or neither of the two cells undergoes a second meiotic division, and on whether one, both or neither are fertilized by a spermatozoon. Austin (1961) quotes an observation by Braden and Edwards of penetration of only one of the two cells by a spermatozoon, as if the egg were about to enter on one of the developmental routes with reference numbers 2, 5, 6 or 9 in the figure.

B. Equal Second Meiotic Division

Here it is the second meiotic division that produces two cells of equal size (Fig. 15.6). The process was called "immediate cleavage" by its discoverers, Braden and Austin, (1954a) (see also Austin, 1961). Further development would depend on whether one, both or neither of the two cells is penetrated by a spermatozoon. Austin (1961) lists several eggs from mated mice in which the presence of two pronuclei in one cell and one pronucleus in the other suggests the operation of the developmental route with reference number 2 in Fig. 15.6, the result being a $1n/2n$ mosaic. Parthenogenetic mosaic haploid development (reference number 3 in the Fig.) has been recorded in the mouse by Tarkowski, Witkowska and Novicka (1970), and such eggs have been cultured to the 4-cell stage by Graham (1970). Their later haploid stages must have originated by the routes numbered 3 or 5 in the Fig., and $1n/2n$ mosaics observed in later development must have come into being by chromosome doubling in one or more blastomeres of an originally haploid egg. Ford (1969), in reviewing human mosaics and chimaeras, refers to several cases where one egg is involved in two separate acts of syngamy. It may be pointed out that implanted diploid/triploid mosaics occur in mice (Fischberg and Beatty, 1951) and men (Schindler and Mikamo, 1970), and may arise by one of the developmental routes in Fig. 15.5, though a more complicated explanation is sometimes necessary.

XVI. DISCUSSION

Most of the cytological accidents that give rise to parthenogenetic or heteroploid development occur in the oocyte and the female gamete. The early embryo and the male germ line are less implicated. I would like to stress the wide range of chromosomal constitutions that can be possessed by the mammalian oocyte and embryo, and sometimes also by the adult. At the same time, I will outline the limits of this variation. In the euploid series, all degrees of chromosome number from haploid to hexaploid have been recognized in embryos, but only the triploid and the normal diploid can come to birth. Among the sex-chromosome aneuploids, the OO constitution is unknown at any stage, but OY can exist as an embryo and XO as an adult. All kinds of complex constitutions such as XXY, XXYY, XXXXXY

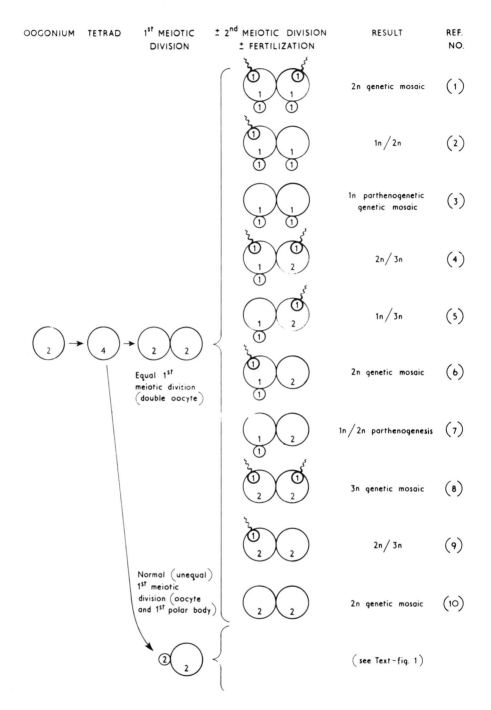

Fig. 15.5 Chromosome distribution in the egg after "equal first meiotic division", with normal or suppressed second meiotic divisions, and with or without fertilization. The symbolism is explained in Fig. 15.1.

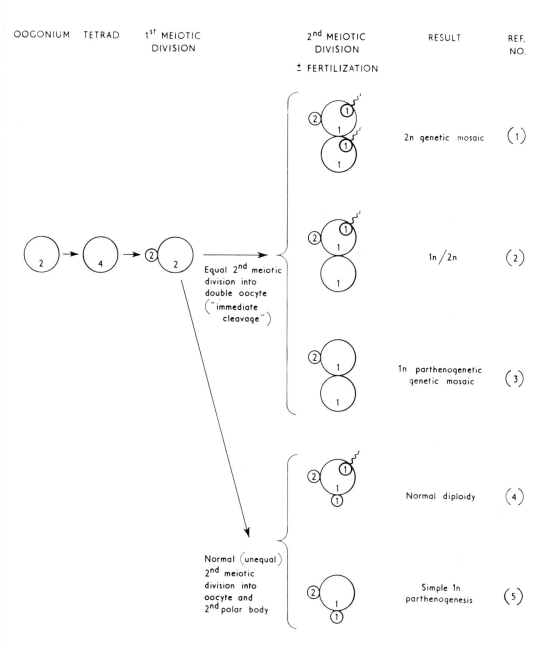

OOGONIUM   TETRAD   1ˢᵗ MEIOTIC   2ⁿᵈ MEIOTIC   RESULT   REF.

DIVISION   DIVISION   NO.

± FERTILIZATION

Fig. 15.6 Chromosome distribution in the egg after normal first meiotic division, with "equal second meiotic division" ("immediate cleavage"), and with or without fertilization. The symbolism is explained in Fig. 15.1.

exist as born human beings and some have been recognized in other mammals. Among the autosomal aneuploids, little is known about embryonic stages, but individuals can exist trisomic for one of a limited number of autosomes. Trisomy of other autosomes must be presumed to arise from time to time, but the resulting individuals are not found, and the constitutions must be lethal.

The gamete also can withstand a wide range of chromosomal abnormality, and yet remain fertile. From this and other evidence it is concluded that the differentiation and fertility of male and female gametes are almost wholly independent of their genetic content from meiosis onwards (see Beatty, 1970a, b).

There is evidence that many of the theoretically possible cytological accidents that might occur, do occur. Much of the literature on parthenogenesis and euploid heteroploidy can be referred to the operation or non-operation, singly or in combination, of four cytological anomalies in the egg: development without fertilization; chromosome doubling at the first maturation division; chromosome doubling at the second maturation division; chromosome doubling at the time of the first cleavage of the embryo. Much of the literature on aneuploid heteroploidy can be referred to non-disjunction in the male or female gamete, or in the early embryo. A major part of the research on mammalian heteroploidy is now concerned with working out these cytological mechanisms in detail.

Adult heteroploidy has now been recognized in all the main vertebrate groups: fish, amphibia, reptiles, birds and mammals. Adult parthenogenesis is known in fish, amphibia, reptiles and birds; in mammals, parthenogenetic development has been scored after implantation, and on comparative grounds it would not be too surprising if adult mammalian parthenogenesis were eventually to be reported.

At the level of cellular biology, research on mammalian parthenogenesis and heteroploidy may afford a means of producing tissue cultures of haploid, triploid and tetraploid cells for use in other fields of study. In experimental mammals, embryos with these degrees of ploidy can be produced to order. The mode of origin of triploid embryos raises a problem concerning the centriole. There is presumably a supernumerary centriole in one or both of the types of triploid produced either by dispermic fertilization or else by suppression of the second polar body. Yet regular cleavage can ensue in both types of triploid embryo, instead of the abortive multipolar divisions that would be expected from the presence of too many centrioles. Therefore, we face the possibility that the fertilized egg can in some way regulate the number of *effective* centrioles within it, or else that a correct number of centrioles arises *de novo* in the fertilized egg.

At the genetic level, heteroploid animals make available gene dosages additional to those available from normal diploidy and give extra points on a gene-dosage curve. By taking advantage of the availability of various chromosomal deviants we can, in the statistical sense, vary independently the number of X chromosomes, the number of Y chromosomes, the number of a limited number of autosomes, or the numbers of all chromosomes, When these variations in chromosome number are compared with the gonadal sex that develops in the animal, it is found that animals without a Y chromosome usually have a female gonad, but definite exceptions with a male gonad can exist. But, virtually without exception, animals with one or more Y chromosomes have male gonadal tissue. It is inferred that the Y chromosome has a powerful and mandatory effect in determining male gonadal tissue. This discovery, made possible by advances in mammalian heteroploidy, disposed of a prevailing but false analogy with *Drosophila,* where the Y chromosome is a dummy from the point of view of determining male gonadal tissue. The mandatory role of the Y chromosome (or, *mutatis mutandis,* of the W chromosome in animals with the ZW/ZZ mechanism) seems to be the same in all vertebrates (Beatty, 1964b).

On the evolutionary side, I would point out that triploid races are known in amphibia and reptiles, and that unisexual reproduction, once a rare exception in vertebrates, is known in fish, amphibia and reptiles (Schultz, 1969; Uzzell, 1970). But unisexual reproduction and polyploid races do not seem to have evolved in mammals.

In the medical field, fundamental, mission-orientated and practical researches are proceeding hand in hand. The chain of cause and effect leading to various syndromes such as Mongolism and Turner's syndrome, whose etiology was a mystery only twelve years ago, is now being described in detail. The immediate cause of many syndromes is quite certainly an imbalance in chromosome number. Advances have also been made in studying a second link in the chain - the cause of the chromosome abnormality itself. This has been traced to non-disjunction, or else to doubling of the chromosome number. Moving back to a third link in the chain - what causes the non-disjunction or chromosome doubling? Here again there have been advances of service in steering the course of further research. For example, work on the mouse would have told us that the main research effort for tracking the causes of non-disjunction would best be applied to the female, since male meiosis is little involved; but in man we already know that either parent can be involved, and both are therefore due for study. And it helps further research to know the time and place of the non-disjunctional event or the chromosome doubling in the oocyte or embryo. Cytological accidents traced by cytogenetical arguments to the first maturation of the oocyte must have taken place in the Fallopian tube near the time of fertilization; events occurring in the embryo must have taken place after fertilization, in the Fallopian tube or the uterus. With autosomal trisomy, such as Mongolism, genetic counselling is possible, because of a known dependence on the age of the mother, and because in familial Mongolism parents in a certain genetic situation will have a high and specifiable probability of producing a Mongol child at each birth. Triploidy is the only euploid state of importance in man. We have a causal explanation in terms of triploidy of about one-fifth of all spontaneous human abortion. Perhaps later research will show how to reduce in both male and female parents the incidence of the cytological accidents giving rise to triploidy. Parthenogenesis seems at present to have little medical relevance in man, but a negative point may be made: although ionizing radiation is a potent cause of gynogenesis and therefore of inviable embryos, the effective dose of radiation is so high that the real matter for concern would be survival of the male rather than survival of his progeny.

Finally, I must emphasize that all kinds of specific and non-specific factors, including mere handling of eggs, are known to result in maldistribution of chromosomes in the oocyte and embryo. Some of the chromosome abnormalities will produce inviable embryos, but other types will certainly lead to abnormal adults. In assessing the dangers of fertility control imposed by blocking oocyte or egg development in man, we need continued fundamental study of the mechanisms producing parthenogenesis or heteroploidy, and detailed cytogenetic evaluation of the effects of each individual blocking technique.

XVII. REFERENCES

Amoroso, E. C. and Parkes, A. S. (1947) Effects on embryonic development of X-irradiation of rabbit spermatozoa *in vitro. Proc. Roy. Soc. Edinb. B.* **134**, 57.

Austin, C. R. (1951) Activation and the correlation between male and female elements in fertilization. *Nature (London).* **168**, 558.

Austin, C. R. (1961) The Mammalian Egg. Oxford, Blackwell Scientific Publ.

Austin, C. R. and Braden, A. W. H. (1954a) Induction and inhibition of the second polar division in the rat egg and subsequent fertilization. *Aust. J. Biol. Sci.* **7**, 195.

Austin, C. R. and Braden, A. W. H. (1954b) Anomalies in rat, mouse and rabbit eggs. *Aust. J. Biol. Sci.* **7**, 537.

Bacsich, P. and Wyburn, G. M. (1945) Parthenogenesis of atretic ova in the rodent ovary. *J. Anat.* **79**, 177.

Balfour-Lynn, S. (1956) Parthenogenesis in human beings. *Lancet.* **2**, 1071.

Bartalos, N. and Baramki, T. A. (1967) Medical Cytogenetics. Baltimore, Williams and Wilkins Co.

Beatty, R. A. (1957) Parthenogenesis and polyploidy in Mammalian Development. Cambridge University Press.

Beatty, R. A. (1964a) Gynogenesis in vertebrates: fertilization by genetically inactivated spermatozoa. *In* Proc. Intern. Symp., "Effects of ionizing radiation on the reproductive system". (W. D. Carlson and F. X. Gassner, eds.), p. 229. Colorado.

Beatty, R. A. (1964b) Chromosome deviations and sex in vertebrates. *In* "Intersexuality in Vertebrates including man". (C. N. Armstrong and A. J. Marshall, eds.), p. 17. London and New York, Academic Press.

Beatty, R. A. (1967) Parthenogenesis in vertebrates. *In* "Fertilization". (C. Metz and A. Monroy, eds.), Vol. 1, p. 413. New York, Academic Press.

Beatty, R. A. (1970a) Phenotype of spermatozoa in relation to genetic content. *J. Anim. Sci.* (In press).

Beatty, R. A. (1970b) The genetics of the mammalian gamete. *Biol. Rev.* **45**, 73.

Beatty, R. A. and Fischberg, M. (1951) Cell number in haploid, diploid and polyploid mouse embryos. *J. Exp. Biol.* **28**, 541.

Beatty, R. A. and Fischberg, M. (1952) Heteroploidy in mammals. III. Induction of tetraploidy in pre-implantation mouse eggs. *J. Genet.* **50**, 471.

Bomsel-Helmreich, O. and Thibault, C. (1962) Developpement d'oeufs triploides experimentaux chez la lapine. *Annls. Biol. anim. Biochim. Biophys.* **2**, 265.

Braden, A. W. H. (1957) Variation between strains in the incidence of various abnormalities of egg maturation and fertilization in the mouse. *J. Genet.* **55**, 476.

Braden, A. W. H. and Austin, C. R. (1954a) Reactions of unfertilized mouse eggs to some experimental stimuli. *Exp. Cell Res.* **7**, 277.

Braden, A. W. H. and Austin, C. R. (1954b) Fertilization of the mouse egg and the effect of delayed coitus and of hot-shock treatment. *Aust. J. Biol. Sci.* **7**, 552.

Bruce, H. M. and Austin, C. R. (1956) An attempt to produce the Hertwig effect by X-irradiation of male mice. *Stud. Fert.* **8**, 121.

Chang, M. C. (1950) Cleavage of unfertilized ova in immature ferrets. *Anat. Rec.* **108**, 31.

Chang, M. C. (1952) Fertilizability of rabbit ova and the effects of temperature *in vitro* on their subsequent fertilization and activation *in vivo*. *J. Exp. Zool.* **121**, 351.

Chang, M. C. (1954) Development of parthenogenetic rabbit blastocysts induced by low temperature storage of unfertilized ova. *J. Exp. Zool.* **125**, 127.

Dempsey, E. W. (1939) Maturation and early cleavage figures in ovarian ova. *Anat. Rec.* **75**, 223.

Edwards, R. G. (1954a) Colchicine-induced heteroploidy in early mouse embryos. *Nature (London).* **174**, 276.

Edwards, R. G. (1954b) The experimental induction of pseudogamy in early mouse embryos. *Experientia.* **10**, 499.

Edwards, R. G. (1957a) The experimental induction of gynogenesis in the mouse. I. Irradiation of the sperm by X-rays. *Proc. Roy. Soc. [Biol.]* **146**, 469.

Edwards, R. G. (1957b) The experimental induction of gynogenesis in the mouse. II. Ultra-violet irradiation of the sperm. *Proc. Roy. Soc. [Biol.]* **146**, 488.

Edwards, R. G. (1958) The experimental induction of gynogenesis in the mouse. III. Treatment of sperm with trypaflavine, toluidine blue, or nitrogen mustard. *Proc. Roy. Soc. [Biol.]* **149**, 117.

Fechheimer, N. S. (1968) Consequences of chromosomal aberrations in mammals. *J. Anim. Sci.* **21**, 27.

Fechheimer, N. S. (1970) Cytogenetics in man. *In* "Genetic Disorders of Man". (R. M. Goodman, ed.), p. 22. Boston, Little Brown and Co.

Fischberg, M. and Beatty, R. A. (1951) Spontaneous heteroploidy in mouse embryos up to mid-term. *J. Exp. Zool.* **118**, 321.

Ford, C. E. (1969) Mosaics and chimaeras. *Brit. Med. Bull.* **25**, 104.

Graham, C. F. (1970) Parthenogenetic mouse blastocysts. *Nature (London).* **226**, 165.

Krafka, J. (1939) Parthenogenetic cleavage in the human ovary. *Anat. Rec.* **75**, 19.

Matthey, R. (1957) Les bases cytologiques de l'heredite "relativement" liee au sexe chez les mammiferes. *Experientia.* **13**, 341.

McFeely, R. A. (1967) Chromosome abnormalities in early embryos of the pig. *J. Reprod. Fert.* **13**, 579.

Mittwoch, U. (1967) Sex Chromosomes. London and New York, Academic Press.

Ohno, S. (1967) Sex Chromosomes and Sex-linked Genes. Berlin, Heidelberg and New York, Springer-Verlag.

Penrose, L. S. (1966) The causes of Down's syndrome. *Adv. Teratologyy.* **1**, 9.

Piko, L. and Bomsel-Helmreich, O. (1960) Triploid rat embryos and other chromosomal deviants after colchicine treatment and polyspermy. *Nature (London).* **186**, 737.

Pincus, G. (1936a) The Eggs of Mammals. New York, Macmillan.

Pincus, G. (1936b) The parthenogenetic activation of rabbit eggs. *Anat. Rec.* **67**, 67.

Pincus, G. (1939a) The breeding of some rabbits produced by recipients of artificially activated ova. *Proc. Nat. Acad. Sci., U.S.A.* **25**, 557.

Pincus, G. (1939b) The comparative behaviour of mammalian eggs *in vivo* and *in vitro*. IV. The development of fertilized and artifically activated rabbit eggs. *J. Exp. Zool.* **82**, 85.

Pincus, G. and Shapiro, H. (1940) The comparative behaviour of mammalian eggs *in vivo* and *in vitro*. VII. Further studies on the activation of rabbit eggs. *Proc. Am. Phil. Soc.* **83**, 631.

Polani, P. E. (1969) Autosomal imbalance and its syndromes, excluding Down's. *Brit. Med. Bull.* **25**, 81.

Race, R. R. and Sanger, R. (1969) Xg and sex-chromosome abnormalities. *Brit. Med. Bull.* **25**, 99.

Russell, L. B. (1962) Chromosome aberrations in experimental mammals. *Progr. Med. Genet.* **2**, 230.

Saglik, S. (1938) Ovaries of gorilla, chimpanzee, orang-utan and gibbon. *Contr. Embryol.* **27**, 181.

Schindler, A. M. and Mikamo, K. (1970) Triploidy in man: Report of a case and a discussion on etiology. *Cytogenetics.* **9**, 116.

Schultz, R. J. (1969) Hybridization, unisexuality, and polyploidy in the teleost *Poeciliopsis* (Poeciludae) and other vertebrates. *Amer. Nat.* **103**, 665.

Shapiro, H. (1942) Parthenogenetic activation of rabbit eggs. *Nature (London).* **149**, 304.

Shaver, E. L. and Carr, D. H. (1967) Chromosome abnormalities in rabbit blastocysts following delayed fertilization. *J. Reprod. Fert.* **14**, 415.

Stern, C. and Austin, C. R. (1958) In discussion of: *Anomalies of fertilization leading to triploidy. J. Cell Physiol.* **56**, 1.

Tarkowski, A. K., Witkowska, A. and Nowicka, J. (1970) Experimental partheogenesis in the mouse. *Nature (London).* **226**, 162.

Thibault, C. (1949) L'oeuf des mammiferes. Son developpement parthenogenetique. *Annls. Sci. nat. (Zool.).* **11**, 136.

Thibault, C. (1952) La fecondation chez les mammiferes et les premiers stades de developpement. *Report, II Intern. Congr. Physiol. and Pathol. Anim. Repr. Art. Insem.* **1**, 7.

Uzzell, T. (1970) Meiotic mechanisms of naturally-occurring unisexual vertebrates. *Amer. Nat.* **104**, 433.

Vickers, A. D. (1969) Delayed fertilization and chromosomal anomalies in mouse embryos. *J. Reprod. Fert.* **20**, 69.

16

OBSERVATIONS ON THE BEHAVIOR OF OOGONIA AND OOCYTES IN TISSUE AND ORGAN CULTURE

Richard J. Blandau
D. Louise Odor

I.   Introduction
II.   Observations on the movements of living primordial germ cells from mice
III.   Motility characteristics of human oogonia
IV.   Observations on tissue culture of fetal mouse and human ovaries
V.   Organ cultures of fetal mouse ovaries
VI.   Zona pellucida formation in organ culture
VII.   Follicular development in organ culture
VIII.   Electron microscopic studies of oocytes in organ culture
IX.   Acknowledgments
X.   References

I. INTRODUCTION

There are few areas in mammalian embryology that have evolved with as rich and interesting controversy as that of the origin of the primordial germ cells. When one studies the historical development of knowledge in this field, one appreciates the importance of the application of new techniques in the solution of complex biological problems; even more, the value of observing and interpreting the experiments that nature often performs for us.

The source of the germ cells has been the primary subject for debate: whether they arise from the germinal epithelium of the gonads (Kingery, 1917, mouse; Simkins, 1923, mouse, rat; Hargitt, 1925, rat; Brambell, 1927, mouse; Nelsen and Swain, 1942, opossum) or extra-gonadally (Rubaschkin, 1912, guinea pig; Fuss, 1912, human; Vanneman, 1917, armadillo; Rauh, 1926, rat; Everett, 1943, mouse;

Witschi, 1948, human; Chiquoine, 1954, mouse; Mintz, 1957, mouse; Mintz and Russell, 1957, mouse; Blandau, White and Rumery, 1963, mouse; and Ozdzenski, 1967, mouse).

The interpretations of the earlier work were based largely on cytomorphological criteria, and the specific identification of the primordial germ cells often presented problems. Investigation in this field was facilitated greatly when it was learned that the germ cells could be stained specifically by virtue of their high alkaline phosphatase content (McKay, Hertig, Adams and Danziger, 1953; Chiquoine, 1954).

In one of the first experimental approaches to the problem, Everett (1943) transplanted the germinal ridges of mouse embryos under the kidney capsules of adults and observed that germ cells developed only if some primordial germ cells had already seeded the genital ridges. The first experimental verification of the extra-gonadal origin of the primordial germ cells in mammals was made by Mintz and Russell (1957) when they evaluated the embryological basis for sterility found in mice of WW, WW^v and W^{v}W^v genotypes. Eight-day-old embryonic mice harboring these mutant genes showed presence of the full complement of primordial germ cells in the yolk sac. Mitotic failure of the germ cells, preventing increases in their numbers, and their inability to migrate to the germinal ridges with resultant cytogenetic failure of the gonads, were the primary manifestations of the mutation.

The preponderance of evidence at this writing affirms that the germ cells arise in or near the yolk sac endoderm and migrate into specific areas overlying the mesonephros that become the gonads.

It was the classical work of Professor Witschi (1948) on the origin and migration of human germ cells that suggested to us the possibility of observing the motility patterns of the living germ cells of various mammals and prompted us to study their characteristics of growth in tissue culture preparations. In this paper we will summarize some of our experiences in looking for and observing the motility characteristics of the living primordial germ cells and some observations on the *in vitro* cultivation of embryonic and fetal mouse and human ovaries.

II. OBSERVATIONS ON THE MOVEMENTS OF LIVING PRIMORDIAL GERM CELLS FROM MICE

Very thin squash preparations were made from small pieces of tissue removed from the base of the allantois, the primitive streak, the regions of the hindgut and its mesentery of 8- and 9-day embryonic mice. Similar procedures were followed in studying the germ cells that had already seeded the gonads in 10- through 16-day mouse fetuses. Owing to the extreme sensitivity of germ cells to changes in their environment, great care was exercised in the preparation to insure an adequate time span in which to observe the ameboid movements. The gentle handling of the delicate embryonic tissues, the special care in the cleaning of the glassware, and other precautionary measures have been described elsewhere in detail (Blandau, White and Rumery, 1963).

It is our impression that equal parts of fresh mouse embryo extract and Waymouth's medium provide a good environment for observing the living germ cells of mice; a modified Eagle's medium to which is added heat-inactivated horse serum in a final concentration of 10% is most satisfactory for observing human oogonia (Blandau, 1969).

Tissues for the squash preparations were cut from the embryo with a finely ground iridectomy scissors and placed immediately on a glass slide in a large drop of the tissue culture medium. An 18 mm round coverglass was lowered gently over the tissue. Excess fluid was withdrawn by touching a piece of No. 1 Whatman filter paper to the edge of the coverglass. The degree of flattening was controlled by following continuously the entire process under a dissecting microscope. The preparation was sealed by ringing the coverglass with paraffin oil. The procedure was carried out in an enclosed chamber at temperatures of either 34°C or 37°C. All observations of cellular behavior were made with Zeiss dark medium-contrast phase objectives. Motility patterns were recorded cinematographically on either Dupont 914-A Panchromatic or Eastman Kodak Plus X negative film.

It takes considerable experience to make the preparation thin enough so that the various cell types in the tissues can be recognized and yet not be damaged. When the cytologic and behavioral characteristics of the living germ cells are once recognized they may be identified readily and distinguished easily from other embryonic cells. Motile germ cells from the region of the hindgut in a 9-day mouse embryo (a) and 12-day mouse gonad (b) are pictured in Fig. 16.1.

The motile germ cells from the extra-gonadal regions usually assume an elongated shape. The leading edge of the cell is often free of organelles although this may change momentarily as the cell changes its direction of movement. The leading edge of the pseudopodium is delimited by a membrane that flows out rapidly and appears as an enlarged bleb. Constrictions of the cytoplasm just beyond the bleb often cause the nucleus to assume an hour-glass figure as it flows forward into the pseudopodium (Fig. 16.1b). Another pseudopodium is usually not formed until the nucleus has completed its forward movement. The relatively few mitochondria in the cytoplasm are globular in shape and often aggregate about the nucleus. In the moving cell they are located in the trailing cytoplasm. Granted that it is difficult to quantitate, the germ cells that will develop into spermatogonia appear to move more slowly than those destined to be oogonia. Even after the germ cells have seeded the gonads, they continue to show considerable activity, differing somewhat from that observed during the migratory phase. Rather than moving forward, the cells move in a circular pattern. Often they are quite immobile; even so, they display almost continuous undulating movements of their peripheral cytoplasm. Nuclear rotation is a common feature of the non-motile cells, complete rotation occurring four to six times per minute. In 16-day fetuses many of the oocytes are in various stages of

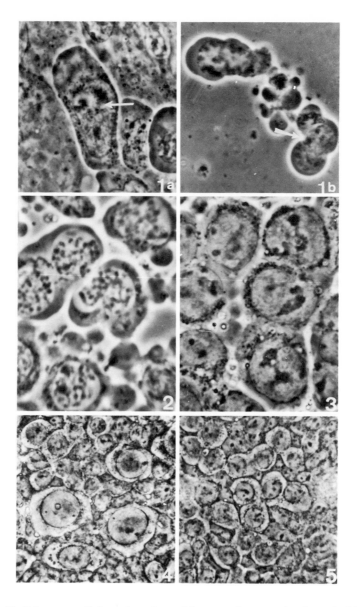

Fig. 16.1a A motile living germ cell (arrow) as observed in a squash preparation from the hindgut of a 9-day mouse embryo (X 875).

Fig. 16.1b Several moving germ cells from the gonad of a 12-day mouse embryo. Note the constriction of the nucleus (arrow) as it flows into the pseudopodium (X 770).

Fig. 16.2 A group of living oocytes in the pachytene stage from a 16-day mouse fetus. The oocytes still move about quite actively (X 860).

Fig. 16.3 A number of human oogonia from an 11-week embryo. The nuclei are large and the numerous organelles are concentrated about them (X 770).

Figs. 16.4 and 16.5 The appearance of living oocytes growing in tissue culture for two weeks. The irregular outline of the plasma membranes in the largest oocytes in Fig. 16.4 is due to their undulations (X 300).

prophase of meiosis; they nevertheless continue to display considerable localized movements (Fig. 16.2).

III. MOTILITY CHARACTERISTICS OF HUMAN OOGONIA

The same technique was applied to living oogonia recovered from the ovaries of 11- and 12-week human embryos. In general the motility characteristics described for the mouse pertain. Human oogonia are larger in size at this stage of development than mouse oogonia. The cytoplasm contains many globular mitochondria (Fig. 16.3). In most cells the mitochondria huddle about the nucleus. The nucleus is large, often irregular in shape, and may contain multiple nucleoli.

There is a remarkable similarity in the pattern of ameboid movement in certain of the human germ cells to that of the mouse. The rate of forward movement, however, differs significantly. The human germ cells seem to move much more slowly. Time lapse cinematography is necessary to demonstrate these forward movements properly. Time lapse cinematography of human oogonia in tissue culture reveals almost continuous undulations of their peripheral membranes often accompanied by nuclear rotations (Fig. 16.4), this despite the fact that they may be surrounded by a continuous layer of follicular cells. In both mouse and human multinucleate oogonia are common. They, too, are capable of independent movement.

We conclude that: (1) the primordial germ cells of the mouse and human have the capacity for independent movement; (2) the pattern of motility is ameboid during the migratory phase and undulatory after they have become located within the genital ridges; (3) the germ cells that will form the spermatogonia are less active in their movement and do not display the continuous undulatory activity of oogonia and oocytes; and (4) that the oocytes in the various stages of prophase of meiosis may continue their ameboid activity but move in a circular pattern.

IV. OBSERVATIONS ON TISSUE CULTURE OF FETAL MOUSE AND HUMAN OVARIES

To date there have been very few attempts to culture embryonic or fetal ovarian tissue. In view of the dramatic changes that occur in oogonia and oocytes during the fetal periods, it would be of value to learn: (1) whether these cells can survive and be induced to grow in culture; (2) whether a zona pellucida will form; (3) what the role of the follicular cells in the growth and development of the oocyte is; (4) whether multilaminar and vesicular follicles can develop; (5) whether exogenous hormones added to the culture medium can influence the growth and/or differentiation of the oocyte; (6) whether oogonia can complete the prophase of meiosis and proceed to the dictyotene stage; and (7) what length of time oocytes can be maintained in culture.

The first successful tissue culture of fragments of adult human ovary removed

by surgery revealed principally an outgrowth of fibroblasts with occasional sheets of
epithelial cells (Zondek and Wolff, 1924; Wolff and Zondek, 1925). No observations
on the oocytes were reported. Asayama and Furusawa (1960) cultured gonads from
12- to 16-day mouse embryos in roller-tubes for 70 to 94 hours and reported
significant differentiation even during this short culture period. Our observations are
based on our experience in culturing ovarian fragments from 16-day mouse fetuses
and human ovaries from embryos and fetuses obtained from therapeutic abortions.
The crown-rump lengths of the human embryos and fetuses varied from 25 to 180
mm.

The ovaries were removed and placed in a large volume of modified Eagle's
medium. They were cut into fragments of approximately 1mm in size with sharply
honed cataract knives with care to avoid excessive squashing, which injures large
numbers of oocytes. Then, too, because of the loose packing of the cells in the
stroma, many will be freed and not survive.

Fifteen to 20 fragments were deposited on a large coverslip that formed the
base of the Rose culture chamber. The excess medium was withdrawn carefully
leaving only sufficient fluid to prevent drying. Immediately a strip of dialysis
membrane was placed over the explant so that the fragments would remain in place
during the assembling and filling of the chamber.

The Rose chamber was filled with a modified Eagle's medium containing
heat-activated horse serum in a final concentration of 10% (see Blandau, Warrick and
Rumery, 1965, for details). The cultures were examined frequently with phase
objectives. Time lapse cinematography was helpful in evaluating the behavior of the
cells *in vitro.*

The great variation in the preservation of the oocytes is a notable feature of the
growth and differentiation of the various ovarian fragments in the culture. In some
fragments the oocytes disappear completely within several days, whereas the stroma
that contained them appears viable and remains so for weeks. In other bits of tissue
the oogonia and oocytes persist; some of them will begin to grow and, at a later
period, even a zona pellucida may develop.

The tissue fragments growing in Rose chambers are often sufficiently thin so
that the cytologic details of the oocytes may be studied (Fig. 16.4 and 16.5). After
several weeks in culture, some fragments still contain large numbers of oocytes that
vary significantly in size (Fig. 16.4). All of them appear to be surrounded by a single
layer of follicular cells. It is usually impossible at the light microscopic level to
outline the full extent of the follicular cells encompassing the smaller oocytes.
Oocytes of 30-50 μm in diameter may be surrounded by a single layer of follicular
cells resting on a definitive basement membrane. The surrounding stromal cells may
be arranged circumferentially, although the extent to which this occurs varies
greatly.

Oocytes of similar size are often grouped in clusters. This was particularly noticeable in an ovarian fragment from an 11-mm human embryo maintained in culture for many weeks. The clones of cells are often in identical stages of mitosis (Fig. 16.6).

In the relatively few observations we were able to make on cultures of human ovaries containing primarily oogonia, transformation into oocytes did not take place irrespective of the length of time in culture. If prophase of meiosis was already in progress in the fragments of mouse and human ovaries when explanted, they did proceed to the dictyotene stage. Several multilaminar follicles were seen in the mouse tissue culture preparations but none of these developed into vesicular follicles (Fig. 16.7). When such follicles were dissected from the fragments they remained intact. In this preparation only a very thin zona encompasses the egg. Definitive zonae pellucidae appeared in oocytes that had attained a size of from 30-50 μm in diameter (Fig. 16.8). The zonae were irregular in thickness in some oocytes; in others their thickness was uniform but relatively thinner than eggs of similar size *in vivo*. With the PAS staining procedures, the zona appeared very red; a useful method for detecting its earliest appearance. As mentioned previously, time lapse cinematography revealed the almost continuous undulatory movements of the egg cytoplasm. When the moving cytoplasm pressed upon the zona pellucida, it indented it momentarily, giving the impression that the zona was quite soft and pliable.

Oocytes in tissue culture may attain a size of 80-100 μm in diameter. All of the larger ones were enclosed in zonae pellucidae. Fully grown mouse ova *in vivo* attain diameters of from 70-85 μm, including the zona pellucida. When the large oocytes were dissected from the ovarian fragments they had an elliptical rather than ovoid shape. This shape was maintained even after the oocytes were freed from the stroma. The elliptical shape was no doubt due to the fact that the oocytes were growing in a confined space beneath the dialysis membrane. With the passage of time, eggs in culture with intact zonae were freed from the stroma. The zonae of these eggs frequently stick to the coverglass even during the rigors of changing the medium (Fig. 16.9). This may indicate that the physical characteristics of the *in vitro* zonae differ from those *in vivo*.

We conclude from these studies that fragments of fetal mouse and human ovarian tissue can be maintained in Rose chambers for more than 50 days (mouse) and 80 days (human); large numbers of oocytes degenerate and disappear from the cultured tissues; only those oocytes which are enclosed in the gonadal stroma survive; some of those remaining grow to a large size and most of them develop zonae pellucidae; and the zonae pellucidae do not appear to be as thick nor as highly polymerized *in vitro* as those formed *in vivo*.

Human oogonia that have not entered the prophase of meiosis at the time they are explanted will not do so *in vitro*. Mouse and human oocytes that have begun the prophase of meiosis may complete this process *in vitro* and attain the dictyotene stage.

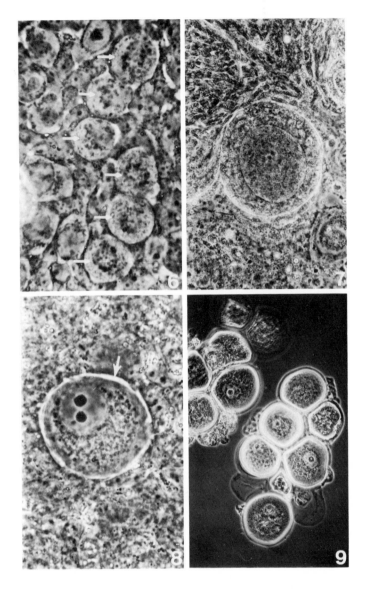

Fig. 16.6 A cluster of living human oogonia from an 11-mm human embryo. Note that all of the cells (arrows) are in the same stage of mitosis (X 670).

Fig. 16.7 A multilaminar follicle in a fragment from a 16-day mouse fetus, in culture for 23 days. Only a very thin zona pellucida is present. Note the definitive basement membrane separating the follicle cells from the stroma (X 310).

Fig. 16.8 A mouse oocyte in tissue culture for 50 days. Note the presence of a continuous but irregular zona pellucida (X 410).

Fig. 16.9 A cluster of oocytes with zonae which have become freed from the tissue culture fragments. The zonae adhere to the coverglass because of their stickiness.

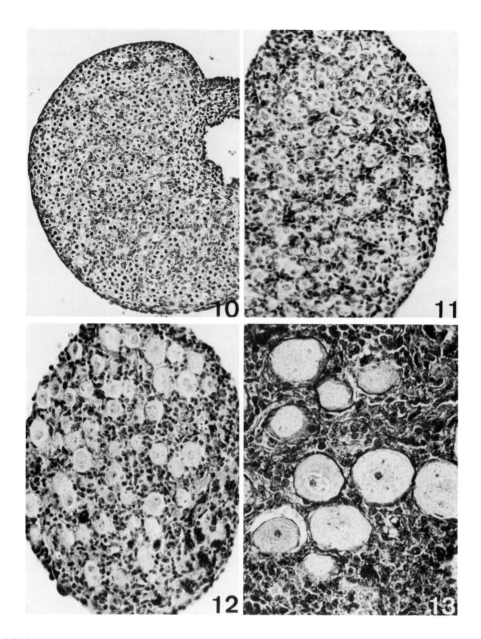

Fig. 16.10   A section through a 16-day fetal mouse ovary. The oocytes are arranged in clusters surrounded by delicate connective tissue. PAS stain   (X 100).

Fig. 16.11   A segment of a 16-day fetal mouse ovary which has been in culture for 6 days. The oocyte clusters are being separated into individual cells or smaller groups of cells by the invasion of follicular cells   (X 120).

Figs. 16.12 and 16.13   A section from a 16-day fetal mouse ovary that had been in culture for 12 days. A large number of the smaller oocytes have disappeared. The larger oocytes are encompassed by zona pellucida material (Fig. 16.13).

V. ORGAN CULTURES OF FETAL MOUSE OVARIES

Organ culture is a better method than tissue culture for evaluating histogenesis and cytodifferentiation of the fetal mouse ovary. In this study 16-day fetal mouse ovaries were dissected from the mesovarium. Six to 8 pairs of whole ovaries were placed on agar strips in a Petri dish. A modified Eagle's medium, to which had been added heat-inactivated horse serum in a concentration of 10%, was pipetted into the dish to a depth sufficient to form only a thin film of medium over the surfaces of the ovaries since ovaries submerged in the medium do not survive. The culture medium was replaced every 2 or 3 days. Incubation at 37° C was carried out in an atmosphere of 5% carbon dioxide and 95% air with a humidity of 70%. Several hundred cultures were harvested between one and 45 days after explantation. Some were fixed in Hsu's fixative (Rumery, 1968), serially sectioned at 8 μm and treated in accordance with the periodic acid-Schiff method, and then counterstained with Delafield's hematoxylin.

When 16-day fetal mouse ovaries were placed in organ culture they consisted basically of numerous small oocytes arranged in irregular groupings. The majority of oocytes were in the prophase of meiosis (Fig. 16.10). When 16-day fetal ovaries were gently squashed by the procedure described earlier, large numbers of oocytes were released individually and arranged themselves in the form of a halo about the more dense medullary tissue. At this stage of organization there were no true follicles present and oocytes were all about the same size.

We were mystified by the great variation from ovary to ovary in the degree of organ differentiation, the number of oocytes that survive and grow, and the extent of degenerative changes. Although every effort was made to follow identical procedures from culture to culture, the oocytes in some ovaries in the same culture dish showed remarkable consistent development, whereas others degenerated rapidly after only a few days on the agar strips. Generally, the oocytes showed relatively little change in arrangement or size during the first 3 days in culture. The chromosomal configuration of most oocytes was either zygotene, diplotene or dictyotene. A few were in the pachytene stage. As we have shown in tissue culture the meiotic prophase apparently can go to completion in organ culture since by the 5th day in culture the majority of the oocytes were in the dictyotene stage.

The oocytes in organ culture did not all grow uniformly. After 14-20 days in culture, the oocytes surviving attained maximum diameters of 46-53 μm and many of them had become separated from one another by a single layer of flattened follicular cells. There was some stratification of oocytic development from the medulla to the cortex. The oocytes near the medullary area increased in size somewhat more rapidly than those in the outer reaches of the cortex. Unilaminar follicles were seen first in this area. The size attained by oocytes in organ culture varied greatly from ovary to ovary and with the length of time in culture (Figs. 16.11 through 16.13). The largest normal appearing oocyte (73 μm) was found in a 45-day culture, a favorable comparison with the size of normally ovulated eggs in *vivo.* The medium or large size oocytes usually had either spherical or oval nuclei and a single large nucleolus. As the oocytes increase in size they contain aggregations of PAS-positive bodies. Some of these assume a juxtanuclear position. The nature of these accumulations is as yet unknown.

In only one of the hundreds of larger oocytes studied did we see a metaphase spindle of an abortive first meiotic division. Definitive follicular cells may be recognized and distinguished readily from the remaining stromal connective tissue cells at the light microscopic level. Mitoses were observed quite frequently in the follicular cells.

We wish to emphasize again the significant variations in the survival of oocytes in the various organ cultures within the same dish. At 20 days in culture some ovaries still had a remarkable number of small, apparently perfectly normal oocytes present. In others, most of the oocytes had been cytolyzed. This was accompanied by an extensive central necrosis of the ovary.

Attempts to accelerate oocytic growth and differentiation in organ culture by adding various concentrations of gonadotrophins and steroid hormones in a variety of combinations have so far not been successful. We believe that the basic reason for this failure may be related to the problem of exposing the organs to adequate concentration of the hormones. Efforts to improve the technique for accomplishing this are receiving attention.

VI. ZONA PELLUCIDA FORMATION IN OOCYTES IN ORGAN CULTURE

Thin, discontinuous zonae pellucidae appear on some enlarging oocytes (approximately 24 μm in diameter) after 3-4 days in organ culture. After six days in culture, all of the large oocytes showed continous zonae pellucidae but they were less thick than on eggs of comparable size *in vivo* (Figs. 16.13 and 16.15).

The fully formed zona pellucida *in vivo* is quite uniform in thickness. In organ cultures the zonae are often quite irregular in thickness and appear to be less polymerized. As the vitellus degenerates in the older cultures, the zona persists. It is often surrounded by macrophages; some of which penetrate the membrane (Fig. 16.16).

Oocytes with intact zonae can be harvested readily from ovaries that have been in culture for 20 or more days (Fig. 16.15). When capacitated spermatozoa are added to such preparations they may penetrate the zona and vitellus. Many spermatozoa may come to rest in the perivitelline space (Fig. 16.14). In none of these preliminary observations have we observed any change in the nuclei of the oocytes after sperm penetration. The only purpose in mentioning these observations is to point out that spermatozoa may penetrate the zona pellucida in eggs harvested from organ cultures.

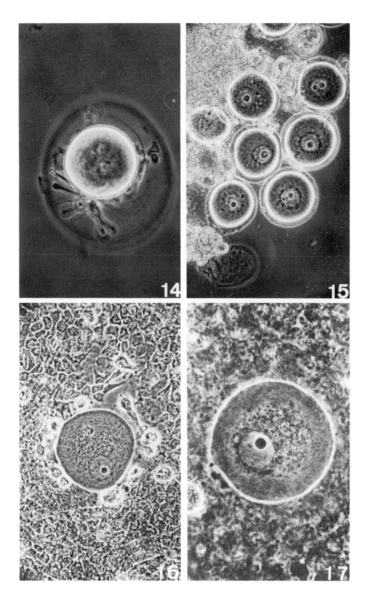

Fig. 16.14 A mouse oocyte harvested from a 20-day organ culture and exposed to capacitated spermatozoa. Numerous spermatozoa have penetrated the zona but none have entered the vitellus. The vitellus has undergone considerable shrinkage.

Fig. 16.15 The appearance of oocytes recovered from a 16-day fetal ovary in organ culture for 22 days. The zonae pellucidae are quite uniform and the eggs appear normal.

Figs. 16.16 and 16.17 Oocytes in squash preparations of organ cultures illustrating the loss of follicular cells and their replacement by macrophages. Oocytes under such conditions may remain normal in appearance for many days.

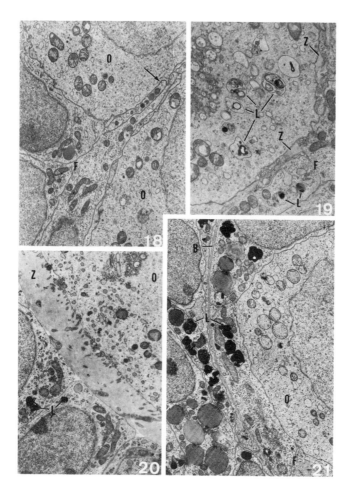

Fig. 16.18 24-day culture. The fetal arrangement of oocytes (O) intermingled irregularly with follicular cells (F) persisted in some cultures. The two oocytes are separated from one another by the processes and cell bodies of three follicular cells (X 4,000).

Fig. 16.19 10-day culture. Zona pellucida substance (Z) has formed between the oocyte and follicular cells (F) at two sites. A follicle had not developed. Lysosomal-like bodies (L) are present in both the follicular cells and the oocyte. We interpret the presence of many of these bodies in an oocyte as indicative of early degenerative changes. However, the structure of the mitochondria, multivesicular bodies and endoplasmic reticulum in the oocyte is normal, as is that of the organelles of the follicular cells (X 6,510).

Fig. 16.20 7-day culture. Portion of an unilaminar follicle having a columnar epithelium and a zona pellucida (Z). Mitochondria, small profiles of endoplasmic reticulum, multivesicular bodies and a small Golgi complex lie in the ooplasm. The organelles of the follicular cells are normal, but two lysosomal-like bodies (L) are present. The zona pellucida is quite wide and contains processes of the follicular cells and microvilli of the oocyte. This is the normal relationship between the zona and the cells of the ovarian follicles (X 5,880).

Fig. 16.21 70-day culture. "Primordial" unilaminar follicles such as this one are found with decreasing frequency in the older cultures. The follicular cells (F) rest on a basal lamina (B), external to which lie stromal connective tissue cells. Some of the oocytic mitochondria are enlarged and have dilated cristae that appear as vesicles. The follicular epithelium and some of the stromal cells contain dense lysosomal-like bodies (L) and lipid (X 4,480).

VII. FOLLICULAR DEVELOPMENT

The structural unit of a follicle is an oocyte surrounded by one or more layers of follicular cells resting on a basement membrane. The majority of follicles in the organ cultures are encompassed by one layer of follicular cells. A few may proceed to the bi- or trilaminar stage. Antral spaces have not been observed in any of the follicles in organ culture. Follicular development is much more irregular in organ culture than *in vivo*. Occasionally, oocytes of different sizes, arranged in a group, will be enclosed by several layers of follicular cells.

A few macrophages are present in cultures only a few days old but their number increases greatly with time. In the older cultures macrophages may completely replace the follicular cells encompassing the larger oocytes (Fig. 16.16). Even though oocytes under these circumstances appear quite normal there is a tendency for them to degenerate more rapidly than those continuously surrounded by follicular cells.

VIII. ELECTRON MICROSCOPIC STUDIES
OF OOCYTES IN ORGAN CULTURES

A number of organ cultures of ovaries from 16-day fetal mice that had been in culture for 2 through 89 days were fixed and sectioned for examination with the electron microscope. These studies were helpful in resolving many questions that could not be answered at the light microscopic level.

No true follicles were present in the ovaries of 16-day fetal mice on the first day of explantation. We reiterate that a follicle is defined as an oocyte surrounded by a follicular epithelium resting on a basal lamina. There was no evidence of zona pellucida formation in any of the oocytes and most of them were in the prophase of meiosis. At this stage the oocytes were grouped in irregular clusters. The cell bodies of follicular cells were interspersed among the oocytic groups and their cell processes surrounded them. A few stromal connective tissue cells lay between the clusters of oocytes and follicular cells, beneath the germinal epithelium and in the medullary area (Odor and Blandau, 1969, a and b).

The length of time this primitive arrangement of cells persists varies considerably, even in ovaries cultured for the same number of days. It remained longest in the most peripheral areas of the cortex. An example of the intermingling of oocytes and follicular cells is shown in a 24-day culture in Fig. 16.18. Note the separation of the two oocytes (O) by thin processes of the follicular cells (↑) and the group of three follicular cells to the left in the photograph. The cytoplasmic organelles in both cell types were normal for this stage of development.

True follicles were seen first on day 4 in the whole organ cultures. The oocytes were surrounded by flattened follicular cells, forming "primordial" unilaminar follicles. A few such follicles persisted even in a 70-day culture (Fig. 16.21). A "primary" unilaminar follicle resulted when the follicular epithelium became cuboidal or columnar. These were noted as late as 64 days *in vitro.* Follicles having two or three layers of epithelium were rarely encountered in the electron microscopic sections.

Zona pellucida material appeared initially as small discontinuous accumulations between the oocyte and follicular cells in 3- and 4-day cultures (Fig. 16.19). Continuous, relatively wide zonae pellucidae were present around medium sized oocytes after 7 days of explantation (Figs. 16.20, 16.22). Between 7 and 20 days the normal relationships between oocytes and follicular cells were observed just as in the case of oocytes of similar size *in vivo.* That is, processes of the follicular cells extended into the zonae (Figs. 16.20, 16.22, 16.23). Some of the processes terminated in juxtaposition to the plasma membranes of the oocytes (Fig. 16.22↑). Oocytic microvilli projected into the zona pellucida also. At this stage the organelles of the oocyte were usually normal in size, number and distribution and were similar to those in growing oocytes *in vivo* examined 3 or more days after birth. As the length of time *in vitro* increased, such normal relationships were observed with decreasing frequency.

The zona appeared to be composed of irregularly oriented groups of very fine filaments (Fig. 16.23). In some instances the outer region of the zona contained small areas of low electron density (Fig. 16.22). These features were more prominent in the organ cultures than in ovaries *in situ,* a possible indication that the formation of the zona pellucida was not entirely normal *in vitro.*

Although some oocytes grew to a normal maximum size, the zonae usually appeared somewhat narrower than those around oocytes that matured *in vivo.* The possibility exists that the replacement of follicular cells with macrophages, as described below, was one of the factors which brought this about.

One of the earliest signs of degeneration of a follicle having a zona pellucida around the oocyte was the disappearance of the follicular cell processes from the zona. The microvilli on the surface of the plasma membrane of the oocyte persisted longer but often became bulbous and disoriented (Fig. 16.24). Eventually, neither oocytic microvilli nor follicular cell processes penetrated the zonae (Fig. 16.25).

Often by the time the processes of the follicular cells had disappeared one or more macrophages had migrated into the follicular epithelium and had become attached to the outer surface of the zona pellucida. In Fig. l6.24 such a macrophage lies on the surface of the zona and, external to it, cytoplasm of two follicular cells is seen. The ultrastructure of the macrophages in the organ cultures is so distinctive that they can be distinguished easily from the follicular cells. The macrophage in Fig. 16.25 contains a teleophase chromosomal mass partially surrounded by cisternal elements of the endoplasmic reticulum from which the nuclear envelope is reconstituted. Numerous large lysosomal-like bodies enclosed by a single membrane and containing very electron dense masses are present. These are considered to be the same structures that appeared as PAS-positive granules in these cells on the light

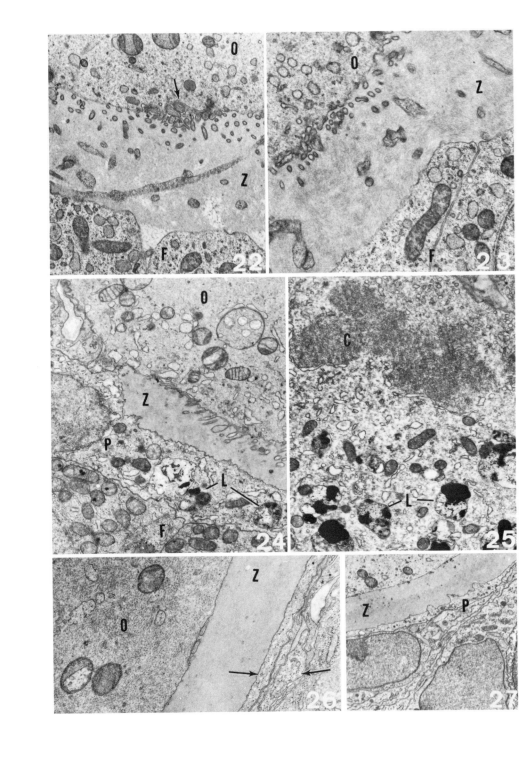

microscopic level. The cytoplasmic matrix is less electron dense, the mitochondria are smaller and there are more small agranular vesicles, fewer ribosomes and less granular endoplasmic reticulum than in the follicular cells (compare Figs. 16.24 and 16.25 with Figs. 16.20 and 16.21).

Eventually, macrophages completely replaced the follicular cells and no basal laminae were observed. Their cytoplasm, adjacent to the zona pellucida, often formed very thin layers almost free of organelles (Figs. 16.26 and 16.27). In Fig. 16.27 three small protrusions of macrophagic cytoplasm had begun to penetrate the substance of the zona pellucida. In other instances the ooplasm had degenerated and the zonae were present as shells with macrophages attached to their exteriors and filling their interiors.

There were other interesting changes in the cells of the explanted ovaries. As early as day 3 or 4 *in vitro*, lysosomal-like bodies appeared in the follicular cells, but they did not produce any injurious effects that could be visualized (Figs. 16.19 and 16.20). In long-termed cultures follicular cells, either in follicles (Fig. 16.21) or in epithelial clusters bounded by basal laminae, contained these bodies while other organelles remained normal. Also some oocytes in both short-and long-termed

Fig. 16.22   7-day culture. Microvilli of the oocyte (O) and processes of the follicular cells (F) lie within the substance of the zona pellucida (Z). One long follicular cell process is obliquely oriented. Other processes are in contact with the plasma membrane of the oocyte (↑). Note the areas of low electron density in the outer part of the zona   (X 14,700).

Fig. 16.23   14-day culture. This follicle has a normal oocyte (O) and follicular cells (F). Of particular interest is the presence of interwoven bundles of fine filaments in the zona pellucida (Z). These are more prominent than those in oocytes of similar size developed *in vivo*. Follicular cell processes and oocytic microvilli lie within the zona (X 15,100).

Fig. 16.24   44-day culture. The structure of this oocyte (O) is not normal. Note the large mitochondrion containing vesicles that are dilated cristae. Some profiles of the endoplasmic reticulum have enlarged. A few of the microvilli within the zona pellucida (Z) are bulbous and others are oriented more obliquely than usual. The zona does not completely surround the oocyte. Where present, it is covered by a macrophage (P) that has several large lysosomal-like bodies (L) and small mitochondria in its cytoplasm. Portions of two follicular cells (F) are located external to the macrophage.

Fig. 16.25   46-day culture. Pictured is part of a macrophage in telophase of mitosis. The chromosomal mass (C) is only partially surrounded by cisternal elements of the endoplasmic reticulum. The cytoplasm is characterized by a matrix of low electron density, small mitochondria, many small agranular vesicles and cisternae and numerous large lysosomal-like bodies (L)   (X9,600).

Fig. 16.26   36-day culture. Most of the oocytic organelles have disappeared, leaving only a few mitochondria and endoplasmic reticulum profiles. The zona pellucida (Z) no longer contains oocytic microvilli or follicular cell processes. It is covered by thin cytoplasmic layers of macrophages (↑). The entire epithelium had been replaced by macrophages   (X 22,640).

Fig. 16.27   50-day culture. The follicular epithelium around this large oocyte had been replaced by macrophages (P). The latter are attached to the zona pellucida and have begun to penetrate it, as evidenced by the three blunt protrusions into the zona. The thin layers of cytoplasm of macrophages are typical of this stage in the degeneration of the follicle   (X 4,580).

explants contained lysosomal-like bodies. Usually their structure was different from those observed in both follicular cells and connective tissue cells (Fig. 16.19). Their single membranes enclosed a matrix of low electron density and structures of high density that often had the form of small myelin figures.

Small oocytes containing many of these bodies, as in Fig. 16.19, were considered to be in an early degenerative stage. The remainder of the ooplasmic organelles appeared normal. Another indication of minimal oocytic abnormality was the enlargement of some mitochondria to the extent that dilated cristae appeared as rounded vesicles (Figs. 16.19 and 16.26). A few large oocytes with vesiculated and enlarged mitochondria were observed in explants cultured for as long as 82 days. Many of the large oocytes in long-term cultures lost most of their characteristic organelles, only a decreased number of mitochondria and profiles of endoplasmic reticulum remaining in the ooplasm (Fig. 16.26 and 16.27).

The following conclusions may be reached from the observations on whole organ cultures of fetal mouse ovaries:
    (1)   There is great variation in the preservation and growth of the oocytes in the various ovaries in the same culture; (2)   Some oocytes may attain a maximum diameter of 73 μm, which is within the normal maximum found *in vivo*; (3)   True follicles, the majority unilaminar, may develop in organ culture and when examined at the ultrastructural level many of these appear normal up to 20 or 25 days in culture; (4)   Some follicles develop to the multilaminar stage, but no more than three layers of follicular cells were counted in any of the organ cultures. (Vesicular follicles were never observed); (5)   The zona pellucida is formed about the oocytes in organ culture and does not appear as thick nor as highly polymerized as on an oocyte of similar size *in vivo*; (6)   In the older cultures the follicular cells may be replaced completely by macrophages which may arrange themselves in a manner simulating follicular cells. (Macrophages have a characteristic ultrastructure that clearly differentiates them from follicular cells); (7)   In the thousands of oocytes examined on the light microscopic level from 2-45 days in culture, we observed but one first metaphase spindle.

IX. ACKNOWLEDGMENTS

Previously unreported work mentioned in this paper was supported by grants HD-00606 and HD-03752 from the National Institutes of Health. A 16 mm motion picture film illustrating various behavioral characteristics of mammalian germ cells is available through the Audio-Visual Services, 110 Lewis Hall, University of Washington, Seattle, Washington 98105. We are grateful to Mr. Roy Hayashi for his photographic skills in the preparation of the film and plates.

X. REFERENCES

Asayama, S. and Furusawa, M. (1960) Culture *in vitro* of prospective gonads and gonad primordia of mouse embryos. *Dobutsugako Zasshi (Zool. Soc. Japan).* **69**, 283.

Blandau, R. J. (1969) Observations on living oogonia and oocytes from human embryonic and fetal ovaries. *Amer. J. Obstet Gynec.* **104**, 310.

Blandau, R. J., Warrick, E. and Rumery, R. E. (1965) *In vitro* cultivation of fetal mouse ovaries. *Fertil. Steril.* **16**, 705.

Blandau, R. J., White, B. J. and Rumery, R. E. (1963) Observations on the movements of the living primordial germ cells in the mouse. *Fertil. Steril.* **14**, 482.

Brambell, F. W. R. (1927) The development and morphology of the gonads of the mouse. I. The morphogenesis of the indifferent gonad and of the ovary. *Proc. Roy. Soc. (Lond.), Series B.* **101**, 391.

Chiquoine, A. D. (1954) The identification, origin, and migration of the primordial germ cells in the mouse embryo. *Anat. Rec.* **118**, 135.

Everett, N. B. (1943) Observational and experimental evidences relating to the origin and differentiation of the definitive germ cells in mice. *J. Exp. Zool.* **92**, 49.

Fuss, A. (1912) Uber die Geschlechtszellen des Menschen und der Saugetiere. *Arch. mikr. Anat.* **81**, 1.

Hargitt, G. T. (1925) The formation of the sex glands and germ cells of mammals. I. The origin of the germ cells in the albino rat. *J. Morph.* **40**, 517.

Kingery, H. M. (1917) Oogenesis in the white mouse. *J. Morph.* **30**, 261.

McKay, D. G., Hertig, A. T., Adams, E. C. and Danziger, S. (1953) Histochemical observations on the germ cells of human embryos. *Anat. Rec.* **117**, 201.

Mintz, B. (1957) Embryological development of primordial germ-cells in the mouse: Influence of a new mutation, WJ[1]. *J. Embryol. Exp. Morph.* **5**, 396.

Mintz, B. and Russell, E. S. (1957) Gene-induced embryological modifications of primordial germ cells in the mouse. *J. Exp. Zool.* **134**, 207.

Nelsen, O. E. and Swain, E. (1942) The prepubertal origin of germ-cells in the ovary of the opossum *(Didelphys virginiana). J. Morph.* **71**, 335.

Odor, D. L. and Blandau, R. J. (1969a) Ultrastructural studies on fetal and early postnatal mouse ovaries. I. Histogenesis and organogenesis. *Amer. J. Anat.* **124**, 163.

Odor, D. L. and Blandau, R. J. (1969b) Ultrastructural studies on fetal and early postnatal mouse ovaries. II. Cytodifferentiation. *Amer. J. Anat.* **125**, 177.

Ozdzenski, W. (1967) Observations on the origin of primordial germ cells in the mouse. *Zool. Poloniae.* **17**, 367.

Rauh, W. (1926) Ursprung der weiblichen Keimzellen und die Chromatischen Vorgange bis zur Entwicklung des Synapsisstadiums. Beobachtet an der Ratte. *Zeit. Anat.* **78**, 637.

Rubaschkin, W. (1912) Zur Lehre von der Keimbahn bei Saugetieren. *Anat. Hefte, Wiesh.* **46**, 343.

Rumery, R. E. (1968) The fetal mouse oviduct in organ and tissue culture. *In* "The Mammalian Oviduct." (E. S. E. Hafez and R. J. Blandau, eds.), p. 445-457. Chicago, The University of Chicago Press.

Simkins, C. S. (1923) On the origin and migration of the so-called primordial germ cells in the mouse and the rat. *Acta Zool.* **4**, 241.

Vanneman, A. S. (1917) The early history of the germ cells in the armadillo. *Tatusia novemcincta. Amer. J. Anat.* **22**, 341.

Wolff, E. K. and Zondek, B. (1925) Die Kultur menschlichem Ovarial und Amniongewebes. *Virchow Arch. [Path. Anat.]* **254**, 1.

Witschi, E. (1948) Migration of the germ cells of human embryos from the yolk sac to the primitive gonadal folds. *Carnegie Inst. Contrib. Embryol.* **32**, 67.

Zondek, B. and Wolff, E. (1924) Uber Zuchtung von menschlichem Ovarialgewebe *in vitro. Zbl. Gynak.* **48**, 2193.

17

FUNCTIONAL INTERACTIONS BETWEEN THE AMPHIBIAN OOCYTE AND GENERAL OVARIAN CELLS

Antonie W. Blackler

I.   Nature, origin and growth of the ovary
II.   Growth and developmental capacity of the oocyte
III.   Commentary
IV.   Acknowledgments
V.   References

I. NATURE, ORIGIN AND GROWTH OF THE OVARY

The amphibian ovary is developmentally established early in tadpole life. At the stage of the neurula, certain cells situated in the endoderm (frogs and toads) or mesoderm (salamanders and newts), termed primordial germ cells, migrate by amoeboid means toward the dorsal midline of the embryonic posterior gut and there become incorporated in the tissue of the dorsal mesentery during the latter's formation. The number of these primordial germ cells at this time varies from one species to another, but eventually all the cells become situated along the dorsal mesentery with the possible exception of a few which may be left in the walls of the posterior intestinal tract (notably the rectum) or at the junction of the gut with the ventral root of the dorsal mesentery. Shortly thereafter, when the tadpole is beginning to feed, the primordial germ cells separate into equal sized right and left groups on each side of the dorsal mesentery and on the medio-ventral faces of the mesonephric kidney rudiments. Cells of the mesoderm in this region (the precise source seems somewhat questionable) now begin to invest the longitudinal columns of the embryonic sex cells, and also to infiltrate them. Thus the gamete precursors originate prior to the appearance of the gonadal rudiments and in an embryonic site distant from those gonadal sites (see Blackler, 1966, for a review).

Until the moment when the primary gonads are established, the total number of sex cells has remained about the same as the number found at the neurula stage. Thus the primordial germ cells have undergone a long period of mitotic arrest since the latter stage. At this time, however, as the mesoderm cells invade their ranks, the embryonic sex cells begin to mitose at an increasing rate. The invasion of cells is rapidly completed, and well-defined, but sexually indifferent, gonads exist from this moment. The mesodermal cells now begin to multiply and the gonads grow rapidly in size, until, by the closing stages of tadpole life, sexual differentiation in the gonadal tissue effectively leads to the formation of histologically distinct ovaries and testes. With the passage of further time the ovary continues to grow and its 15 - 20 lobes expand in volume to fill the abdominal space not otherwise occupied by the organs of the gastrointestinal tract (Fig. 17.1). These simple observations

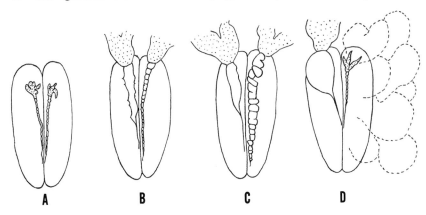

A   B   C   D

Fig. 17.1 Growth of the gonad in *Xenopus laevis*. The kidney length of each animal has been adjusted to a common length to facilitate comparison of the gonads. Male gonads are shown at left, female at right. a. Gonads of the middle tadpole. Fat bodies are small and the gonadal appearance is the same in each sex. b. Gonads at the time of metamorphosis. The testis becomes shorter and more compact, and the ovary shows growth of the lobes. c. Gonads two weeks after metamorphosis. The testis becomes elliptical and tubule formation is much advanced. The ovary shows continued growth of the lobes as growing oocytes fill their walls. d. Gonads ten weeks after metamorphosis. Testes are almost spherical and the first mature sperm appear. The ovary (dashed outline) consists of voluminous lobes whose anterior-posterior arrangement is obscured. The other gonad to the right (solid outline) is representative of a completely sterile ovary at this time, except that the fat body may be normal in size; completely sterile testes usually have a smoother outline.

demonstrate the basic dualism of the amphibian ovary, a dualism composed of the reproductive cells (the primordial germ cells) and the somatic cells (the cells of the ovarian supporting tissue and follicles). Moreover, although there are differences in detail, the pattern of ovarian formation is essentially the same throughout the vertebrates.

A number of studies, experimental as well as descriptive, have established the fact that only the primordial germ cells in amphibia have the ability to form gametes (Witschi, 1929; Blackler, 1962; Smith, 1966). The mesodermal cells of the ovary form the follicular tissue and general stroma of the ovary, but neither in the absence of the embryonic sex cells nor in their presence, are able to assume a gametogenetic role. This feature applies also to the male. The possibility of somatic cells adopting a gamete-forming role has been warmly argued in the past, particularly in the case of mammals (see reviews of Everett, 1945; Zuckerman, 1951), but at least in the amphibia this possibility must be discarded (Blackler, 1966).

There is evidence that the proper and normal growth of the gonad depends on a reciprocal interaction between its germinal and somatic elements. The most anterior part of the amphibian gonad is devoid of reproductive cells and forms the fat body (late tadpole life and adult); this is present whether or not the rest of the gonad contains embryonic sex cells. It is probable, in consequence, that the instructional event for the formation of the fat body has nothing to do with the embryonic sex cells. Those primordial germ cells that fail to become incorporated into the gonad do not undergo mitosis when the germ cells within the gonad begin to divide. It is likely, although never demonstrated by experiment, that the germ cells require some stimulus from the investing mesoderm, which is mediated through direct physical contact, in order to divide. All such extra-gonadal sex cells gradually disappear and have not been seen after metamorphosis. It has been proposed from time to time by workers in the vertebrate field that extra-gonadal sex cells may be responsible for a significant proportion of teratomas, embryomas, ectopic pregnancies and tumors which may be found in the posterior part of the adult body, but while such proposals may remain arguable for warm-blooded animals, it seems not to be the case for the South African clawed toad *Xenopus,* in which a search for extra-gonadal cells shows that their frequency is extremely low (Blackler, unpublished).

The stimulus for cell division given by the somatic elements to the sex cells seems to be accompanied by a reciprocal arrangement. In the complete absence of primordial germ cells, whether by spontaneous natural sterility or as a result of experimental castration (e.g. by irradiation of the egg or by surgical removal of embryonic sex cells from the neurula), a purely "somatic" gonad is formed. This latter consists of a columnar mass bearing a fat body at the anterior end (Fig. 17.1). As general development proceeds, such sterile gonads may increase in mass slowly, but present an almost unchanged appearance throughout tadpole life and thereafter. When sexual differentiation occurs in fertile animals, the sterile ovaries remain unchanged in appearance from the indifferent gonad, but the sterile testes receive the influx of rete cells from the mesoderm and present a somewhat thickened and solid appearance. Such testes, however, are always insignificant in size as compared to fertile testes, and show no seminiferous tubule formation.

It is possible to arrange matters such that the initial gonads become populated by a smaller number than usual of primordial germ cells, and in the ensuing period of gonad growth rather curious phenomena are manifest. Even the existence of one or two germ cells in the entire length of a very young gonad has a marked effect on the mitotic rate in the somatic cells. The gonad is less affected in the female than in the male, in which the testis may grow to about 30% of the mass of a fertile testis and exhibit the formation of seminiferous tubules. Such testes, however, also develop large vacuoles, excessive vascularization and unusual numbers of melanophores. When the initial number of primordial germ cells is larger, one of two consequences can occur in the developing ovary, which always shows increased growth as compared with totally sterile or almost-sterile ovaries. Either the primordial germ cells apparently fail to enter mitosis in recognizable numbers but instead develop directly through oocyte stages to fully formed ova (replete with yolk and pigment), or mitosis occurs at rates in excess of those found in fertile animals such that rapid restoration of normal (fertile) germ cell counts is achieved (see Bounoure, 1964). It is possible that the decision to undergo regulation to normality or to continue to manifest the syndromes of complete or partial sterility is based upon some physiological threshold at the cellular level which in turn is based on some quantitative relationship between the somatic and germinal elements of the ovary. It should be added that this decision is taken and implemented during larval and immediate post-metamorphic life; adult animals are either completely sterile or normally fertile, although it is not uncommon in *Xenopus* to encounter animals which have one fully fertile ovary and the other absolutely sterile.

II. GROWTH AND DEVELOPMENTAL CAPACITY OF THE OOCYTE

Another evident effect of the somatic cells of the gonad upon the germinal cells occurs during the period of sexual differentiation. It can be experimentally demonstrated that up to this moment in development the germ cells are entirely plastic in respect to their ability to adopt permanently sperm-forming or egg-forming differentiative pathways (Humphrey, 1957; Blackler, 1965). Experimental evidence can also be mustered to substantiate the view that the differentiative pathway is elected for the primordial germ cell by the sexual nature (i.e. genetic sex constitution) of the somatic elements of the gonad; these studies are based mainly upon the induction of sex reversal. Again, however, the generally reciprocal nature of gonadal events is shown in cases where the gonad is naturally or artificially deprived of all sex cells, whereby the sterile gonads fail to show sex differentiation in the somatic structure apart from the appearance of the rete tissue. Sex differentiation of the gonadal mesoderm is itself evoked by the presence of reproductive cells.

Primary oocytes begin to grow shortly after metamorphosis. In view of the inter- relationships of the germinal and somatic elements of the gonads reviewed above, one is given pause to inquire if such reciprocity extends to the actual differentiation of the ova. The problem is especially apt in the case of oogenesis, because the formation of the egg involves not only the genetic adjustments involved in the preparations for sexual syngamy, but also the creation of an extraordinarily large cell which contains the nutriment to support early development and definite information for that development. In particular, one may ask if the follicle cells are purely nutritive in their physiology, affording non-instructional sustenance to the growing oocyte, or if they might also transfer some molecules whose presence is vital for successful early development. In some sense, ovarian events evidently influence the course of early development. The basic egg polarity in amphibian eggs seems associated with the orientation of the growing oocyte relative to the ovarian wall, and this polarity is later on crucial in determining the anterior-posterior axis of the future embryo. The egg polarity is revealed to the observer in a number of ways, not the least of which is the distribution of yolk. Amphibian yolk has drawn some considerable research attention in recent years, and a number of studies (Wallace and Dumont, 1968) have led to the general conclusion that amphibian yolk is not synthesized in the oocyte at all, as one might reasonably have surmised, but in fact is manufactured in the maternal liver from whence it is shunted via the bloodstream and the follicle cells into the egg. The distribution of the yolk, in addition to influencing the anterior- posterior axis also has, as classical studies have repeatedly shown, marked effects on the pattern and rate of cleavage, and the type of gastrulation shown in the different species. Developmental instruction, as seen in these examples, is limited to a crude 'roughing out' of the embryo plan, but one is left with the consideration that molecules other than yolk may also be transferred to the oocyte and that these may have greater developmental consequence, even to the extent of being responsible for some fraction of cellular differentiation and morphogenesis.

It is simple to pose the problem, but not easy to conceive experiments to elucidate it. One may look for the passage of specific chemicals across the follicle cell-oocyte boundary, but even if detected, they have to be demonstrated to have developmental significance. The rest of this presentation is devoted to a description of some experimental work in the African toad genus *Xenopus* which has attempted to test the problem, not by the means of assembling a list of chemical candidates, but by subjecting the developmental 'purity' of the oocyte to a purely developmental test.

In a previous publication (Blackler, 1962) I reported on the results of the surgical transfer of primordial germ cells between the South African clawed toad *Xenopus laevis laevis* and another subspecies *X. l. victorianus* at the neurula stage. The object of the study was to determine whether the germ cells of one subspecies could function effectively, in terms of gamete differentiation and zygote development, in the gonad of the other subspecies. Since the gonads are not yet formed at the neurula stage, the donor primordial germ cells migrate to the endodermal crest of the host and thence participate in the formation of a gonad the somatic cells of which are entirely supplied by the host. The results were that females laid eggs whose size and pigmentation were identical with those normally laid by the donor species, and that experimental males produced sperm which functioned genetically in accord with the donor genotype. Since the tests of progeny used in the analysis yielded only hybrid or host-type offspring (the latter the result of faulty surgery), the examination of donor character (other than the egg characters mentioned above) was limited to the time of appearance of the melanophores in the epidermis overlying the tadpole rectal tube, which is clearly different in the two species and seems to be inherited as a simple dominant-recessive trait. Notwithstanding the limitations of the experiment, it did not appear that the development of the donor-type eggs had been influenced by either the kind of host materials transferred to them while they were still oocytes, or by any kind of developmental 'information' imparted to the egg cortex by the host follicular cells.

Unfortunately, transfer of germ cells and their 'transmission' between subspecies may not be considered a fair test. It could be argued that the differences between the subspecies are slight from a genetic viewpoint, and that even if polymorphic forms of the same protein existed between the South African and Ugandan *Xenopus laevis,* the differences are not enough to influence the course of ontogeny. Therefore, I have set out to repeat this work using distinct species of *Xenopus* which not only display distinctive characteristics separating them developmentally but also in which the hybrid is lethal at a specific stage of development. If germinal cell lines may be subjected to transmission between species, and the eggs laid by the experimental female are altered in structure or behavior through residence of the growing oocyte in a foreign follicle, or if the eggs laid may be used in hybrid studies in which the hybrid survives or at least develops beyond the usual stage of arrest, then some functional molecular relationship between follicle and egg would be established and further studies involving the search for specific transferred molecules would be justified. In fact, as we shall see, there is no evidence in the intersubspecific study that is at variance with the new results.

The species used were *X. l. laevis* from South Africa and *X. tropicalis* from Camaroun. As Fig. 17.2 shows, the species are allopatric. Moreover, their ecology is different, *laevis* being a toad of open country while *tropicalis* is a denizen of forested regions. There is no difficulty in distinguishing the two species at any stage in their life histories, as may be seen from the list of characters used in the study (Table 17.1). Forty-two neurula transfers were performed, according to the technique of Blackler and Fischberg (1961), and from these 16 adult experimental *laevis* were obtained. Eight of these animals carried the Oxford nuclear marker (deletion of the nucleolar organizer site on the chromosome) in the heterozygous form; the use of this marker enables clear distinction to be made between donor and host cells (see Blackler and Fischberg, 1961, for details).

All transfers were made from *tropicalis* to *laevis*. There were three reasons for making the transfers in only one interspecific direction. *Tropicalis* does not respond readily to mammalian gonadotropins (customarily used to provoke mating in *Xenopus),* whereas *laevis* females will almost always lay eggs after receiving 300 I.U. of chorionic gonadotropin (human) and males clasp females after a dose of 50 I.U.; the erratic response of *tropicalis* could have made the analysis of *tropicalis* females carrying *laevis* sex cells a frustrating experience. Second, the egg of *tropicalis* is smaller than that of *laevis* and there is a corresponding difference in size of their neurula: thus, while a fragment of the *tropicalis* neurula large enough to contain a reasonable number of primordial germ cells may be transplanted to *laevis* without difficulty at this stage, the reverse graft is extremely difficult to bring off - a quart cannot be put into a pint pot. A final consideration was that it was hoped to be able to carry out an analysis of experimental males by mating them with females of the other species; the male *tropicalis* is so small that he is unable to enter into correct amplex with the much larger *laevis* female.

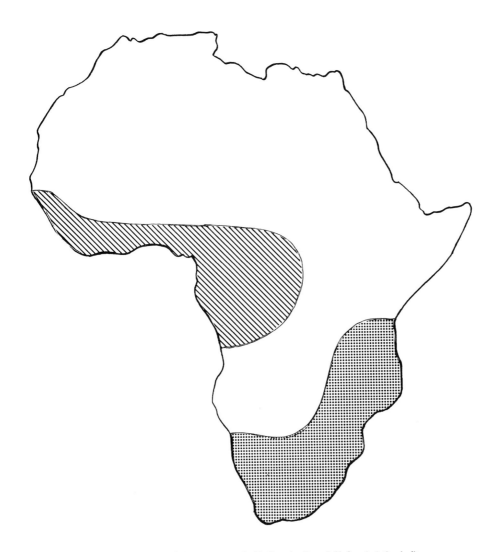

Fig. 17.2   Geographical distribution of *Xenopus tropicalis* (hatched) and *X. laevis* (stippled).

Gurdon (1962, 1967) has reported that species of *Xenopus* will not hybridize, either in a natural mating or by means of artificial fertilization. In his nuclear transplantation study of the two species used by me, he found that the transplant hybrids (diploid nucleus of one species in the cytoplasm of the other) arrest in both transfer directions, albeit at two different developmental stages. I tried to mate a female *tropicalis* with a *laevis* male three times, without success, although the male clasped the female on each occasion and she laid some eggs. However, it seems that the lack of success is due to the impossibility, given the size differences of the animals, of bringing the cloacae into correct approximation, since it was found possible to get some development if eggs were taken directly from the cloacal lips of

Table 17.1 **Phenotypic characteristics of** *Xenopus tropicalis* **and** *Xenopus laevis*

Phenotype	Character	*tropicalis*	*laevis*
Adult	Female size[1]	Not exceeding 65 mm.	May exceed 100 mm.
	Eyes	Small	Large
	Suboptic tentacle	Prominent	Usually absent or very reduced
	Dorsal color[2]	Yellowish 'flash' between eyes, greenish overall color with slightly darker markings	No flash. Finely reticulate or large solid patches of green, black and deep yellow
	Ventral color[3]	Finely stippled in blackish-brown	Immaculate white or thighs smudged in gray
	Hind foot claws[4]	Four	Three
	Handling	Docile	Nervous
	Response to chorionic gonadotropin[5]	Erratic	Usually responds
	Oviposition	About 4 hours after injection	7-10 hours after injection
Egg	Diameter	0.7 - 0.8 mm.	1.3 - 1.5 mm.
	Color[6]	Coarsely speckled brown-gray animal zone, cream marginal zone, very pale brown vegetal zone	Chocolate-brown animal zone, white or pale yellow vegetal zone
	Cleavage at set temperature[7]	Faster than *laevis*	Slower than *tropicalis*

Larva	Adhesive organ[8]	Pointed and protuberant	Blunted and less protuberant
	Head melanophores	Evenly distributed	Scarce between eye and brain
	Rectal tube melanophores	Appear in early larval life	Appear at end of larval life
	Maximum size[9]	80 - 130 mm.	70 - 85 mm.
	Anal fin[10]	Strongly arched	Weakly arched
	Tail axis when swimming[10]	Sharply bent to body axis	Smoothly bent to body axis
	Abdomen	Circular in profile, bulbous	Oval in profile, smooth contour
	Overall color	Sandy brown	Black and white
Metamorphosis	Duration	13 days	8 - 9 days
	Tail resorption	Tail cylindrical in cross-section, drooping between legs	Tail laterally flattened, strongly curled at end
	Length at end of metamorphosis[1]	17 - 25 mm.	14 - 16 mm.
	Growth to sexual maturity	7 - 8 months	10 - 12 months

Notes:
1. Snout-vent length
2. Fig. 17.4, a and b
3. Fig. 17.3, a and d
4. Fig. 17.3, a
5. *X. tropicalis* receive 150 I.U. HCG (♀) or 50 I.U. (♂),
   X. laevis receive 350 I.U. (♀) or 150 I.U. (♂)
6. Fig. 17.4, c and d
7. At 22°C, gastrulation begins in 5 hours in *tropicalis*,
   in 7 hours in *laevis*
8. Fig. 17.5, b
9. Tadpoles raised in water at density of 1 tadpole per 2 liters
   water, and fed with a suspension of nettle powder *(Urtica)*
10. As viewed with naked eye

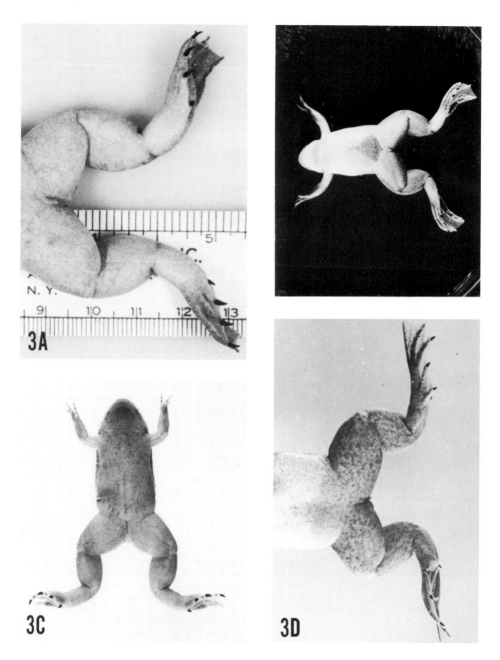

Fig. 17.3 a. Hindlegs of *X. tropicalis.* Note the four claws; it is the one on the inside of the foot that is missing in *X. laevis.* b. Young experimental *laevis* (length 50 mm.) showing pigmented patch of skin in the ventral midline which marks the presence of *tropicalis* material. c. Young *tropicalis* (length 30 mm.) showing the customary four claws and pigmentation of the species. This animal, however, resulted from a cross between a *tropicalis* female and an experimental *laevis* male. d. Hindlegs of experimental *laevis* (length 65 mm.) showing the mottled thigh pattern of *laevis* but the four claws on each foot typical of *tropicalis.*

the female, rapidly dried with filter paper, and then dabbed with small pieces of a minced *laevis* testis: about a third of the treated eggs cleaved and died uniformly at the beginning of gastrulation. The reverse hybridization has been attempted once only, principally because very few *tropicalis* males were available. Less than 5% of the eggs cleaved, but their development was largely normal except for some difficulties during gastrulation until the late neurula stage, at which time they died uniformly. It is interesting that these stages are the same as those reported by Gurdon (1962) as the arresting stages in his transplant embryos.

In the present study the 16 adult experimental animals turned out to consist of 4 females, 3 males and 1 sterile male with the Oxford nuclear marker, and 6 females and 2 males unmarked. The unmarked animals were not used in the analysis because the progeny of the males could not be screened for as many characters as female-derived progeny, and the females either laid strictly *laevis* eggs (the result of graft failure and poor removal of host sex cells at the time of the transfer operation) or laid such a small percentage of *tropicalis* eggs among the *laevis* eggs that getting enough of them for a quantitative analysis would have been difficult. Nonetheless, the fact that some indisputable *tropicalis* eggs were obtained among the larger and differently colored *laevis* eggs afforded encouragement to analyze the marked individuals.

Before considering the analysis, it should be mentioned that the experimental *laevis* already showed some *tropicalis* characters. When the graft is made, a small piece of donor mesoderm and epidermis is also transferred to promote rapid healing. Shortly after metamorphosis the *tropicalis* epidermis can be recognized as a brownish-black and finely granular patch in the posterior belly region of the host *laevis* (see Fig. 17.3b). As the animals grew, *laevis* and *tropicalis* influences could be seen in the nature of the hindlimbs, which were of either species form in size, contour and pigmentation of skin. However, there was no guarantee that a *tropicalis*-like limb should be armed with the species number of four claws, or a *laevis* limb should bear three, as is indicated in Table 17.2 and Fig. 17.3b.

Table 17.3 summarizes the results of the mating analysis. Since the number of experimental animals available was small, and one may mate animals with optimal fertilization rates twice a year only, a systematic analysis would have been difficult. Therefore, experimental females were either mated with *laevis* males or with experimental males, while experimental males were also mated with *tropicalis* females. Little attempt was made to examine large numbers of eggs in detail since the qualitative aspects of the analysis were considered to be more important. Nonetheless, at least a hundred *laevis*-like eggs were followed in their development, and as many *tropicalis*-like eggs as could be obtained.

Table 17.3 shows that two experimental females produced some eggs that were indistinguishable from eggs laid by wild-type *tropicalis* females. When these eggs were fertilized by sperm from experimental males, some of them commenced development through to the close of metamorphosis (Fig. 17.5a and b), and

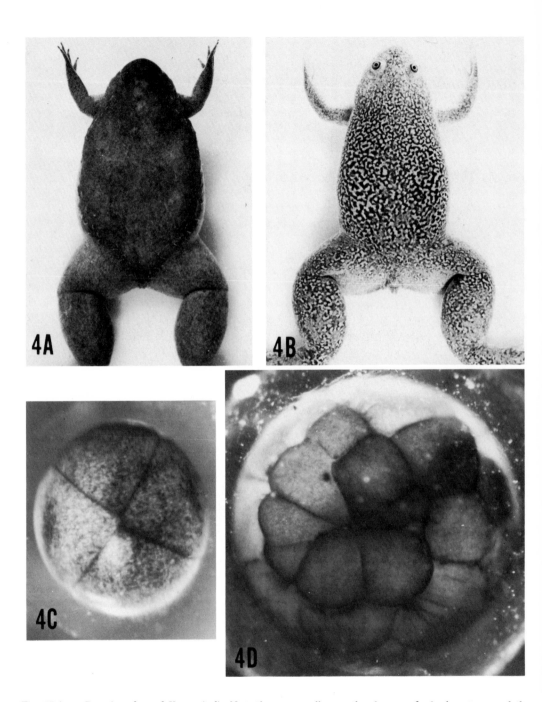

Fig. 17.4 a. Dorsal surface of *X. tropicalis*. Note the very small eyes, the absence of a back pattern, and the body shape. Length 65 mm. b. Dorsal surface of *X. laevis*. Note the prominent eyes, reticulate back pattern, and the body shape. Length 95 mm. c. Egg of *X. tropicalis* viewed at animal pole. The speckled pigmentation is characteristic. Diameter 0.8 mm. d. Egg of *X. laevis* viewed at animal pole. The fine pigmentation of the animal hemisphere is typical. Diameter 1.3 mm.

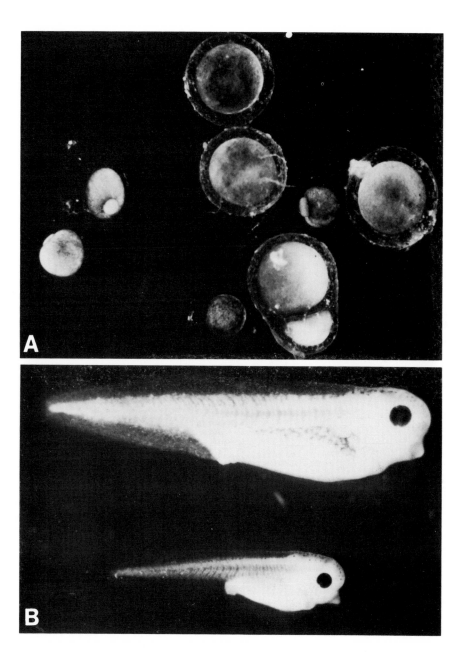

Fig. 17.5a *Tropicalis* and *laevis* eggs laid by an experimental *laevis* female mated to an experimental *laevis* male. Some eggs of both species will develop to metamorphosis, while others will show interspecific hybrid arrest. The *tropicalis* eggs (0.8 mm.) are at the late gastrula stage; the *laevis* eggs (1.3 mm.) are somewhat "younger" in stage. Note the inclusion of both kinds of eggs within a single jelly capsule at bottom center.

Fig. 17.5b Newly hatched larvae from the same mating as Fig. 17.5a. Above is the *laevis* type (length 6.0 mm.) showing the blunt adhesive organ, and below it is the *tropicalis* type (length 3.6 mm.) with pointed sucker. Note that the *tropicalis* larva is more developmentally advanced than the *laevis* although both have the same developmental age.

Table 17.2 **Hindlimb Type and Claw Number in the Adult Experimental Toads**

Toad	Sex	Leg Type	Left Leg No. of Claws	Leg Type	Right Leg No. of Claws
1	♀	laevis	3	laevis	3
2	♀	laevis	3	laevis	3
3	♀	laevis	4	laevis	4
4	♀	laevis	3	laevis	3
9	♀	tropicalis	4	laevis	3
10	♀	laevis	3	laevis	3
11	♀	laevis	3	laevis	3
12	♀	tropicalis	3	laevis	4
13	♀	laevis	4	tropicalis	3 + bud
14	♀	tropicalis	3	tropicalis	4
5	♂	laevis	3	laevis	3
6	♂	laevis	3	tropicalis	3
7	♂	laevis	3	laevis	4
8	♂	laevis	3	laevis	3
15	♂	laevis	3	tropicalis	4
16	♂	tropicalis	3	laevis	3

throughout development no deviation was observed from the characteristics of *tropicalis* as listed in Table 17.1. Confirmation of gametic purity was obtained from a mating of an experimental male with a *tropicalis* female in which all the eggs developed strictly in accordance with *tropicalis* characteristics. Hybrid arrests were also noted in some matings, - or rather arrested embryos that were indistinguishable from the arrested embryos in natural hybrid matings. Unfortunately there is always some abnormality in 'natural' *Xenopus* matings (5-15%) and there was no way of distinguishing between truly hybrid arrests and intraspecific abnormal embryo

Table 11.3 Results of ... mating Xenopus laevis ... (L = laevis, X = ? ♀/♂)

X= experimental *laevis* eggs

Frog	Sex	Mate	Egg type	Development	Graft success
X1	♀	L♂	L	Typically L development.	−ve
X2	♀	X6♂	93% L, 7% T	T eggs arrested as early gastrulae, L (except anucleolate forms) developed to metamorphosis.	+ve
X2	♀	L♂	94% L, 6% T	No T eggs developed, but L develop to metamorphosis.	+ve
X3	♀	X6♂	L	Typically L development (except anucleolate forms) with no unusual incidence of arrested development.	−v
X4	♀	X7♂	86% T, 14% L	L eggs obtained metamorphosis, or died either as anucleolate forms, arrested gastrulae, or late neurulae.	+ve
X4	♀	X7♂	83% T, 17% L	See also under X7 below. T eggs developed to metamorphosis or died as early gastrulae.	
X4	♀	L♂	85% T, 15% L	L eggs developed to metamorphosis, T eggs died as early gastrulae.	
X5	♂	T♀	T	Few eggs developed (15) but all showed typical T development to metamorphosis.	+ve
X6	♂	X2♀	see above	see above	−ve
X7	♂	X4♀	see above	65% T eggs developed to metamorphosis, 35% T arrested as gastrulae. 25% L eggs developed to metamorphosis, 9% died as anucleolate larvae, 30% arrested as gastrulae and 36% as late neurulae.	+ve
X8	♂	T♀	T	No eggs developed. Dissection revealed sterility.	−ve

arrests at the same moment in development; this was particularly true in the case of arrests at the beginning of gastrulation. It must also be admitted that attempts to resolve this matter by examining nucleolar number, by means of squashing the embryo under a coverglass and examining the nuclei by phase contrast microscopy, proved inconclusive: not only are nucleoli difficult to see at the beginning of gastrulation, but the fact that many of the arrested gastrulae were in the process of degeneration rendered the nuclear contents ambiguous, to say the least. Be that as it may, and given that this eventuality introduced an unwelcome subjectivity into the analysis, the results of the matings did not signify that the time and character of interspecific hybrid arrest were altered by the residence and growth of *tropicalis* oocytes in *laevis* gonads.

The most extraordinary result was the obtaining of eggs, embryos and metamorphosed toads of both species from a mating of an experimental female *laevis* with an experimental *laevis* male. It demonstrates that *tropicalis* oogonia can embark on oogenesis and yield a fertilizable egg that will develop in a manner typical of its species, in spite of being forced to accomplish this in the environment of a foreign follicle and ovary. The degree of autonomy on the part of the growing oocyte is indeed impressive and bespeaks a high level of cellular integrity on the part of reproductive cells in general.

III. COMMENTARY

The extent of interaction between the reproductive and the somatic elements of the amphibian gonad throughout larval life seems to be of a fairly high order up to the moment when the reproductive cells settle into a sperm-forming or egg-forming developmental pathway. Thereafter it appears that, in ovaries, the general ovarian cells may supply much aid and comfort to the growing oocytes but that the latter employ this beneficence strictly for their specific advantage. Perhaps the most striking example of gametogenic single-mindedness is the case (very rare) in which the gonad of a mature *Xenopus* turns out to be a true hermaphrodite, with oocytes in vitellogenesis placed in follicles immediately adjacent to seminiferous tubules filled with apparently normal spermatozoa. Such cases are not freely available for experimental study, but the ability of "transmission" of one species through the gonad of a second species is almost as striking. The fact that this ability exists strongly suggests in turn that instructional molecules necessary for development are not transmitted during oogenesis. Moreover, it would appear that the growing oocyte is either non-selective for incoming molecules, or else accepts them but chemically re-manufactures them to its specific requirements. More research in this area of study is evidently needed, especially in the light of current speculation about the absorption of drugs, etc. in the oocyte cytoplasm and their effect on later development.

How more generalized, beyond the amphibia, the findings reported in this paper happen to be, is unknown. However, just as amphibian studies have led to reassessments of mammalian studies of maturation and ovulation (see Schuetz, Smith, this symposium), the results of species transmission in *Xenopus* indicate that it may be possible to devise new techniques of contraception involving oogenesis without the concomitant danger of introducing a range of developmental abnormalities into fertilized eggs when the contraceptive embargo is lifted. Finally, there is a strong possibility of maintaining species which are rare, or difficult to maintain under laboratory conditions, by carrying their reproductive cell lines through the somas of related, and easily maintained, species.

IV. ACKNOWLEDGMENTS

Aided by a grant from the United States Public Health Service (HD-01663).

V. REFERENCES

Blackler, A. W. and Fischberg, M. (1961) Transfer of primordial germ cells in *Xenopus laevis*. *J. Embryol. Exp. Morph.* **9**, 634.

Blackler, A. W. (1962) Transfer of primordial germ cells between two subspecies of *Xenopus laevis*. *J. Embryol. Exp. Morph.* **10**, 641.

Blackler, A. W. (1965) Germ cell transfer and sex ratio in *Xenopus laevis*. *J. Embryol. Exp. Morph.* **13**, 51.

Blackler, A. W. (1966) Embryonic sex cells of amphibia. *Adv. Reprod. Physiol.* **1**, 9.

Bounoure, L. (1964) La lignee germinale chez les batraciens anoures. *In* "L'Origine de la Lignee Germinale." (E. Wolff, ed.), Paris, Masson et Cie.

Everett, N. (1945) The present status of the germ cell problem in vertebrates. *Biol. Rev.* **20**, 45.

Gurdon, J. B. (1962) The transplantation of nuclei between two species of *Xenopus*. *Develop. Biol.* **5**, 68.

Gurdon, J. B. (1967) African clawed frogs. *In* "Methods in Developmental Biology." (F. H. Wilt and N. K. Wessells, ed.), New York, Crowell.

Humphrey, R. R. (1957) Male homogamety in the Mexican axolotl: a study of the progeny obtained when germ cells of a genetic male are incorporated in a developing ovary. *J. Exp. Zool.* **134**, 91.

Smith, L. D. (1966) The role of a 'germinal plasm' in the formation of primordial germ cells in *Rana pipiens*. *Develop. Biol.* **14**, 330.

Wallace, R. A. and Dumont, J. N. (1968) The induced synthesis and transport of yolk proteins and their accumulation by the oocyte in *Xenopus laevis*. *J. Cell Physiol. Suppl. 1.* **72**, 73.

Witschi, E. (1929) Studies on sex differentiation and sex determination in amphibians. *J. Exp. Zool.* **52**, 235.

Zuckerman, S. (1951) The numbers of oocytes in the mature ovary. *Recent Progr. Hormone Res.* **6**, 63.

18

**THE ROLE OF PROTEIN UPTAKE IN
VERTEBRATE OOCYTE GROWTH
AND YOLK FORMATION**

Robin A. Wallace

I.   Introduction
II.   Definition of experimental conditions and terms
        A.   Animals
        B.   Yolk protein
III.   Synthesis and turnover of vitellogenin
IV.   Incorporation of vitellogenin by the ovary
        A.   Selectivity
        B.   Transformation
V.   Protein incorporation by the isolated oocyte
        A.   Experimental procedure
        B.   Protein incorporation as a function of oocyte size
        C.   Progress and localization of protein incorporation
        D.   Temporal extent of HCG stimulation
        E.   Other observations
VI.   Relevance to the mammalian oocyte
VII.   Acknowledgments
VIII.   References

I. INTRODUCTION

Research in oogenesis has traditionally focused on early chromosomal events or the terminal stages of oocyte maturation, while the intermediate period of oocyte development and particularly the growth of the oocyte have not been considered as carefully. This is somewhat surprising because one of the prime characteristics of the oocyte is its size; it is generally the largest cell found in most animals and is only occasionally rivaled by nerve cells in this respect. Since one of the purposes of this conference is to define and examine some of the unique aspects of oogenesis, I would like to consider the question of oocyte size and present a personal, admittedly inadequate, but exploratory tour of the operant mechanisms of oocyte growth in vertebrates.

We have used the amphibian oocyte for our initial studies on oocyte growth because it can be readily obtained in large numbers and cultured under controlled experimental conditions and because the macromolecular components of the amphibian oocyte are relatively simple when compared, for example, to those from most fish, birds, and reptiles. Hopefully, what we find out about the amphibian oocyte will have some application to the mammalian oocyte as well, and we view the process of oocyte growth and protein deposition in the amphibian oocyte as a conveniently exaggerated model of similar events occurring during mammalian oocyte development.

In most amphibians, then, after oogonia divide and give rise to oocytes, cell division arrests for up to three years, while a period of seasonal oocyte growth ensues. Grant (1953) has diagrammatically illustrated this process for *Rana temporaria* (Fig. 18.1), and his figure indicates that at a given time three groups of

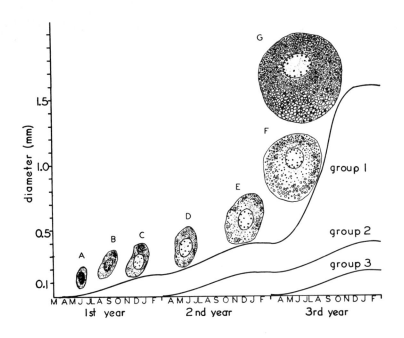

Fig. 18.1 Seasonal growth of the *Rana temporaria* oocyte [From Grant, 1953].

oocytes are present in the ovary and that the major portion of oocyte growth occurs during the middle of the third year. During this limited period, the diameter of the oocyte enlarges by a factor of four so that the volume increases some 11-fold and about 90% of the oocyte is formed. The major event which occurs during this growth phase is vitellogenesis or the deposition of protein material (Wallace and Dumont, 1968), so that a remarkable amount of protein is packaged within a single cell over a relatively short period of time. We have been particularly interested in this vitellogenic period and over the past several years have attempted to define some of the physiological processes which are taking place.

II. DEFINITION OF EXPERIMENTAL CONDITIONS AND TERMS

A. Animals

Our experimental animal has generally been *Xenopus laevis* imported from South Africa. These animals have ovaries which, under natural conditions, apparently undergo seasonal enlargement due to vitellogenesis in the oldest group of oocytes (Gitlin, 1939), as has been depicted for *R. temporaria* (Fig. 18.1). Animals arriving in the laboratory, however, have generally been subjected to a variety of environmental conditions, sometimes for a considerable length of time, so that most females contain a large number of mature oocytes in addition to immature oocytes in an infinite variety of stages. The immature oocytes from such animals are also extremely variable with respect to the rate at which they are developing. We have therefore found it necessary to provide a uniform diet and a 12-hour light cycle for at least two months before any experimental manipulation. Animals treated in this manner are described as "normal", and all but the largest oocytes from normal females still undergo a very slow but measurable rate of growth. As we shall see, oocyte growth in normal females can be stimulated by a single, large dose of human chorionic gonadotropin (HCG).

B. Yolk Protein

The principal component of the mature oocyte is what has generally been described in amphibians as the yolk platelet (Fig. 18.2). Yolk platelets comprise better than 80% of the protein-nitrogen (Gregg and Ballentine, 1946) and more than 90% of the protein-phosphorus (Grant, 1953) of the amphibian egg and are the primary components elaborated during the terminal growth phase of the oocyte (Wittek, 1952; Grant, 1953; Kemp, 1953). The yolk platelet itself consists of a central, crystalline "main body" (Karasaki, 1963) containing the proteins lipovitellin and phosvitin (Wallace, 1963b), a surrounding "superficial layer" which may contain serum-like antigens (Glass, 1959), phosphatases (Brown and Millington, 1968), and polysaccharide substances (Ohno, Karasaki and Takata, 1964; Tandler and La Torre, 1967), and an investing membrane (Karasaki, 1963). Lipovitellin and phosvitin together comprise about 98% of the yolk platelet protein (Wallace, 1963a) and represent the molecular elements of what we shall define as "yolk" (Wallace and Dumont, 1968).

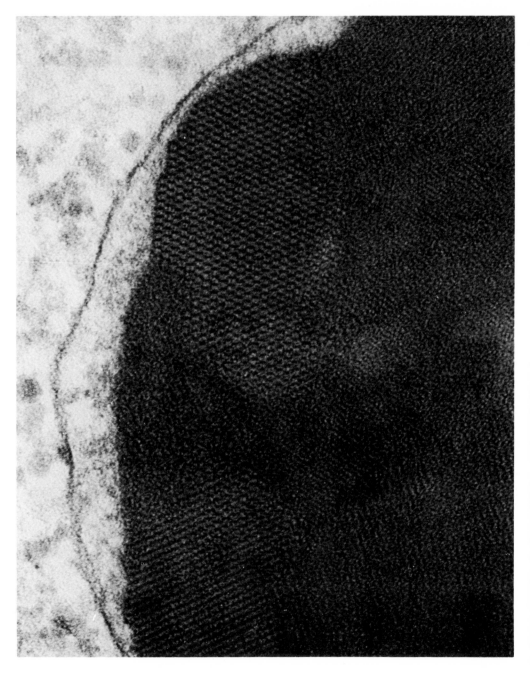

Fig. 18.2 Electron micrograph of part of a small yolk platelet from an early *X. laevis* morula (before yolk utilization has begun) depicting an outer membrane, peripheral "superficial layer," and a central, electron-dense "main body." A variety of crystalline planes, including portions with no apparent ordered structure can be seen in the main body. Fixed in veronal-buffered 1% OsO_4, embedded in Epon, and stained with uranyl acetate and lead citrate   [Courtesy of S. Karasaki]   (X 256,000).

III. SYNTHESIS AND TURNOVER OF VITELLOGENIN

Many female vertebrates display seasonal changes involving elevated levels of serum components, particularly lipid, calcium, and phosphoprotein, and such changes appear to be under the control of estrogen. Immunological and biochemical similarities between certain serum components and the proteins of the oocyte have also been noted, and the transfer of serum proteins to the developing oocyte has been demonstrated in a variety of animals (for literature, see Follett and Redshaw, 1968; Wallace and Jared, 1968). Similar observations have been made on *X. laevis* (Zwarenstein and Shapiro, 1933; Follett and Redshaw, 1968; Wallace and Jared, 1968).

The macromolecular components involved in these processes have been poorly understood, however, and as an initial step to define such compounds in *X. laevis*, we prepared TEAE-cellulose chromatograms of sera derived from animals treated with various hormones and injected with $[^{32}P]NaH_2PO_4$ and $[^{14}C]$leucine 20 hours previous to bleeding (Fig. 18.3). The pattern found for the normal male (Fig. 18.3a) indicates that the serum proteins are slightly labeled by $[^{14}C]$leucine, but not by $[^{32}P]NaH_2PO_4$, and that there is virtually no alkali-labile protein-phosphorus in any of the eluant fractions. In the normal female (Fig. 18.3b), a new, small peak is present at elution position 0.64 and is well-resolved from the other serum components. Although present in small amounts, this normally sex-limited material appears to be synthesized at about the same rate as most other serum proteins, as judged by the extent of $[^{14}C]$leucine labeling. It is also the only component labeled by $[^{32}P]NaH_2PO_4$ and appears to be associated with a small amount of protein-phosphorus. We have isolated this component, found it to be a lipophosphoprotein (12% lipid, 1.4% protein-phosphorus) with a molecular weight of 460,000, and have designated it as vitellogenin (Wallace, 1970a). Experiments with tissue slices have also shown that it is synthesized by the liver (Wallace and Jared, 1969).

If normal female *X. laevis* are injected with HCG under controlled laboratory conditions, they ovulate many of their mature oocytes and at the same time become what we call "vitellogenic", *i.e.*, engaged in the production of yolk protein necessary for the growth of new oocytes (Wallace and Dumont, 1968). A chromatogram of the serum of such a vitellogenic female (Fig. 18.3c) indicates an increased amount of vitellogenin relative to the normal female. This increase is evident in both the amount present and especially the amount labeled. The production of vitellogenin appears to be mediated by estrogen, for if estradiol-17β is administered to the male, vitellogenin is found in the serum (Fig. 18.3d). The specific activity of vitellogenin in females is higher than that found for the estrogen- treated male, suggesting that it is turning over more rapidly. This fact was confirmed by the observation that the physiological half-life for vitellogenin in the serum of vitellogenic females and estrogen-treated males is approximately 2 days and 40 days respectively (Wallace and Jared, 1968). Once the ovary is removed from the vitellogenic females, however,

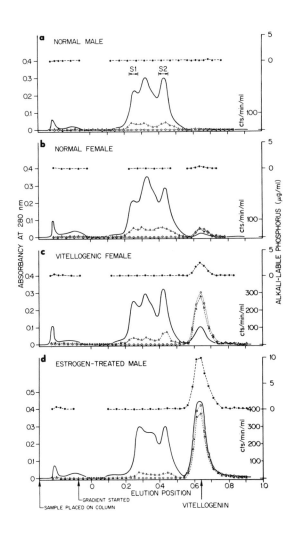

Fig. 18.3 Chromatography on TEAE-cellulose of sera derived from normal males (a), normal females (b), vitellogenic females (c), and estrogen-treated males (d). The absorbancy of the effluent is indicated by a solid line, and effluent fractions were analyzed for alkali-labile protein-phosphorous (●——●) and for protein labeling associated with ^{14}C (△——△) and ^{32}P (○- · -○) [From Wallace and Jared, 1969].

the physiological half-life changes to 40 days, so that the ovary appears to be the tissue which normally removes most of the vitellogenin from the bloodstream.

As an initial step in defining the conditions necessary for this ovarian uptake, two groups of females were injected with estradiol-17β and HCG respectively. Protein-phosphorus and ^{32}P-labeled protein, as specific markers for vitellogenin (see

Fig. 18.3), were then determined at various times in the serum in order to indicate the course of vitellogenin synthesis and turnover (Fig. 18.4). Estrogen-treated females immediately began to synthesize vitellogenin, which accumulated in the serum until around day 12 (Fig. 18.4a; see also Follett and Redshaw, 1968). When $[^{32}P]NaH_2PO_4$ was injected into estrogen- treated females on day 5, the amount of labeled vitellogenin observed in the serum during the following week was constant (Fig. 18.4b), thus indicating that no significant amounts were being removed from the circulation. During the same period, the specific activity of protein- phosphorus dropped to some extent because of a dilution of labeled phosphoprotein by additionally synthesized, unlabeled phosphoprotein (Fig. 18.4c). By day 12 after the administration of estrogen, vitellogenin synthesis ended, and both its absolute amount in the serum and its specific activity reached a relatively constant level. When HCG was given to the estrogen- treated females on day 13, they ovulated their mature oocytes within 24 hours; subsequently, all values for protein-phosphorus, protein labeling, and specific activity decreased with time. The decline in the amount of labeled protein indicates that vitellogenin was removed from the circulation as a response to HCG administration, whereas the decrease in specific activity indicates that additional, unlabeled vitellogenin was added to the circulation. The decline in protein-phosphorus values suggests that even though additional vitellogenin was introduced into the serum compartment, it was removed from this compartment at a faster rate.

For comparison, a second group of females was given HCG on day 0, after which they ovulated and began a period of vitellogenesis. The total amount of protein-phosphorus in the serum rose only slightly after HCG administration (Fig. 18.4a), and when $[^{32}P]NaH_2PO_4$ was injected on day 8, the vitellogenin subsequently labeled was found to disappear from the serum at a rate of $t_{1/2}$ = 1.8 days (Fig. 18.4b). The specific activity of vitellogenin dropped at an identical rate (Fig. 18.4c), indicating that during the period examined, a steady state existed in which the amount of vitellogenin entering the circulation was balanced exactly by the amount leaving the circulation. This amount can be calculated to be about 7.7 mg vitellogenin/day (Wallace and Jared, 1969).

The conclusion we may draw from this experiment is that the injection of estrogen or the mere presence of vitellogenin in the circulation does not promote an uptake of vitellogenin by the ovary, and in fact there is evidence that estrogens may actually suppress this process (Follett, Nicholls and Redshaw, 1968). Instead, ovarian uptake is promoted in some manner by HCG, which also induces ovulation and apparently the ovarian secretion of estrogens (Gallien and Chalumeau-Le Foulgoc, 1960; Follett *et al.,* 1968).

IV. INCORPORATION OF VITELLOGENIN BY THE OVARY

A. Selectivity

In order to determine whether or not the uptake of vitellogenin by the ovary is

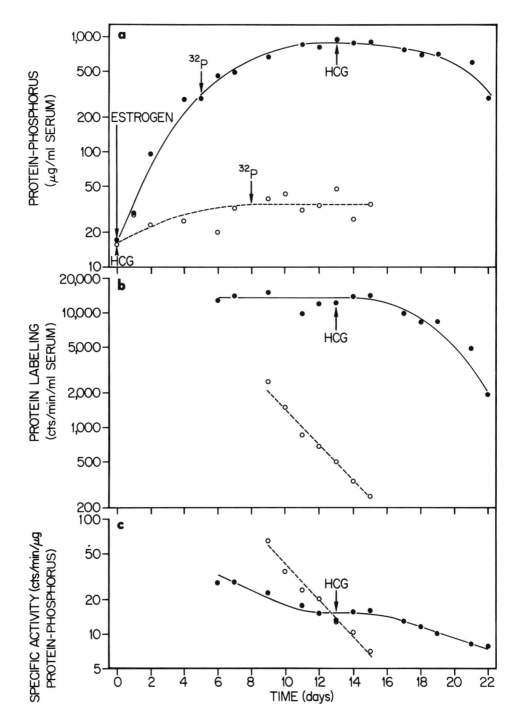

Fig. 18.4 Patterns of serum phosphoprotein synthesis and turnover as a function of time after hormone injections in two groups of *X. laevis* females as indicated by (a) the presence of alkali-labile protein-phosphorous in the serum, (b) the amount of labeled protein present after the injection of $[^{32}P] NaH_2PO_4$, and (c) the specific activity of the serum protein- phosphorus [From Wallace and Jared, 1969].

a selective process, we injected equivalent amounts of [³H] vitellogenin or serum protein fractions S1 or S2 (indicated in Fig. 18.3a) into vitellogenic females and assayed for labeled protein present in the serum and in the ovary as a function of time in each of the three cases. The results (Fig. 18.5) indicated that vitellogenin left the circulation and became associated with ovarian protein much more rapidly than

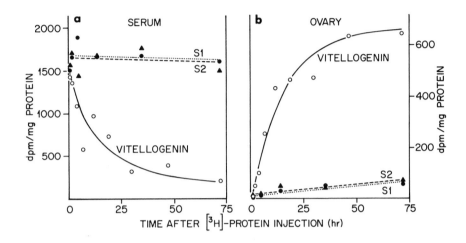

Fig. 18.5 Loss of labeled protein from the serum (a) and its incorporation into the ovary (b) following the injection of equivalent amounts of [³H] vitellogenin (o ——o) or serum protein fractions [³H]S1 (•. . .•) and [³H]S2 (△——△) into vitellogenic females [From Wallace and Jared, 1969].

the other two serum protein fractions. Initially, when the relative amounts of the three labeled components in the serum were similar, the incorporation of vitellogenin into ovarian protein appeared to be about 50 times more rapid than the incorporation of either S1 or S2. Thus, although the incorporation of protein into the ovary would appear to be a selective process, it is important to note that the discrimination is not absolute and the data suggest that any macromolecular component may become incorporated to some extent. This conclusion is supported by the previously observed localization of serum-like antigens (Glass, 1959) and both injected iron-dextran (Wartenberg, 1964) and Trypan blue (Wallace and Dumont, 1968) within amphibian oocytes undergoing vitellogenesis.

B. Transformation

The protein of the mature amphibian oocyte or egg is localized primarily in the yolk platelet inclusions, which are the crystalline structures comprised almost exclusively of lipovitellin and phosvitin (Wallace, 1963b, 1965; Wallace and Dumont, 1968). Vitellogenin itself has not been observed within the oocyte or egg. We therefore undertook a study of the fate of vitellogenin during its incorporation into ovarian protein.

As a first comparative step, a vitellogenic female was injected with [^{32}P]NaH$_2$PO$_4$, and the main-body components from the ovarian yolk platelets were subsequently isolated and chromatographed. The resulting chromatogram (Fig. 18.6a) indicated a major component analogous to lipovitellin (see Wallace, 1965) associated with a small amount of labeled protein-phosphorus at position 0.40; the

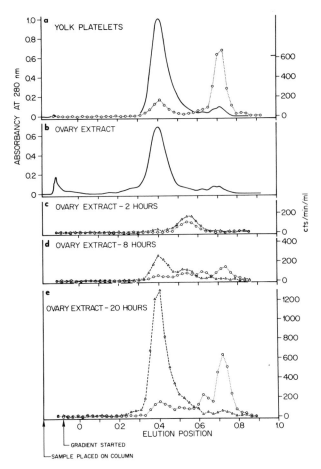

Fig. 18.6 Chromatography on TEAE-cellulose of (a) crystalline yolk platelet protein isolated from the ovary of a vitellogenic female injected with [^{32}P]NaH$_2$PO$_4$ 24 hours previous to isolation and of ovarian extracts prepared (b,c) 2 hours, (d) 8 hours and (e) 20 hours after the injection of [^{3}H,^{32}P] vitellogenin into vitellogenic females. Absorbancy (a,b) is indicated by a solid line, [^{32}P] protein labeling (a, c-e) by open circles (o—o), and [^{3}H] protein labeling (c-e) by open triangles (△- - -△) [From Wallace and Jared, 1969].

chromatogram also indicated a minor component analogous to phosvitin around position 0.72 and with which most of the labeled protein-phosphorus was associated. Lipovitellin contains 98% of the leucine in the yolk platelet crystal, and phosvitin contains 79% of the protein-phosphorus (Wallace and Jared, 1969), so that isotopic markers for these two components serve as fairly specific indicators for lipovitellin and phosvitin respectively.

As a second step, therefore, vitellogenin doubly labeled with [^{3}H] leucine and [^{32}P] NaH$_2$PO$_4$ was injected into vitellogenic females. At various time intervals thereafter, ovarian extracts containing a maximum amount of protein were made and chromatographed on TEAE-cellulose. A typical chromatogram (Fig. 18.6b) of such ovarian extracts indicates that most of the protein is comprised of the crystalline yolk platelet components (compare Fig. 18.6a). The labeling found in an extract several hours after injection of [^{3}H, ^{32}P] vitellogenin indicates the presence of a single component, eluting between 0.50 and 0.60, which has a ^{3}H/^{32}P ratio similar to the injected [^{3}H, ^{32}P] vitellogenin (Fig. 18.6c). Eight hours after the injection the labeling pattern appears heterogeneous, with most of the ^{3}H eluting in position 0.40 and most of the ^{32}P eluting at position 0.72 (Fig. 18.6d). By 20 hours (Fig. 18.6e), the amounts of ^{3}H-label in position 0.40 and ^{32}P-label in position 0.72, corresponding to lipovitellin and phosvitin respectively, is even more pronounced. Several other labeled components indicated in Fig. 18.6d and e, particularly in positions 0.55 and 0.63, may represent intermediates in the formation of lipovitellin and phosvitin. Follett *et al.* (1968) have also observed a similar distribution of phosphorus label in yolk platelet preparations after the injection of [^{32}P] vitellogenin. These observations, taken together with the physical, chemical and immunological similarities between vitellogenin and the yolk proteins lipovitellin and phosvitin (Wallace, 1970a), thus suggest that vitellogenin is a complex comprised of the precursor molecules for lipovitellin and phosvitin and that once the complex is transported to the ovary, an intra-molecular rearrangement occurs whereby it is transformed into the two proteins of the yolk platelet crystal.

V. PROTEIN INCORPORATION BY THE ISOLATED OOCYTE

A. Experimental Procedure

The developing amphibian oocyte is surrounded by three cellular layers, described by Wischnitzer (1963) in *Notophthalmus viridescens* as the: (a) *follicular epithelium,* the compact layer of follicle cells immediately adjacent to the oocyte; (b) *theca,* the middle layer of connective tissue in which can be found fibroblasts and elements of a capillary network; and (c) *surface epithelium,* the outermost layer of squamous cells which represents, over most of the oocyte, the inner ovarian epithelium (Fig. 18.7). A similar stratification has been observed for *X. laevis* oocytes (Wallace and Dumont, 1968).

In the intact vitellogenic female, we have seen that vitellogenin is rapidly incorporated by ovarian tissue (Fig. 18.5) and transformed into the yolk proteins of the oocyte (Fig. 18.6). During this process, vitellogenin must enter the capillary network in the theca and from there pass through the connective tissue and follicular epithelium to the oocyte surface, where it is presumably sequestered by a pinocytotic mechanism (Wallace and Dumont, 1968). The factors involved during these steps are difficult to study using the intact animal. We have therefore developed a culture procedure for oocytes which employs 50% labeled serum protein equilibrated against a simple saline solution (Jared and Wallace, 1969). The

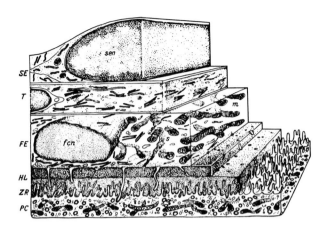

Fig. 18.7 A three dimensional illustration of the layers enveloping yolk forming oocytes in *Notophthalmus (Triturus) viridescens.* SE, surface epithelial layer; T, thecal layer; FE, follicular epithelial layer; HL, homogeneous layer; ZR, zona radiata; PC, peripheral cytoplasm of the oocyte; sen, surface epithelial cell nucleus; fcn, follicle cell nucleus. In *X. laevis*, the structure of the zona pellucida (HL + ZR) appears somewhat different [Wallace and Dumont, 1968], but other aspects of the enveloping layers are similar to *N. viridescens* [From Wischnitzer, 1963].

preparation of the labeled serum is such that at least 98% of the labeled protein incorporated by the oocytes is represented by vitellogenin (Wallace, Jared and Nelson, 1970).

As an initial consideration, then, we investigated the influence of the cellular layers on protein incorporation by isolated oocytes. The results are provided in Table 18.1 and indicate that protein is poorly incorporated into oocytes surrounded

Table 18.1 **Effect of Investing Cellular Layers on Protein Incorporation by Vitellogenic Oocytes** *in vitro**

Cellular layers present during culture	dpm·mm^{-2} ± S.E.	
	Exp. 1	**Exp. 2**
All	22 ± 3	11 ± 2
Follicular epithelium	320 ± 6	373 ± 15
None (EDTA-treated)	22 ± 1	52 ± 5

* Each value represents the average incorporation found for 12 oocytes cultured for 24 hours. (From Wallace *et al.*, 1970).

by all the cellular layers, but if the outer two layers are manually removed, protein is readily sequestered. This observation suggests, simply, that the outermost cellular layers are impermeable to protein.

The follicular epithelium is difficult to remove manually but can be dissociated from the oocyte by EDTA treatment (Masui, 1967). When this is done, again relatively little incorporation of labeled protein into the oocyte is observed (Table 18.1). This lack of incorporation is more difficult to interpret, because although it suggests that the integrity of the investing follicular epithelium is necessary for protein uptake, the possibility remains that binding sites on the oocyte surface were altered during the period of EDTA treatment. For the observations which follow, then, oocytes were cultured with the follicular epithelium intact but with the outer cellular layers removed.

B. Protein Incorporation as a Function of Oocyte Size

To determine which oocytes are capable of sequestering protein from the medium, we isolated oocytes of various sizes from the ovary and cultured them for 24 hours. Oocytes derived from a normal female ranged up to about 1.30 mm diameter and became labeled in culture only to a slight extent (Fig. 18.8a, open circles). Oocytes in the size range of 0.80-1.20 mm appeared most active in this respect. When the donor female was injected with HCG on the previous day, however, protein uptake by isolated oocytes was greatly enhanced (Fig. 18.8b, open circles). In this case, a peak of activity was observed for oocytes 1.00-1.15 mm in diameter, whereas the activity for the smallest and largest oocytes tapered off.

The extent of protein uptake observed for each oocyte in Fig. 18.8 has also been divided by the surface area of the oocyte (closed circles). An interesting aspect of this data is that oocytes between 0.6-1.2 mm diameter and particularly between 0.8-1.1 mm diameter appear to incorporate protein both maximally and at about the same rate per unit surface area. "Surface area" in this context simply defines the ideal delimitation of the oocyte derived from the observed diameter, whereas the true surface area of the oolemma is considerably larger since it is greatly convoluted at this time (Wallace and Dumont, 1968; Dick, Dick and Bradbury, 1970). However, since the inclusion of the "surface area" term allows us to normalize the data over a certain size range, it may have some realistic basis which more accurately reflects a physiological rather than physical surface, *i.e.,* that area immediately adjacent to the oocyte which can be considered as a physiological compartment into which protein passes from a distal source and from which the oocyte sequesters protein. We should note that the cultured oocytes which incorporate protein maximally per unit surface area correspond to those from intact vitellogenic females which have been previously observed to undergo marked pinocytotic activity and to be actively engaged in yolk formation (Wallace and Dumont, 1968).

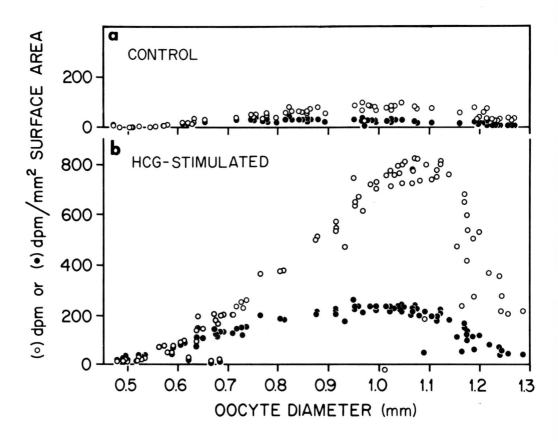

Fig. 18.8 Extent of protein incorporation as a function of oocyte size in oocytes from (a) normal females and (b) vitellogenic females. The incorporation of labeled protein is expressed both as dpm/oocyte (○) and dpm/oocyte divided by the surface area of the oocyte (●) [From Wallace *et al.*, 1970].

C. Progress and Localization of Protein Incorporation

Oocytes (0.8-1.1 mm) derived from normal or vitellogenic females were also cultured for up to 6 days in order to follow the progress of protein uptake. The results (Fig. 18.9) indicated that the incorporation of protein generally follows linear kinetics and that oocytes from vitellogenic females are approximately 10 times more active.

Since the cultured oocytes were invested by a follicular epithelium, it was important to ascertain whether the oocytes or the surrounding cells primarily incorporated the protein. Therefore, radioautographs were made of sections from oocytes which had been cultured for 4 hours in ^{3}H-labeled medium (Fig. 18.10). Virtually all the labeled material was found to be associated with the interior of the oocyte, so that protein is not appreciably incorporated by the follicle cells and does not accumulate at the oocyte surface. Protein uptake also appears to occur more

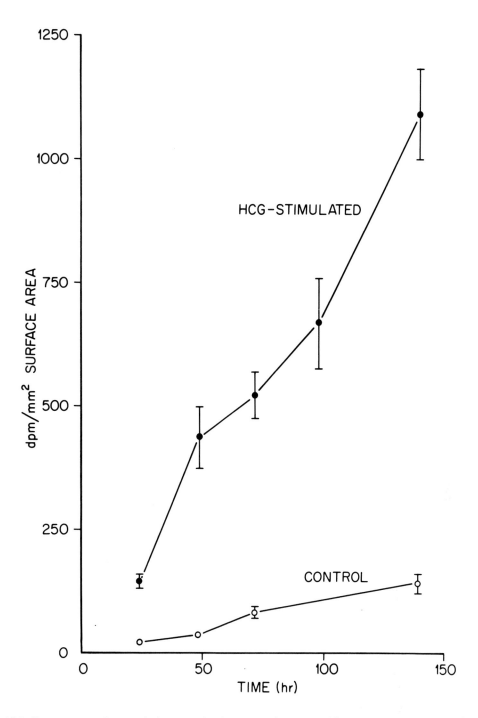

Fig. 18.9 Progress curve for protein incorporation by oocytes from normal females (○) and vitellogenic females (●). Each point represents the average value found for 8-12 oocytes and the bars represent the standard errors [From Wallace *et al.*, 1970].

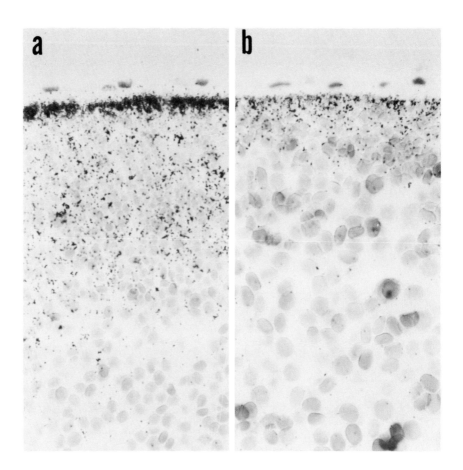

Fig. 18.10 Radioautography of the animal pole (a) and vegetal pole (b) from the same midsection of a vitellogenic oocyte (approx. 0.9 mm diameter) cultured for four hours in medium containing [^{3}H] vitellogenin as the only labeled component (25,000 dpm/μl). A thick layer of pigment granules is located just beneath the animal pole surface of the oocyte (a) obscuring some of the photographic detail; the follicular epithelium surrounds the oocyte. Freeze-substituted preparation sectioned at 5μ, coated with Kodak NTB3 emulsion, incubated for 90 days, developed and stained (yolk platelets, cell nuclei) with Mayer's haemalum [Unpublished data]   (X 550).

extensively at the animal pole (Fig. 18.10a) than at the vegetal pole (Fig. 18.10b) of the oocyte.

D. Temporal Extent of HCG Stimulation

It is apparent, then, that HCG administration to the intact animal stimulates oocytes within a certain size range to greatly increase their protein-sequestering activity as tested *in vitro*. We next determined how long oocytes remain activated after HCG injection. Accordingly females were given one or two injections of HCG and oocytes were isolated and cultured for 24 hours at various time intervals thereafter. The data for individual donors are provided in Fig. 18.11a and serve primarily to indicate the extent of individual variation, and that an increased level of protein incorporation by oocytes appears to be reached within 24 hours after HCG injection. If the data for individual females are grouped according to weeks after HCG injection, the general trend emerges more clearly (Fig. 18.11b): a single dose of HCG promotes an increased ability of oocytes to sequester protein for about 3 weeks, after which activity tapers off. This length of time is similar to that derived from turnover studies *in vivo* (Wallace and Jared, 1968). A second dose given 5 weeks later also promotes another period of sequestering activity.

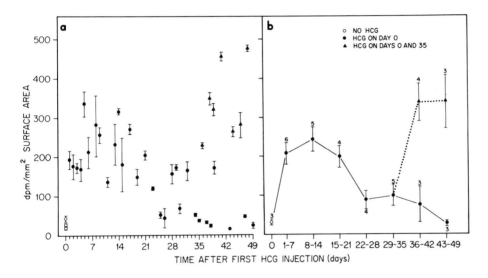

Fig. 18.11   Extent of protein incorporation by oocytes as a function of time after HCG injections: (a) data from individual females (the bars represent the standard errors for the average values derived from 8-12 oocytes); (b) data from females grouped according to weeks (the numbers over the standard error bars represent the females in each group) [From Wallace *et al.*, 1970].

E.   Other Observations

Much of what has just been described is of a preliminary observational nature and was performed in order to define the experimental system. We eventually hope, with the help of the isolated oocyte, to understand some of the basic controlling mechanisms in oocyte growth, selective membrane transport of macromolecules, and crystalline protein assembly. To this end we have begun experiments which:

(a) Attempt to induce protein uptake *in vitro*. Over 20 compounds (including HCG) which have been described to enhance macromolecular incorporation in other systems have been tested. However, the only substance we have found to be effective thus far is insulin, which promotes a 4-5 fold increase in sequestering activity at concentrations of 10 μg/ml or greater (Wallace, 1970b).

(b) Define the metabolic requirements for protein incorporation and transformation. In general, inhibitors of respiration (cyanide, antimycin A) or oxidative phosphorylation (oligomycin) have little effect on protein incorporation as opposed to inhibitors of glycolysis (fluoride, iodoacetate). Actinomycin D and puromycin (1 μg/ml) also appear to have no effect, even after several hours of preincubation, whereas cycloheximide (0.5-5.0 μg/ml) inhibits protein incorporation by up to 50%.

(c) Examine the selective mechanism of protein uptake. In an initial experiment, 5-6 times more vitellogenin (by weight) was incorporated by oocytes per unit time than normal serum protein, even though normal serum protein represented two-thirds of the total protein present in the culture medium (Wallace *et al.,* 1970). This selective bias was not as great as has been found in the intact animal (Fig. 18.5) and, in fact, may indicate no selectivity at all if it is assumed that the isolated oocyte randomly incorporates surrounding protein molecules and that vitellogenin is about 5-6 times larger than the average serum protein.

(d) Quantitate the relative roles of heterosynthetic *vs* autosynthetic protein assembly in the oocyte. In this regard, doubly labeled vitellogenin incorporated into the cultured oocyte has been found to undergo a transformation into lipovitellin and phosvitin in a manner analogous to what has been observed *in vivo* (Fig. 18.6). During this transformation, the small molecular weight pools remained essentially unlabeled, so that yolk protein formation appears to be due to the direct conversion rather than breakdown of vitellogenin and resynthesis. Conversely, most of the free [^{3}H]leucine taken up by the oocyte in culture appeared to become incorporated into a protein (membrane?) fraction which was rapidly turning over; no specific synthesis of labeled lipovitellin was observed.

VI. RELEVANCE TO THE MAMMALIAN OOCYTE

Changes in serum calcium, phosphorus, or protein levels as a response to season or estrogen administration have not been observed in mammals (Urist and Schjeide, 1961). Further, crystalline yolk inclusions suggestive of a lipovitellin-phosvitin complex have not been described from mammalian eggs or oocytes; however, a systematic search for these proteins has not yet been made. Nevertheless, it would appear that placental mammals have lost the capacity to synthesize some of the yolk components, which are characteristic for the eggs of lower vertebrates.

Inclusion bodies which may be related to yolk material have been observed in certain mammalian oocytes, however, and the suggestion has been made that the observable pinocytosis taking place at the surface of the guinea pig oocyte could carry nutritive substances into the ooplasm (Anderson and Beams, 1960). Adams and Hertig (1964) have also observed in the guinea pig that "the complex development and the peripheral location of the cytoplasmic organelles of the primary oocyte suggest that it is equipped for the absorption, utilization and intracellular transport of materials delivered to its surface membrane."

The most direct evidence for the incorporation of protein material into mammalian oocytes has been obtained by Glass, who demonstrated the localization of autologous and systemically injected heterologous serum antigens and proteins in the developing mouse oocyte by immuno-histological and radioautographic methods (Glass, 1961; Glass, 1966; Glass and Cons, 1968; see also Anderson, this conference). These observations have followed her earlier study on the localization of serum-like antigens in frog oocytes (Glass, 1959). Based in part on this study, it would appear that much of the serum protein, other than vitellogenin, which is incorporated into the adult amphibian oocyte, becomes associated with the outer noncrystalline portion of the yolk platelet after lipovitellin and phosvitin are formed (Wallace and Jared, 1969). In the absence of vitellogenin, as is apparently the case for mammals, serum proteins would thus be incorporated into inclusion bodies which lack a central, crystalline core. This suggestion, of course, needs to be documented by further experimentation, but it would nevertheless appear that mammalian oocytes do undergo a definite but more limited period of growth and that during this period proteinaceous material is incorporated. The significance of this process and the control mechanisms involved remain challenging subjects for future research.

VII. ACKNOWLEDGMENTS

Research sponsored by the U. S. Atomic Energy Commission under contract with Union Carbide Corporation.

VIII. REFERENCES

Adams, E. C. and Hertig, A. T. (1964) Studies on guinea pig oocytes I. Electron microscopic observations on the development of cytoplasmic organelles in oocytes of primordial and primary follicles. *J. Cell Biol.* **21**, 397.

Anderson, E. and Beams, H. W. (1960) Cytological observations on the fine structure of the guinea pig ovary with special reference to the oogonium, primary oocyte and associated follicle cells. *J. Ultrastruct. Res.* **3**, 432.

Brown, A. C. and Millington, P. F. (1968) Electron microscope studies of phosphatases in the small intestine of *Rana temporaria* during larval development and metamorphosis. *Histochemie.* **12**, 83.

Dick, E. G., Dick, D. A. T., and Bradbury, S. (1970) The effect of surface microvilli on the water permeability of single toad oocytes. *J. Cell Sci.* **6**, 451.

Follett, B. K., Nicholls, T. J., and Redshaw, M. R. (1968) The vitellogenic response in the South African clawed toad *(Xenopus laevis* Daudin). *J. Cell. Physiol.* **72**, (Suppl.) 91.

Follett, B. K. and Redshaw, M. R. (1968) The effects of oestrogen and gonadotrophins on lipid and protein metabolism in *Xenopus laevis* Daudin. *J. Endocrinol.* **40**, 439.

Gallien, L. and Chalumeau-Le Foulgoc, M. T. (1960) Mise en evidence de steroides oestrogenes dans l'ovaire juvenile de Xenopus laevis Daudin, et cycle des oestrogenes au cours de la ponte. *Compt. Rend. Acad. Sci.* **251**, 460.

Gitlin, G. (1939) Gravimetric studies of certain organs of *Xenopus laevis* (the South African clawed toad) under normal and experimental conditions I. The oviducts. *S. Afr. J. Med. Sci.* **4**, (Suppl.) 41.

Glass, L. E. (1959) Immuno-histological localization of serum-like molecules in frog oocytes. *J. Exp. Zool.* **141**, 257.

Glass, L. E. (1961) Localization of autologous and heterologous serum antigens in the mouse ovary. *Develop. Biol.* **3**, 787.

Glass, L. E. (1966) Dissimilar localization of serum albumin and globulins in mouse ovaries. *Fertil. Steril.* **17**, 226.

Glass, L. E. and Cons, J. M. (1968) Stage dependent transfer of systemically injected foreign protein antigen and radiolabel into mouse ovarian follicles. *Anat. Rec.* **162**, 139.

Grant, P. (1953) Phosphate metabolism during oogenesis in *Rana temporaria. J. Exp. Zool.* **124**, 513.

Gregg, J. R. and Ballentine, R. (1946) Nitrogen metabolism of *Rana pipiens* during embryonic development. *J. Exp. Zool.* **103**, 143.

Jared, D. W. and Wallace, R. A. (1969) Protein uptake *in vitro* by amphibian oocytes. *Exp. Cell Res.* **57**, 454.

Karasaki, S. (1963) Studies on amphibian yolk I. The ultrastructure of the yolk platelet. *J. Cell Biol.* **18**, 135.

Kemp, N. E. (1953) Synthesis of yolk in oocytes of *Rana pipiens* after induced ovulation. *J. Morph.* **92**, 487.

Masui, Y. (1967) Relative roles of the pituitary, follicle cells, and progesterone in the induction of oocyte maturation in *Rana pipiens. J. Exp. Zool.* **166**, 365.

Ohno, S., Karasaki, S., and Takata, K. (1964) Histo- and cytochemical studies on the superficial layer of yolk platelets in the *Triturus* embryo. *Exp. Cell Res.* **33**, 310.

Tandler, C. J. and La Torre, J. L. (1967) An acid polysaccharide in the platelets of *Bufo arenarum* oocytes. *Exp. Cell Res.* **45**, 491.

Urist, M. R. and Schjeide, O. A. (1961) Partition of calcium and protein in the blood of oviparous vertebrates during estrus. *J. Gen. Physiol.* **74**, 495.

Wallace, R. A. (1963a) Studies on amphibian yolk III. A resolution of yolk platelet components. *Biochim. Biophys. Acta.* **74**, 495.

Wallace, R. A. (1963b) Studies on amphibian yolk IV. An analysis of the main-body component of yolk platelets. *Biochim. Biophys. Acta.* **74**, 505.

Wallace, R. A. (1965) Resolution and isolation of avian and amphibian yolk-granule proteins using TEAE-cellulose. *Anal. Biochem.* **11**, 297.

Wallace, R. A. (1970a) Studies on amphibian yolk IX. *Xenopus* vitellogenin. *Biochim. Biophys. Acta.* **215**, 176.

Wallace, R. A. (1970b) Some factors involved in protein incorporation and transformation by isolated amphibian oocytes. *J. Cell Biol.* **47**, 219a.

Wallace, R. A. and Dumont, J. N. (1968) The induced synthesis and transport of yolk proteins and their accumulation by the oocyte in *Xenopus laevis. J. Cell. Physiol.* **72**, (Suppl.) 73.

Wallace, R. A. and Jared, D. W. (1968) Studies on amphibian yolk VII. Serum phosphoprotein synthesis by vitellogenic females and estrogen-treated males in *Xenopus laevis. Canad. J. Biochem.* **46**, 953.

Wallace, R. A. and Jared, D. W. (1969) Studies on amphibian yolk VIII. The estrogen- induced synthesis of a serum lipophosphoprotein and its selective uptake by the ovary and transformation into yolk platelet proteins in *Xenopus laevis. Develop. Biol.* **19**, 498.

Wallace, R. A., Jared, D. W. and Nelson, B. L. (1970) Protein incorporation by isolated amphibian oocytes 1. Preliminary Studies. *J. Exp. Zool.* **175**, 259.

Wartenberg, H. (1964) Experimentelle Untersuchungen uber die Stoffaufnahme durch Pinocytose wahrend der Vitellogenese des Amphibienoocyten. *Z. Zellforsch. Mikroskop. Anat.* **63**, 1004.

Wischnitzer, S. (1963) The ultrastructure of the layers enveloping yolk-forming oocytes from *Triturus viridescens. Z. Zellforsch. Mikroskop. Anat.* **60**, 452.

Wittek, M. (1952) La vitellogenese chez les amphibiens. *Arch. Biol.* **62**, 133.

Zwarenstein, H. and Shapiro, H. A. (1933) Metabolic changes associated with endocrine activity and the reproductive cycle in *Xenopus laevis* III. Changes in the calcium content of the serum associated with captivity and the normal reproductive cycle. *J. Exp. Biol.* **10**, 372.

19

FOLLICLE GROWTH IN THE MOUSE OVARY

Torben Pedersen

I. Introduction
II. Methods
III. Materials
IV. Results
    A. Differential counts of follicles
    B. Labelling pattern and labelling index
    C. Duration of the S-phases
    D. Doubling times and transit times
    E. Number of follicles starting to grow per unit time
    F. Follicle kinetics
V. Discussion
VI. References

I. INTRODUCTION

The ovaries of mice contain throughout life a mixed population of follicles in different stages of development. The composition of this population may be determined by differential counting, and it can be demonstrated that the composition of the population varies, not only from one age to another, but at certain ages also with the stage of the estrous cycle (Mandl and Zuckerman, 1951; Jones and Krohn, 1961; Pedersen, 1970b). Differential counting can, however, only reveal a picture of the growth of the follicles at the precise moment when the animal is killed; it is not possible by this method alone to give a complete description of the dynamics of follicle growth, including the time it takes a follicle to grow from one stage of development to another.

With the introduction of autoradiography after pulse-labelling with tritiated thymidine, it has become possible to study the kinetics of the granulosa cell populations in the follicles and thereby also to determine the kinetics of whole follicles (Pedersen, 1969, 1970a, b).

The ovarian follicles can be divided into two groups, the non-proliferating and the proliferating follicles. The first group consists of almost all the small follicles; they form a pool which the follicles leave when they start their development. The second group contains all the larger follicles. When a follicle has started its development, it continues to grow until it degenerates or ovulates (Pedersen, 1970b). A full description of the dynamics of follicle growth will then include a) a determination of the size of the two groups (i.e. a differential count), b) the flow of follicles from the first to the second group per time unit, c) the growth rate of the follicles in the second group, and d) the outflow from the second group through degeneration or ovulation.

It has been the aim to describe in this way the dynamics of follicle growth at different ages from birth to the end of reproductive life, during the estrous cycle and in pregnancy.

II. METHODS

The methods used have been described in detail earlier (Pedersen, 1969, 1970a), and they will therefore only be briefly mentioned.

The follicles were classified according to the number of granulosa cells in the largest cross section of each follicle as proposed by Pedersen and Peters (1968) (Fig. 19.1). From the number of granulosa cells in the largest cross section, the diameter of the granulosa cells and of the whole follicle, the total number of granulosa cells in a follicle may be calculated. This has been done for a number of follicles of different types, and the relation between the number of granulosa cells in the largest cross section and the corresponding number of cells in the whole follicle was estimated (Fig. 19.3). To calculate the time it takes a follicle to grow from one stage of development to another, it is necessary to know the total number of granulosa cells in these stages and the time it takes the granulosa cells to double their number, i.e., the doubling time (T_D). T_D can be investigated in autoradiographs after pulse-labelling with tritiated thymidine. In ovaries prepared shortly after labelling (i.e. one hour), the autoradiographs reveal the cells which at that time are in the DNA-synthesis phase (S-phase). The distribution of the labelled cells in the population represents its labelling pattern, while the percentage of labelled cells in a cell population is expressed as its labelling index (LI). In autoradiographs prepared at different time-intervals after pulse-labelling it is possible to follow the appearance and disappearance of labelled mitoses. The results (the percent labelled mitoses) are plotted against time after labelling (PLM-curve). On the curve it is possible to measure the duration of the S-phase (t_s) and of the G_2-phase plus half of the mitosis phase (t_2).

The doubling time (T_D) may then be calculated from the following formula

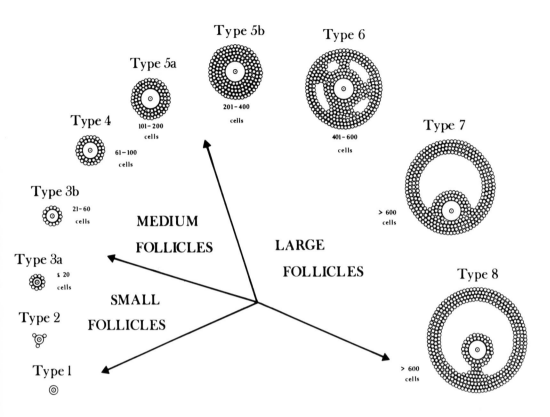

Fig. 19.1 Classification of follicles in the mouse ovary according to the number of granulosa cells in the largest cross section of each follicle.

(Cleaver, 1965):

$$LI = (\exp[t_s \ln 2/T_D] - 1)(\exp[t_2 \ln 2/T_D]).$$

The time it takes a follicle to grow from one type to another, or the time it takes to grow through a certain type is defined as the transit time of that type of follicle (T_F). It may be calculated from the following formula:

$$T_F = T_D(\ln N_o - \ln N_i/\ln 2)$$

where N_i and N_o represent the minimum and maximum number of granulosa cells in that type.

Table 19.1 Differential Counts of Follicles in Ovaries of Mice at Different Ages

Age	Type of follicles												Total number of medium + large follicles
	Type 3b		Type 4		Type 5a		Type 5b		Type 6		Type 7		
	Mean	± S.E.	Mean	± S.E.	Mean	± S.E.	Mean	± S.E.	Mean	± S.E.	Mean	± S.E.	
Immature period													
7 days	141	45	—		—		—		—		—		141
14 days	155	9	125	0	31	8	—		—		—		311
21 days	163	15	113	18	54	8	38	3	13	3	—		381
28 days	142	20	94	13	60	7	22	4	19	9	—		337
35 days	105	12	55	10	33	7	13	3	12	3	—		218
Mature period													
3 months cyclic mice	139	12	105	6	45	3	23	2	9	2	10	2	331
3 months pregn. mice	100	6	79	5	35	2	17	2	6	1	10	1	247
9 months	70	10	50	5	28	2	20	5	8	2	—		176
12 months	68	7	53	17	20	10	18	7	10	—	8	2	177
16 months	38	19	37	18	15	8	12	7	8	6	5	3	115
24 months	—		—		—		—		—		—		0

III. MATERIAL

Mice of the Bagg strain were used. Five infant or immature mice were used at each age for the determination of LI. Four mice, 3 months of age, were killed at each stage of the estrous cycle. At other ages investigated only two mice were used at two different stages in the cycle. Two mice were killed at different times during pregnancy. The curves of labelled mitoses were worked out at the ages of 14, 21, and 28 days. In order to test whether the duration of the S-phases was different at other ages, double-labelling experiments using two markedly differing doses of [³H] thymidine were performed at the ages of 3, 12, and 16 months, at the 12th day of pregnancy and at the age of 28 days.

IV. RESULTS

A. Differential Counts of Follicles

The differential counts of the follicles (Table 19.1) reflect the composition of the follicle population at different ages and during pregnancy. Only the medium and large follicles, both of which are growing, will be discussed. The table shows that the number of medium and large follicles is greatest in the last half of the immature period, at the ages of 21 and 28 days. At 35 days the numbers have decreased. In the mature mouse after the first ovulation, the numbers of medium and large follicles increase again during the first months of reproductive life. At the age of 3 months their numbers again start diminishing, and at the age of 2 years no follicles are left in the ovaries. It will furthermore be seen that during pregnancy there are fewer follicles of all types than in the cyclic mice of the same age. The number of medium follicles did not vary during the cycle, but the number of large follicles, especially types 5b and 6, does vary with the cycle, reaching a maximum at ovulation and in the first half of the cycle. It may furthermore be seen that although the total number of follicles varies, the proportion between the different types of follicles is almost constant up to the age of 12 - 13 months. Hereafter, there are relatively more of the large follicles than of the medium follicles.

B. Labelling Pattern and Labelling Index

In autoradiographs it is possible to distinguish between proliferating and non-proliferating cell populations, as only the first contain labelled cells in autoradiographs prepared shortly after pulse-labelling. The labelling pattern of medium and large follicles is constant throughout life: All the follicles contain labelled granulosa cells in their largest cross sections. This fact means that all follicles contain a proliferating granulosa cell population, i.e., they are all growing. It may therefore be concluded that when a follicle has reached the developmental stage of type 3b, it continues to grow until it degenerates or ovulates; it cannot remain 'inactive' in the ovary.

The small follicles of type 3a constitute a transitional group between 'resting' and 'growing' follicles. Some of these do not grow, and are therefore unlabelled in the autoradiographs, while others have started their development, and therefore contain one or more labelled granulosa cells. The percentage of the type 3a follicles which contain one or more labelled cells in their largest cross section will not give an exact count, but it will give a good estimate of how many type 3a follicles have started to grow. In immature mice about 74% of type 3a follicles are labelled at the age of 7 days. This value decreases and reaches 35% in the last half of the immature period. On the average this value is constant the rest of the life, although it changes with the stage of the cycle, reaching 60% at estrus and about 30% at metestrus.

The medium and large follicles all contained labelled granulosa cells but not to the same degree. The labelling indices (LI) of the different types of follicles are shown in Table 19.2. Medium follicles of types 3b and 4 have less than 10% labelled

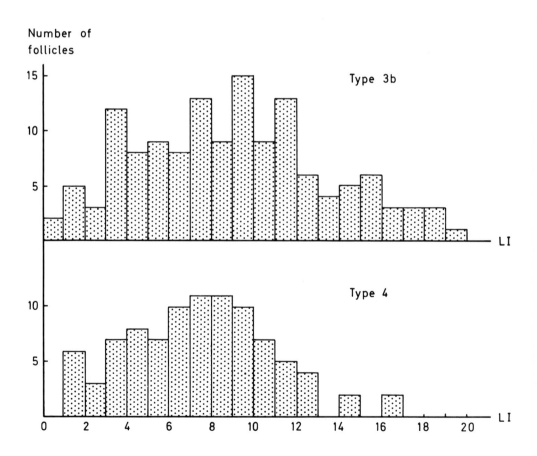

Fig. 19.2   The distribution of the labelling indices of type 3b and 4 follicles in ovaries from 28-day-old mice.

Table 19.2 **MEAN LABELLING INDEX OF FOLLICLES IN THE MOUSE OVARY AT DIFFERENT AGES**

Age		Type 3b	Type 4	Type 5a	Type 5b	Type 6	Type 7
		%	%	%	%	%	%
Immature period	7 days	20.8	—	—	—	—	—
	14 —	13.2	10.3	14.3	—	—	—
	21 —	8.5	8.4	17.3	30.5	30.0	—
	28 —	7.4	7.4	17.2	38.7	35.0	—
	35 —	8.3	8.3	16.4	35.4	32.6	—
Mature period	3 months cyclic mice at oestrus	11.2	8.4	20.1	42.8	38.9	35.6
	— metoestrus	7.0	6.6	13.7	34.2	35.6	29.1
	— dioestrus	8.6	7.0	14.2	32.2	35.4	29.7
	— early pro.	9.0	7.2	14.5	34.5	26.4	23.2
	— late pro.	9.6	7.2	15.2	32.5	38.2	18.3
	3 months old pregn. mice mean for a pregnancy	8.9	6.9	14.8	32.5	31.1	19.2
	9 months old at oestrus	13.8	10.2	32.0	39.5	28.4	—
	at metoestrus	8.5	7.1	15.8	29.1	32.9	—
	12 months old at oestrus	13.4	14.6	24.5	44.3	32.9	—
	at metoestrus	8.7	12.7	17.2	—	—	—
	16 months old at oestrus	12.4	10.7	22.9	45.2	46.3	—
	at metoestrus	5.2	7.2	7.9	28.3	—	—

granulosa cells while the large follicles have about 35 to 40%. These values are means of a number of determinations in different follicles from several ovaries. The question arises whether the follicles of a certain type under given circumstances constitute a homogenous group with respect to their LI, or whether there are several populations within the same type. Figure 19.2 shows two examples of the distribution of the LI of a single type of follicle at a certain age. Due to the relatively few cells which were counted in each follicle, namely the number of cells in the largest cross section, which in type 3b never exceeds 60, and in type 4, 100, there must be a considerable spread in the results, and the broad distribution seen in this figure was thus expected. On the other hand, the results do not show two distributions, which supports the assumption that the follicles at a given developmental stage are homogenous with respect to LI.

C. Duration of the S-phases

To be able to calculate the doubling times of the granulosa cells it was necessary to know not only the LI of the granulosa cells, but also the duration of the S-phase of the granulosa cell. In immature mice at the age of 28 days these values were determined by the method of labelled mitoses (Pedersen, 1970a; Hartmann and Pedersen, 1970; Pedersen and Hartmann, in press) which gives accurate and reliable values (Table 19.3). There is a considerable shortening in the duration of the S-phase of the granulosa cells during the development of the follicles. On the curve of labelled mitosis it was also possible to measure the duration of the G_2-phase plus half of the mitosis phase of the granulosa cells (Table 19.3).

Table 19.3 **Duration of the S-Phase (t_s) and of the G_2 + Half of the Mitosis Phase (t_2) in Granulosa Cells from Different Types of Follicles**

	type 3b	type 4	type 5a	type 5b	type 6	type 7
t_s	9.8 hrs	11.2 hrs	10.3 hrs	6.8 hrs	6.8 hrs	6.8 hrs
t_2	2.0 hrs	2.0 hrs	2.0 hrs	1.8 hrs	1.8 hrs	1.8 hrs

It has not been possible to apply this method to all other ages and circumstances in the life of the mouse, and therefore another method to measure the duration of the S-phase had to be introduced, namely the double-labelling technique. Double-labelling experiments were performed at the different ages and also at the age of 28 days, where the exact measurements by the labelled mitoses method were performed. The results of the double-labelling experiments in mice of different ages were not different from those found at 28 days, and it was concluded that the duration of the S-phase of the granulosa cells was not dependent on the age of the mouse, but only on the stage of development of the follicle in which they resided.

D. Doubling Times and Transit Times

From the LI of a given follicle type and the duration of the S-phase and the G_2-phase of the granulosa cells in that type, it is then possible to calculate the doubling time of the granulosa cells. This calculation has been done for each type of follicle at the ages under investigation. Each stage of follicle development was characterized not only by the number of granulosa cells in the largest cross section, but also by the total number of granulosa cells (Fig. 19.3). When the doubling times of these cells are known, it is possible to calculate the growth rate of the follicles expressed as their transit times (T_F), i.e., the time it takes to grow from one stage of

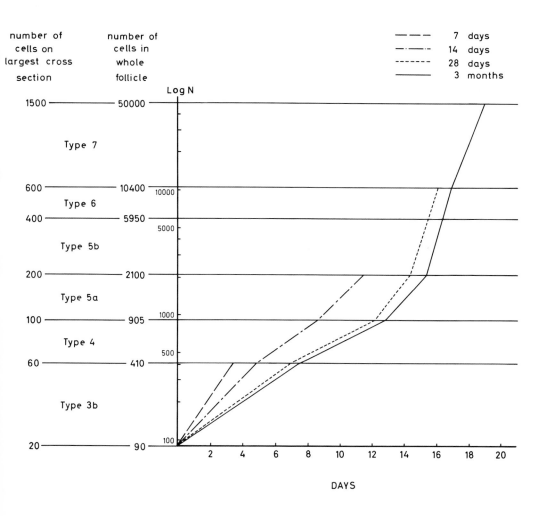

Fig. 19.3 a) The relation between the number of granulosa cells in the largest cross sections and the corresponding number of cells in a whole follicle, and b) The time it takes a follicle to grow from one stage of development to another at different ages.

Table 19.4 **TRANSIT TIMES IN HOURS OF FOLLICLES IN OVARIES OF MICE AT DIFFERENT AGES**

		Type 3b	Type 4	Type 5a	Type 5b	Type 6
Immature mice	7 days	83	–	–	–	–
	14	119	91	67	–	–
	21	185	13	56	29	15
	28	167	126	56	23	14
	35	145	115	58	25	14
Mature mice	3 months old at oestrus	142	113	48	21	12
	3 months old at metoest	221	139	69	26	13
	3 months old at dioest	183	133	67	27	13
	3 months old at early pro	177	129	62	27	13
	3 months old at late pro	166	129	62	27	13
	3 months old cyclic mice mean for a cycle	178	129	62	25	14
	3 months old pregnant mice mean for a pregnancy	178	129	62	25	14
	9 months old at oestrus	117	93	32	22	16
	at metoest	184	132	60	29	14
	12 months old at oestrus	120	66	41	20	14
	at metoest	181	76	55	–	–
	16 months old at oestrus	129	89	43	20	11
	at metoest	295	129	115	30	–

development to another (Table 19.4). The transit times of the medium follicles are always much longer than those of large follicles. The shortest transit times of types 3b and 4 are found early in immaturity. In the rest of the immature period their growth rate is almost the same as in cyclic mice. In types 5a, 5b, and 6, the transit times show minor variations from one stage to another. It may be seen that the growth rate of all types of follicles is independent of pregnancy. In the 9, 12, and 16 month-old mice, the transit times of types 4 and 5a are somewhat shorter than in the 3 month-old mice, while the transit times of the other types of follicles are unchanged.

The transit times of type 7 are not listed here, as they depend to some extent on the length of the estrous cycle. In mice with an estrous cycle of 5 - 6 days, it takes somewhat longer for a type 6 follicle to develop to an ovulating follicle than in mice with a cycle-length of only 4 days (Pedersen, 1970b). Figure 19.3 shows the time it takes a follicle to grow from one stage to another at different ages. In 3 month-old mice it takes about 19 days from the time a follicle starts its development as type 3b until ovulation. In old mice (12 - 17 months) it takes probably a shorter period of time as the transit times of some of the types are shorter; the exact values however, have not be determined.

E. Number of Follicles Starting to Grow per Unit Time

Type 3b follicles are never, or at least very seldom, seen in degeneration. Furthermore, they can not remain in the ovaries as non-growing follicles. Under these circumstances the number of follicles starting to grow as type 3b per hour can be calculated as the exact number present divided by the transit time of type 3b. This value is defined as the inflow in type 3b and is expressed per 24-hour period. The results are shown in Table 19.5. In the first week of life more follicles start to grow than at any other period in the life span of the mouse. In the cyclic 3 month-old mice the values vary somewhat, but, on the average, the number of follicles starting to grow is lower than in immature mice. In pregnancy fewer follicles start to grow than in cyclic mice. The inflow decreases as the mouse grows older.

F. Follicle Kinetics

In the first 5 weeks of life (the immature period) the decrease in the number of small follicles amounts to about 4000 - 5000 (Peters, 1969). The question arises of how this number compares to the number of follicles leaving the pool of small follicles due to follicle development. The differential counts show that at day 7 there are about 140 3b follicles in one ovary, and as the inflow was determined at that and the following ages (Table 19.5), it is possible to calculate the total inflow into type 3b in the whole immature period (from day 0 to day 35). This amounts to about 850. It may then be seen that only a minor part of the decrease in the population of small follicles is due to development, and that the major reason for the decrease in the number of small follicles in the immature period is degeneration of the small follicles.

Table 19.5 **Number of Follicles Starting to Grow as Type 3b Follicles**
Per 24 Hour Period at Different Ages

IMMATURE PERIOD	7 days	14 days	21 days	28 days	35 days	Average
	41	31	21	20	17	26

MATURE PERIOD	3 months	3 months	9 months	12 months	16 months	24 months
	cyclic mice	pregnant mice	cyclic mice	cyclic mice	cyclic mice	
	19	14	10	10	3	0

From 1 to 12 months also it is possible to calculate the number of follicles leaving the pool of non-proliferating follicles. The inflow-values found at 3 months of age are taken as representative for the period from 1 to 6 months, and the inflow-values at the age of 9 months are taken as representative values for the ages from 6 to 12 months. This calculation does not pretend to be exact, but gives an idea of the number leaving the pool between the ages of 1 and 12 months. The number amounts to 4700. It has been shown that the number of small follicles in mature mice decreases gradually with age following an exponential line (Pedersen, 1970b):

$$\log y = 3.8118 + 0.0031x,$$

where x = time in days
and y = number of follicles.

From this formula the decrease in the pool of small follicles in the same period of time is calculated to be about 4450. The fine agreement between these figures justifies the conclusion that the decrease in the pool of small follicles in the mature period is solely due to follicle development.

V. DISCUSSION

The mouse ovary contains non-proliferating ('resting') and proliferating ('growing') follicles. The small follicles constitute the pool of non- growing follicles. The size of the pool is largest at birth and thereafter decreases with age. In the first 3 weeks of life the decrease in the pool is, for the most part, due to degeneration of some of the small follicles, while the rest of the decrease is due to outflow of follicles which start on their development. After the onset of maturity, the decrease in the pool seems to be solely due to follicles leaving the pool when they start to grow, while degeneration of small follicles does not take place any more. This conclusion is in contrast to what has been suggested by Jones and Krohn (1961).

At all times during the life of the mouse, new follicles start to develop, but the number of starting follicles varies from one age to another. In the first two weeks of life more follicles start than at any other age. The number of starting follicles decreases with age. The question arises: what factor determines the number of starting follicles? Krarup, Pedersen and Faber (1969) have shown that it is not the age of the animal alone, but that there exists a relation between the number of non-proliferating and proliferating follicles. Moreover, the number of growing follicles present would determine the number of follicles which at the same time start to grow. This factor cannot, however, provide the only mechanism which influences the number of starting follicles as pregnancy involves a decrease before any change in the number of growing follicles has taken place.

Several investigators have found a large number of growing follicles about the age of 21-28 days (described as the prepubertal peak) (Jones and Krohn, 1961). This phenomenon can be explained by the variations in the number of starting follicles. In the first two weeks of life more follicles start than later. These follicles develop into large follicles within 14 days, and it is therefore not surprising that we find more large follicles at the age of 21 - 28 days than at any other age of the life.

As fewer follicles start to grow during pregnancy than in the cycle, it would be expected that multiparous mice have a larger number of small follicles than nulliparous mice. In rats (Shelton, 1959) this was not observed, while in mice forced-breeding resulted in a somewhat, although not significant, slower depletion in the population of small follicles (Jones, 1957). The variations in the number of small follicles from one animal to another, however, completely overshadow a possible difference between multi- and nulliparous mice.

When follicles have started to grow as type 3b follicles, they do not stop. Degeneration of medium follicles is very seldom seen in the ovaries, and as they cannot remain in the ovary as resting follicles, it is concluded that a follicle which has started development as type 3b, continues to grow until it has become type 5b. Hereafter, its fate varies. In the immature mouse follicles cannot develop beyond type 6 or an early type 7, and as they do not remain in the ovaries as non-growing follicles, they degenerate. In cyclic mice, only those follicles which at a certain stage in the cycle have reached a certain size can continue to develop into ovulating follicles (Pedersen, 1970b), namely, only those follicles present as type 5b or 6 at metestrus. All follicles reaching this size at other times degenerate. It follows that in mature mice, all the small follicles present at the onset of maturity sooner or later start to grow. They continue their development until they have reached a given size, and thereafter, the majority degenerate and a few are ovulated.

The growth rate of the follicles expressed as their transit times varies relatively little in the large follicles. The growth rate of the medium follicles changes more; they grow fastest in the first two weeks of life, but thereafter the variations in growth rate are of minor importance. It may as a whole, be concluded that, within a given type of follicle, the growth rate shows only small and unimportant variations at different ages and in pregnancy. The changing hormonal status seems to have little influence on the growth rate.

The growth of the follicles of different types varies enormously in the single ovary. In the medium follicles only 10% of the granulosa cells are in DNA-synthesis phase, and they have a relatively long S-phase, while in types 5b and 6, the LI increases to about 35 - 40%, while the duration of the S-phase is shortened. It is thus the milieu, i.e., the stage of development of the follicle in which the granulosa cells reside, that determines their growth rate. In the medium follicles few of the granulosa cells proliferate, while the others do not proliferate or proliferate at a very slow speed. During the development of the follicles, more granulosa cells proliferate and the mitotic cycle is shortened. In the large follicles all the granulosa cells proliferate at a high speed. The question arises: what is the cause of the acceleration of the granulosa cells? It is not a stimulus outside the follicle, as this acceleration takes place whatever the stage of the cycle or the age of the animal. It must be the follicle itself which stimulates the granulosa cells. The acceleration coincides in time with the formation and proliferation of the theca cells around the follicle. It might be incipient hormone production in the follicle itself, e.g., estrogen-formation, which accelerates the granulosa cells. That estrogen has such an effect on cells in target organs (e.g. the epithelium of the uterus) has been reported several times (Beato and Dienstbach, 1968; Bresciani, 1964; Epifanova, 1966).

Another question is: what protects some follicles from degeneration and allows further growth leading to ovulation? As it is only at a certain stage of the cycle from where follicles can continue their development, it appears most likely that it is a shift in the hormonal status, which determines when a follicle will be brought to ovulation or not, and not the follicle itself.

The growth of follicles is a continuous process. Follicles start to grow at all times, attain a certain development and then degenerate or ovulate. The whole process must be under both exogenous as well as endogenous control (Hisaw, 1947). The endogenous control is the control system residing in the follicle itself, and the exogenous is the hormonal influence the follicle receives from the other part of the ovary and outside the ovary, e.g., influence of FSH and LH. What controls the number of starting follicles is the number of growing follicles present at the particular time, but most likely modified somewhat by exogenous factors. After a follicle has started on its development, it is the endogenous system which takes over the control and which is responsible for the acceleration of the growth. Little control is exerted by exogenous factors. The prevention of degeneration and further development to ovulation seems, however, to be subject only to exogenous control.

VII. REFERENCES

Beato, M. and Bienstbach, F. (1968) Effect of estrogens and gestagens on the duration of DNA-synthesis in the genital tract of ovariectomized mice. *Virchow Arch. [Zellpath.]* **1**, 197.

Bresciani, F. (1964) DNA synthesis in alveolar cells of the mammary gland, acceleration by ovarian hormones. *Science.* **146**, 653.

Cleaver, J. E. (1965) The relationship between the duration of the S-phase and the fraction of cells which incorporate ^{3}H-thymidine during exponential growth. *Exp. Cell Res.* **39**, 697.

Epifanova, O. I. (1966) Mitotic cycles in estrogen-treated mice: A radioautographic study. *Exp. Cell Res.* **42**, 562.

Hartmann, N. R. and Pedersen, T. (1970) Analysis of the kinetics of granulosa cell populations in the mouse ovary. *Cell and Tissue Kinetics.* **3**, 1.

Hisaw, F. L. (1947) Development of the Graafian follicle. *Physiol. Rev.* **27**, 95.

Jones, E. C. (1957) The aging ovary. Thesis. University of Birmingham.

Jones, E. C. and Krohn, P. L. (1961) The relationship between age, numbers of oocytes and fertility in virgin and multiparous mice. *J. Endocr.* **21**, 469.

Krarup, T., Pedersen, T. and Faber, M. (1969) Regulation of oocyte growth in the mouse ovary. *Nature (London).* **224**, 187.

Mandl, A. M. and Zuckerman, S. (1951) Cyclical changes in the number of medium and large follicles in the adult rat ovary. *J. Endocr.* **8**, 341.

Pedersen, T. (1969) Follicle growth in the immature mouse ovary. *Acta Endocr.* **62**, 117.

Pedersen, T. (1970a) Determination of follicle growth rate in the ovary of the immature mouse. *J. Reprod. Fertil.* **21**, 81.

Pedersen, T. (1970b) Follicle kinetics in the ovary of the cyclic mouse. *Acta Endocr.* **64**, 304.

Pedersen, T. and Peters, H. (1968) Proposal for a classification of oocytes and follicles in the ovary of the mouse. *J. Reprod. Fertil.* **17**, 555.

Pedersen, T. and Hartmann, N. R. (1971) The kinetics of granulosa cells in developing follicles in the mouse ovary. *Cell and Tissue Kinetics.* (In press).

Peters, H. (1969) The development of the mouse ovary from birth to maturity. *Acta Endocr.* **62**, 98.

Shelton, M. (1959) A comparison of the population of oocytes in nulliparous and multiparous senile laboratory rats. *J. Endocr.* **18**, 451.

20

GONADOTROPHIN-INDUCED MATURATION
OF MOUSE GRAAFIAN FOLLICLES
IN ORGAN CULTURE

T. G. Baker
P. Neal

I.   Introduction
II.   Materials and methods
        A.   Animals
        B.   Organ culture procedures
        C.   Hormone treatment
        D.   Observations on 'fresh' (unfixed) material
        E.   Histology
        F.   Quantitative studies
III.   Results
        A.   Observations on 'fresh' (unfixed) oocytes
        B.   Histology
        C.   Quantitative results
IV.   Discussion
V.   Acknowledgments
VI.   References

I. INTRODUCTION

It is now well established that gonadotrophic hormones play an important role in the regulation of the processes of follicular growth, meiotic maturation of oocytes, ovulation, and atresia (see Young, 1961; Hartman and Leatham, 1961; Ingram, 1962; Mauleon, 1969; Baker, 1970b). The majority of the early studies of these processes involved such procedures as hypophysectomy and the injection of various steroid and protein hormones: they were limited in their interpretation by the fact that many endocrine systems and target organs were affected. It was soon realized, however, that the ovaries of hypophysectomized animals contain large follicles for only a limited period, after which follicular growth beyond about four layers of granulosa cells is prevented. Treatment of such animals, or of immature mammals, with pregnant mares serum gonadotrophin (PMSG) alone induces follicular growth without the ensuing ovulation, unless pre-ovulatory follicles are in the ovary at the time of injection (Rowlands and Williams, 1943; Rowlands, 1944). However, the injection of human chorionic gonadotrophin (HCG) into mice some 40 - 48 hours after priming with PMSG induces a resumption of meiosis in oocytes which were previously arrested at the dictyate stage, and ovulation some 13 hours later (Edwards, 1966). The number of oocytes that are ovulated is dependent on the doses of PMSG and HCG employed and may be at least double that found naturally occurring at estrus (Fowler and Edwards, 1957; Edwards and Gates, 1959; Zarrow and Wilson, 1961).

The precise mechanism of hormonal action *in vivo* is difficult to investigate due to the multiplicity of action and the various target organs and feed-back involved. With the first successful organ culture of mammalian ovaries by Martinovitch (1937, 1938), a technique became available whereby the precise effects of hormones, either individually or in groups, could be identified for each parameter of their effects (see Fainstat, 1968). It has been shown, for example, that the addition of PMSG or follicle stimulating hormones (FSH) to organ cultures of the ovaries of pre-pubertal rats and mice, increases the number of granulosa cells undergoing mitosis, and also the number of medium-sized follicles (Fainstat, 1968; Crooke and Ryle, 1968; Ryle, 1969a, b, c). The effects of tthe hormones in culture may well be similar to that recorded from studies *in vivo*.

The results of some of the *in vitro* studies are difficult to evaluate, however, since many variations of culture technique have been employed, using a variety of hormone preparations, and animals of different age and strain. It would seem that one of the most reliable parameters for studying the action of gonadotrophins on the ovary is the maturation and ovulation of oocytes under conditions comparable to the superovulation technique (Fowler and Edwards, 1957). It has already been indicated by means of a preliminary communication that the technique of superovulation can be used in studies of organ culture; mouse oocytes undergo meiosis from dictyate to metaphase II if HCG is added to the nutrient medium; some eggs undergo 'ovulation' *in vitro* (Fritz, Cho and Biggers, 1965).

The present experiments were designed to re-investigate the effects of superovulatory doses of gonadotrophins on the mouse ovary *in vivo* and *in vitro* by means of quantitative histological and cytological studies. The rationale behind this study was that a technique that would allow follicular growth and a resumption of meiosis in oocytes in organ culture would prove of great value in studies of the growth and maturation of the oocyte and follicle, and of the hormonal dependence of these processes up to the time of ovulation. It would also provide a supply of oocytes for studies of *in vitro* fertilization particularly from species where egg recovery following hormone treatment is impracticable (human, and domestic animals of agricultural importance). The present communication deals with the treatment of ovaries with PMSG and HCG at specific dose-levels and time intervals.

II. MATERIAL AND METHODS

A. Animals

Mice of the Schofield albino strain aged 6 - 8 weeks were selected randomly for stage of the estrous cycle, since it has been shown previously that superovulation in the mouse, at the doses used in this study, is largely unaffected by the day of the cycle when gonadotrophin treatment is commenced (Edwards, Wilson and Fowler, 1963). Animals were killed by cervical dislocation and their reproductive tracts removed for histology with or without prior culture.

B. Organ Culture Procedures

The technique of organ culture used is similar to that of Fritz, Cho and Biggers (1965) and Foote and Thibault (1969), and was a modification to that which we have published previously (Baker and Neal, 1969). Immediately after removal from the animal, the ovaries were dissected free of connective tissue and cut into 6 - 10 pieces, each about 1-2 mm^3. The ovarian fragments were placed on a square of 'Millipore' membrane filter supported on a stainless steel platform in a vented disposable petri dish (Sterilin Ltd., Richmond, U.K.). Each petri dish contained all the explants derived from only one ovary: in this way the effects of hormone treatment could be compared with other ovaries which had not received gonadotrophins. Five ml of Eagle's minimal essential medium (Earle's salts; Flow Laboratories, Irvine, U.K.), supplemented with glutamine and calf serum (20% v/v; Flow Laboratories Ltd.), was added to each petri dish such that the level of the medium was slightly lower than the top of the stainless steel platform. The ovarian fragments were thus only in contact with medium which diffused through the 'Millipore' membrane (see Baker and Neal, 1969). The addition to the nutrient medium of 40 I.U. Ampicillin ('Penbritin'; Beecham Labs., Brentford, U.K.), and 10μg/ml Amphotericin B ('Fungizone', Squibb Pharmaceuticals), minimized the chance of infection. Four petri dishes containing the ovarian explants were placed in each modified 'Kilner' assembly (see Baker and Neal, 1969), which was then gassed with 5% CO_2 in air at a pressure of 0.703 kgm/cm^2 (10 lbs/in.^{2}). The high gas pressure, which was maintained in the culture vessel, prevented necrosis within the explants; lower pressures are associated with damage to the central region of the tissue, while high pressures affect the surface of the fragments. Incubation was carried out at 36°C for varying intervals depending upon the treatment employed (see Fig. 20.1; also below).

C. Hormone Treatment

The gonadotrophins used throughout these experiments were PMSG ('Gestyl';

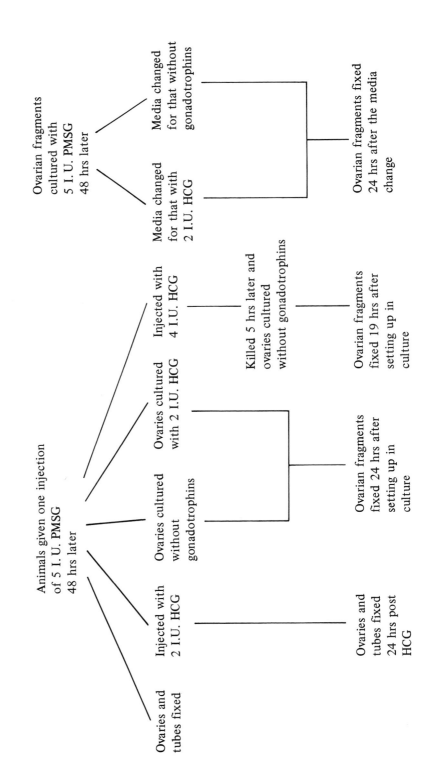

Figure 20.1 **Synopsis of Gonadotrophin Treatments Employed**

Organon Laboratories, London), and HCG ('Pregnyl'; Organon), since these preparations were usually selected for superovulation *in vivo* (see Edwards and Gates, 1958). The hormones were dissolved in Eagle's medium to a concentration of 20 I.U./ml (PMSG) or 10 I.U./ml (HCG), shortly before use. For superovulation experiments *in vivo,* the gonadotrophins were administered by intraperitoneal injection. The addition of the same quantities of hormone to the petri dishes at the onset of culture (or during a change of nutrient medium) facilitated the study of the response *in vitro.* The doses used in the present study were those which have been shown to induce a suitable quantitative response in terms of superovulation in intact adult mice (5 I.U. PMSG, followed 48 hours later by 2-4 I.U. HCG; see Fowler and Edwards, 1957; Edwards and Gates, 1958; Zarrow and Wilson, 1961; Whittingham, 1968). Modifications to the basic technique of superovulation are shown in Fig. 20.1.

D. Observations on 'Fresh' (unfixed) Material

Ovarian explants were removed from culture at varying times after the addition of HCG and compared with similar tissue recovered from mice after the injection of HCG. The oocytes were released from 'large' follicles by puncture with a hypodermic needle and examined using the techniques of Austin (1961). The chromosomes were brought into optical focus under a phase contrast microscope after rolling the oocytes until the nucleus and polar body were clearly seen. Preparations were stained with acetic lacmoid or orcein and lightly pressed to reduce their optical thickness for photography.

E. Histology

Ovaries derived from organ culture, and also those recovered directly from the animal, were fixed in Bouin's aqueous fixative, dehydrated through a graded ethanol series, and embedded in paraffin wax. Serial sections cut at 5 - 8 μm were stained either with Harris's haematoxylin and eosin, or by means of a modification of the 'Azan' technique (Neal, unpublished).

The morphological appearance of the ovarian explants was assessed and those showing necrotic changes discarded. The histology of the follicles, and the pre-ovulatory maturation changes within their oocytes, was compared with those seen in the ovaries of mice during the various phases of the estrous cycle and following superovulation. In the latter studies, the Fallopian tubes of the animals were serially sectioned along with the ovaries and were scanned for the presence of eggs.

F. Quantitative Studies

Serial sections of the cultured and control ovaries were scanned under the microscope and counts made of oocytes at the various stages in their maturation. For the purposes of compiling Tables 20. 1-3, the oocytes were classified as follows:

Table 20.1 **Superovulation in intact Schofield white mice; injection of 5 I.U. PMSG followed 48 hrs later by HCG**

(Figures in parenthesis denote range)

No. ovaries	HCG treatment		Eggs maturing			
	Dose (I. U.)	Time (hrs)	Meta I*	Meta II*	Ovulated†	Total
11	-	-	11.8	2.8	3.4	18.6
			(4-21)	(1-8)	(0-11)	(7-35)
8	2	24	14.8	2.8	7.8	25.0
			(8-19)	(1-6)	(5-9)	(19-28)

* see text
† tubal eggs

Table 20.2 **No. of eggs maturing *in vitro* after the administration of 5 I.U. PMSG *in vivo* followed 48 hrs later by HCG treatment *in vivo* or *in vitro***

(Figures in parenthesis denote range)

No. ovaries	HCG treatment		Culture*	Eggs maturing			
	Dose (I. U.)	Time (hrs)		Meta I+	Meta II+	'Free'†	Total
6	-	-	24	10.8	5.8	-	16.5
				(5-17)	(0-15)		(5-31)
9	4	5*	19	14.3	6.6	1.5	22.5
				(3-24)	(4-11)	(0-4)	(9-33)
8	2	24	-	18.5	4.3	-	22.6
				(14-23)	(2-7)		(21-28)

* duration of *in vitro* culture without gonadotrophins added (hrs)
+ see figure 20.1
† tubal eggs as for table 20.1 (eliminated from culture)

Table 20.3 **No. of eggs maturing after treatment with 5 I.U. PMSG (48 hrs) and HCG** *in vitro*

(Figures in parenthesis denote range)

No. ovaries	HCG treatment		Culture*	Eggs maturing			
	Dose (I. U.)	Time (hrs)		Meta I	Meta II	'Free'	Total
6	-	-	24	3.2	0.8	1.5	5.3
				(0-8)	(0-3)	(0-4)	(1-11)
12	2	24	19	12.8	4.2	0.0	17.3
				(3-20)	(0-9)	-	(6-29)

* Duration of *in vitro* culture without added gonadotrophins (hrs)

Metaphase I - includes all oocytes which do not possess a polar body (metaphase I through anaphase I), but excluding those at the dictyate stage; Metaphase II - only those oocytes which possess a polar body and with chromosomes on a metaphase spindle.

This distinction was made since oocytes at anaphase I and telophase I were only rarely encountered, and the polar body could be clearly seen in at least one of the serial sections of each oocyte at metaphase II.

Following the removal of the explants from the petri dishes, the 'Millipore' rafts and culture medium were scanned and counts made of the oocytes which were seemingly 'ovulated' ('free eggs'). Similarly, serial sections of the Fallopian tubes of animals who had received hormone treatment *in vivo,* were examined and counts made of the eggs which had been ovulated. In each, the stage of meiosis attained by the oocyte was recorded, together with the status of the egg (normal, degenerate, or 'aged').

III. RESULTS

A. Observations on 'Fresh' (unfixed) Oocytes

Oocytes recovered from the cultured explants are indistinguishable from those collected from mice at autopsy; they show well-defined chromosomes and spindle

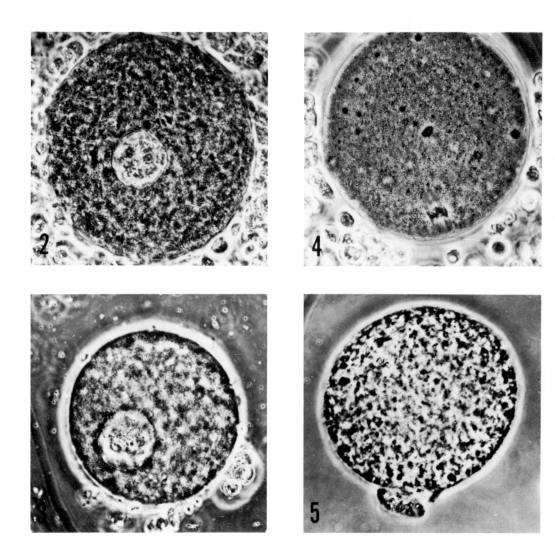

Fig. 20.2 'Living' oocyte at the dictyate stage recovered from a Graafian follicle 12 hrs after the onset of culture with HCG. Phase-contrast microscopy (X 475).

Fig. 20.3 Similar preparation to Fig. 20.2; early stage of chromatin condensation (X 480).

Fig. 20.4 Oocyte at metaphase I recovered from an ovary treated with PMSG *in vivo* and HCG *in vitro*. The chromosomes were stained with acetic lacmoid during observation with phase-contrast illumination (X 475).

Fig. 20.5 Oocyte at metaphase II (↗) and associated polar body. The egg was recovered from a Graafian follicle 24 hrs after the onset of culture with HCG (X 510).

formation (Figs. 20.2-5). The majority of the oocytes from both the control and cultured ovaries are at the dictyate stage (stage of arrested development; Figs. 20.2 and 20.3), but those from the medium-sized and Graafian follicles frequently show signs of pre-ovulatory maturation (metaphase I or II; Figs. 20.4 and 20.5).

Normal oocytes are usually surrounded by a dense layer of cumulus cells which often renders observation with the phase contrast microscope difficult. The application of the staining solutions (acetic lacmoid or orcein) usually disperses the cells sufficiently to permit photography of the nuclear structures within the oocyte. The absence of a complete corona radiata usually indicates that the oocytes are undergoing degeneration. Their cytoplasm is often vacuolated or fragmented, such oocytes being a constant feature of both the cultured and control ovaries.

Some of the oocytes observed during the course of the present study appeared to contain no visible nuclear structures. They are found in the 'fresh' preparations made from ovaries in all the treatment groups employed (see Fig. 20.1), but are never encountered in histological preparations. It would thus seem that the failure to detect chromosomes in a proportion of the eggs studied with the phase contrast microscope is due to a deficiency in the manipulation and staining techniques.

The 'free eggs' recovered from the petri dishes at the end of the culture period are invariably at metaphase II and are usually 'normal', resembling those which have undergone ovulation in controls. Although it appears likely that a proportion of these oocytes have been shed by a process similar to ovulation *in vivo,* the possibility cannot at present be discounted that some resulted from mechanical damage to the follicle during handling.

B. Histology

The histological appearance of the ovary is essentially the same irrespective as to whether the gonadotrophins are administered *in vivo* or *in vitro* (see Figs. 20.6 and 20.7). The ovaries of mice which have been subjected to superovulation contain more medium and large follicles than those of untreated mice at estrus Similarly, the appearance of the ovary after 9 days of culture in the absence of hormones resembles that of mice following hypophysectomy; the small follicles persist while some of the medium sized and antral follicles degenerate. The addition of PMSG at the onset of culture allows the maintenance of the large follicles for longer periods (Figs. 20.6 and 20.7).

The appearance of oocytes undergoing pre-ovulatory maturation is similar irrespective of the treatment group from which the ovary is recovered. Consequently, the selection of photographs illustrating meiotic changes in the oocytes includes eggs from a variety of procedures, as well as from controls (Figs. 20.10-20.15 and Fig. 20.18). The period of arrested development (dictyate or dictyotene stage; Fig. 20.10), generally persists until shortly after treatment with HCG. Thereafter, the oocyte proceeds through metaphase (Fig. 20.11), anaphase,

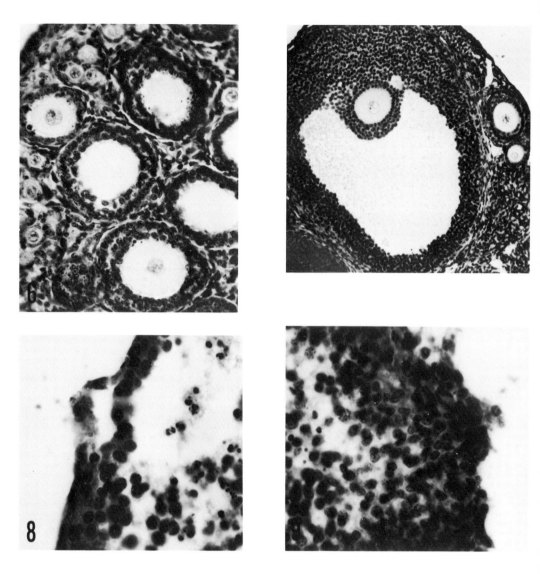

Fig. 20.6   Histological appearance of the ovary of a 3-week-old mouse after culture for a further 9 days. The primordial and 'growing' follicles, and also their oocytes, are similar in appearance to those in controls   (X 380).

Fig. 20.7   Ovary treated with PMSG *in vivo* and subsequently cultured in medium containing HCG. The Graafian follicle is normal in appearance although larger than those which are seen to ovulate. Oocyte at the dictyate stage   (X 130).

Fig. 20.8   The surface of a Graafian follicle showing what may be the formation of an ovulatory stigma *in vitro*. Same treatment as for Fig. 20.7   (X 510).

Fig. 20.9   Graafian follicle which contained no oocyte and appeared to be undergoing luteinization. The raised area at the surface of the follicle may be the remains of the ovulatory stigma. Same treatment as for Figs. 20.10 and 20.11   (X 445).

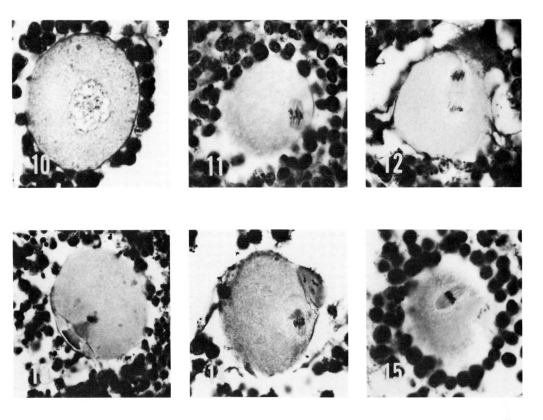

Fig. 20.10 Oocyte at the dictyate stage within a Graafian follicle which had been treated with PMSG and HCG *in vitro*. Note mitosis of cumulus cell (X 580).

Fig. 20.11 Oocyte at metaphase I after treatment with PMSG *in vivo* and HCG *in vitro* (X 510).

Fig. 20.12 Oocyte at anaphase of the first meiotic division. Same treatment as for Fig. 20.11 (X 550).

Fig. 20.13 Polar body abstriction in an oocyte maturing within a Graafian follicle treated with PMSG and HCG in organ culture (X 480).

Fig. 20.14 Oocyte with a polar body and chromosomes at metaphase II. Treatment as for Fig. 20.16 (X 510).

Fig. 20.15 Oocyte containing the spindle of the second meiotic division and with a polar body. PMSG by injection into the animal; HCG added in culture 48 hrs later (X 560).

and telophase (Fig. 20.12) of the first meiotic division, and the first polar body is abstricted (Fig. 20.13). The extrusion of the polar body occurs after the rotation of the spindle through 90° (cf. Figs. 20.11, 20.12 and 20.13). The second metaphase rapidly ensues, during which the first polar body can clearly be identified alongside the egg (Figs. 20.14, 20.15 and 20.18). Further development of the oocyte appears to be 'blocked', although in the abnormal conditions of 'parthenogenetic' development structures resembling cleavage stages are occasionally seen (Fig. 20.20). It should be pointed out, however, that such eggs can also result from the resumption of meiosis in each of two oocytes within a common zona pellucida. Some of the oocytes are seen to contain two nuclei (Fig. 20.19), and a few of the follicles contain two oocytes which are usually at the same stage of meiosis. However, one of the follicles recovered following treatment with HCG *in vitro* contained one oocyte at the dictyate stage, the other having matured to metaphase II.

The follicles in ovaries treated with PMSG contain more granulosa cells undergoing mitosis than in the untreated group. The effect is less marked, however, when the hormone is administered *in vitro,* than it is *in vivo.* Preliminary observations indicate that the follicles in explants are smaller than those *in vivo,* although precise measurements have not as yet been completed. PMSG is less effective in inducing pre-ovulatory maturation than the combined superovulatory treatment. Nevertheless, some of the oocytes are ovulated; they are initially of normal appearance but subsequently lose their covering of cumulus cells and degenerate.

Treatment of either the explants or of intact mice, with HCG some 48 hours following a stimulatory dose of PMSG leads to the maturation of many eggs in Graafian follicles. The oocytes in the largest follicles often remain at the dictyate stage until degeneration sets in (Fig. 20.7), while those in preantral follicles appear to be blocked at metaphase I (Fig. 20.16). The remaining follicles with antra possess eggs that undergo maturation to metaphase II and *in vivo,* some 13 hours after treatment with HCG, large numbers of eggs are recovered from the Fallopian tubes. On the other hand, oocytes which reach metaphase II *in vitro* are usually retained within their follicles, although a few 'free eggs' are found in the culture medium or on the 'Millipore' rafts. Where such eggs are found, the corresponding explants contain follicles devoid of oocytes which possess structures resembling post-ovulatory stigmata (Fig. 20.9). The 'ovulatory' process *in vitro* appears to consist of a thinning of the follicle wall at a point on its surface. This localized elevated area is associated with pyknosis of a few granulosa cells after which the follicle is split open and granulosa cells and oocytes are seen to be released (Fig. 20.8). Similar changes are observed in follicles undergoing ovulation *in vitro.*

C. Quantitative Results

The number of eggs ovulated from Schofield strain mice following the superovulation treatment is approximately double that found during the estrous cycle (Baker and Neal, unpublished results). The treatment with PMSG and HCG used during the present study results in the resumption of meiosis and ovulation such that four types of eggs are encountered during quantitative procedures; namely:

(1)   those which do not proceed beyond metaphase I,
(2)   oocytes which reach metaphase II without undergoing ovulation,
(3)   those at metaphase II recovered from the oviducts,
(4)   seemingly 'aged' eggs in the Fallopian tubes; they are devoid of cumulus cells and nuclear structures and have been excluded from the counts.

When the superovulatory doses of gonadotrophins are administered *in vitro,* only the first and second types of eggs are encountered in the same proportion as in the controls; type three is represented by the 'free eggs' and type four is seemingly eliminated prior to culture (Tables 20.1-3).

The numbers of oocytes undergoing maturation varies considerably between individuals within each group, irrespective of whether the PMSG alone or PMSG and HCG are administered *in vivo* or *in vitro.*

IV. DISCUSSION

The results of the present study demonstrate that the meiotic maturation of oocytes within their follicles is dependent on the presence of gonadotrophic hormones. HCG induces the resumption of meiosis beyond the dictyate stage only in those oocytes which are of an appropriate size, whereas PMSG appears to function by promoting the growth of medium sized follicles (which constitute a precursor 'pool') to this pre-ovulatory size. The optimum period of treatment of the ovaries of Schofield mice, whether *in vivo* or *in vitro,* appears to be about 48 hours, after which HCG induces maturation in the oocytes to metaphase I or II within about 13 hours in the intact mouse and slightly longer in organ culture (*ca,* 16 hours).

The present observations confirm and extend those of Fritz, Cho and Biggers (1965), who treated mouse ovaries in organ culture with HCG, after stimulating follicular growth with PMSG *in vivo.* In a preliminary communication, they reported that the oocytes within the follicles resume meiosis to metaphase II and 'ovulate' if HCG is present. The 'free eggs' found in our experiments could also have been naturally ovulated. The possibility remains, however, that at least a proportion of such oocytes result from mechanical damage to the follicle during handling of the explants. The fact that the 'free oocytes' are invariably at metaphase II is no 'proof' of ovulation unless the process is actually observed, since it is well known that eggs cultured in the absence of their follicles spontaneously mature to this stage (see Schuetz, 1969).

Perhaps the best evidence from the present study that oocytes are occasionally ovulated *in vitro* derives from the histological observations of pre- and

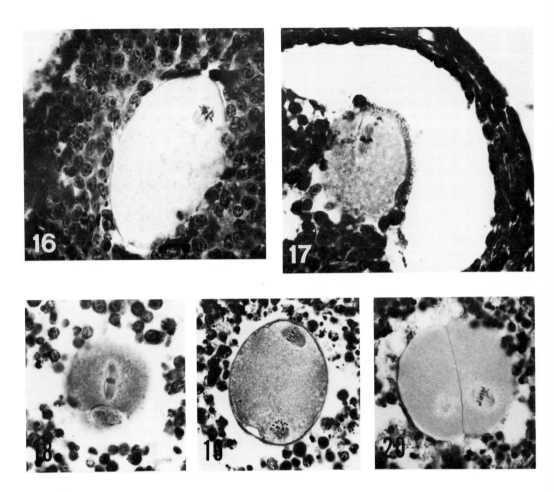

Fig. 20.16 Normal oocyte at metaphase I enclosed within a pre-antral follicle. PMSG by injection into the intact animal; HCG treatment in organ culture   (X 450).

Fig. 20.17 Graafian follicle enclosing an oocyte at metaphase II; treatment with PMSG and HCG *in vitro*. The oocyte shows early degenerative changes   (X 425).

Fig. 20.18 Pre-ovulatory oocyte at metaphase II in the ovary of a mouse which had been given super-ovulating injections of PMSG and HCG. Compare morphology with oocytes shown in Figs. 20.17, 20.18, and 20.20   (X 510).

Fig. 20.19 Apparently normal oocyte containing two metaphase spindles; ovary treated *in vivo* with PMSG and *in vitro* with HCG   (X 440).

Fig. 20.20 Parthenogenetic, or possibly binovular, oocyte which is similar in appearance to a 2-cell cleavage stage. The egg is enclosed within a Graafian follicle which had been matured *in vitro* but similar eggs have been seen in the ovaries of mice subjected to super-ovulating procedure *in vivo*   (X 415).

post-ovulatory changes in the follicle wall. These changes, which include pyknosis of a localized region of granulosa cells around an elevated stigma-like structure, and in some cases the rupture of the follicle, are similar to those described for ovulatory follicles *in vivo* (see Blandau, 1966). Other follicles contained no oocytes and possessed a post-ovulatory plaque of cells over the remains of the stigma.

It should be pointed out, however, that the proportion of oocytes at metaphase II which seemingly 'ovulate' *in vitro* is much smaller than following comparable treatment in the intact animal. The lower incidence of 'free oocytes' in culture, and also the smaller size of the follicles, may well be due to the lack of a blood supply or to a deficiency of steroid hormones and their precursors. Noyes, Clewe and Yamate (1961) found a similar reduction in the number of ovulated eggs in rat ovaries which had been transplanted to the anterior chamber of the eye of male recipients. These authors pointed out that the mileu of steroids was probably inadequate to sustain normal follicular growth and ovulation, and that an inadequate blood supply prevented the final growth of the Graafian follicle to the size found in controls.

It is generally accepted, however, that the size of the Graafian follicle is no indication of the nuclear status of the oocyte that it contains. Thus Walton and Hammond (1928) observed that the largest follicles do not necessarily show pre-ovulatory changes; they often fail to ovulate. Chang (1953) showed that abnormally large follicles are particularly common following treatment with gonadotrophins, as in the case of the present study. The oocytes in these follicles persist at the dictyate stage until degeneration sets in. It would thus seem that the ability of the oocyte to respond to the maturing effects of HCG is largely mediated by the follicle which surrounds it.

The present results also demonstrate that oocytes fail to complete their meiotic maturation if the follicle is smaller than those of ovulatory size. Thus pre-antral follicles, and those in which the formation of the antrum is at an early stage, usually contain oocytes whose maturation beyond metaphase I appears to be 'blocked'. It remains to be established whether, if sufficient time elapsed, such follicles would grow to a size where meiotic maturation could be completed, or whether degeneration subsequently occurs (see Ingram, 1962).

A similar 'block' during the first meiotic division was recorded by Edwards (1965) for oocytes matured in a chemically defined medium. He found that a variable proportion of the eggs derived from a variety of mammalian species progressed only to anaphase I, while other oocytes of similar appearance reached metaphase II. Edwards' conclusion that inadequate conditions of culture prevented the maturation of some oocytes may well be true, especially in the light of studies which indicate that a high proportion of such oocytes are abnormal (e.g., Chang, 1968). It may well be, however, that the 'block' during meiosis is due to the considerable variation in size of the follicles from which the oocytes are derived.

The process of follicular growth has recently been investigated using quantitative autoradiographic techniques (see Peters and Levy, 1966; Pedersen, 1970). These authors have shown that the ovaries of mice experiencing normal estrous contain groups of follicles destined to ovulate or degenerate during a number of ensuing cycles. Follicles of type 5b (*ca.* 5 layers of granulosa cells; Pedersen, 1970), appear to constitute a precursor 'pool' of relatively constant size. A porportion of these follicles proceed through the final stages of growth and antrum formation under the stimulus of FSH from the pituitary gland. It has been shown that the process from type 5b to type 8 (mature Graafian follicles) takes about 5 days, while the entire period of follicular growth from type 3b (single layer of granulosa cells) to ovulation takes a minimum of 19 days (see Pedersen, 1970). The results of the present study indicate that superovulatory doses of PMSG and HCG can induce pre-ovulatory changes in follicles at any stage beyond type 5b, but only those of type 8 possess oocytes at metaphase II. It would seem that if the time interval following HCG treatment was sufficiently long, the oocytes in follicles of types 6 and 7 would also reach metaphase II, otherwise the increased number of eggs which are ovulated *in vivo* could not be accounted for. Pedersen's (1970) results show that there are insufficient pre-ovulatory follicles in the ovaries of mature mice to account for the superovulatory response unless more of the type 5b are stimulated to proceed to stages 6 - 8.

The response of ovarian explants to PMSG and HCG is less effective than following treatment of intact animals with similar doses of gonadotrophins. Thus only some 17 oocytes undergo maturation to metaphase I and II *in vitro* compared with about 25 *in vivo*. The impaired ability of PMSG to promote follicular growth in organ culture could be due to the lack of a blood supply and of endogenous endocrines. The importance of steroid hormones has been clearly demonstrated in experiments involving hypophysectomy where PMSG is less effective than in controls (Williams, 1945a). But the response is improved if ovarian atrophy is prevented by pre-treatment with estrogen (Williams, 1945b; Mauleon, 1969). Moreover, oestrogens have been shown to exert an FSH-like effect both *in vivo* (Williams, 1956; Croes-Buth, Paesi and de Jongh, 1959; Meyer and Bradbury, 1960), and *in vitro* (Lostroh, 1959; Fainstat, 1968). It may well be that the addition of estrogen and PMSG to the explants will induce a synergistic effect and increase the number of oocytes undergoing pre-ovulatory changes to the level seen in controls (see Hisaw, 1961).

It is conceivable that a deficiency of steroid hormones was responsible for the failure by Foote and Thibault (1969) to obtain maturation of cow and pig oocytes within follicles *in vitro,* although species differences may also play an important part. The techniques of culture that they employed, and the composition of the nutrient media, were similar to those in the present study, except that purified preparations of ovine FSH and LH were substituted for PMSG and HCG. The resumption of meiosis beyond the dictyate stage did not occur unless the oocytes were removed from the follicles, although the histological appearance of the cultures appeared to be satisfactory.

We have found that the histology of the follicles and of the oocytes within them is similar in explants to those in the intact animal. It is clear, however, that the capacity of the eggs to sustain fertilization, undergo cleavage, and develop into normal offspring needs to be tested. Such tests have recently been initiated and so far only a few cleaved eggs have been recovered following fertilization *in vitro*. The possibility remains that these eggs are undergoing parthenogenetic development, a phenomenon reported for cytologically normal eggs cultured in the absence of their follicles. Chang (1955a) has shown that the number of such oocytes which sustain cleavage is very small and few offspring can be obtained following transfer of eggs to suitable recipients. The number of cleavage stages recovered is considerably increased, however, if the oocytes are recovered from Fallopian tubes or from follicles in which maturation is well advanced (Chang, 1955b; Suzuki and Mastroianni, 1968; Whittingham, 1968). These results indicate that the follicular environment is essential for the normal progression of all but the last stages of oocyte maturation.

Studies with the electron microscope have consistently shown that from an early stage in follicular growth, microvilli emanating from the surface of the egg interdigitate within the zona pellucida with those from the follicle cells (e.g., Franchi, 1960; Adams and Hertig, 1964). The delicate relationship between the oocyte and granulosa cells is clearly important in terms of the nutrition of the oocyte and also the passage of informational material (see Baker, 1970a). Indeed, one of the earliest histological signs of atresia affecting 'growing' follicles is the withdrawal of the processes from within the zona pellucida: the oocyte rapidly becomes eosinophilic and its nucleus pyknotic (see Ingram, 1962). It is likely that similar changes occur in some *in vitro* systems for culturing eggs in the absence of their follicles. But such factors are unlikely to affect the oocytes in the present experiments since the morphological integrity of the oocyte/follicle relationship is maintained until pre-ovulatory maturation is completed.

V. ACKNOWLEDGMENTS

Some of the expenses incurred in this study were defrayed out of grants to Dr. B. M. Hobson by Organon Laboratories Ltd., who also provided the gonadotrophins. One of us (T.G.B.) gratefully acknowledges the provision of travel grants by the National Institutes of Health, Bethesda.

VI. REFERENCES

Adams, E. C. and Hertig, A. T. (1964) Studies on guinea pig oocytes. 1. Electron microscopic observations on the development of cytoplasmic organelles in oocytes of primordial and primary follicles. *J. Cell Biol.* **21**, 397.

Austin, C. R. (1961) The Mammalian Egg. Oxford, Blackwell.

Baker, T. G. and Neal, P. (1969) The effect of X-irradiation on mammalian oocytes in organ culture. *Biophysik.* **6**, 39.

Baker, T. G. (1970a) Electron microscopy of the primary and secondary oocyte. *In* "Advances in the Biosciences 6." Oxford, Pergamon Press.

Baker, T. G. (1970b) Oogenesis and ovarian development. *In* "Reproductive Biology," (H. Balin and S. Glasser, eds.) New York, Excerpta Medica Foundation. (In press).

Blandau, R. J. (1966) The mechanism of ovulation. *In* "Ovulation". (R. B. Greenblatt, ed.), p. 3, Philadelphia, J. B. Lippincott.

Chang, M. C. (1953) *In* "Ciba Foundation Symposium: Mammalian germ cells." (G. E. W. Wolstenholme, M. P. Cameron and J. S. Freeman, eds.), London, J. and A. Churchill Ltd.

Chang, M. C. (1955a) The maturation of rabbit oocytes in culture and their maturation, activation, fertilization and subsequent development in the Fallopian tube. *J. Exp. Zool.* **128**, 379.

Chang, M. C. (1955b) Fertilization and normal development of follicular oocytes in the rabbit. *Science.* **121**, 867.

Chang, M. C. (1968) *In vitro* fertilization of mammalian eggs. *J. Anim. Sci.* **27**, Suppl. 1, 15.

Croes-Buth, S., Paesi, F. J. A. and Jongh, S. E. (1959) Stimulation of ovarian follicles in hypophysectomized rats by low doses of oestradiol benzoate. *Acta Endocr.* **32**, 399.

Crooke, A. C. and Ryle, M. (1968) *In vitro* activity of human follicle stimulating hormone. *In* "Gonadotrophins 1968." (E. Rosemberg, ed.), p. 193, Los Altos, Geron-X Inc.

Edwards, R. G. and Gates, A. H. (1959) Timing of the stages of the maturation divisions, ovulation, fertilization and the first cleavage of eggs of adult mice treated with gonadotrophins. *J. Endocr.* **18**, 292.

Edwards, R. G., Wilson, E. D. and Fowler, R. E. (1963) Genetic and hormonal influences on ovulation and implantation in adult mice treated with gonadotrophins. *J. Endocr.* **26**, 389.

Edwards, R. G. (1965) Maturation *in vitro* of mouse, sheep, cow, pig, rhesus monkey and human ovarian oocytes. *Nature (London).* **208**, 349.

Edwards, R. G. (1966) Mammalian eggs in the laboratory. *Sci. Amer.* **215**, 72.

Fainstat, T. (1968) Organ culture of postnatal rat ovaries in chemically defined medium. *Fertil. Steril.* **19**, 317.

Foote, W. D. and Thibault, C. (1969) Recherches experimentales sur la maturation *in vitro* des ovocytes de Truie et de Veau. *Ann. Biol. Anim. Bioch. Biophys.* **9**, 329.

Fowler, R. E. and Edwards, R. G. (1957) Induction of superovulation and pregnancy in mature mice by gonadotrophins. *J. Endocr.* **15**, 374.

Franchi, L. L. (1960) Electron microscopy of oocyte-follicle cell relationships in the rat ovary. *J. Biophys. Biochem. Cytol.* **7**, 397.

Fritz, H. I., Cho, W. K. and Biggers, J. D. (1965) Ovulation from whole ovaries of mice in organ culture. *J. Cell Biol.* **27**, 31A. (abstract).

Hartmann, C. G. and Leathem, J. H. (1963) Oogenesis and Ovulation. *In* "Mechanisms Concerned with Conception." (C. G. Hartmann, ed.), p. 205, Oxford, Pergamon Press.

Hisaw, F. L. (1961) *In* "Control of Ovulation." (C. A. Villee, ed.), p. 48, Oxford, Pergamon Press.

Ingram, D. L. (1962) Atresia. *In* "The Ovary." (S. Zuckerman, A. M. Mandl and P. Eckstein, eds.), Vol. 1, p. 247, New York, Academic Press.

Lostroh, A. J. (1959) The response of ovarian explants from post-natal mice to gonadotrophins. *Endocrinology.* **65**, 124.

Martinovitch, P. N. (1937) Development *in vitro* of the mammalian gonad. *Nature (London).* **139**, 413.

Martinovitch, P. N. (1938) Development *in vitro* of the mammalian gonad - ovary and ovogenesis. *Proc. Roy. Soc. [Biol.].* **125**, 232.

Mauleon, P. (1969) Oogenesis and Folliculogenesis. *In* "Reproduction in Domestic Animals." (H. H. Cole and P. T. Cupps, eds.), 2nd ed., p. 187, New York, Academic Press.

Meyer, J. E. and Bradbury, J. T. (1960) Influence of stilbesterol on the immature rat ovary and its response to gonadotrophin. *Endocrinology.* **66**, 121.

Noyes, R. W., Clewe, T. H. and Yamate, A. M. (1961) Follicular development, ovular maturation and ovulation in ovarian tissue transplanted to the eye. *In* "Control of Ovulation." (C. A. Villee, ed.), p. 24, Oxford, Pergamon Press.

Pedersen, T. (1970) Follicle kinetics in the ovary of the cyclic mouse. *Acta Endocr.* **64**, 304.

Peters, H. and Levy, E. (1966) Cell dynamics of the ovarian cycle. *J. Reprod. Fertil.* **11**, 227.

Rowlands, I. W. and Williams, P. C. (1943) Production of ovulation in hypophysectomized rats. *J. Endocr.* **3**, 310.

Rowlands, I. W. (1944) The production of ovulation in the immature rat. *J. Endocr.* **3**, 384.

Ryle, M. (1969a) A quantitative *in vitro* response to follicle stimulating hormone. *J. Reprod. Fertil.* **19**, 87.

Ryle, M. (1969b) The duration of an FSH effect *in vitro. J. Reprod. Fertil.* **19**, 349.

Ryle, M. (1969c) Morphological response to pituitary gonadotrophins by mouse ovaries *in vitro. J. Reprod. Fertil.* **20**, 307.

Schuetz, A. W. (1969) Oogenesis: processes and their regulation. *In* "Advances in Reproductive Physiology." (A. McLaren, ed.), Vol. 4, p. 99, London, Logos Press.

Suzuki, S. and Mastroianni, L. (1968) *In vitro* fertilization of rabbit follicular oocytes in tubal fluid. *Fertil. Steril.* **19**, 716.

Walton, A. and Hammond, J. (1928) Observations on ovulation in the rabbit. *Brit. J. exp. Biol.* **6**, 190.

Whittingham, D. G. (1968) Fertilization of mouse eggs *in vitro*. *Nature (London)*. **220**, 592.

Williams, P. C. (1945a) Studies of the biological action of serum gonadotrophins. 1. Decline in ovarian response after hypophysectomy. *J. Endocr.* **4**, 127.

Williams, P. C. (1945b) Studies of the biological action of serum gonadotrophins. 2. Ovarian response after hypophysectomy and oestrogen treatment. *J. Endocr.* **4**, 131.

Williams, P. C. (1956) The history and fate of redundant follicles. *In* "Ciba Foundation Colloquia on Aging." (G. E. W. Wolstenholme and E. C. P. Miller, eds.), Vol. 2, p. 59, London, J. and A. Churchill Ltd.

Young, W. C. (1961) The mammalian ovary. *In* "Sex and Internal Secretions." (W. C. Young, ed.), 3rd ed., Vol. 1, p. 449, London, Bailliere, Tindall and Cox Ltd.

Zarrow, M. X. and Wilson, E. D. (1961) The influence of age on superovulation in the immature rat and mouse. *Endocrinology*. **69**, 851.

21

FINAL STAGES OF MAMMALIAN OOCYTE MATURATION

Charles G. Thibault

I.   Introduction
II.   *In vitro* studies on the initiation of the meiotic prophase
III.   When do oocytes become capable of resuming meiosis?
IV.   Role of the granulosa cells in the maintenance of the dictyate stage
V.   Final maturation *in vitro*
        A.   Chronology of *in vitro* nuclear maturation
        B.   Spindle abnormalities
        C.   *In vitro* matured rabbit oocytes are able to cleave after parthenogenetic activation
        D.   *In vitro* matured rabbit oocytes are fertilizable
        E.   Failure of sperm head to swell during fertilization of *in vitro* matured oocytes
VI.   Summary
VII.   References

I. INTRODUCTION

The role of endocrine factors in mammalian germ cell maturation appears very different in the male and the female. FSH, LH, or both initiate the meiotic processes in the testis at puberty, and then support them permanently. On the contrary, the meiotic prophase in the mammalian ovary generally appears during a relatively short period of time of embryonic life, and oocytes reach the dictyate stage. The endocrine stimuli which induce meiotic prophase in the female mammal remain unknown, although many experiments have been done in order to explain the sexual differentiation of the gonocytes, and hermaphrodism in invertebrates and lower vertebrates. It is also well known that oocytes from prepuberal or mature females can resume meiosis *in vivo* up to formation of the second metaphase spindle, if an exogenous or endogenous stimulus of gonadotropic hormone is given and if the oocytes are present within large-size Graafian follicles.

The resumption of meiosis may also be observed in at least two other circumstances:

(1) In oocytes removed from Graafian follicles of different sizes and cultured *in vitro,* or transferred *in vivo* into the Fallopian tube or uterus.

(2) In degenerating Graafian follicles as shown by Benoit and Mauleon (personal communication) in adult female rats 3 - 9 days after hypophysectomy. These observations seem contradictory since meiosis is resumed, in the one case after gonadotropic stimulation and in the other cases when gonadotropic hormones are absent. Moreover, under *in vitro* conditions many oocytes which resume meiosis belong to small or middle-size Graafian follicles, and would be unable to mature during the normal estrous period.

Using mainly *in vitro* procedures, we have tried to answer the following questions:

(1) What endocrine support is required for the initiation of meiotic prophase in the female embryo?

(2) At what time in development do oocytes achieve nuclear maturation?

(3) How are oocytes maintained in the dictyate stage?

(4) What is the physiological significance of preovulatory gonadotropic stimulation on the oocyte itself, and on its corona cells?

II. *IN VITRO* STUDIES OF THE INITIATION OF THE MEIOTIC PROPHASE

In vitro studies of rat, mouse, and human embryonic ovaries have shown that different steps of the meiotic prophase appear without any endocrine support (Rat - Martinovitch, 1938; Mouse - Borghese and Venini, 1956; Van de Kerckhove, 1956; Blandau, Warrick and Rumery, 1965; Human - Blandau, 1969). However, in all experiments it is highly probable that meiotic prophase had already commenced at the beginning of the culture at least in some parts of the ovaries.

In the sheep ovary, meiotic prophase begins around the 50th - 51st day of pregnancy. The appearance of the first leptotene figures is rapidly followed by the other meiotic stages, and at 75 days most of the ovocytes are formed and have reached the final stage of meiotic prophase. Experiments conducted with sheep ovaries put in culture before or after the beginning of the meiotic prophase, indicate that the initiation of meiotic prophase needs a trigger stimulus. When ovaries of 54-day-old embryos are cultured under rigid conditions of pH and oxygen tension, the successive stages of meiotic prophase appear normally during culture. But, if ovaries of younger embryos that are 47, 48, and 49 days old are cultured, the meiotic prophase does not appear for as long as 12 days after the beginning of the culture (Figs. 21.1 and 21.2). All stages of meiotic prophase are visible after the same duration in the control embryos.

All attempts to initiate *in vitro* meiosis in ovaries of embryonic sheep, 48 days old, were unsuccessful, including culture of the embryonic ovary with fragments of the pituitary of older fetuses or of maternal pituitary, and culture in presence of pure ovine gonadotropic hormones. Only a few abnormal leptotene stages were observed. These results indicate that meiosis does not appear spontaneously in sheep. However, the trigger stimulus is not pituitary dependent and its nature remains to be discovered. This trigger stimulus is probably of short duration since the meiotic processes can progress *in vitro* without any other support, if they are initiated *in vivo*. We can only know if such a trigger stimulus is generally significant by new *in vitro* studies of rat, mouse, or human ovaries from embryos which are too young to have started meiotic prophase.

III. WHEN DO OOCYTES BECOME CAPABLE OF RESUMING MEIOSIS?

Pincus and Enzmann (1935), Chang (1955a, b), Edwards (1965a, b), and more recently many others (see Thibault, 1971) have shown that oocytes from prepuberal or mature mammals can resume meiosis when cultured outside the follicle. However, a variable proportion of oocytes remain in the dictyate stage or at different stages between dictyate and the second metaphase spindle - usually in metaphase I. Even when a high proportion of oocytes reach metaphase II, as in Donahue's experiments on the mouse oocyte, there is asynchrony among the oocytes.

We have tried to find out when oocytes become competent to mature completely *in vitro*. We used prepuberal rabbits between 70 and 90 days old and adult does in oestrus from our New Zealand colony (Thibault and Gerard, 1971). All

Table 21.1 In vitro **Maturation of Oocytes from Larger Follicles of Immature and Adult Estrous Rabbits**

	Duration of culture (in hrs.)	No. of rabbits	No. of oocytes	D	PM	1st MS	A-T	M2	2nd MS
ADULT	7.51 - 8.28	3	18				1	17	
IMMATURE	7.41 - 8.15	4	86	6	22	19	9	23	7
						47%		46%	

D	Dictyate stage
PM	Prometaphase
1st MS	First metaphase spindle
A-T	Anaphase-telophase
M2	Second metaphase plate
2nd MS	Second maturation spindle always visible

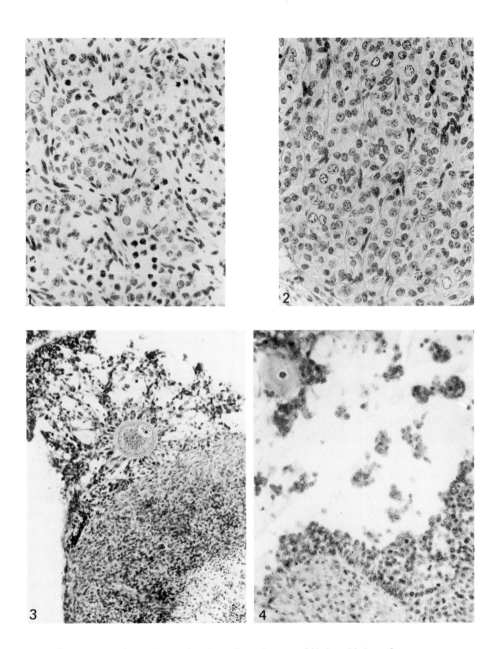

Fig. 21.1   Different stages of meiotic prophase in embryonic ovary of 62-day-old sheep foetus.

Fig. 21.2   Absence of meiotic prophase in 50-day-old ovary after 12 days of culture. Note clusters of oogonia; many of them are dividing.

Fig. 21.3   Cow oocyte grafted on the granulosa cell (G.C.) of a foreign follicle. The nucleus remains in the dictyate stage.

Fig. 21.4   Oocyte remains a part of a healthy granulosa layer (G.C.) and resumes meiosis (Metaphase I).

the oocytes from preovulatory follicles resumed meiosis up to the second metaphase when maintained in culture for 8 hours. There was a striking difference in the results obtained with oocytes from immature rabbits, using the same technique since 42% only reach metaphase II (Table 21.1). This difference offers a good explanation for the variable percentage of oocytes reaching the second metaphase spindle, as reported in the literature for different species. The results indicate that oocytes which reached their maximum size in growing follicles have not yet developed or synthesized the chemical machinery permitting germinal vesicle breakdown and the achievement of metaphases I and II, including formation of the polar bodies. All the oocytes from preovulatory follicles reached maturity, but in small or medium-size follicles, oocyte maturity is not always attained. At the present time, we cannot give a criterion of oocyte maturity based on follicle size. We can speculate that active synthesis on lampbrush chromosomes, as described by Baker and Franchi (1967a, b) and Baker (1970) is implicated in this second stage of oocyte maturation.

IV. ROLE OF THE GRANULOSA CELLS IN THE MAINTENANCE OF THE DICTYATE STAGE
(Foote and Thibault, 1969)

From the preceding experiments, it is obvious that a large proportion of oocytes from Graafian follicles of different sizes have attained the maturity necessary for resumption of meiosis. However, those oocytes within the ovary remain in the dictyate stage until the follicles are able to respond to an ovulatory gonadotropic stimulus. In order to understand why they remain in the dictyate stage, we have cultured (a) cow and pig oocytes inside the follicle, and (b) oocytes from medium-size follicles in contact with granulosa cells or theca cells.

In the first part of this work, medium-sized cow or pig follicles included theca cells, or they were limited to granulosa cells and their surrounding membrane. All follicles were cultured under the same conditions of medium, pH and temperature. After 48 - 72 hours of culture, all oocytes remained in the dictyate stage as long as the granulosa cells stayed healthy. These conclusions have been recently confirmed by Foote, Matley and Tibbits (1970), with heifer oocytes, who compared simultaneously the behavior in culture of free oocytes and oocytes inside the follicle: 47% of free oocytes matured up to metaphase II, whereas all oocytes in the follicles remained in the dictyate stage.

In the second part of the work, oocytes from medium-size follicles were cultured either in contact with the theca cells of isolated small follicles or with granulosa cells of experimentally opened follicles. Oocytes grafted on the theca cells invariably resumed meiosis as free oocytes whereas oocytes grafted on the granulosa cells remained in the dictyate stage. In these experiments, when the grafting between foreign corona cells and granulosa cells was prevented by pycnosis or by mechanical disturbance, resumption of meiosis was found (Figs. 21.3 and 21.4, and Table 21.2).

The persistence *in vitro* of the dictyate stage in oocytes cultured in their own

Table 21.2 **Culture of Pig Oocytes on the Granulosa Cells or Theca Cells of Foreign Follicles**

Experiment	No. of follicles	No. of grafted oocytes	Nuclear Stages of Grafted Oocytes		
			Dictyate	Met. 1	Met. 2
Grafts on theca cells outside the follicle	3	5	0	I	4
Grafts on granulosa cells inside the follicle	10	19	16	I*	2*

follicle, or grafted on to the granulosa cells, and on the other hand the resumption of meiosis in oocytes cultured outside the follicle whether or not in contact with theca cells, constitute presumptive evidence that the granulosa cells are responsible for maintenance of the dictyate stage in mature oocytes from Graafian follicles.

V. FINAL MATURATION *IN VITRO*
(Thibault and Gerard, 1970, 1971)

A. Chronology of *in Vitro* Nuclear Maturation

Only oocytes from preovulatory follicles of estrous rabbits were used in this study to compare the duration of the final stage of meiosis *in vitro* and *in vivo* on similar oocytes. This comparison showed the following facts:

(1) For all *in vitro* and *in vivo* oocytes, the length of time was exactly the same between first modification of the germinal vesicle and formation of the second metaphase spindle (Table 21.3). However, maturation begins approximately one hour earlier *in vitro* than *in vivo,* if time zero is respectively the beginning of the culture and the moment of coitus or HCG injection. This one-hour difference remains throughout the whole nuclear maturation and represents the time necessary *in vivo* for the release of the gonadotropins and the commencement of their effect.

(2) All the oocytes from preovulatory follicles matured *in vitro* up to the second maturation spindle in 9 - 9.5 hours, and we did not observe asynchrony among the oocytes as reported by Donahue (1968) for mouse oocytes.

(3) The nuclear maturation of rabbit oocytes *in vivo* does not appear to be DNA-RNA dependent. The intrafollicular injection of 1 ng of actinomycin D, 10 - 31 minutes after HCG injection, did not prevent the resumption of meiosis within a

Table 21.3 Comparison between *in vivo* and *in vitro* Maturation
of the **Rabbit Oocyte**

Control time after coitus (in hrs)	*In vitro* duration of culture (in hrs)	No. of rabbits	No. of oocytes	PM	M1	1st MS	A-T	M2	2nd MS
3.00 - 3.30		4	13	13					
	3.16 - 3.37	2	15	2	13				
4.00 - 4.50		4	14	4	10				
	6.17 - 6.24	3	17			4	10	3	
6.58 - 7.15		3	21			19	2		
	8.05 - 8.28	2	14				1	13	
9.00 - 9.30		3	22					21	6
	9.00 - 9.30	4	31					12	19
	15.20 - 18.00	19	114						114

For explanation, see Table 21.1.

normal time. In these experiments, oocytes were studied only from follicles in which ovulation had been prevented by actinomycin injection (Pool and Lipner, 1964; Espey and Lipner, 1965) (Table 21.4).

Table 21.4 **Effect of Intrafollicular Injection of Actinomycin D on Nuclear Oocyte Maturation**
in vivo

No. of rabbits	Oocytes from injected follicles	Control oocytes from untreated follicles	
		non-ovulated	ovulated
7	24 Met.2*/24	6 Met.2/6	12 Met.2/12

*Met. 2 = metaphase 2

B. Spindle Abnormalities

In rabbit oocytes from preovulatory follicles, the formation of the first metaphase spindle, the extrusion of the first polar body and the formation of the second metaphase spindle appear perfectly normal using cytological criteria. However, in most cases, the spindle does not lie tangential to the cytoplasmic membrane of the oocyte, but is orientated radially (Thibault and Gerard, 1971) (Figs. 21.3 and 21.4).

In calf or cow oocytes taken from unripe follicles, many abnormalities were observed when follicular fluid was included in the medium, such as formation of one or two nuclei which resemble pronuclei. A characteristic abnormality of those reaching metaphase II was always visible; this was the presence of a membranous structure surrounding the spindle. Clumping of the spindle fibers sometimes occurs in normally ovulated oocytes during aging, but the appearance is not similar to the abnormality just described.

C. *In Vitro* Matured Rabbit Oocytes are Able to Cleave after Parthenogenetic Activation (Thibault, 1971)

Oocytes matured *in vitro* were cooled at 10°C for 24 hours in the same media and then put back at 37°C (Chang, 1954). After a further 20 hours of culture, 12 out of 16 oocytes from two rabbits had divided into two blastomeres; after 48 hours 13 out of 16 from two other rabbits were at the 8-cell stage. These results give proof that *in vitro* matured rabbit oocytes possess all the components required for cleavage.

D. *In Vitro* **Matured Rabbit Oocytes are Fertilizable**
(Thibault and Gerard, 1971; Thibault, 1971)

Oocytes recovered after 7, 10, 12, and 15 hours of culture were mixed *in vitro* with capacitated sperm recovered from the uterus of rabbits mated 12 hours previously. The results are summarized in Table 21.5. The percentage of fertilization

Table 21.5 **Duration of** *In Vitro* **Maturation and Fertilization Rate**

In vitro maturation (in hrs)	No. of Rabbits	No. of Oocytes	Eggs Fertilized	
			No.	%
7	16	116	42	36
10	7	47	23	49
12	6	33	10	30
15.5	4	29	14	48
Total	33	225	91	40

did not increase with the duration of the preliminary culture. It was lower (40% fertilization) than the percentage of *in vitro* fertilization obtained when naturally ovulated oocytes were used (65%). It seems that this relatively low fertilization rate has little physiological significance, since in other experiments, with *in vitro* matured oocytes, proportions higher than 65% were observed. The numerous manipulations undergone by the oocytes could be an explanation.

Trials to fertilize *in vivo* matured rabbit oocytes after germinal vesicle breakdown have shown that sperm penetration does not occur before first polar body formation. Thus, after nuclear maturation oocytes become fertilizable.

E. **Failure of Sperm Head to Swell During Fertilization**
of *in Vitro* **Matured Oocytes** (Thibault and Gerard,
1970; Thibault, 1971)

Fertilization begins as usual but the sperm head remains morphologically unchanged although migrating to the center of the egg. The absence of sperm swelling is visible immediately after sperm head engulfment in the cytoplasm. Whereas normally a few hours after fertilization (Thibault, 1967) two pronuclei come into contact, in this situation the female pronucleus joins to a sperm head and not a male pronucleus. Two or three hours after fertilization a membrane appears, surrounding an unmodified sperm head. Then, the chromatin of the sperm head

Table 21.6 **Evolution of the Sperm Head from 3.5 hrs. to 9 hrs. after Introducing Capacitated Sperm**

Time after mixing egg and sperm	No. of eggs fertilized	No. modification of sperm head		Membrane formation but no swelling		Small male pronucleus		Sub-normal male pronucleus		Cleaved eggs 2-cell	
		No.	%	No.	%	No.	%	No.	%	No.	%
3.30	25	23	92	2	8						
6.	12	8	66	3	25			1	9		
7.30	29	9	31	12	41	5	17	3	10		
15.00	42	3	7			10	24	8	19	21	50

Table 21.7 **Time Required for the Presence of the MPGF into the Oocyte During In Vivo Maturation**

Time between HCG or coitus and culture	In vitro maturation (in hours)	No. of rabbits	No. of oocytes	Fertilized eggs		Sperm head	
				No.	%	no or abnormal	normal
2 hrs.	12.00	4	31	20	64	100%	0
3 hrs.	12.00	2	15	8	53	100%	0
5 hrs.	12.00	3	14	9	64	100%	0
7 hrs.	5.00	3	25	11	44	20%	80%

undergoes fragmentation and some fragments swell, giving progressively a small male pronucleus and finally, in a few cases, a subnormal male pronucleus (Table 21.6; Figs. 21.5 - 21.8).

The fact that the sperm head does not evolve into a male pronucleus during the first hours after fertilization, and that the formation of the male pronucleus is abnormal and delayed, reveals that during natural oocyte maturation in the follicle a factor appears in the cytoplasm which is not synthesized *in vitro,* or at least not to a sufficient degree. We suggest calling this factor the male pronucleus growth factor (MPGF).

We tried to find out when this substance first appears in the oocyte during *in vivo* maturation, and if addition of different hormones to the culture medium stimulates the production of MPGF in the *in vitro* matured oocyte. Oocytes from rabbit preovulatory follicles were recovered 2, 3, 5 hours after HCG injection or 7 hours after coitus. They were cultured for 12, 12, 12, and 5 hours respectively and mixed *in vitro* with capacitated sperm. The results are summarized in table 21.7, and they suggest that MPGF only becomes available *in vivo* in oocytes 6.5 - 7 hours after the gonadotropic ovulatory stimulus.

The finding of this late passage into the oocyte fits well with Chang's conclusion (1955a, b) that normal development after *in vivo* fertilization is only possible with oocytes recovered from Graafian follicles 7 hours after mating or HCG injection. As the nucleus of the oocyte has reached metaphase I two hours earlier, it seems that the MPGF is not synthesized by the oocyte itself, but rather by either the corona cells or the follicular cell after gonadotropic stimulation.

In order to check if corona cells can synthesize the MPGF *in vitro* when they receive an adequate hormonal stimulation, we added crude pituitary extract, LH, FSH, prolactin, estradiol, progesterone, or 20 α-OH progesterone to the culture medium. Oocytes were examined between 3.5 and 6 hours after introduction of capacitated sperm. None of the hormones studied was able to stimulate the synthesis of MPGF at a normal level. There is a tendency with prolactin alone, or in combination with FSH and LH, or with crude pituitary extracts to hasten pronucleus formation, but at the beginning of sperm penetration the normal swelling of the sperm head does not occur. It is of interest that there is a high percentage of digyny when oocytes are matured in the presence of prolactin (Thibault, 1971).

The expansion of the corona cell processes, so characteristic of the freshly ovulated rabbit oocyte, is observed when FSH is present in the medium. This cellular reaction is not so strong with LH, and does not appear without hormones or in the presence of prolactin. The same response of the corona cells was obtained with calf oocytes after 36 hours of culture (Figs. 21.9 - 21.12). In spite of the FSH action on the calf corona cells, abnormalities of the second maturation spindle were not prevented (Figs. 21.11 and 21.12). As FSH stimulates corona cells in such a way

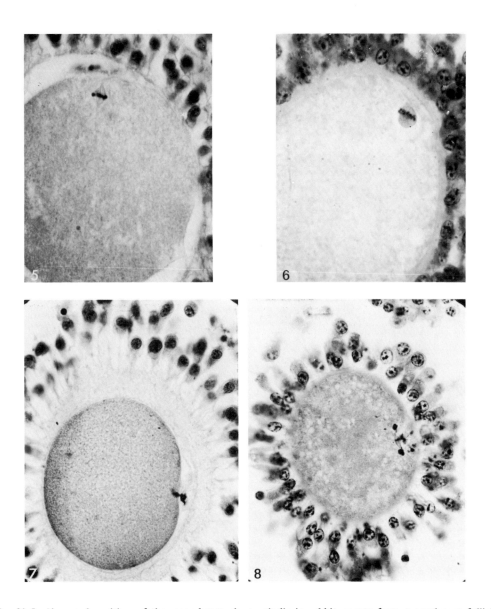

Fig. 21.5 Abnormal position of the second metaphase spindle in rabbit oocyte from preovulatory follicles cultured for 12 hours.

Fig. 21.6 Abnormal second metaphase spindle of *in vitro* matured cow oocyte: round-shape spindle surrounded by a nuclear membrane.

Fig. 21.7 Oocyte and its corona cells cultured for 12 hours in presence of 10 μg of prolactin. Corona cells remain aggregated around the oocyte. Note the two chromosome sets in the oocyte during telophase I (the first polar body is on the left) (rabbit).

Fig. 21.8 2 μg of FSH stimulate expansion of the corona cell processes as it occurs *in vivo* in the Graafian follicle before ovulation (rabbit).

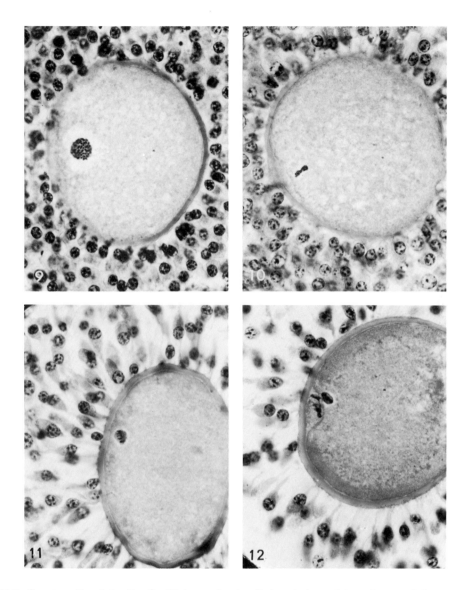

Fig. 21.9 Oocyte cultured *in vitro* for 30 hours showing first metaphase plate and aggregated corona cells (cow).

Fig. 21.10 Oocyte cultured *in vitro* for 35 hours in presence of LH (2 µg/ml) showing the second metaphase plate and aggregated corona cells (cow).

Fig. 21.11 Oocyte cultured *in vitro* for 35 hours in presence of FSH (2 µg/ml) showing expanded corona cells and second metaphase spindle. In spite of the normal appearance of the corona cells, the second metaphase spindle is also enveloped by a membrane, regularly found in cow oocytes matured without gonadotrophic hormones (cow).

Fig. 21.12 Oocyte cultured *in vitro* for 35 hours in presence of FSH (2 µg) and LH (2 µg/ml) showing expanded corona cells and abnormal second metaphase spindle (cow).

that *in vitro* matured oocytes and their corona cells cannot be distinguished from naturally ovulated oocytes, but is unable to promote the synthesis of MPGF, we must conclude that corona cells are not responsible for the synthesis of this factor.

VI. SUMMARY

Rabbit oocytes from preovulatory follicles always resume meiosis *in vitro* up to the second metaphase spindle with the same timing as *in vivo* after gonadotropic ovulatory stimulation.

By contrast, in the smallest follicles, oocyte ability to resume meiosis is not always attained, as shown by the variable percentage of oocytes from medium-size follicles which are able to reach the second metaphase spindle in culture.

The results of *in vitro* grafting of oocytes from cow and pig onto theca or granulosa cells of foreign follicles suggest that granulosa cells are responsible for the maintenance of the dictyate stage during the final stages of follicular growth.

In studying fertilization of *in vitro* matured rabbit oocytes we have shown that a substance is present *(in vivo)* in the rabbit oocyte a few hours before ovulation, but not during *in vitro* maturation. This substance, which we have called male pronucleus growth factor, is responsible for the immediate swelling of the sperm head after penetration. Its absence, after *in vitro* maturation of oocytes surrounded only by corona cells, and in the presence of pituitary or steroid hormones, indicates that granulosa cells or theca cells must be implicated in its synthesis rather than corona cells.

VII. REFERENCES

Baker, T. G. (1971) Electron microscopy of the primary and secondary oocytes. 6th Shering Symp., Venice, April 1970. *Adv. Biosci.* **6**, (In press).

Baker, T. G. and Franchi, L. L. (1967a) The structure of the chromosomes in human primordial oocytes. *Chromosoma.* **22**, 358.

Baker, T. G. and Franchi, L. L. (1967b) The fine structure of chromosomes in bovine primordial oocytes. *J. Reprod. Fertil.* **14**, 511.

Blandau, R. J. (1969) Observations on living oogonia and oocytes from human embryonic and fetal ovaries. *Am. J. Obstet. Gynec.* **104**, 310.

Blandau, R. J., Warrick, E. and Rumery, R. E. (1965) *In vitro* cultivation of fetal mouse ovaries. *Fertil. Steril.* **16**, 705.

Borghese, E. and Venini, M. A. (1956) Culture *in vitro* di gonadi embryonali di *mus musculus.* *Symp. genetica.* **5**, 69.

Chang, M. C. (1954) Development of parthenogenetic rabbit blastocysts induced by low temperature storage of unfertilized ova. *J. Exp. Zool.* **125**, 127.

Chang, M. C. (1955a) The maturation of rabbit oocytes in culture and their maturation, activation, fertilization and subsequent development in the Fallopian tubes. *J. Exp. Zool.* **128**, 379.

Chang, M. C. (1955b) Fertilization and normal development of follicular oocytes in the rabbit. *Science.* **121**, 867.

Donahue, R. P. (1968) Maturation of the mouse oocyte *in vitro*. 1. Sequence and timing of nuclear progression. *J. Exp. Zool.* **169**, 237.

Edwards, R. G. (1965a) Maturation *in vitro* of mouse, sheep, cow, pig, rhesus monkey and human ovarian oocytes. *Nature (London).* **208**, 349.

Edwards, R. G. (1965b) Maturation *in vitro* of human ovarian oocytes. *Lancet.* **2**, 926.

Espey, L. L. and Lipner, H. (1965) Enzyme rupture of rabbit Graafian follicle. *Am. J. Physiol.* **298**, 208.

Foote, W. D., Matley, D. L. and Tibbits, F. D. (1970) Follicular inhibition of oocyte maturation *in vitro*. Proc. Western Sect., *Am. Soc. Anim. Sci.* **21**, 45.

Martinovitch, P. N. (1938) The development *in vitro* of the mammalian gonad ovary and ovogenesis. *Proc. Roy. Soc. [Biol.]* **125**, 232.

Pincus, G. and Enzmann, E. V. (1935) The comparative behavior of mammalian eggs *in vivo* and *in vitro*. *J. Exp. Med.* **62**, 665.

Pool, W. R. and Lipner, H. (1964) Inhibition of ovulation in the rabbit by actinomycin D. *Nature (London).* **203**, 1385.

Thibault, C. (1971) *In vitro* maturation and fertilization of rabbit and cattle oocytes. *In* "Regulation of Mammalian Reproduction". (S. J. Segal, R. Crozier and P. A. Corfman, eds.), Springfield, Illinois, Charles C.Thomas.(In press).

Thibault, C. and Gerard, M. (1970a) Facteur cytoplasmique necessaire a la formation du pronucleus male dans l'ovocyte de Lapine. *C. R. Acad. Sci. (Paris).* **270**, 2025.

Thibault, C. and Gerard, M. (1970b) La maturation et la fecondation *in vitro* de l'ovocyte de quelques Mammiferes. Symp. "Les malformations congenitales des Mammiferes", Pfizer, Amboise mai 1970. (In press).

Van de Kerckhove, D. (1959) L'ovaire prenatal de la Souris blanche en culture organotypique. *C. R. Ass. Anat.* 754.

22

THE RELATION OF OOCYTE MATURATION TO OVULATION IN MAMMALS

Roger P. Donahue

I.   Introduction
II.   Methods
        A.   Oocyte collection during period of ovulation
        B.   Examination of ovarian oocytes
III.   Control by gonadotrophins
IV.   Maturation *in vitro*
V.   Ovulation of immature stages
VI.   Maturation without ovulation
VII.   Exceptional species
VIII.   Summary
IX.   Acknowledgments
X.   References

I. INTRODUCTION

The term oocyte maturation will be used here to refer to the progression of the nucleus from the germinal vesicle stage (Fig. 22.1) to second meiotic metaphase (metaphase II - Fig. 22.8). A mature oocyte will be at metaphase II and an immature one will be at or beyond the germinal vesicle stage but one which has not yet reached metaphase II. In the original description of this phase in the history of the mammalian oocyte, Van Beneden (1875) applied the term "la maturation" and the expression (in German, die Reifung) has been used, although not universally, to the present day. Since the whole process of oocyte formation, growth, and preparation for post-fertilization development involves maturing of the oocyte, alternative terms have been offered for the progression from germinal vesicle to metaphase II. "Initiation of meiosis" is inappropriate since the first meiotic division begins in the fetal ovary and is interrupted about the time of diplonema in late prophase when the interphase-like germinal vesicle forms. Neither is "meiotic maturation" entirely accurate, for when the nucleus leaves the germinal vesicle stage and progresses through metaphase I (Fig. 22.2), anaphase I (Fig. 22.3) and telophase I (Fig. 22.4) with first polar body formation, there is an arrest at metaphase of the second meiotic division without an intervening prophase II. Meiosis is completed only after the second polar body is expelled, normally brought about by sperm penetration inducing the oocyte nucleus to leave metaphase II and pass through anaphase and telophase II with subsequent second polar body formation. Maturation thus applies to the interval between two periods of natural arrest. It might be better to retain the term maturation, considering it an abbreviation for "completion of the first maturation (meiotic) division".

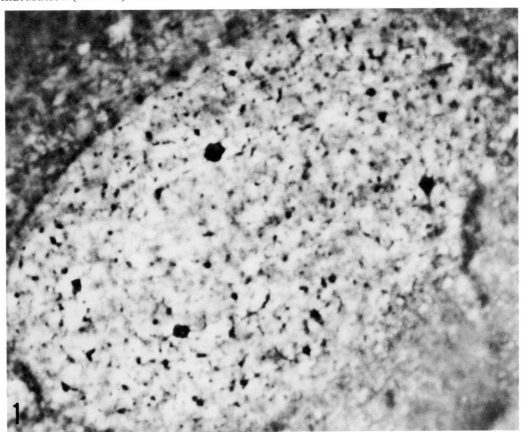

Fig. 22.1 Germinal vesicle. 32 hrs after HCG. Although no nucleolus is evident, the fixative is not well suited for visualization of the usually single nucleolus (X 90).

Oocyte maturation has been investigated in twenty-one mammals (Table 22.1); in a few cases more than one species may be included under one mammal. As would be expected, oocytes of the small laboratory animals (the mouse, rat, rabbit, and guinea pig) have been the most extensively studied. Major areas of interest have been morphology and timing, control by gonadotrophins, influence of the follicular cells, maturation *in vitro* and, more recently, energy metabolism of oocytes during maturation. The present account reviews the relation between maturation and ovulation, and provides further information on ovulation of immature stages and maturation without subsequent ovulation.

II. METHODS

A. Oocyte Collection During Period of Ovulation

Unbred 8-10 week old CF-1 female mice (Carworth, Inc.) were superovulated with intraperitoneal injections (15 units each) of pregnant mare's serum (Equinex, Ayerst) followed 48 hours later with human chorionic gonadotrophin (APL, Ayerst) at the end of the 12-hour light period (Fowler and Edwards, 1957). Females were sacrificed at 11, 11½, 12, 13, and 14 hours after HCG and the Fallopian tubes torn so as to free ova located in the ampulla. Three separate experiments, each consisting of oocytes pooled from three to four females, were carried out on separate days for each of the five collection times.

After liberation of the oocytes into a Krebs-Ringer salt solution containing pyruvate, they were transferred to a mixture of this medium and hyaluronidase to remove the cumulus cells (Biggers, Whittingham and Donahue, 1967). Oocytes were then transferred to a micro slide, fixed with ethanol-acetic acid (3:1), stained with 1% aceto-orcein and examined with phase contrast optics (Donahue, 1968). Between 30-60 minutes elapsed between sacrifice of females and fixation of oocytes.

B. Examination of Ovarian Oocytes

Ovaries from unbred 8-10 week old **CF-1** female mice (Carworth, Inc.) selected without regard to the phase of the estrous cycle were removed and placed in embryological watch glasses containing about 0.5 ml of the Krebs-Ringer salt solution with pyruvate under 1 ml paraffin oil. Oocytes were liberated by puncturing follicles with a 25 gauge hypodermic needle, washed once, then fixed, stained, and examined (Donahue, 1968). Almost all oocytes liberated in this way appear to be full-sized, and it is presumed they come from follicles with antra. After one washing most of the cumulus cells had separated from the oocytes (in contrast, after administration of gonadotrophins, the cumulus is not easily separated from the oocyte).

III. CONTROL BY GONADOTROPHINS

Previous investigations have led to the conclusion that gonadotrophins, endogenous or injected, induce not only ovulation but also stimulate the oocyte to mature, resulting in metaphase II formation shortly before ovulation. Thus in rabbit oocytes, chromatin is seen condensing in the germinal vesicle within 2 hours after

Table 22.1 Reports of Mammalian Oocyte Maturation In Vivo and In Vitro

Order	Mammal	In Vivo	In Vitro
Insectivora	Mole	Heape, 1886	
	Ericulus	Strauss, 1939	
	Shrew	Pearson, 1944	
Chiroptera	Bat	Van Beneden and Julin, 1880	
		Van der Stricht, 1906, 1923	
		Guthrie and Jeffers, 1938	
Edentata	Armadillo	Newman, 1912	
Lagomorpha	Rabbit	Van Beneden, 1875	Pincus and Enzmann, 1935
		Rein, 1883	Chang, 1955a, b
		Flemming, 1885	Robertson and Baker, 1969
		Friedgood and Pincus, 1935	
		Pincus and Enzmann, 1935	
		Pincus, 1936	
		Pincus and Enzmann, 1937	
		Moricard and Gothie, 1953	
		Zamboni and Mastroianni, 1966	
Rodentia	Mouse	Sobotta, 1895, 1907	Moricard and Fonbrune, 1937
		Kirkham, 1907a, b	Gothie and Tsatsaris, 1939
		Lams and Doorme, 1908	Edwards, 1962, 1965a
		Long and Mark, 1911	Biggers *et al.*, 1967
		Van der Stricht, 1923	Donahue, 1968, 1970
		Branca, 1925	Donahue and Stern, 1968
		Engle, 1927	
		Gresson, 1941	
		Makino, 1941	
		Nakamura, 1957	
		Edwards and Gates, 1959	
		Jagiello, 1965, 1967, 1968, 1969a, b	
		Jagiello and Ohno, 1966	
		Kuhlmann, 1966	
		McGaughey and Chang, 1969	
		Zamboni, 1970	
Rodentia	Rat	Coe, 1908	Edwards, 1962
		Sobotta and Burckhard, 1910	
		Kirkham and Burr, 1913	
		Boling *et al.*, 1941	
		Blandau, 1945	
		Noyes, 1952	

	Rat	Odor, 1955, 1960 Clewe *et al.,* 1958 Odor and Renninger, 1960 Mandl, 1963 Jagiello, 1965	
	Hamster	Ward, 1948	Edwards, 1962
	Vole	Breed and Clark, 1970	
	Guinea pig	Rein, 1883 Rubaschkin, 1905 Moore and Tozer, 1908 Lams, 1913 Van der Stricht, 1923 Branca, 1925	Jagiello, 1969c
Carnivora	Dog	Van der Stricht, 1908, 1923 Evans and Cole, 1931	
	Cat	Longley, 1911 R. Van der Stricht, 1911 O. Van der Stricht, 1923 Dawson and Friedgood, 1940	
	Ferret	Robinson, 1918	
	Fox	Pearson and Enders, 1943	
Perissodactyla	Horse	Hamilton and Day, 1945	
Artiodactyla	Pig	Corner, 1917 Spaulding *et al.,* 1955 Hunter and Polge, 1966	Edwards, 1965a, 1968 Foote and Thibault, 1969
	Cow		Edwards, 1965a Foote and Thibault, 1969 Robertson and Baker, 1969
	Sheep	Dziuk, 1965	Edwards, 1965a
Primates	Monkey	Hartman and Corner, 1941	Edwards, 1962, 1965a Suzuki and Mastroianni, 1966
	Man	Thomson, 1919 Hoadley and Simons, 1928 Dixon (cited by Allen *et al.*) Allen *et al.,* 1930 Baca and Zamboni, 1967 Jagiello *et al.,* 1968 Steptoe and Edwards, 1970	Pincus and Saunders, 1939 Edwards *et al.,* 1966 Edwards, 1965a, b, 1968 Jagiello *et al.,* 1968 Edwards *et al.,* 1969 Kennedy and Donahue, 1969 Mastroianni and Noriega, 1970 Suzuki and Iizuka, 1970

copulation (an event which stimulates luteinizing hormone release from the pituitary), metaphase I is first found at 4 hours, metaphase II at 9 hours, and ovulation occurs between 9½ and 10½ hours (Pincus and Enzmann, 1935). Rat oocytes reach metaphase I by 3-4 hours after the onset of heat (behavioral estrus) with metaphase II and ovulation occurring 6 hours later (Odor, 1955). Upon administration of an ovulation- inducing dose of gonadotrophin, usually human chorionic gonadotrophin (HCG), the germinal vesicle begins to break down after 2 hours in the mouse (Edwards and Gates, 1959), 18 hours in the pig (Hunter and Polge, 1966) and an estimated 25 hours in the human female (Edwards, 1965b). Metaphase II and ovulation occur at 11-14 hours in the mouse (Edwards and Gates, 1959), 37 hours in the pig (Hunter and Polge, 1966), and 35-48 hours after HCG in man (Jagiello, Karnicki, and Ryan, 1968; Steptoe and Edwards, 1970). In general then, the entire sequence of maturation takes between 10 to 24 hours once maturation starts. Most rapid of the individual stages are those occurring between metaphase I and II (Table 22.2).

Although precise time relationships are not established, maturation is accompanied by an increase in follicular size and quantity of follicular fluid, a decrease in the density of cells (cumulus) surrounding the oocyte, and retraction of the cytoplasmic processes of the cumulus which have traversed the zona pellucida. A separation of the cumulus from those cells (granulosa) lining the follicle also occurs and the cumulus- clad ovum is released from the ovary (Pincus and Enzmann, 1935; Boling, Blandau, Soderwall and Young, 1941; Zamboni, 1970). But maturation and ovulation are not inseparable events since (1) maturation usually occurs after manual removal of oocytes from ovarian follicles; (2) ovulation of immature oocytes can occur; (3) mature oocytes, apparently not destined to ovulate, are found in ovaries; and (4) ovulation of oocytes in the germinal vesicle stage normally occurs in at least two mammals and, in one case, it is known that fertilization precedes ovulation.

IV. MATURATION *IN VITRO*

First observed in rabbit oocytes by Pincus and Enzmann (1935) and now known in nine other species (Table 22.1), maturation can occur in a suitable cell culture medium upon release of full-sized oocytes from ovarian follicles. Pincus and Enzmann (1935) suggested that maturation both *in vivo* and *in vitro* commences upon removal of oocytes from the influence of the somatic cells of the follicle. According to this hypothesis, follicular cells (particularly the granulosa) maintain the oocyte at the germinal vesicle stage by producing a maturation inhibitor or by withholding essential nutrients required for maturation. Recent experimental evidence supports such a role for some component within the follicle, possibly the follicular cells themselves. When individual follicles were separated from the ovary and cultured with the follicular wall intact, oocytes did not mature in the three species studied, the mouse (Edwards, 1962) and the cow and pig (Foote and Thibault, 1969). Maturation also failed to occur in pig and cow oocytes even when the follicle was broken open, provided that the oocyte or the cumulus remained in

Table 22.2 Duration in Hours (+ S.E.) of Individual Stages of the Maturation Division in Mouse, Rat, and Pig Oocytes

Stage	MOUSE, *In Vivo*[1] Induced Maturation (Edwards & Gates, 1959)	MOUSE, *In Vitro*[2] (Talbert, 1969)	RAT, *In Vivo*[3] Spontaneous Maturation (Mandl, 1963)	PIG, *In Vivo* Induced Maturation (Hunter & Polge, 1966)
Chromatin Condensation	—	4.3 ± 0.7	3.0	4
Prometaphase I	—	3.2 ± 0.3	1.0	4
Metaphase I	6.0 ± 0.3	3.8 ± 0.3	2.3	9
Anaphase I	1.2 ± 0.3	0.4 ± 0.3		
Telophase I	0.3 ± 0.2	0.2 ± 0.1	1.0	
Chromatin Mass	—	0.6 ± 0.1	—	2
Prometaphase II	—	0.5 ± 0.1	0.5	—
Metaphase II to Ovulation	—	—	1.0	—

1. Duration calculated by probit analysis.

2. Duration calculated by recursive multiple regression using the data of Donahue (1968).

3. Duration estimated using peak incidence of successive stages.

contact with the granulosa (Foote and Thibault, 1969). (For further relationships between oocytes and follicular cells, see chapters by T. G. Baker, J. D. Biggers and A. V. Nalbandov.) In addition, fewer rabbit oocytes matured (26%) when cultured in follicular fluid from unmated rabbits than when oocytes were cultured in the control medium (serum) where 46% matured (Chang, 1955a).

Not all oocytes placed into culture will mature. Between 35 and 73% of cultured human oocytes obtained at different times of the menstrual cycle reached metaphase II and another 5-25% underwent an apparent arrest at metaphase I (Table 22.3); the remainder were composed mostly of those that stayed at the germinal

Table 22.3 **Nuclear Stages Attained by Human Follicular Oocytes After 43 or more Hours of Culture in Various Media**

No. Oocytes	MET. I	MET. II	Reference
65	5%	71%	Edwards, 1965b.
30*	27%	73%	Jagiello, Karnicki, and Ryan, 1968
87	11%	45%	Kennedy and Donahue, 1969
72	25%	35%	Mastroianni and Noriega, 1970

*Oocytes obtained 24 hours after HCG *in vivo* then cultured for 25-28 hours.

vesicle stage or showed signs of degeneration. In two studies using human oocytes matured *in vitro* for subsequent fertilization *in vitro*, 26/104 or 25% (Edwards, Donahue, Baramki and Jones, 1966) and 20/56 or 36% (Edwards, Bavister, and Steptoe, 1969) remained at the germinal vesicle stage. About 5% of mouse oocytes showed no sign of maturation and another 5% arrested at metaphase I. (Both the arrested germinal vesicle and metaphase I were morphologically normal when examined by light microscopy.) Except for metaphase I, arrest at a stage between germinal vesicle and metaphase II is rare in mouse oocytes (Donahue, 1968). *In vitro* maturation progresses at about the same rate as maturation *in vivo* in pig and human (Edwards, 1968) and rabbit (Chang, 1955a) oocytes. In a large population of mouse oocytes, an asynchrony of up to 6 hours was observed both in the breakdown of the germinal vesicle and the formation of metaphase II (Donahue, 1968).

The extent to which these observations accurately reflect *in vivo* maturation is not known, but it has now been demonstrated that oocytes matured *in vitro* are capable of normal development. Mouse oocytes matured *in vitro* can be fertilized *in vitro* and subsequently develop into normal 15-day fetuses when transferred to foster mothers although the number of recovered fetuses was less than the number obtained from oocytes matured *in vivo* (Cross, 1970).

V. OVULATION OF IMMATURE STAGES

The frequency of tubal oocytes possessing an intact germinal vesicle was 1.9% (28/1472) in pigs induced to ovulate (Polge and Dziuk, 1966), 0.2% (3/1677) in naturally mated pigs (Hancock, 1961), 1.3 and 0.08% in (mostly) naturally ovulating mice and rats, respectively (Austin and Braden, 1954). The present writer found one oocyte with an apparently normal germinal vesicle (Fig. 22.1) out of 2700 (0.04%) examined within 24 hours of ovulation in superovulated mice. In addition, Longley (1911) found one oocyte at the germinal vesicle stage in the Fallopian tube of a cat. Pearson (1944) finding five such oocytes in the tube and periovarial space of a short-tailed shrew suggested they may have been released prematurely by "cataclysmic" ovulation in adjacent follicles.

When adult mice are given exogenous gonadotrophins, many oocytes leave the ovary before maturation is complete and the final stages of maturation take place in the Fallopian tube. In these mice (see **METHODS**), ovulation has begun by 11 hours after HCG when a mean of 6.8 oocytes were collected from each female previously injected with gonadotrophins. Ovulation appears to be completed by 14 hours (mean of 45 oocytes per female -- see Table 22.4). This period of ovulation is in agreement with that found by Edwards and Gates (1959).

Between 11 and 14 hours, five tubal oocytes out of 1313 (0.4%) were at metaphase I (Fig. 22.2) and nine (0.7%) at anaphase I (Fig. 22.3) or telophase I (Fig. 22.4). Those in the chromatin mass stage, which is formed after telophase and characterized by a fully formed polar body along with a crescent-shaped chromatin group in the oocyte (Donahue, 1968), represented 21% of all tubal oocytes collected at 11 hours (Table 22.4, Fig. 22.7). The percentage of oocytes at the chromatin mass stage decreased to 10% at 11½ hours, 3% at 12 hours, less than 0.5% (2/430) at 13 hours, and none were found at 14 hours. From the crescent-shaped chromatin mass, discrete chromosomes formed, first without apparent organization (early prometaphase II, Fig. 22.5) but subsequently were organized onto a spindle (later prometaphase II, Fig. 22.6). The prometaphase II stage was found in 29% of the tubal eggs collected at 11 hours but in only 3% of the oocytes at 14 hours. During the period of ovulation there was corresponding increase of oocytes at metaphase II, when all chromosomes lie on the metaphase plate (Fig. 22.8). At 11 hours, only 10% of all tubal oocytes were at metaphase II but this increased to 79% at 13 hours and 71% at 14 hours (Table 22.4).

Talbert (1969) has estimated the duration of the chromatin mass and prometaphase II stages to be 0.6 and 0.5 hours, respectively, in mouse oocytes maturing *in vitro* (Table 22.2). These times may be valid *in vivo,* with ovulation of

Table 22.4 **Nuclear Stages of Mouse Tubal Oocytes Collected During the Period of Ovulation***

Hours After HCG	No. Females	No. Eggs Collected	Scored	No. Eggs Collected Per Female **	Met I	Ana 1/ Telo I	Chromatin Mass	Promet II	Met II	Deg Met II	Deg Chromatin	Frag.	Other
11	9	61	58	6.8	0	0/0	12 (21%)	17 (29%)	6 (10%)	5 (9%)	10 (17%)	6 (10%)	2 (3%)
11½	10	116	105	11.6	3 (3%)	3/4 (3%/4%)	11 (10%)	29 (28%)	31 (30%)	14 (13%)	3 (3%)	5 (5%)	2 (2%)
12	12	388	327	32.3	0	0/0	11 (3%)	63 (19%)	175 (54%)	36 (11%)	12 (4%)	22 (7%)	8 (2%)
13	12	430	403	35.8	1 (0%)	1/0 (0%)	2 (0%)	28 (7%)	318 (79%)	45 (11%)	1 (0%)	4 (1%)	3 (1%)
14	10	450	420	45.0	1 (0%)	1/0 (0%)	0	12 (3%)	298 (71%)	43 (10%)	16 (4%)	36 (9%)	13 (3%)
TOTAL	53	1445	1313										

* Oocytes from 3-4 females were pooled at each collection time.

** Mean number of oocytes collected per female injected with gonadotrophins.

Deg-Degenerated

Frag-Fragmented

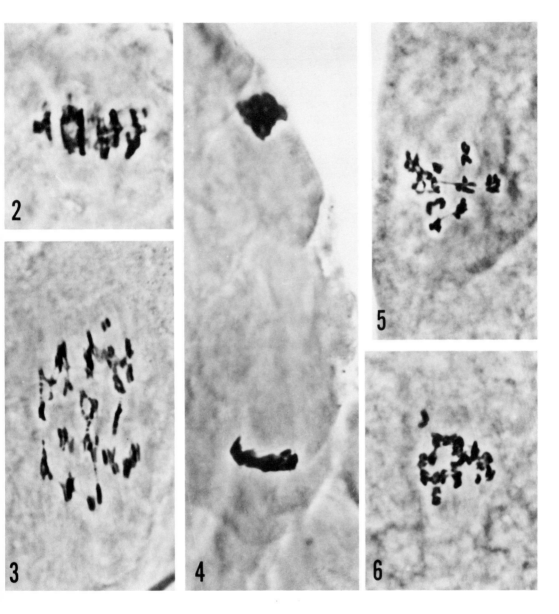

Fig. 22.2   Metaphase I. 12 hrs after HCG   (X 80).

Fig. 22.3   Anaphase I. 13 hrs after HCG   (X 80).

Fig. 22.4   Late telophase I or early chromatin mass. 12 hrs after HCG   (X 80).

Fig. 22.5   Early prometaphase II. 12 hrs after HCG   (X 80).

Fig. 22.6   Late prometaphase II. 12 hrs after HCG   (X 80).

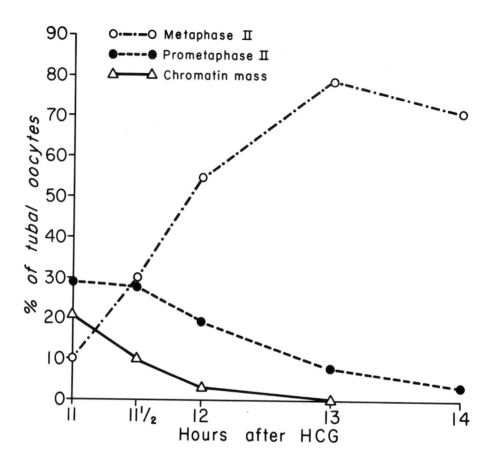

Fig. 22.7 Nuclear stages of mouse oocytes during the period of ovulation. Oocytes were examined 30-60 minutes after the indicated collection times. Data taken from Table 22.4.

chromatin mass and prometaphase II stages still occurring at 12 and 13 hours, respectively (Table 22.4). However, one-half to one hour elapsed between sacrifice of the female and oocyte fixation, allowing additional time for maturation in the liberation medium. For this reason, the number of immature stages reported at each time is an underestimate of the degree of immaturity in tubal oocytes.

Immature stages thus comprised 83% (29/35) of the non-degenerated ova at 11 hours, and 62, 30, 9 and 4% of the ova at 11½, 12, 13, and 14 hours after HCG, respectively. It therefore appears that of the first oocytes to ovulate, a high proportion have not yet reached metaphase II but complete maturation in the

Fallopian tube, while those ovulated at progressively later times have continued to mature in the follicle with the result that more have attained metaphase II by the time of ovulation. (The occurrence of degenerated ova is discussed in the following section.) Whether ovulation of premetaphase II stages occurs naturally in the absence of exogeneous gonadotrophins is not known. However, in an ultrastructural study, Zamboni (1970) has shown that recently ovulated mouse oocytes have elongated chromosomes (presumably in a chromatin mass or early prometaphase II stage) and typical dyads form somewhat later. In naturally mated mice not induced to ovulate, Braden and Austin (1954) found that fertilization did not begin as soon as expected in the presence of competent sperm, a fact which led them to suggest that "maturation" of egg membranes after ovulation was necessary before sperm penetration could occur. Tubal oocytes with immature nuclear stages might, at least in part, account for this delay in fertilization.

There is some evidence that spermatozoa cannot penetrate the oocyte cytoplasm until maturation has progressed to metaphase I. Spermatozoa were found in the cytoplasm of some pig tubal oocytes at the germinal vesicle stage, but since sperm head transformation had not occurred, further development was considered unlikely (Polge and Dziuk, 1965). In other ovulated germinal vesicle oocytes of the pig (Dziuk and Dickmann, 1965) as well as those of mice and rats (Austin and Braden, 1954), sperm had passed through the zona pellucida but remained in the perivitelline space and failed to enter the egg cytoplasm. Development (to term) of rat follicular oocytes was dependent upon their being at metaphase I or beyond before transfer to mated foster mothers (Noyes, 1952). Sperm apparently were unable to enter the cytoplasm of rabbit oocytes until metaphase I or a somewhat later stage (Chang, 1955a), and the probability of subsequent development was greatly enhanced if follicular oocytes were close to the time of first polar body formation before transfer to mated females (Chang, 1955b).

VI. MATURATION WITHOUT OVULATION

Friedgood and Pincus (1935) reported that stimulation of the cervical sympathetics caused rabbit ova to mature, but most of these failed to ovulate. Maturation without ovulation also occurred in rabbits injected with one-fourth to one-half the amount of gonadotrophins required to bring about ovulation (Pincus and Enzmann, 1935). On the basis of these studies, Pincus (1936) concluded that gonadotrophins at levels too low to induce ovulation can nevertheless lead to oocyte maturation, an interpretation still prevalent today, e.g., see Breed and Clark (1970). Indeed, a cyclic variation in the number of mouse ova containing a meiotic spindle (either metaphase I or II) but which presumably are not destined to ovulate has been reported. Engle (1927) found that ova with "pseudomaturation" spindles were most prevalent the first day of diestrum (about one day after ovulation) with a mean of 28.4 per ovary, and reached their lowest number the second day of diestrum (11.2 per ovary). Following induced or natural ovulation in guinea pigs, all medium and large follicles remaining in the ovaries contained ova with meiotic spindles

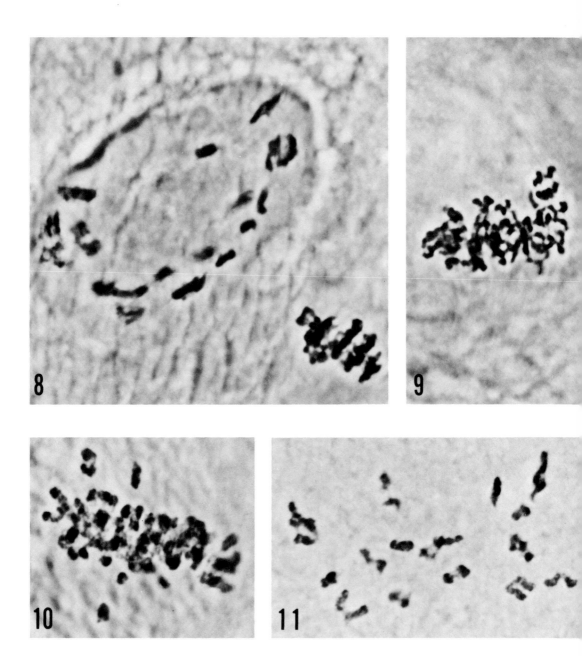

Fig. 22.8   Metaphase II and the first polar body with its scattered chromosomes. 14 hrs after HCG   (X 90).

Fig. 22.9   Degenerating metaphase II. Liberated from ovary   (X 90).

Fig. 22.10   Degenerated metaphase II. Liberated from ovary   (X 90).

Fig. 22.11   Degenerated and scattered metaphase II chromosomes. 34 hrs after HCG   (X 90).

(metaphase I or II) and the follicular cells were undergoing lutein changes (Dempsey, 1939). However, since meiotic spindles were also found following hypophysectomy of guinea pigs, Dempsey (1939) concluded that "pituitary hormone are not essential for the initiation of maturation changes in ova". In the rabbit (Pincus and Enzmann, 1937) and mouse (Engle, 1927; Pincus, 1936), the earliest meiotic spindles are seen in full-sized ova from multilayered follicles just prior to antrum formation.

In man, few mature ova are found in the ovaries at any time during the menstrual cycle. Among oocytes freshly recovered from large follicles, Allen, Pratt, Newell and Bland (1930) found only one of the more than 200 oocytes examined to be at metaphase II (at mid-cycle and perhaps destined to ovulate), and Edwards (1965b) found one out of 32 with degenerated metaphase II chomosomes, the remainder of oocytes in these studies being at the germinal vesicle stage. Similarly, Kennedy and Donahue (1969) detected no sign of maturation in 69 ova upon liberation from the ovary.

At least some mature oocytes which fail to ovulate are capable of normal development. When rat ovaries were transplanted to the anterior eye chamber of host rats, HCG led to ova maturation but only a limited amount of ovulation. Forty-five of the mature follicular oocytes were transferred to mated females and eleven were carried to term and weaned without sign of abnormality (Clewe, Yamate and Noyes, 1958). These authors noted that non-ovulating follicles formed corpora lutea around degenerating ova.

When mouse oocytes are released from ovarian follicles without regard to the estrous stage and examined immediately (see **METHODS**), some mature but degenerated ova are found (Table 22.5). While the majority (79% of 533 oocytes) of ova were at the germinal vesicle stage, the following degenerated stages were found: 1% metaphase I, 12% metaphase II, 3% degenerated chromatin of unknown stage, 2% unknown, 2% fragmented, and 1% miscellaneous stages. Degeneration is characterized by either a fusing of adjacent chromosomes or a swelling of chromosomes leading to a loosely arranged spindle (Fig. 22.9), loss of sharp chromosomal outline (Fig. 22.10), and chromosome scattering (Fig. 22.11). (Changes in metaphase II chromosomes from unfertilized mouse eggs at different times after ovulation have been described by McGaughey and Chang (1969)). Fragmented ova (Fig. 22.12) usually have several larger cytoplasmic pieces along with smaller fragments of varying number and typically one or two "nuclei" per ovum. Fragmentation of oocytes within the follicle has been noted frequently and was thought by some to be a form of parthenogenesis (Pincus, 1936). Whether fragmented oocytes arise from ova that have completed maturation is not known.

Of the degenerated follicular oocytes, 55% were metaphase II, 24% unknown, 8% fragmented, 6% metaphase I, and 5% miscellaneous. It therefore appears that if an oocyte starts to mature it is much more likely to proceed all the way to

Table 22.5 Nuclear Stages of Mouse Oocytes upon Liberation from Ovarian Follicles

Exp. No.	No. Females	No. Eggs	G.V.	Deg. Met I	Deg. Met II	Deg. Chromatin	Frag.	Unknown	Other*
1	4	172	130	5	25	0	3	8	1
2	3	176	141	1	20	5	4	3	2
3	3	185	150	1	19	10	2	1	2
Total	10	533	421	7	64	15	9	12	5*
			(79%)	(1%)	(12%)	(3%)	(2%)	(2%)	(1%)

G. V. = germinal vesicle
Deg. = degenerated
Frag. = fragmented

*Other: 1 deg. anaphase I; 3 chromatin mass, probably deg.; 1 "pronucleus" and a polar body

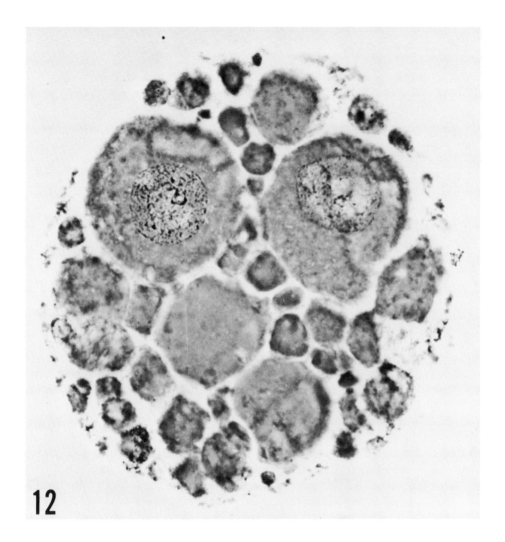

Fig. 22.12   Fragmented oocyte. 32 hrs after HCG.

metaphase II rather than arrest at an intermediate stage. This same conclusion was previously reached with oocytes matured *in vitro,* and even the occurrence of 6% at metaphase I closely parallels the 5% found to arrest at metaphase I *in vitro* (Donahue, 1968).

Are any of the degenerated follicular oocytes ovulated? In Table 22.6, the number of degenerated metaphase I and II, degenerated chromatin, fragmented, unknown, and "other" oocytes recovered from the Fallopian tube between 11 and 14 hours after HCG are compared with the number of degenerated oocytes

Donahue

Table 22.6 **Comparison of the Number of Ovarian and Tubal Oocytes with Degenerated Nuclear Stages**

Source	Hours After (HCG)	No. Eggs Scored	No. Females	No. Eggs Scored Per Female	Degenerated Eggs*		
					No.	%	No. Per Female
ovary	- -	533	10	53.3	112	21	11.2
tube	11	58	9	6.4	23	40	2.6
tube	11½	105	10	10.5	24	23	2.4
tube	12	327	12	27.3	78	24	6.5
tube	13	403	12	33.6	53	13	4.4
tube	14	420	10	42.0	108	26	10.8

* Includes deg. Met I, deg. Met II, deg. chromatin, frag., unknown, and other categories from Tables 22.4 and 22.5

recovered from the ovary. Between 2.4 and 10.8 degenerated eggs per female occur in the Fallopian tubes whereas 11.2 degenerated eggs per female were recovered from ovaries. It is thus possible that some of the degenerated ovarian oocytes are ovulated in response to the administered HCG. If this is so, they appear to be the first ovulated since they comprise 40 per cent of all tubal oocytes at 11 hours but only 23 per cent at 11½ hours. Indeed, the last observation suggests that many degenerated ova are already present in the tubes prior to the beginning of ovulation. Edwards and Gates (1959) found degenerated ova in the Fallopian tubes of mice induced to ovulate and the "number per female rarely exceeded the number of eggs normally expected after natural mating". They suggested these ova might have arisen either by ovulation in response to the prior PMS injection and retained in the tube, or else matured in the ovary in response to endogenous hormones, degenerated under the influence of the PMS and induced to ovulate by the HCG. PMS is known to induce ovulation within 24 hours (Runner, 1950).

If the degenerated tubal oocytes were indeed ovulated much earlier than 11 hours, they nevertheless remained in the vicinity of the collection site, the ampulla. Some eggs in the ampulla were without cumulus but, since these usually adhered to the cumulus of the other eggs, a separation of those with and without cumulus was not usually possible. When they could be separated, those without cumulus were usually, but not always, degenerated while ova with cumulus were occasionally degenerated. The exact origin(s) of the degenerated tubal ova remains unknown.

VII. EXCEPTIONAL SPECIES

Of the 21 mammals studied (Table 22.1), the dog, fox, and possibly the horse are exceptional in that their oocytes remain at the germinal vesicle stage until after ovulation. The nature of the maturation-inducing stimulus in these species remains unknown. In dog oocytes, the first polar body was found a day or two after ovulation (Evans and Cole, 1931), and Van Der Stricht (1923) concluded that sperm penetration can occur at any time during maturation. Fox oocytes are believed to reach metaphase II about one day after ovulation and remain at this stage until sperm penetration occurs (Pearson and Enders, 1943). Of six horse tubal oocytes recovered following induced ovulation, three were degenerated and three (collected 1, 3, and 7 days after ovulation) contained a large vesicular nucleus, presumably a germinal vesicle (Hamilton and Day, 1945).

In the insectivore *Ericulus,* found only in Madagascar, spermatozoa enter the ovary, pass into the follicle and there fertilize the oocyte that has already formed the first polar body. The ovum has two distinct nuclei (pronuclei) by the time ovulation occurs, and the first and subsequent cleavage divisions take place in the Fallopian tube (Strauss, 1939). In a later note, Strauss (1950) states that intra-follicular fertilization occurs in the sub-family *Centetinae* (which includes *Ericulus*), and that sperm occasionally penetrate immature eggs, inducing maturation and cleavage but not ovulation. Intra-follicular fertilization has also been reported to take place in the short-tailed shrew (Pearson, 1944).

VIII. SUMMARY

In most of the twenty-one mammals previously studied, oocytes mature, i.e. progress from late prophase (germinal vesicle stage) of the first meiotic division to metaphase of the second meiotic division, in ovarian follicles with ovulation occurring shortly after metaphase II formation. Although gonadotrophins, directly or indirectly, induce both maturation and ovulation, the two events are not inseparably linked. In oocytes of ten mammals, maturation can take place outside of the follicle in a suitable cell culture medium and, in the mouse, at least some of the resulting mature oocytes are capable of normal fertilization and development. Ovulation of immature stages from germinal vesicle to prometaphase II can occur, but it is doubtful whether spermatozoa can penetrate stages earlier than metaphase I. Maturation without ovulation may occur routinely as a result of sub-optimal gonadotrophic stimulation, and in one case some of the mature follicular oocytes proved capable of normal development after transfer to a foster mother. Dog, fox and possibly horse oocytes are normally ovulated in the germinal vesicle stage and maturation takes place in the Fallopian tube. Ovulation of pronuclear zygotes, a consequence of fertilization within the ovarian follicle, occurs in a group of insectivores from Madagascar.

New information presented here indicates that the majority of the first oocytes to ovulate in superovulated mice are immature and complete maturation in the Fallopian tube. Immature stages, mostly chromatin mass and prometaphase II, accounted for 83, 62, 30, 9 and 4% of the nondegenerated ova collected at 11, 11½, 12, 13, and 14 hours, respectively, after injection of HCG. It was also found that 79% of the oocytes liberated from the mouse ovary without regard to the estrus cycle were at the germinal vesicle stage while 21% were in various stages of degeneration with signs of maturation. Most of the latter were at metaphase II, suggesting that if an oocyte starts to mature, it is likely to complete maturation rather than arrest at an intermediate stage.

IX. ACKNOWLEDGMENTS

This research was supported by grants from the Ford Foundation and the National Institutes of Health (GM 15253).

X. REFERENCES

Allen, E., Pratt, J. P., Newell, Q. U., and Bland, L. J. (1930) Human ova from large follicles; including a search for maturation divisions and observations on atresia. *Amer. J. Anat.* **46**, 1.

Austin, C. R. and Braden, A. W. H. (1954) Anomalies in rat, mouse, and rabbit eggs. *Aust. J. Biol. Sci.* **7**, 537.

Baca, M. and Zamboni, L. (1967) The fine structure of human follicular oocytes. *J. Ultrastruct. Res.* **19**, 354.

Biggers, J. D., Whittingham, D. G., and Donahue, R. P. (1967) The pattern of energy metabolism in the mouse oocyte and zygote. *Proc. Nat. Acad. Sci. (U.S.A.).* **58**, 560.

Blandau, R. J. (1945) The first maturation division of the rat ovum. *Anat. Rec.* **92**, 449.

Boling, J. L., Blandau, R. J., Soderwall, A. L., and Young, W. C. (1941) Growth of the Graafian follicle and the time of ovulation in the albino rat. *Anat. Rec.* **79**, 313.

Braden, A. W. H. and Austin, C. R. (1954) Fertilization of the mouse egg and the effect of delayed coitus and of hot-shock treatment. *Aust. J. Biol. Sci.* **7**, 552.

Branca, A. (1925) L'ovocyte atresique et son involution. *Arch. Biol.* **35**, 325.

Breed, W. G. and Clarke, J. R. (1970) Ovulation and associated histological changes in the ovary following coitus in the vole *(Microtus agrestis). J. Reprod. Fertil.* **22**, 173.

Chang, M. C. (1955a) The maturation of rabbit oocytes in culture and their maturation, activation, fertilization, and subsequent development in the Fallopian tubes. *J. Exp. Zool.* **128**, 379.

Chang, M. C. (1955b) Fertilization and normal development of follicular oocytes in the rabbit. *Science.* **121**, 867.

Clewe, T. H., Yamate, A. M. and Noyes, R. W. (1958) Maturation of ova in mammalian ovaries in the anterior chamber of the eye. *Int. J. Fertil.* **3**, 187.

Coe, W. R. (1908) The maturation of the egg of the rat. *Science N.S.* **27**, 444.

Corner, G. W. (1917) Maturation of the ovum in swine. *Anat. Rec.* **13**, 109.

Cross, P. C. (1970) The fertilization and post-implantation development of oocytes matured *in vitro*. *In* "Program of the Society for the Study of Reproduction, 3rd annual meeting." Columbus, Ohio.

Dawson, A. B. and Friedgood, H. B. (1940) The time and sequence of preovulatory changes in the cat ovary after mating or mechanical stimulation of the cervis uteri. *Anat. Rec.* **76**, 411.

Dempsey, E. W. (1939) Maturation and cleavage figures in ovarian ova. *Anat. Rec.* **75**, 223.

Donahue, R. P. (1968) Maturation of the mouse oocyte *in vitro*. I. Sequence and timing of nuclear progression. *J. Exp. Zool.* **169**, 237.

Donahue, R. P. (1970) Maturation of the mouse oocyte *in vitro*. II. Anomalies of first polar body formation. *Cytogenetics.* **9**, 106.

Donahue, R. P. and Stern, S. (1968) Follicular cell support of oocyte maturation: production of pyruvate *in vitro*. *J. Reprod. Fertil.* **17**, 395.

Dziuk, P. J. (1965) Timing of maturation and fertilization of the sheep egg. *Anat. Rec.* **153**, 211.

Dziuk, P. J. and Dickmann, Z. (1965) Failure of the zona reaction in five pig eggs. *Nature (London).* **208**, 502.

Edwards, R. G. (1962) Meiosis in ovarian oocytes of adult mammals. *Nature (London).* **196**, 446.

Edwards, R. G. (1965a) Maturation *in vitro* of mouse, sheep, cow, pig, rhesus monkey and human ovarian oocytes. *Nature (London).* **208**, 349.

Edwards, R. G. (1965b) Maturation *in vitro* of human ovarian oocytes. *Lancet.* **2**, 926.

Edwards, R. G. (1968) The beginnings of human development. *In* "Proceedings of the Eighth International Planned Parenthood Federation." International Planned Parenthood Federation.

Edwards, R. G. and Gates, A. H. (1959) Timing of the stages of the maturation divisions, ovulation, fertilization, and the first cleavage of eggs of adult mice. *J. Endocr.* **18**, 292.

Edwards, R. G., Bavister, B. D. and Steptoe, P. C. (1969) Early stages of fertilization *in vitro* of human oocytes matured *in vitro*. *Nature (London).* **221**, 632.

Edwards, R. G., Donahue, R. P., Baramki, T. A. and Jones, H. W. (1966) Preliminary attempts to fertilize human oocytes matured *in vitro*. *Amer. J. Obstet. Gynecol.* **96**, 192.

Engle, E. T. (1927) A quantitative study of follicular atresia in the mouse. *Amer. J. Anat.* **39**, 187.

Evans, H. M. and Cole, H. H. (1931) An introduction to the study of the oestrous cycle in the dog. *Mem. Univ. Calif.* **9**, 65.

Friedgood, H. B. and Pincus, G. (1935) Studies on conditions of activity in endocrine organs. XXX. The nervous control of the anterior hypophysis as indicated by maturation of ova and ovulation after stimulation of cervical sympathetics. *Endocrinology.* **19**, 710.

Gothie, S. and Tsatsaris, B. (1939) Etudes de la premiere mitose de maturation dans des ovocytes de souris cultives in vitro (cytologie nucleaire et cinematographique). *Ann. Physiol. Physicochim. Biol.* **Pt. 2, 15,** 837.

Gresson, R. A. R. (1941) A study of the cytoplasmic inclusions during maturation, fertilization and the first cleavage division of the egg of the mouse. *Quart. J. Microscop. Sci.* **83**, 35.

Guthrie, M. J. and Jeffers, K. R. (1938) The ovaries of the bat *Myotis lucifugus lucifugus* after injection of hypophyseal extract. *Anat. Rec.* **72**, 11.

Hamilton, W. J. and Day. F. T. (1945) Cleavage stages of the ova of the horse, with notes on ovulation. *J. Anat.* **79**, 127.

Hancock, J. L. (1961) Fertilization in the pig. *J. Reprod. Fertil.* **2**, 307.

Hartman, C. G. and Corner, G. W. (1941) The first maturation division of the macaque ovum. *Contrib. Embryol. Carnegie Institution.* **29**, 1.

Heape, W. (1886) The development of the mole (*Talpa europea*), the ovarian ovum, and segmentation of the ovum. *Quart. J. Microscop. Sci.* **26**, 157.

Hoadley, L. and Simons, D. (1928) Maturation phases in human oocytes. *Amer. J. Anat.* **41**, 497.

Hunter, R. H. F. and Polge, C. (1966) Maturation of follicular oocytes in the pig after injection of human chorionic gonadotrophin. *J. Reprod. Fertil.* **12**, 525.

Jagiello, G. M. (1965) A method for meiotic preparations of mammalian ova. *Cytogenetics.* **4**, 245.

Jagiello, G. M. (1967) Streptonigrin: effect on the first meiotic metaphase of the mouse egg. *Science.* **157**, 453.

Jagiello, G. M. (1968) Action of phleomycin on the meiosis of the mouse ovum. *Mutat. Res.* **6**, 289.

Jagiello, G. M. (1969a) Meiosis and inhibition of ovulation in mouse eggs treated with actinomycin D. *J. Cell Biol.* **42**, 571.

Jagiello, G. M. (1969b) Inhibition of reduction in female mouse meiosis. *Experientia.* **25**, 695.

Jagiello, G. M. (1969c) Some cytologic aspects of meiosis in female guinea pig. *Chromosoma.* **27**, 95.

Jagiello, G. M. and Ohno, S. (1966) Isopycnotic behavior of the X-univalent in the XO mouse ovum. *Exp. Cell Res.* **41**, 459.

Jagiello, G. M., Karnicki, J. and Ryan, R. J. (1968) Superovulation with pituitary gonadotrophins. Method for obtaining meiotic metaphase figures in human ova. *Lancet.* **1**, 178.

Kennedy, J. F. and Donahue, R. P. (1969) Human oocytes: maturation in chemically defined medium. *Science.* **164**, 1292.

Kirkham, W. B. (1907a) The maturation of the mouse egg. *Biol. Bull.* **12**, 259.

Kirkham, W. B. (1907b) Maturation of the egg of the white mouse. *Trans. Conn. Acad.* **13**, 65.

Kirkham, W. B. and Burr, H. S. (1913) The breeding habits, maturation of eggs and ovulation of the albino rat. *Amer. J. Anat.* **15**, 291.

Kuhlmann, W. (1966) Cited by A. B. Griffen. *In* "Biology of the Laboratory Mouse." 2nd edition. (Earl L. Green, ed.), p. 51-85. New York, McGraw Hill.

Lams, H. (1913) Etude de l'oeuf de cobaye aux premiers stades de l'embryogenese. *Arch. Biol.* **28**, 229.

Lams, H. and Doorme, J. (1908) Nouvelles recherches sur la maturation et la fecondation de l'oeuf des mammiferes. *Arch. Biol.* **23**, 259.

Long, J. A. and Mark, E. L. (1911) The maturation of the egg of the mouse, Pub. No. 142, p. 72, Washington, D. C., Carnegie Institution of Washington.

Longley, W. H. (1911) The maturation of the egg and ovulation in the domestic cat. *Amer. J. Anat.* **12**, 139.

Makino, S. (1941) Studies on the murine chromosome. I. Cytological investigations of mice included in the genus *Mus. J. Fac. Sci. Hokkaido Univ.* **7**, 305.

Mandl, A. M. (1963) Pre-ovulatory changes in the oocyte of the adult rat. *Proc. Roy. Soc. [Biol.]* **158**, 105.

Mastroianni, L. and Noriega, C. (1970) Observations on human ova and the fertilization process. *Amer. J. Obstet. Gynec.* **107**, 682.

McGaughey, R. W. and Chang, M. C. (1969) Meiosis of mouse eggs before and after sperm penetration. *J. Exp. Zool.* **170**, 397.

Moore, J. E. S. and Tozer, F. (1908) On the maturation of the ovum in the guinea-pig. *Proc. Roy. Soc. [Biol.]* **80**, 285.

Moricard, R. and De Fonbrune, P. (1937) Nouvelles etudes experimentales sur les mecanismes de la formation du premier globule *in vitro* chez les mammiferes. *Arch. Anat. Microscop.* **33**, 113.

Moricard, R. and Gothie, S. (1953) Hormonal mechanisms of the first polar body formation in the follicle. *In* "Mammalian Germ Cells." (G. E. W. Wolstenholme, ed.), Boston, Little Brown.

Nakamura, T. (1957) Cytological studies on abnormal ova in mature ovaries of mice observed at different phases of oestrus cycle. *J. Fac. Fish. Anim. Husb. Hiroshima Univ.* **1**, 343.

Newman, H. H. (1912) The ovum of the nine-banded armadillo. Growth of the ovocytes, maturation and fertilization. *Biol. Bull.* **23**, 100.

Noyes, R. W. (1952) Fertilization of follicular ova. *Fertil. Steril.* **3**, 1.

Odor, D. L. (1955) The temporal relationship of the first maturation division of rat ova to the onset of heat. *Amer. J. Anat.* **97**, 461.

Odor, D. L. (1960) Electron microscopic studies on ovarian oocytes and unfertilized tubal ova in the rat. *J. Biophys. Biochem. Cytol.* **7**, 567.

Odor, D. L. and Renniger, D. F. (1960) Polar body formation in the rat oocyte as observed with the electron microscope. *Anat. Rec.* **137**, 13.

Pearson, O. P. (1944) Reproduction in the shrew (*Blarina brevicorda* Say). *Amer. J. Anat.* **75**, 39.

Pearson, O. P. and Enders, R. K. (1943) Ovulation, maturation, and fertilization in the fox. *Anat. Rec.* **85**, 69.

Pincus, G. (1936) The Eggs of Mammals. New York, Macmillan.

Pincus, G. and Enzmann, E. V. (1935) The comparative behavior of mammalian eggs *in vivo* and *in vitro*. I. The activation of ovarian eggs. *J. Exp. Med.* **62**, 665.

Pincus, G. and Enzmann, E. V. (1937) The growth, maturation and atresia of ovarian eggs in the rabbit. *J. Morph.* **61**, 351.

Pincus, G. and Saunders, B. (1939) The comparative behavior of mammalian eggs *in vivo* and *in vitro*. VI. The maturation of human ovarian ova. *Anat. Rec.* **75**, 537.

Polge, C. and Dziuk, P. (1965) Recovery of immature eggs penetrated by spermatozoa following induced ovulation in the pig. *J. Reprod. Fertil.* **9**, 357.

Rein, G. (1883) Beitrage zur Kenntniss der Reifungserscheinungen und Befruchtungsvorgange am Saugethierei. *Archiv. Mikroskop. Anat.* **22**, 233.

Robertson, J. E. and Baker, R. D. (1969) Role of female sex steroids as possible regulators of oocyte maturation. (Abstract). "Program of the Society for the Study of Reproduction, 2nd annual meeting." Davis, California.

Robinson, A. (1918) The formation, rupture, and closure of ovarian follicles in ferrets and ferret-polecat hybrids, and some associated phenomena. *Trans. Roy. Soc. Edinb.* **52**, 303.

Rubaschkin, W. (1905) Uber die Reifungs und Befruchtungsprozesse des Meerschweincheneies. *Anat. Hefte.* **29**, 507.

Runner, M. N. (1950) Induced ovulations in immature mice as a source of material for studies on mammalian eggs. *Anat. Rec.* **106**, 313.

Sobotta, J. (1895) Die Befruchtung und Furchung des Eies der Maus. *Arch. Mikroskop. Anat.* **45**, 15.

Sobotta, J. (1907) Die Bildung der Richtungskorper bei der Maus. *Anat. Hefte.* **35**, 493.

Sobotta, J. and Burckhard, G. (1910) Reifund und Befruchtung des Eies der weissen Ratte. *Anat. Hefte.* **42**, 433.

Spaulding, J. F., Berry, R. O. and Moffit, J. G. (1955) The maturation process of the ovum of swine during normal and induced ovulations. *J. Anim. Sci.* **14**, 609.

Steptoe, P. C. and Edwards, R. G. (1970) Laparoscopic recovery of pre-ovulatory human oocytes after priming of ovaries with gonadotrophins. *Lancet.* **1**, 683.

Strauss, F. (1939) Die Befruchtung und der Vorgang der Ovulation bei *Ericulus* aus der Familie der Centetiden. *Biomorphosis.* **1**, 281.

Strauss, F. (1950) Ripe follicles without antra and fertilization within the follicle; a normal situation in a mammal. *Anat. Rec.* **106**, 251. (abstract).

Suzuki, S. and Iizuka, R. (1970) Maturation of human ovarian follicular oocytes *in vitro.* *Experientia.* **26**, 640.

Suzuki, S. and Mastroianni, L. (1966) Maturation of monkey ovarian follicular oocytes *in vitro.* *Amer. J. Obstet. Gynec.* **96**, 723.

Talbert, A. J. (1969) Recursive multiple regression with a polychotomous response variable: An application to stagewise biological development data. M.S. (Biostatistics) thesis. Johns Hopkins University.

Thomson, A. (1919) The maturation of the human ovum. *J. Anat.* **53**, 172.

Van Beneden, E. (1875) La maturation de l'oeuf, la fecondation et les premieres phases du developpement embryonnaire des mammiferes d'apres les recherches faites chez le lapin. *Bull. Acad. Roy. Med. Belg.* **40**, (2nd Series) 686.

Van Beneden, E. and Julin, C. (1880) Observations sur la maturation, la fecondation et la segmentation de l'oeuf chez les cheiropteres. *Arch. Biol.* **1**, 551.

Van der Stricht, O. (1906) Mitoses de maturation de l'oeuf de chauvesouris *(V. noctula). Compt. Rend. Assoc. Anat. Huitieme reunion.* 51.

Van der Stricht, O. (1908) La structure de l'oeuf de chienne et la genese du jaune corps. *Compt. Rend. Assoc. Anat. Dixieme reunion.* 1.

Van der Stricht, O. (1923) Etude comparee des ovules de mammiferes aux differentex periodes de l'ovogenese d'apres les travaux du laboratore d'histologie et d'embryologie de l'universite de Gand. *Arch. Biol.* **33**, 229.

Van der Stricht, R. (1911) Vitellogenese dans l'ovule de chatte. *Arch. Biol.* **26**, 365.

Ward, M. C. (1948) The maturation divisions of the ova of the golden hamster *Cricetus auratus. Anat. Rec.* **101**, 663.

Zamboni, L. (1970) Ultrastructure of mammalian oocytes and ova. *Biol. Reprod. Suppl.* **2**, 44.

Zamboni, L. and Mastroianni, J. R. (1966) Electron microscopic studies on rabbit ova. I. The follicular oocyte. *J. Ultrastruct. Res.* **14**, 95.

23

MATURATION AND FERTILIZATION OF HUMAN OOCYTES *IN VITRO*

Joseph F. Kennedy

I.   Introduction
II.   Review of literature
        A.   Maturation *in vitro*
        B.   Maturation *in vivo*
        C.   Fertilization *in vitro*
                1.   After maturation *in vitro*
                2.   After maturation *in vivo*
III.   Recovery of human oocytes for culture
IV.   Discussion
V.   Acknowledgments
VI.   References

I. INTRODUCTION

Very little is known about the earliest stages of human development mainly because of the lack of sufficient material for study and inadequate methodology. The only stage at which human ova are available in sufficient numbers for proper study is the germinal vesicle stage found in follicular oocytes. Only a few oocytes recovered from follicles will have resumed meiosis (Allen, Pratt, Newell and Bland, 1930; Kennedy and Donahue, 1969b). In order to study oocytes in the first or second meiotic divisions they must be obtained during maturation *in vitro,* after gonadotrophin induced superovulation or from the ovaries of women after gonadotrophin stimulation.

II. REVIEW OF LITERATURE

A. Maturation *in Vitro*

Oocyte maturation *in vitro* has been studied in a number of mammals including

Table 23.1 **Maturation of Human Oocytes** *In Vitro*

Author	Hrs.	No. Oocytes	G.V./Deg.	Promet. I	Met. I	Telo. I	Met. II with or without P.B.
Pincus and Saunders (1939)	0	34	34				
	8½ - 72	110					27 (16)
Edwards (1965 a & b)	0	43	42				1
	25	19	8	10	1		
	27 - 28	6	2	3	1		
	28½ - 36	23	2	1	18	1	1
	40 - 43	12	5		2		5
	48 - 70	85	25		4		56
Jagiello et al. (1968)*	0	25	17		8		
	25 - 28	30			8		22
Kennedy and Donahue (1969)	0	69	69				
	40 - 43	466	279		71		116
Mastroianni Noriega (1970)	0	7	7				
	24	4	4				
	46 - 50	68	27		18		23
	72	4	2				2
Steptoe and Edwards (1970)*	0	87	55	12	18		9
	3 - 4	9					9

G.V./Deg.: Germinal vesicle or degenerating.
Promet. I, Met. I, Telo. I: Prometaphase, Metaphase and Telophase of First Meiotic Division
Met. II: Metaphase of second Meiotic Division
P. B.: First Polar Body
( ): Number of meiotic spindles
*: Oocytes recovered following gonadotrophic injections

man (Table 23.1) since Pincus and Enzman (1935) first observed nuclear maturation of rabbit oocytes cultured *in vitro*. Pincus and Saunders (1939) obtained 27 oocytes with polar bodies and 16 with meiotic spindles from 144 cultured human oocytes. Oocytes were cultured in human serum for a period of 8.5-72 hours, and 30 - 50% of oocytes were activated as evidenced by finding spindles or polar bodies. Exposure of oocytes to cytolyzed sperm at 46°C or 1.8% NaCl for brief periods of time did not affect the number of oocytes resuming maturation. Edwards (1965b) obtained 80% resumption of maturation in oocytes cultured in medium 199 with added fetal calf serum. He observed that germinal vesicle breakdown does not occur until after approximately 25 hours in culture, metaphase I between 28 and 35 hours, and metaphase II between 36 and 43 hours. The stage of the menstrual cycle, pregnancy, and various clinical states of patients from whose ovaries the oocytes were recovered had no effect on maturation *in vitro*. Yuncken (1968), using Edwards' methods, was able to confirm this sequence and timing and obtained good metaphase spreads for chromosome study. Kennedy and Donahue (1969b) (Table 23.2) matured human oocytes in various media and between 50% and 63% of oocytes matured *in vitro* in defined media containing pyruvate and bovine serum albumin; these results were similar to the results in media supplemented with fetal calf serum. Of 466 oocytes cultured, 71 went to metaphase I and 116 to metaphase II. Some media supported maturation poorly or not at all. When follicle cells were removed prior to culture, fewer oocytes matured. They found that the energy requirements for meiosis of human oocytes *in vitro* may be similar to that of the mouse, with pyruvate being the key nutrient. Mastroianni and Noriega (1970) studied 83 human oocytes and obtained results similar to those of Kennedy and Donahue when F10 medium containing human serum was used as the culture medium. Many oocytes remained at the germinal vesicle stage or degenerated.

Chandley (1971) cultured human oocytes in various media for 22-24 hours. Seventy-nine out of 267 oocytes reached diakinesis, metaphase I or metaphase II in culture but she was unable to confirm Edwards' timing, finding just as many oocytes at metaphase II at 19 hours as at 26, and conversely just as many in the dictyate stage at 26 hours as at 19. She found wide variation between ovaries in the percentage of oocytes resuming maturation but the age of the patient and the indication for surgery appeared to have no effect on oocyte maturation in culture.

B. Maturation *in Vivo*

Jagiello, Karnicki and Ryan (1968) collected oocytes from ovaries of women who were given pituitary follicle-stimulating hormone (FSH) and human chorionic gonadotrophin (HCG). Twenty hours after HCG a third of the oocytes were at metaphase I, the remainder being at the germinal vesicle stage. When these oocytes were then cultured for 25-28 hours, all resumed meiosis with 66% reaching metaphase II. Excellent chromosome spreads were obtained of both first and second meiotic metaphases either immediately after release from the follicle or after 24 hours incubation. These spreads compared favorably to chromosomes obtained when oocytes were cultured directly from the germinal vesicle stage without

Table 23.2 **Maturation of Human Oocytes in Defined and Undefined Media With and Without Attached Cumulus Oophorus**

No. of Patients	Medium	Cumulus	No. Oocytes Cultured*	Metaphase I No. (%)	Metaphase II No. (%)	Total (I and II) No. (%)
21	F10	Attached	156	26 (16.7)	52 (33.2)	78 (50.0)
10	F10	Removed	72	9 (12.4)	4 (5.6)	13 (18.1)
2	Whitten's - Biggers'	Attached	13	4 (30.8)	2 (15.4)	6 (46.2)
4	0.25 mM Pyruvate	Attached	28	1 (3.6)	2 (7.1)	3 (10.7)
1	1.0 mM Pyruvate +	Attached	25	4 (16.0)	0 (0)	4 (16.0)
		Removed	22	8 (36.3)	4 (18.2)	12 (54.5)
6	199 with‡	Attached	52	7 (13.5)	21 (38.5)	28 (53.8)

* Includes those reported in Kennedy & Donahue (1969b)

+ In Krebs-Ringer salt solution modified to 14.28 mM NaHCO$_3$ and 128.99 mM NaCl

‡ The only undefined medium

stimulation with FSH or HCG.

Steptoe and Edwards (1970) recovered pre-ovulatory human oocytes by laparoscopy 29-31 hours after HCG injection in women primed with human menopausal gonadotrophin (HMG). One hundred thirty-three oocytes were recovered; of 97 oocytes classified microscopically, 55 were at the germinal vesicle stage or degenerate; 18 were at metaphase I; 3 were at metaphase II. Twelve oocytes were in diakenesis or prometaphase and 9 reached metaphase II after culture. Three injections of 225 I.U. of HMG between days 2 and 9 and 5,000 I.U. of HCG on days 9-11 gave the best response but only a third of aspirated follicles yielded oocytes.

C. Fertilization *in Vitro*

Considerable doubt must be cast upon some reports of successful *in vitro* fertilization of human oocytes (Table 23.3). Many experiments suffer from a lack of uniform methodology, inadequate evidence of fertilization, and possible parthenogenesis or anomalous forms of maturation or "fertilization". Although the criteria for identifying fertilization may appear quite precise, such as two polar bodies, two pronuclei, and a sperm tail in the perivitelline space, these might well occur through mechanisms other than normal fertilization.

1. After maturation in vitro. Menkin and Rock (1944, 1948) were the first to report fertilization of human oocytes *in vitro.* Of over 800 oocytes studied, they attempted fertilization of 138 and obtained 2 two-celled and 2 three-celled "embryos", one of which was grossly abnormal. It is difficult to evaluate these results since oocytes were cultured for only 27 hours or less before being placed with washed spermatozoa, and the stage of maturation was not known. According to Edwards' (1965b) timing, these oocytes would have just begun to undergo germinal vesicle breakdown at the time of insemination.

Of over 1,000 oocytes studied by Shettles (1953-1962), approximately 200 were subjected to insemination with 6 reported to have undergone cleavage, one all the way to morula. Again, the stage of maturation at the time of fertilization was not known and many oocytes were inseminated while still in the germinal vesicle stage.

The work of Petrucci (1961) is difficult to interpret since the stage of oocyte maturation was not reported, the photographs did not easily reveal what was said to be shown, and the incidence of successful culture, fertilization or parthenogenesis was not given.

Hayashi (1963) observed fertilization of more than 30% of the eggs he studied *in vitro,* stressing the importance of hormones in the media and cumulus around the oocytes, but said little about capacitation, maturation or parthenogenesis. Penetration of spermatozoa, formation of pronuclei and cleavage to the morula stage were observed but normal cleavage was indistinguishable from that due to

Table 23.3 **Fertilization and Cleavage of Human Oocytes** *in Vitro*

Author	No. oocytes	Medium	Non-ovu-latory	Unf.	F.	Others	1-celled	Pro-nu-cleate & cleaved	Cleaved, Pro-nuclei not seen	Sperm	Stage of oocytes
Menkin & Rock (1948)	138 (800)	Human serum							4	W	Unm.
Shettles (1953-1962)	200* (1000)	Foll. fl. serum							6*	Unw.	Unm. (?)
Petrucci (1961)	?	Various		?					?	Unw.	Unm. (?)
Hayashi (1963)	70 (160)	Various			?				20	Unw.	Unm.
Edwards, Bavister & Steptoe (1969)	70	Bavister, Foll. fl.	26	20	13†			11+		W	Matured *in vitro* (?)
Edwards, Steptoe & Purdy (1970)	393	Various Foll. fl.	65	58	12†	23	150	45	40	W	Matured *in vitro/in vivo(?)* ("Preovular")

*Estimated from several papers
( ) Numbers in parenthesis are number of oocytes studied
+ Pronucleate eggs removed from culture
† Spermatozoa in zona, perivitelline space or vitellus

Foll. fl. = Follicular fluid
F. = Fertilized
Unf. = Unfertilized
W = Washed
Unw. = Unwashed
Unm. = Unmatured

parthenogenesis.

In two reports, Edwards, Bavister and Steptoe (1969) observed eleven pronucleate eggs, five of which had two pronuclei, the remainder up to as many as five. They cultured follicular oocytes for 36 hours, then inseminated the culture with freshly ejaculated, washed spermatozoa and examined oocytes at various times thereafter for the presence of pronuclei. No pronuclei were seen in control eggs. They emphasized that extra bicarbonate and follicular fluid was added to the medium, and that its pH was raised above that previously used for *in vitro* fertilization of human oocytes.

Seitz, Rocha, Brackett and Mastroianni (1970) studied 16 human oocytes cultured for 24 hours in Ham's F10 containing serum. Oocytes were inseminated with washed spermatozoa which were placed in the uterine cavity of mid-cycle rhesus monkeys. Six ova had reached the 2- to 8-cell blastomere stage after various intervals in culture following insemination.

2. After maturation in vivo. Based upon Jagiello's and their own experience in stimulating meiosis *in vivo* with HMG and HCG, Edwards, Steptoe and Purdy (1970) obtained 393 oocytes at laparoscopy 30-32 hours after HCG injection. A high pH and added bicarbonate were used in the culture media, based upon the earlier work of Bavister, Edwards and Steptoe (1969), and Edwards *et al.* (1969). After culture for 1-4 hours, oocytes were inseminated and examined periodically for evidence of fertilization or cleavage. They speculated that follicular fluid or granulosa cells may have been important in achieving fertilization through an effect on the sperm acrosome by steroids in the follicular fluid or synthesized by the granulosa cells. Ham's F10 modified in this manner gave the best results, but because of the small numbers no significant comparisons between media were possible. Forty-five pronucleate ova underwent subsequent cleavage.

III. RECOVERY OF HUMAN OOCYTES FOR CULTURE

We have recovered over 1500 oocytes from more than 100 patients. Most of the oocytes were cultured for experiments on maturation *in vitro* (Figs. 23.1-23.12). Some were used for fertilization *in vitro* (Figs. 23.13, 23.14). Of 1113 oocytes recovered from 78 patients (97 ovaries) (Tables 23.4, 23.5), the mean number of oocytes per ovary was 11.5 and patients ranged in age from 16 to 46 years. Eleven patients were pregnant. Thirty-eight had a proliferative endometrium, 29 were in the secretory phase. The number of oocytes obtained varied with age from a low of 5 in the 41-45 years group to 19 in the 21-25 years group. A few more oocytes per ovary were obtained from the pregnant patients' ovaries than from non-pregnant patients in all age groups.

The number of oocytes recoverable also varied with the stage of the menstrual cycle, with a peak of 13.5 oocytes per ovary occurring between 13 and 16 days after the last menstrual period (LMP) and a three-fold increase from this peak 33 or more

Table 23.4 Numbers of Follicular Oocytes Recovered from Human Ovaries According to the Stage of Menstrual Cycle, Endometrial Histology and Age

Age	Proliferative Endometrium and 1-14 days from L.M.P.		Secretory Endometrium and 15 days from L.M.P.		Pregnant		Proliferative Endometrium and 14 days from L.M.P.	
	No. of Ovaries	Mean No. of Oocytes	No. of Ovaries	Mean No. of Oocytes	No. of Ovaries	Mean No. of Oocytes	No. of Ovaries	Mean No. of Oocytes
16 - 20	22	15.0±	0	0	2	22.5±	0	0
21 - 25	1	8.0±	4	22.2±				
26 - 30	7	8.1±	12	11.1±	4	14.7±	8	33.4±
31 - 35	5	10.2±	9	9.9±	3	15.7±	1	9.0±
36 - 40	7	6.1±	9	6.6±	2	6.5±	0	0
41 - 45	11	6.2±	3	2.3±	0	0	0	0
46 +	0	0	1	12.0±	0	0	0	0

Table 23.5 **Number of Oocytes Per Ovary According to Age and Parity**

	Parity		
Age	0	1 - 5	6+
16 - 20	15		
21 - 25		19.4	
26 - 30	11.6	20.9	12.3
31 - 35	3.0	9.3	14.3
36 - 40	4.0	6.8	
41 - 45	2.3	1.8	15.3
46+	12.0		

days after the LMP, excluding pregnant patients. The largest number of oocytes, 123, was recovered from two ovaries of a 27-year-old (para five) patient with the Chiari-Frommel syndrome who had last menstruated two years before. This patient and several with anovulatory cycles probably account for the larger number of oocytes recovered per ovary from patients more than 32 days since their LMP. When the stage of the menstrual cycle was correlated with the stage of the endometrium and age of the patient (Table 23.4), a few more oocytes were obtained from patients with secretory endometrium and more than 14 days from the LMP in the age groups 26-30 and 31-35. Differences between other age groups were not significant or there were too few patients to make a comparison.

Table 23.5 shows the number of oocytes recovered according to age and parity. The largest number of oocytes was obtained from patients with a parity of 1 - 5 in the age group 21-30 with fewer oocytes generally being obtained in the nulliparous women. Twelve oocytes were recovered from a 47-year-old nulliparous patient whose ovary had many visible follicles. Usually few or no oocytes were obtained from women over age 44 and attempts to recover oocytes from these patients was abandoned later in this study if patients were post-menopausal or the ovary had no visible follicles.

The success or failure of oocytes to mature *in vitro* in the various media could not be correlated with age, parity, endometrial histology, time since LMP, clinical indications for surgery, or ovarian pathology.

IV. DISCUSSION

Although human oocytes, when released from the follicle and incubated in appropriate media, will resume meiotic maturation, a finely integrated system of follicle, oocyte, follicular fluid and follicle cells is thereby disrupted. Many of the cytologic events subsequently observed in culture thereby may be artifactual or degenerative. In fact, numerous oocytes do undergo obvious degeneration and in addition we have observed a number of anomalous forms in culture. Structures such as binuclear oocytes (Kennedy and Donahue, 1969a), large polar bodies, tetraploid number of chromosomes in the oocyte, extrusion of most of the oocyte chromatin into two polar bodies, an interphase-like nucleus instead of metaphase II chromosomes with a polar body, and fragmentation mimicking parthenogenesis have been observed (Kennedy and Donahue, unpublished).With ordinary light and phase contrast microscopy or even with stained specimens, oocytes such as these may be mistakenly identified as fertilized or cleaved. Whether these abnormal cytologic events are representative of those occurring *in vivo* is unknown, but oocytes such as these could account for the 60% abnormal eggs observed by Hertig, Rock, Adams and Menkin (1959) out of a total of 34 fertilized human eggs.

Recovery of pre-ovulatory oocytes by laparoscopy from women who have received HMG and HCG, as has been done extensively by Steptoe and Edwards produces more physiological material as evidenced by their apparent greater success in *in vitro* fertilization of these oocytes. In addition, the demonstration that mouse oocytes matured and fertilized *in vitro* and transferred to foster mothers developed (3.2%) into 15-day-old fetuses, indicates that normal oocyte maturation can occur *in vitro* (Cross and Brinster, 1971). However, the parthenogenetic development of mouse embryos by stimulating the ampulla of the mouse oviduct with an electrical current (Tarkowski, Witkowska and Nowicka, 1970) and mouse blastocysts by shaking hyaluronidase crystals over superovulated mouse eggs (Graham, 1970) leaves the issue in doubt.

Certainly it must be premature to consider implanting human blastocysts obtained by *in vitro* culture into the uterus of a woman regardless of how normal the embryo may appear. Many experiments must be done to evaluate the nutrition of the maturing oocyte and the culture requirements for fertilization and cleavage. Many blastocysts obtained in this way must be shown to be at least chromosomally normal and there must be a compelling medical need before the development of a newborn human, with the attendant risk of monstrous consequences, can be contemplated in this manner.

V. ACKNOWLEDGMENTS

The author thanks Dr. Roger P. Donahue for his assistance in performing this research and Drs. John D. Biggers and Allen C. Barnes for their advice and support. This work was supported by a grant from the Ford Foundation and by grants GM 10189, 5-T01-HD 00109-04 and 5-T01-HD 00023-07 from the National Institutes of Health.

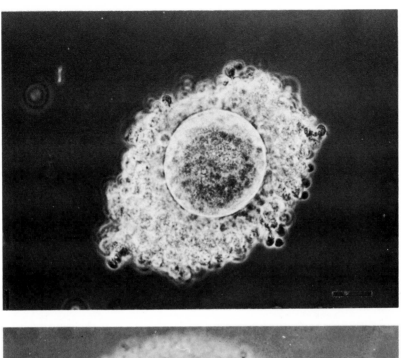

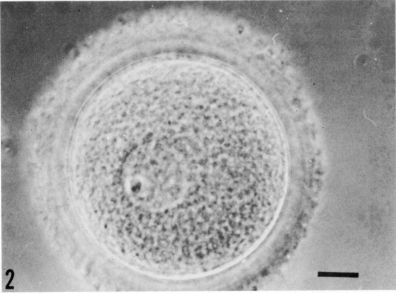

Fig. 23.1   Living human oocyte upon liberation from a large ovarian follicle. The oocyte is surrounded by the dense mass of the cumulus oophorus (follicle cells). The surface of the zona is visible but no cytoplasmic or nuclear detail can be seen. Scale: 40 μm.

Fig. 23.2   Living human oocyte upon liberation from a large ovarian follicle. The surface of the zona pellucida is irregular from mechanical removal of the cumulus. The oocyte typically fills the entire space enclosed by the zona. The nucleus is eccentrically located in the granular cytoplasm as is the single nucleolus in the less granular nucleoplasm. Characteristic germinal vesicle (dictyate) nuclear stage. Scale: 20 μm.

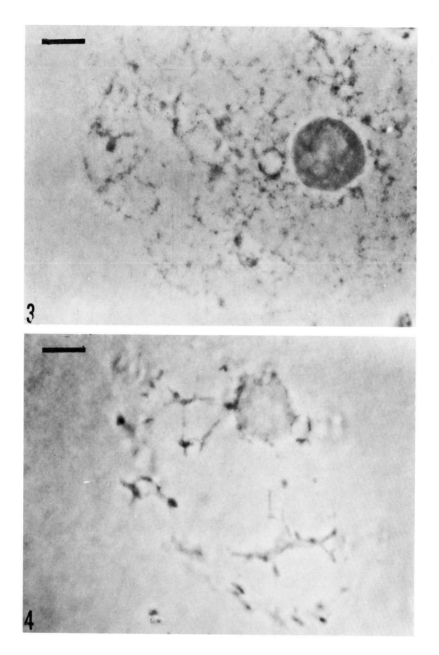

Fig. 23.3 A germinal vesicle, fixed in acetic acid-alcohol, stained with orcein and squashed, immediately after removal from an ovarian follicle. The chromatin is dispersed in a lattice-work of fine filaments. The dark staining nucleolus is sharply defined but the nuclear membrane is not visualized with this technique. Scale: 10 μm.

Fig. 23.4 Nucleus of human oocyte cultured for 43 hours. Arrested at early stage of germinal vesicle breakdown. The thicker and branched chromatin filaments are condensed about the nuclear and nucleolar periphery. Scale: 10 μm.

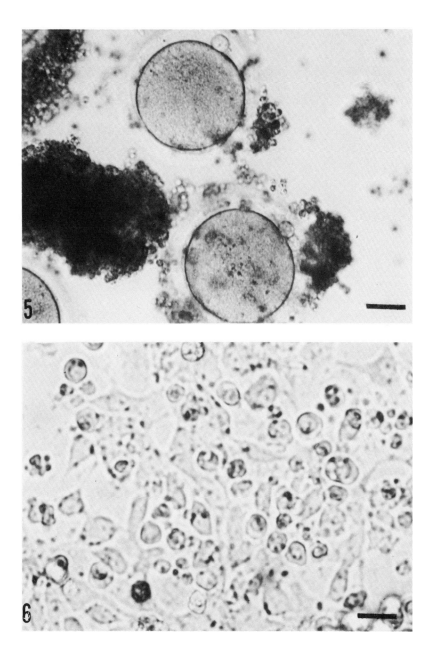

Fig. 23.5 Living human oocytes after culture with intact cumulus in medium with pyruvate for 48 hours. In two oocytes the first polar body has formed and is clearly visible in the perivitelline space. Most of the cumulus has been removed to show the polar bodies. Scale: 40 μm.

Fig. 23.6 Cells of the cumulus growing on the surface of the culture dish from oocytes cultured in medium 199 with serum (48 hours culture). The stellate and spindle shaped forms are more characteristic of cultures in medium containing serum than of serum free medium. Scale: 20 μm.

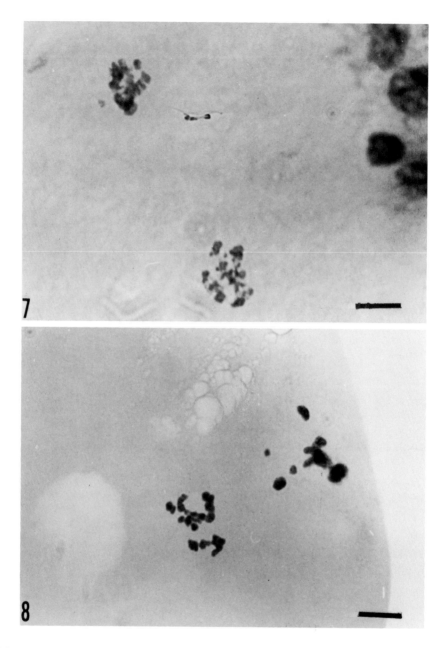

Fig. 23.7 Human oocyte after 45 hours culture in medium 199 with serum with two chromosome groups at telophase of the first meiotic division. The midbody of the spindle is visible near the upper group. Some adherent cumulus cells are in the upper right. Scale: 10 μm.

Fig. 23.8 A similar human oocyte but the polar body chromatin on the right shows early signs of degenerative clumping. The vacuole-like structures are artifacts of squashing. Scale: 10 μm.

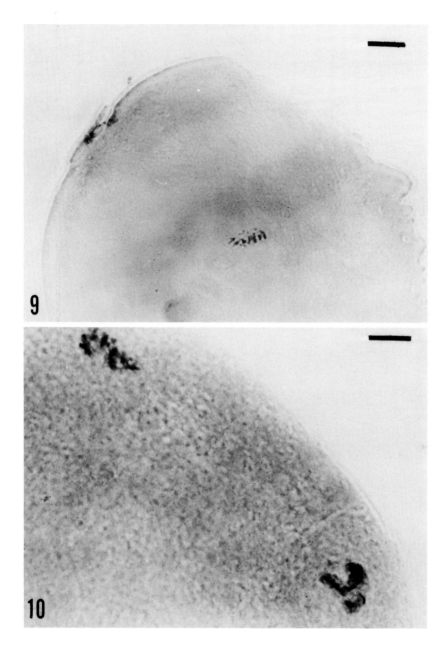

Fig. 23.9 Human oocyte showing the first polar body and a deeply placed second metaphase group of chromosomes. The central location of the metaphase chromosomes could cause equal division of the secondary oocyte with a large second polar body resulting. Scale: 20 μm.

Fig. 23.10 A more typical second metaphase - first polar body human oocyte with clumping of the polar body chromatin and a peripheral location of the second metaphase chromosomes. Scale: 10 μm.

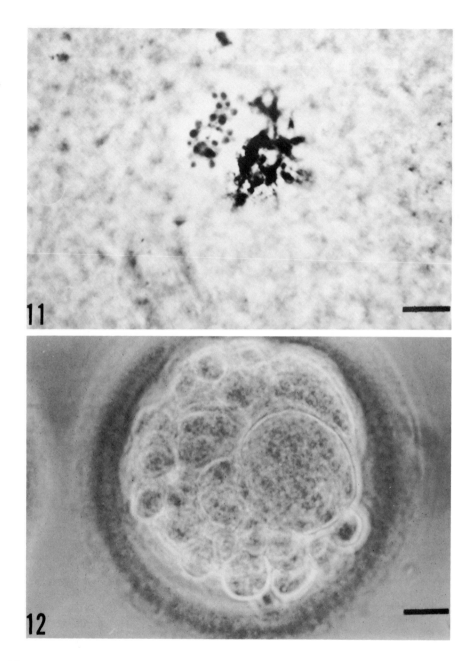

Fig. 23.11 Typical degenerating germinal vesicle found in human oocytes that fail to resume maturation in culture. Scale: 10 μm.

Fig. 23.12 A fragmenting human oocyte after 60 hours of culture. The inequality of the cytoplasmic fragments distinguishes it from normal cleavage. Scale: 10 μm.

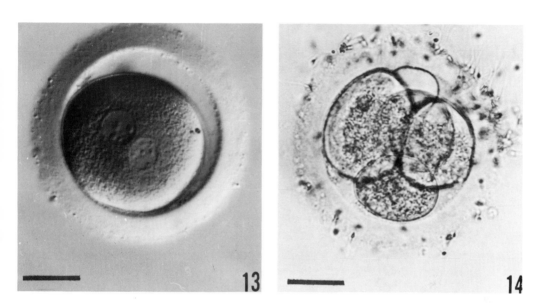

Fig. 23.13 A living human oocyte at the pronuclear stage, matured *in vitro* and inseminated after 36 hours culture and incubated for an additional 54 hours in medium F10. A polar body with two adjacent smaller cytoplasmic fragments is out of the plane of the photograph. Interference contrast (Nomarski) microscopy. Scale: 40 μm.

Fig. 23.14 A living human four-cell egg recovered from the same culture as the oocyte in Fig. 23.13. Numerous spermatozoa are adherent to or have partially penetrated the zona pellucida. Scale: 40 μm.

VI. REFERENCES

Allen, E., Pratt, J. P., Newell, Q. U. and Bland, L. J. (1930) Human ova from large follicles, including a search for maturation divisions and observations on atresia. *Am. J. Anat.* **46**, 1.

Bavister, B., Edwards, R. G. and Steptoe, P. C. (1969) Identification of the midpiece and tail of spermatozoon during fertilization of human eggs *in vitro. J. Reprod. Fert.* **20**, 159.

Chandley, A. C. (1971) Culture of mammalian oocytes. *J. Reprod. Fert.,* Suppl. **14**, 1.

Cross, P. C. and Brinster, R. L. (1971) *In vitro* development of mouse oocytes. *Biol. Reprod.* **3**, 298.

Edwards, R. G. (1965a) Maturation *in vitro* of mouse, sheep, cow, pig, rhesus monkey and human ovarian oocytes. *Nature (London).* **208**, 349.

Edwards, R. G. (1965b) Maturation *in vitro* of human ovarian oocytes. *Lancet.* **2**, 926.

Edwards, R. G., Bavister, B. D. and Steptoe, P. C. (1969) Early stages of fertilization *in vitro* of human oocytes matured *in vitro. Nature (London).* **221**, 632.

Edwards, R. G., Steptoe, P. C. and Purdy, J. M. (1970) Fertilization and cleavage *in vitro* of preovular human oocytes. *Nature (London).* **227**, 1307.

Graham, C. F. (1970) Parthenogenetic mouse blastocysts. *Nature (London).* **226**, 165.

Hayashi, M. (1963) Fertilization *in vitro* using human ova. *In* "Proceedings of the VIIth International Conference of the International Planned Parenthood Federation, Singapore", p. 505.

Hertig, A. T., Rock, J., Adams, E. C. and Menkin, M. C. (1959) Thirty-four fertilized human ova, good, bad and indifferent, recovered from 210 women of known fertility. A study of biologic wastage in early human pregnancy. *Pediatrics.* **23**, 202.

Jagiello, G., Karnicki, J. and Ryan, R. J. (1968) Superovulation with pituitary gonadotrophins. Method for obtaining meiotic metaphase figures in human ova. *Lancet.* **1**, 178.

Kennedy, J. F. and Donahue, R. P. (1969a) Binucleate human oocytes from large follicles. *Lancet.* **1**, 754.

Kennedy, J. F. and Donahue, R. P. (1969b) Human oocytes: maturation in chemically defined medium. *Science.* **164**, 1292.

Mastroianni, L. and Noriega, C. (1970) Observations on human ova and the fertilization process. *Am. J. Obst. Gynec.* **107**, 682.

Menkin, M. F. and Rock, J. (1948) *In vitro* fertilization and cleavage of human ovarian eggs. *Am. J. Obst. Gynec.* **55**, 440.

Petrucci, D. (1961) Producing transplantable human tissue in the laboratory *Discovery.* **22**, 278.

Pincus, G. and Enzman, E. V. (1935) The comparative behavior of mammalian eggs *in vivo* and *in vitro*. I. The activation of ovarian eggs. *J. Exp. Med.* **62**, 665.

Pincus, G. and Saunders, B. (1939) The comparative behavior of mammalian eggs *in vivo* and *in vitro*. VI. The maturation of human ovarian ova. *Anat. Rec.* **75**, 537.

Rock, J. and Menkin, M. F. (1944) *In vitro* fertilization and cleavage of human ovarian eggs. *Science.* **100**, 105.

Seitz, H. M., Rocha, G., Brackett, B. G. and Mastroianni, L. (1971) Cleavage of human ova *in vitro*. *Fertil. Steril.* **22**, 255.

Shettles, L. B. (1953) Observations on human follicular and tubal ova. *Am. J. Obst. Gynec.* **66**, 235.

Shettles, L. B. (1954) Studies on living human ova. *Trans. N. Y. Acad. Sci.* Series II, **17**, 99.

Shettles, L. B. (1955) Further observations on living human oocytes and ova. *Am. J. Obst. Gynec.* **69**, 365.

Shettles, L. B. (1958) The living ovum. *Am. J. Obst. Gynec.* **76**, 398.

Shettles, L. B. (1962) Human fertilization. *Obst. Gynec.* **20**, 750.

Steptoe, P. C. and Edwards, R. G. (1970) Laparoscopic recovery of preovulatory human oocytes after priming of ovaries with gonadotrophins. *Lancet.* **1**, 683.

Tarkowski, A. K., Witkowska, A. and Nowicka, J. (1970) Experimental parthenogenesis in the mouse. *Nature (London).* **226**, 162.

Yuncken, C. (1968) Meiosis in the human female. *Cytogenetics.* **7**, 234.

24

ADENINE DERIVATIVES AND OOCYTE MATURATION IN STARFISHES

Haruo Kanatani

I.   Introduction
II.   Production of meiosis-inducing substance under the influence of gonad-stimulating substance
III.   Purification and identification of the meiosis-inducing substance
IV.   Chemical structure responsible for the induction of oocyte maturation and spawning
V.   1-methyladenosine ribohydrolase in the ovary
VI.   Site of action of 1-methyladenine in inducing oocyte maturation
VII.   Cytoplasmic maturation induced by 1-methyladenine in enucleated oocytes
VIII.   Effect of puromycin on 1-methyladenine-induced maturation
IX.   Concluding remarks
X.   References

I. INTRODUCTION

When an oocyte within an ovary reaches the end of its growth period it contains a single large nucleus called the germinal vesicle. To attain the stage of physiological maturation, the oocyte undergoes meiosis, which is initiated with the breakdown of the germinal vesicle. In vertebrates such as frogs and fishes, meiotic maturation, along with ovulation, can be induced by the injection of pituitary homogenate.

Recently in *Rana pipiens* it has been shown that pituitary gonadotropins do not act directly on the oocyte to induce maturation but affect them indirectly through the follicular tissue (Schuetz, 1967; Masui, 1967; Smith, Ecker and Subtelny, 1968). Further it has been clearly demonstrated that steroid hormones such as progesterone induce maturation of isolated frog oocytes (Schuetz, 1967;

Masui, 1967; Smith *et al.,* 1968). It is therefore probable that pituitary protein hormones act on the follicular tissue to produce steroid hormones or to accelerate the release of steroids.

The presence of progesterone in the ovaries of the starfish, *Pisaster ochraceus,* has been suggested by Botticelli, Hisaw and Wotiz (1960). Recently we confirmed the presence of progesterone in ovaries of a Japanese starfish, *Asterias amurensis*, by means of gas chromatography and mass spectroscopy (Ikegami, Shirai and Kanatani, 1971). However, progesterone failed to induce oocyte maturation in isolated oocytes of *Asterina pectinifera.* The physiological role of steroid hormones such as estradiol-17β (Botticelli *et al.,* 1960) and progesterone present in the starfish ovary, remains to be elucidated.

On the other hand, in invertebrates Chaet and McConnaughy (1959) first reported that injection of a water extract of starfish radial nerves induces shedding of gametes. The active substance has been shown to be a small peptide, called the "radial nerve factor" (Schuetz and Biggers, 1967) or "gonad-stimulating substance" (Kanatani, 1969a), which is released at the time of spawning from the nervous tissue into the coelomic cavity where the gonads are suspended (Kanatani and Noumura, 1962; Chaet, 1966, 1967; Kanatani, 1967, 1969a; Kanatani and Ohguri, 1966; Schuetz, 1969a). However, it has recently been demonstrated that this peptide hormone of the starfish, like the frog gonadotropins, acts indirectly on the gonads. Schuetz and Biggers (1967) and Kanatani and Shirai (1967) have shown that the peptide hormone of the starfish radial nerve acts on the ovary to produce a second substance which directly induces oocyte maturation and spawning *in vitro.* This substance, called the "ovarian factor" (Schuetz and Biggers, 1967) or "meiosis-inducing substance" (Kanatani and Shirai, 1967), has recently been isolated from ovaries of *Asterias amurensis* and identified as 1-methyladenine (Kanatani, Shirai, Nakanisha and Kurokawa, 1969). Although 1-methyladenine invariably induced meiotic maturation in all the starfish species tested, and is believed to be a general meiosis-inducing substance in Asteroidae, such action of a nucleic acid base is known only in starfish (Kanatani, 1969b and unpublished; Stevens, 1970).

In this article I would like to deal with some recent studies on the action of 1-methyladenine as well as its production and identification in relation to the induction of oocyte maturation and spawning in starfishes.

II. PRODUCTION OF MEIOSIS-INDUCING SUBSTANCE UNDER THE INFLUENCE OF GONAD-STIMULATING SUBSTANCE

The presence of a neural substance responsible for meiotic maturation and shedding of gametes is not confined to the radial nerve-circumoral nerve ring system. The nervous tissue of the body wall and tube feet seems also to contain this substance (Kanatani and Ohguri, 1966). However, in the coelomic fluid the neural substance is detectable only when the starfish is undergoing natural spawning (Kanatani and Ohguri, 1966; Kanatani and Shirai, 1970). This fact suggests that the substance is released at the time of spawning from the nervous tissue into the coelomic fluid which bathes the gonads; that is, that the substance is a hormone. Since the substance cannot enter the inside of the starfish body from the outside medium (Kanatani and Shirai, 1969) the original suggestion of Chaet (1966) that it acts as a pheromone between individuals is not tenable.

The clue to elucidating the mechanism by which the peptide hormone contained in the radial nerves induces breakdown of the germinal vesicle was found in 1967 in *Asterias forbesi* (Schuetz and Biggers, 1967) and *Asterina pectinifera* (Kanatani and Shirai, 1967). When oocytes of *Asterias forbesi,* isolated in calcium-free seawater to prevent spontaneous maturation, were placed in calcium-free seawater together with an ovarian fragment and then radial nerve factor was added, the oocytes nearest the fragment began to lose their germinal vesicles. Later, the germinal vesicles of the oocytes farther away from the ovarian fragment also broke down. These observations suggested that a substance (ovarian factor) diffusing from the ovarian fragment was the agent responsible for the breakdown of the germinal vesicles. The percentage of germinal vesicle breakdown was dependent on the amount of the crude ovarian factor, whereas the addition of radial nerve factor alone to immature oocytes failed to induce germinal vesicle breakdown (Schuetz and Biggers, 1967). A similar finding was also made independently with isolated oocytes of *Asterina pectinifera,* which do not usually undergo spontaneous maturation when isolated in seawater (Kanatani and Shirai, 1967). Oocytes expelled from ovarian fragments of *A. pectinifera* within less than one hour after treatment with the nerve extract were undergoing the first meiotic division at the time of spawning, whereas isolated oocytes treated in the same way failed to undergo meiosis. However, when ovarian fragments were placed in seawater containing the nerve extract for 1-3 hours and then the supernatant of this mixture was added to isolated oocytes, all of them underwent breakdown of the germinal vesicles within 30 minutes. The control supernatant lacking the nerve extract had no such effect. These observations also suggested that, as in the case of *Asterias forbesi,* an active substance was produced in the ovary under the influence of the nerve extract.

Since the nerve extract by itself has no effect in inducing oocyte maturation, we have renamed it the "gonad-stimulating substance (GSS)", and the ovarian factor, the "meiosis- inducing substance (MIS)" (Kanatani and Shirai, 1967; Kanatani, 1969a).

MIS has been successfully separated from the GSS contained in the original mixture by gel-filtration through columns of Sephadex G-25, G-15, or G-10 (Schuetz and Biggers, 1967; Kanatani and Shirai, 1967). As shown in Fig. 24.1, MIS eluted much more slowly than GSS. On the other hand, similar gel-filtration of nerve extract and ovary extract failed to produce active fractions capable of inducing oocyte maturation. Further, it was of interest that fractions showing meiosis-inducing activity also had the capacity to induce oocyte shedding. Schuetz (1969a, b) has also demonstrated that MIS of *Pisaster ochraceus* induces oocyte shedding *in vitro.*

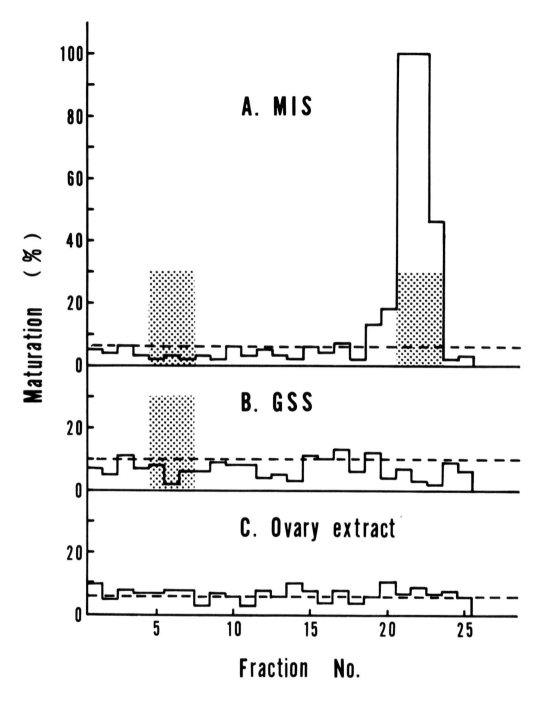

Fig. 24.1 Meiosis-inducing activities of *Asterina pectinifera* fractions obtained by gel-filtration of (a) supernatant of a mixture of ovary and nerve extract (MIS), (b) nerve extract (GSS), and (c) ovary extract, on Sephadex G-15 columns. Broken lines represent percentages of oocyte maturation in seawater (control). Dotted area represents the fractions having gamete-shedding activity. Column size, 1.4 x 42 cm; sample size, 3 ml; eluent, artificial seawater (pH 8.2 - 8.3); fraction size, 5 ml; flow rate, 30 ml/hr. Observations were made after 1 hr [After Kanatani and Shirai, 1967].

The course of breakdown of the follicular envelope and germinal vesicle of isolated oocytes of *Asterina pectinifera* on treatment with MIS is shown in Fig. 24.2.

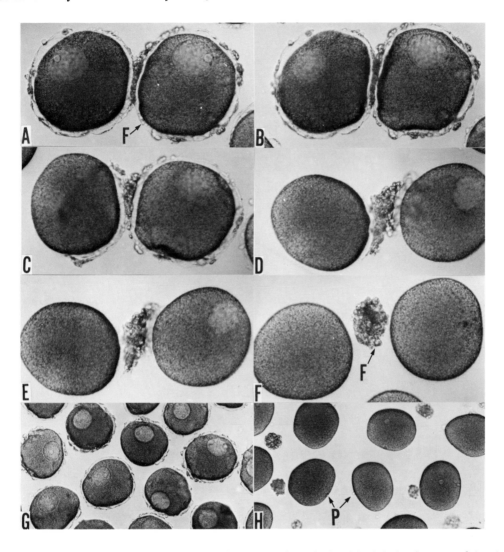

Fig. 24.2   Effect of MIS in causing breakdown of follicles and germinal vesicles in isolated oocytes of *Asterina pectinifera*. Isolated oocytes were transferred to seawater containing MIS. F, follicular envelope; P, polar body. (a) 2 min; (b) 18 min; (c) 20 min; (d) 21.5 min; (e) 23 min; (f) 60 min; (g) control in seawater after 130 min;   (h) 130 min after treatment with MIS [After Kanatani, 1969a].

The follicular envelope ruptures at some points, shrinks, and is gradually stripped off, within approximately 30 minutes, from the oocyte as it contracts to form a small clump of material. Germinal vesicle breakdown is usually completed within the same period, although the time required for its disappearance varies with the concentration of MIS and the state of the ovary during the breeding season. A concentrated MIS solution induces breakdown within 15 minutes, whereas a dilute one requires a much longer time.

At the time of its discovery, MIS was reported to be rather heat stable, dialyzable, insoluble in acetone, ether, petroleum ether and benzene, and not destroyed by treatment with proteolytic enzymes such as trypsin, chymotrypsin, pepsin and pronase. These data suggested that MIS was not a peptide (Schuetz and Biggers, 1967; Kanatani and Shirai, 1967).

The production of MIS in the ovary seems to begin immediately after the addition of GSS. Treatment of ovarian fragments of *Asterina pectinifera* with GSS-seawater for only five minutes is sufficient to induce the production of an active solution which induces 100% maturation. The amount of MIS produced increases with the concentration and duration of treatment with GSS. This fact suggests that production of MIS occurs continuously in the ovary under the influence of GSS (Kanatani and Shirai, 1970). In *Pisaster ochraceus,* production of MIS seems to occur more slowly; Schuetz (1969b) has reported that MIS activity is detectable in the incubation medium approximately one hour after the addition of GSS.

MIS is also produced in the testis under the influence of GSS, although the amount seems to be smaller than that produced in the ovary (Kanatani and Shirai, 1970).

In addition to *Asterias forbesi, Pisaster ochraceus* and *Asterina pectinifera,* production of MIS has been shown in the following starfish species: *Asterias amurensis, Aphelasterias japonica, Coscinasterias acutispina, Luidia quinaria, Astropecten scoparius* and *Patiria miniata* (Kanatani, 1969a and unpublished data).

III. PURIFICATION AND IDENTIFICATION
OF MEIOSIS-INDUCING SUBSTANCE

In order to ascertain the chemical characteristics of MIS, purification of this material was first performed with *Asterias amurensis* material (Kanatani *et al.,* 1969). A total of 20 kg of fresh ovarian fragments was incubated with 100 liters of artificial seawater containing GSS derived from 20 g of lyophilized radial nerves for 6 hours at 20°C to produce MIS. The incubation mixture was centrifuged and concentrated to 14 liters. Inorganic salts were partially precipitated with potassium phosphate and ethanol and concentration was continued to give 1 liter of active solution. This solution was washed with chloroform and ether and the active aqueous phase was concentrated. MIS activity was assayed with isolated oocytes of *Asterina pectinifera.* The sample (118 ml) was gel-filtrated on a large Sephadex G-15 column (5.2 x 109 cm, 0.1M NaCl in 0.05M borate buffer, pH 8.5). The active fractions were pooled and concentrated, and the sample (10 ml, pH 8.7) was further gel-filtrated on a Sephadex G-15 column (Fig. 24.3). The active fractions were

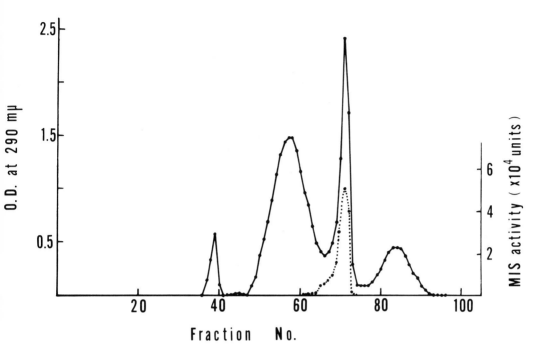

Fig. 24.3 Further gel-filtration of the MIS sample, obtained from previous gel-filtration, on a Sephadex G-15 column (2.5 x 110 cm) in 0.1M γ-collidine-0.1M pyridine acetate buffer (pH 8.7). Fraction size, 10 ml; flow rate, 45 ml/hr; 3°C. Elution curve was obtained by measuring the optical density at 290 mμ using the buffer as a blank.

pooled and lyophilized. The sample (11.6 mg) was dissolved in 0.09M ammonium acetate buffer (pH 6.8) and fractionated on a CM-Sephadex C-25 column, equilibrated with the same buffer, by step-wise elution (Fig. 24.4). MIS was eluted at 0.25M ammonium acetate. The active fractions were concentrated and gel-filtrated again on a Sephadex G-15 column (0.5M pyridine acetate buffer, pH 8.4). The active fractions were lyophilized. This procedure gave about 8.5 mg of purified MIS, which was active at 0.02 μg/ml in inducing 100% breakdown of the germinal vesicles.

Purified MIS showed maximum ultraviolet absorption at 261 mμ in ammonium acetate buffer (pH 6.8), and suggested that it may be a nucleic acid base. Its ultraviolet absorption was therefore remeasured at different pHs. The ultraviolet absorption curves and the infrared absorption spectrum of purified MIS was in good agreement with that of 1-methyladenine (Fig. 24.5). Furthermore, in the high resolution mass spectrum of MIS, a peak assignable to a molecular ion was present at m/e 149.0725 ($C_6 H_7 N_5$), suggesting that MIS is 1-methyladenine itself. The melting

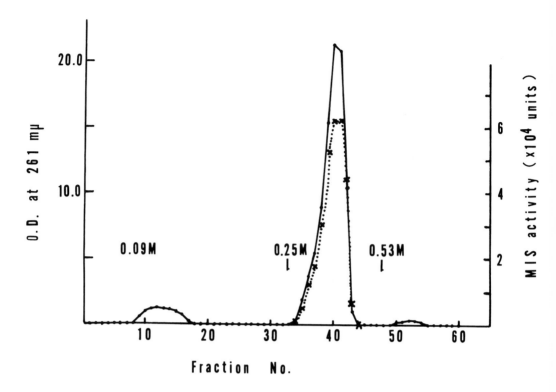

Fig. 24.4 Chromotography of MIS sample, purified by previous gel-filtrations, on a CM-Sephadex column (1.5 x 25 cm) in ammonium acetate buffer. Fraction size, 5 ml; flow rate, 18 ml/hr; 3°C [After Kanatani *et al.*, 1969].

point of MIS (301° - 303°C) was identical with that of 1-methyladenine. Finally, synthetic 1-methyladenine was very effective in inducing oocyte maturation and its minimum effective dose was the same as that of purified MIS. From these results we can conclude that MIS is 1-methyladenine (Kanatani *et al.*, 1969).

Since the discovery of the presence of MIS in the starfish ovary, it has been thought that this substance acts in two ways: one is to induce oocyte maturation and the other is to induce spawning. This was clearly proved by using synthetic 1-methyladenine in the following starfish species: *Asterias amurensis, Asterias forbesi, Asterina pectinifera, Marthasterias glacialis, Astropecten aurantiacus, Ceramaster plàcenta, Patiria miniata, Pisaster brevispinus, Pisaster giganteus, Pisaster ochraceus, Pycnopodia helianthoides, Mediaster aequalis,* and *Leptasterias hexactis* (Kanatani, 1969b and unpublished; Bryan and Sato, 1970; Stevens, 1970). These data indicate that 1-methyladenine is a general inducer of meiosis among starfishes.

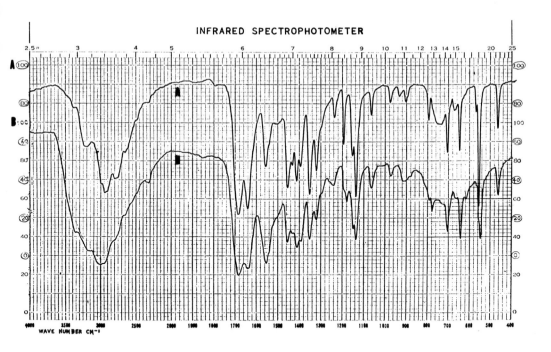

INFRARED SPECTROPHOTOMETER

Fig. 24.5 Infrared absorption spectra of 1-methyladenine (a) and MIS of *Asterias amurensis* (b).

IV. CHEMICAL STRUCTURE RESPONSIBLE FOR THE INDUCTION OF OOCYTE MATURATION AND SPAWNING

In an effort to determine the particular chemical structure which is responsible for inducing oocyte maturation and spawning the effects of various adenine derivatives on such biological processes were investigated (Kanatani, Kurokawa and Nakanishi, 1969; Kanatani and Shirai, 1971). Among twenty-eight compounds tested, on oocytes within ligated ovarian fragments, for breakdown of germinal vesicles and on ovarian fragments for spawning in *Marthasterias glacialis, Asterias forbesi, Patiria miniata* and *Asterina pectinifera,* 1-methyladenine, 1-ethyladenine and 1-methyladenosine were found to be very effective. On isolated oocytes 1-methyladenosine had little effect in inducing maturation. The following related compounds had no effect on either oocyte maturation or spawning: adenine, 3-methyladenine, 7-methyladenine, 9-methyladenine, 1-methylguanine, 1-methylhypoxanthine, 6-methylpurine, N_6-methyladenine, 3-methylcytidine, 5-methylcytosine, N_6-dimethyladenine, N_6-benzyladenine, kinetin, adenosine, 5′-adenylic acid, adenosine 3′, 5′-monophosphate, adenosine triphosphate, hypoxanthine, inosine, 5′-inosinic acid, guanine, 5′-guanylic acid, xanthine and xanthosine. These results are shown in Table 24.1 (see also Fig. 24.6).

Usually 1-methyladenine induces 100% oocyte maturation at 0.3 μM,

Table 24.1 **Effect of 1-methyladenine and Related Compounds** *In vitro* **on Spawning and Oocyte Maturation in Starfishes**

Compounds	Marthasterias glacialis	Asterias forbesi	Patiria miniata	Asterina pectinifera
Adenine	—	—	—	—
3-Methyladenine				—
7-Methyladenine				—
9-Methyladenine	—	—	—	—
1-Methyladenine	+++	+++	+++	+++
1-Methyladenosine	+	+	+	+
1-Methylguanine		—	—	—
1-Methylhypoxanthine				—
6-Methylpurine		—	—	
N_6-Methyladenine		—	—	
3-Methylcytidine				—
5-Methylcytosine				—
N_6-Dimethyladenine				—
N_6-Benzyladenine	—	—		—
N_6-Furfuryladenine (Kinetin)	—	—	—	—
1-Ethyladenine	++	++	++	++
Adenosine	—	—	—	—
5'-Adenylic acid	—	—		—
Adenosine 3',5'-monophosphate (Cyclic AMP)	—	—	—	—
Adenosine triphosphate	—	—		—
Hypoxanthine	—	—		—
Inosine	—	—		—
5'-Inosinic acid	—	—		—
Guanine	—	—		—
Guanosine	—	—		—
5'-Guanylic acid	—	—		
Xanthine			—	
Xanthosine			—	

1-ethyladenine at 1 μM and 1-methyladenosine at 10 - 30 μM, although the minimum effective dose varies to a large extent depending upon the reactivity of the oocytes: late in the spawning season oocytes generally react to lower concentrations. One of the features common to the active compounds is a short alkyl radical such as a methyl or ethyl radical which is combined at the N_1 site of the purine nucleus. Adenine derivatives methylated at other sites such as 3-methyladenine, 7-methyladenine, 9-methyladenine, N_6-methyladenine and N_6-dimethyladenine had no biological activity. However, since neither 1-methylguanine nor 1-methylhypoxanthine was effective, methylation at the N_1 site of the purine nucleus does not seem to be sufficient to induce oocyte maturation. That 1-methylhypoxanthine failed to induce maturation also suggests that the presence of an imino radical combined at the C6 site of the purine nucleus is also important. That 1-ethyladenine is less effective than 1-methyladenine indicates that the presence of such an imino radical at the C6 site of the purine nucleus is not the sole structural characteristic responsible for induction of maturation. Therefore it is

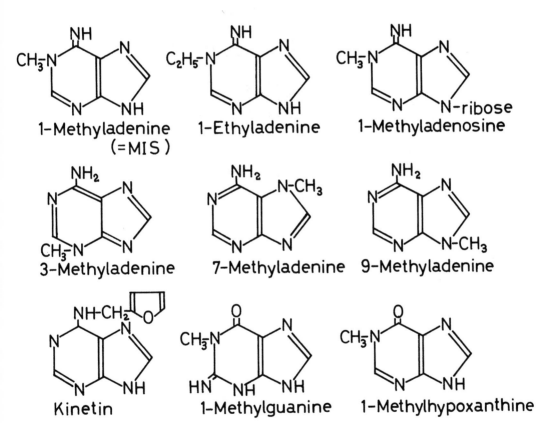

Fig. 24.6 Chemical structure of 1-methyladenine and related compounds tested for biological activity.

tentatively concluded that the presence of a short alkyl radical at the N1 site and an imino radical at the C6 site of the purine nucleus is important for inducing oocyte maturation and shedding of gametes in starfishes, although the effect of certain derivatives of 1-methyladenine, in which the hydrogen atom at the N_6 site is replaced by other atoms or radicals, has not yet been examined.

V. 1-METHYLADENOSINE RIBOHYDROLASE IN THE OVARY

The precise biochemical pathway involved in the production of 1-methyladenine in the gonads under the influence of GSS is not known. That heating of the incubation mixture of ovary and GSS arrests the production of MIS suggests that some enzymatic activity is involved in its production (Kanatani and Shirai, 1970).

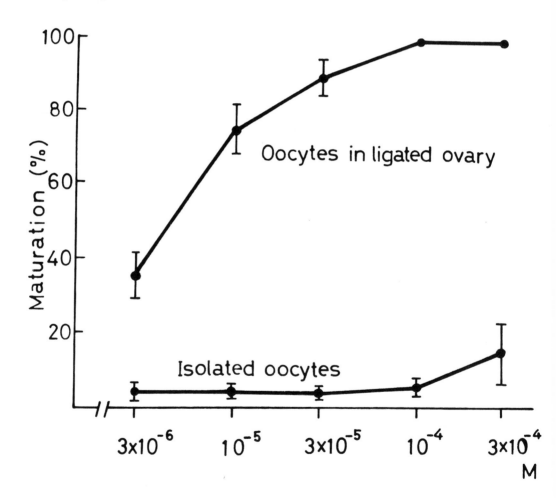

Fig. 24.7 Effect of 1-methyladenosine on maturation of oocytes within ligated ovarian fragments and isolated oocytes in *Patiria miniata*.

Although 1-methyladenosine (1-MAR) is very effective in inducing maturation of oocytes within ligated ovarian fragments, it fails to induce breakdown of germinal vesicles of isolated oocytes in the absence of ovarian tissue (Fig. 24.7). The supernatant of an incubation mixture of ovarian fragments and seawater containing 1-MAR was found to be effective in inducing maturation of isolated oocytes in *Patiria miniata* and *Asterina pectinifera*. This finding suggested that some enzyme which acts on 1-MAR to hydrolyze it into 1-methyladenine and ribose was present in ovarian tissue. Preparation of this enzyme was performed by the following method. Ovarian fragments, from which the oocytes had been removed as completely as possible by treatment with 1-methyladenine for 1 - 2 hours, were washed with seawater and homogenized in deionized water. The enzyme was obtained as a precipitate by adding ammonium sulfate to the supernatant (60,000 g, 50 min) of the homogenate to 0.45 saturation.

The enzyme is not ordinary adenosine ribohydrolase which splits adenosine into adenine and ribose, since it does not hydrolyze adenosine into adenine and ribose. Further, addition of a large amount of adenosine (ten times of 1-methyladenosine) to a reaction mixture of 1-methyladenosine and 1-methyladenosine ribohydrolase does not result in the production of 1-methyladenine when assayed on isolated oocytes. 1-methyladenosine ribohydrolase of *Asterina pectinifera* is stable when frozen and its optimum pH is 7.5. However, since the addition of GSS to the reaction mixture had no effect on its meiosis-inducing activity, the hormonal peptide (GSS) is believed to act on some reaction other than the hydrolysis of 1-methyladenosine into 1-methyladenine and ribose.

The addition of methionine to the mixture of ovarian fragments and GSS-seawater somewhat enhanced the production of 1-methyladenine as assayed by its biological activity. Therefore, although a suitable cell-free reaction system to demonstrate the role of GSS in the production of 1-methyladenine has not yet been devised, GSS seems to accelerate some enzyme involved in methylation at the N1 site of the purine nucleus of a precursor of 1-methyladenosine, or some enzyme involved in the liberation of 1-methyl AMP from ovarian RNA. Although these data are preliminary, it is undoubtedly important to determine the precise role of GSS in 1-methyladenine production, not only to elucidate the precise mechanism of meiotic maturation and spawning in starfish, but to understand the action of such peptide hormones in general from the biochemical point of view.

VI. SITE OF ACTION OF 1-METHYLADENINE IN INDUCING OOCYTE MATURATION

In an effort to elucidate the mechanism of action of 1-methyladenine in inducing oocyte maturation, we have recently used isolated oocytes of *Asterina pectinifera* to investigate the site of action of this substance (Kanatani and Hiramoto, 1970). When a small amount of 1-methyladenine (30 - 100 fmoles) was microinjected into single oocytes with intact germinal vesicles, the substance failed to induce oocyte maturation. Since these injected oocytes invariably underwent meiosis when transferred to seawater containing 1 μM 1-methyladenine, the failure of the injected 1-methyladenine to induce maturation cannot be ascribed to the detrimental effect of the injection itself. It is to be noted that diffusion of 1-methyladenine within the egg cytoplasm seems to be fairly rapid, since aqueous fluorescein having a molecular weight greater than that of 1-methyladenine diffused evenly throughout the egg cytoplasm within a fraction of a minute following microinjection into sea urchin eggs (Hiramoto, 1968).

The quantity of 1-methyladenine required to induce germinal vesicle breakdown in incubated oocytes was compared to that required following microinjection. As a result it was found that when treated with 10 fmoles of 1-methyladenine per oocyte (calculated from the total amount of the substance divided by the number of oocytes in the test solution), all oocytes underwent breakdown of their germinal vesicles within 35 minutes. In this case, the concentration of 1-methyladenine in the test solutions was 0.33 μM, and the minimum effective concentration required to induce maturation in isolated oocytes from the same individuals placed in a large amount of test solution was about 0.3 μM. Therefore, meiotic maturation of *Asterina pectinifera* is induced by the external application of 1-methyladenine provided that the concentration is 0.33 μM or higher, even when the amount of the substance per oocyte is as small as 10 fmoles, which was only one-tenth of the amount of the same substance that was ineffective when microinjected. From these results it is concluded that the site where 1-methyladenine acts to induce meiotic maturation in the starfish is the surface of the oocyte.

Smith and Ecker (1969) have also proposed in *Rana pipiens* that progesterone induces maturation of isolated oocytes by acting at the egg surface. These two examples thus provide evidence that direct triggers of meiotic maturation such as 1-methyladenine and progesterone generally act on the surface of the oocyte, from the outside, in both invertebrates and vertebrates.

VII. CYTOPLASMIC MATURATION INDUCED BY 1-METHYLADENINE IN ENUCLEATED OOCYTES

It has been generally believed that breakdown of the germinal vesicle must occur in marine invertebrate oocytes before they can be fertilized. That is, mixing of the germinal vesicle material with the egg cytoplasm seems to be a necessary prerequisite for acquisition of fertilizability (see Wilson, 1928). In starfish, oocytes with intact germinal vesicles and non-nucleated oocytes or non-nucleated oocyte fragments, whose germinal vesicles had been intact at the time of operation, failed to be fertilized. However, if the breakdown of the germinal vesicle has already begun at the time of enucleation, oocytes or oocyte fragments can be fertilized and continue to develop (Delage, 1901; Chambers, 1921).

We have recently used such enucleation techniques to reinvestigate the role of the germinal vesicle in oocyte maturation as induced by 1-methyladenine in *Asterina pectinifera* (Hirai, Kubota and Kanatani, unpublished). As described in the preceding section, 1-methyladenine seems to act on the surface of the oocyte rather than directly on the germinal vesicle to induce oocyte maturation. This fact suggests that the events leading to breakdown of the germinal vesicle are mediated by some surface change induced by 1-methyladenine. Therefore our interest has been directed to determining whether this substance, in the absence of germinal vesicle material, can induce cytoplasmic maturation as revealed by the formation of a fertilization membrane.

Fifty-eight isolated oocytes from which the germinal vesicles were removed remained unchanged and failed to elevate a fertilization membrane on insemination, thus confirming the results of previous investigators. Another group of 113 enucleated oocytes treated in the same way also failed to form fertilization membranes. However, when these enucleated ooctyes were treated with 10 μM 1-methyladenine for 30-35 minutes and then reinseminated, 84% of them formed a fertilization membrane. Continued observation for an additional 4 hours revealed that these oocytes failed to undergo cleavage, although some of them showed signs of abortive cleavage during this period.

In similar experiments, in which oocytes were cut into nucleated and non-nucleated fragments and treated with 10 μM 1-methyladenine for 30 minutes or longer, 81% of the non-nucleated fragments and 89% of the nucleated fragments formed fertilization membranes on insemination. Fertilized nucleated fragments underwent cleavage after a certain period, whereas their non-nucleated counterparts did not cleave. These results are in good agreement with those obtained recently by Smith and Ecker (1969) on progesterone-induced oocyte maturation in *Rana pipiens*. According to them, mixing of germinal vesicle material with egg cytoplasm is not necessary for artificial activation but is required for cleavage. Therefore it is highly probable that triggers of oocyte maturation such as progesterone and 1-methyladenine act on the surface of the oocyte and induce cytoplasmic maturation, as revealed by fertilizability, without any intervention of germinal vesicle material. Further, the time required for cytoplasmic maturation induced by 1-methyladenine seems to be equivalent to that required for the breakdown of the germinal vesicle (Stevens, 1970). Moreover, Iwamatsu (1966), using the enucleated oocytes of the fish, *Oryzias latipes,* reported that cytoplasmic maturity was induced by some physiological changes in the cytoplasm, which are independent of germinal vesicle material. In this case the enucleated oocytes became fertilizable after incubation in a balanced salt solution for 10-15 hours, and underwent cleavage through normal cortical changes.

VIII. EFFECT OF PUROMYCIN
ON 1-METHYLADENINE-INDUCED MATURATION

According to investigators of oocyte maturation in amphibians, puromycin completely suppresses breakdown of the germinal vesicle of isolated oocytes treated with progesterone (Schuetz, 1967; Smith and Ecker, 1969; Dettlaff and Skoblina, 1969). Dettlaff and Skoblina (1969) reported that this is also the case in sturgeon oocytes. In starfish, Brachet and Steinert (1967) reported that when isolated oocytes from *Asterias rubens* were treated with puromycin (50 μg/ml) spindle formation and migration of the chromosomes towards the animal pole were inhibited. We have also found that treatment with puromycin at a high concentration (1 mM) arrests the breakdown of the germinal vesicle in *Luidia quinaria* whose oocytes usually undergo spontaneous maturation in seawater. However, treatment with 1-methyladenine (1 μM) was found to induce breakdown of germinal vesicle even in the presence of puromycin (Hirai, unpublished). For this reason we have re-examined the effect of puromycin on 1-methyladenine-induced oocyte maturation using the oocytes which do not usually undergo spontaneous maturation when isolated in seawater. Isolated oocytes of *Astropecten aurantiacus* were pretreated in seawater containing puromycin (100 μg/ml, 50 μg/ml and 20 μg/ml) for 4.5 hours and then transferred to seawater containing puromycin (same concentrations) together with 0.6 μM 1-methyladenine. Isolated oocytes kept in seawater, and those transferred to a fresh medium containing puromycin alone after treatment with puromycin for 4.5 hours served as controls. As shown in Table 24.2 such treatment with puromycin had no effect on the capacity of 1-methyladenine to induce breakdown of the germinal vesicle. Breakdown of the germinal vesicle was not retarded in the presence of puromycin when compared with that in the oocytes treated with 1-methyladenine alone. Polar bodies were also observed in many cases.

Table 24.2 **Effect of External Application of Puromycin on 1-methyladenine-induced Oocyte Maturation in** *Astropecten aurantiacus*

SW	Puromycin			1-MA	Puromycin +1-MA		
	100μg	50μg	20μg		100μg	50μg	20μg
5±1%*	3±1%	3±1%	3±1%	100±0%	100±0%	100±0%	100±0%

* mean ± S.E. of 5 experiments

Since it was uncertain whether puromycin entered the oocytes, the inhibitor was microinjected into oocytes of *Asterina pectinifera*. Amounts of puromycin dissolved in deionized water (20 mg/ml) equivalent to 1/200 - 1/50 of the egg volume were microinjected into isolated single oocytes with germinal vesicles and then transferred to seawater containing 1 μM 1-methyladenine. Control oocytes were injected with the same amounts of deionized water and then were transferred to 1 μM 1-methyladenine. Isolated oocytes treated with 1-methyladenine, without injection, also served as controls. Injected puromycin, the concentrations of which, in the oocytes, were as high as 100 μg to 400 μg/ml, had no effect on the breakdown of the germinal vesicle (Table 24.3). Most of the germinal vesicles broke down

Table 24.3 **Effect of Microinjection of Puromycin on 10^{-6}M 1-methyladenine-induced oocyte Maturation in** *Asterina pectinifera*

Injected puromycin			
Volume* (ml)	Final concentration (μg/ml)	Number of examples	breakdown of germinal vesicle
5×10^{-8}	400	10	10
2.5×10^{-8}	200	16	16
1.2×10^{-8}	100	6	6
Control**	0	5	5

*   Average volume of oocyte was 2.5×10^{-6} ml; concentration of puromycin injected was 20 mg/ml.

**   Four out of 5 oocytes were injected with 5×10^{-8} ml of deionized water and the remaining one, with 2.5×10^{-8} ml of water.

within 30 minutes, and puromycin did not retard this process. Formation of polar bodies was also observed. When eight of these oocytes injected with puromycin (final concentration in the egg: 200 μg/ml) were inseminated, they all elevated the fertilization membrane (Hiramoto and Kanatani, unpublished).

Since these experiments did not include direct measurement of protein synthesis or extensive histological examination, the results must be interpreted with caution. Nonetheless, it appears that the role of protein synthesis in the meiotic maturation of starfish is different from that in amphibians and sturgeon. In starfish the process of meiotic maturation proceeds very quickly after the oocyte reaches the full-grown state, generally taking only about 2 hours from the beginning of 1-methyladenine treatment to the discharge of the second polar body.

IX. CONCLUDING REMARKS

Although it is well established that the action of 1-methyladenine triggers the breakdown of the germinal vesicle and the subsequent process of meiotic maturation in starfish, such an activity of the nucleic acid base has been recognized only in Asteroidea. Investigations on the effect of nucleic acid bases in relation to oocyte maturation in other groups of animals are eagerly anticipated not only from the comparative aspect but from the viewpoint which aims to understand the mechanism of meiotic maturation in general.

The results of investigations performed hitherto on the induction of meiotic maturation in starfish have revealed that the site of action of 1-methyladenine is the surface of the oocyte and that the substance does not act directly on the germinal vesicle. The next problem to be elucidated is the nature of the events which occur in the cortex of the oocyte following treatment with 1-methyladenine. Of two major possibilities which are being considered, one concerns possible changes in permeability to environmental ions. Although lack of calcium ions prevents breakdown of the germinal vesicle in those isolated oocytes which undergo spontaneous maturation in seawater, treatment with a sufficient amount of 1-methyladenine can induce meiotic maturation in calcium-free seawater (Shirai and Kanatani, 1970). The other possibility is that 1-methyladenine, the second substance of hormone action, acts on the oocyte surface and there produces a third substance which is referred to as the direct inducer of germinal vesicle breakdown. Although the existence of such a substance has not yet been established, efforts to find it should certainly be made.

At present the analysis of the mechanism of meiotic maturation induced by 1-methyladenine is just beginning. A precise understanding of the biochemical properties of the oocyte surface and their changes during the process of meiotic maturation must be of the greatest importance for the analysis of this phenomenon, as well as of fertilization. Further, persistent analysis of the properties of the egg cortex and their changes during subsequent development will afford the basic data for the better understanding of mechanism of cleavage and differentiation of the embryo.

X. REFERENCES

Botticelli, C. R., Hisaw, Jr., F. L. and Wotiz, H. H. (1960) Estradiol-17β and progesterone in ovaries of starfish *(Pisaster ochraceus). Proc. Soc. Exp. Biol. Med.* **103**, 875.

Brachet, J. and Steinert, G. (1967) Synthesis of macromolecules and maturation of starfish ovocytes. *Nature (London).* **216**, 1314.

Bryan, J. and Sato, H. (1970) The isolation of the meiosis in spindle from the mature oocyte of *Pisaster ochraceus. Exp. Cell Res.* **59**, 371.

Chaet, A. B. (1966) Neurochemical control of gamete release in starfish. *Biol.Bull.* **130**, 43.

Chaet, A. B. (1967) Gamete release and shedding substance of sea-stars. *Symp. zool. Soc. Lond.* **20**, 13.

Chaet, A. B. and McConnaughy, R. A. (1959) Physiological activities of nerve extracts. *Biol. Bull.* **117**, 407.

Chambers, R. (1921) Microdissection studies, III. Some problems in the maturation and fertilization of the echinoderm egg. *Biol. Bull.* **41**, 318.

Delage, Y. (1901) Etudes experimentales sur la maturation cytoplasmique et sur la parthenogenese artificielle chez les echinodermes. *Arch. de Zool. Exp. et Gen. 3 ser.* **9**, 285.

Dettlaff, T. A. and Skoblina, M. N. (1969) The role of germinal vesicle in the process of oocyte maturation in *Anura* and *Acipenseridae. Annal. Embryologie et de Morphogenese suppl.* **1**, 133.

Hiramoto, Y. (1968) The mechanics and mechanism of cleavage in the sea-urchin egg. *XXII Sympos. Soc. Exp. Biol.* 311.

Ikegami, S., Shirai, H. and Kanatani, H. (1971) On the occurrence of progesterone in ovary of the starfish, *Asterias amurensis. Zool. Mag. (Tokyo).* **80**, 26.

Iwamatsu, T. (1966) Role of germinal vesicle materials on the acquisition of developmental capacity of the fish oocyte. *Embryologia.* **9**, 205.

Kanatani, H. (1967) Neural substance responsible for maturation of oocytes and shedding of gametes in starfish. *Gunma Symp. Endocrinology.* **4**, 65.

Kanatani, H. (1969a) Mechanism of starfish spawning: Action of neural substance on the isolated ovary. *Gen. Comp. Endocr. Suppl.* **2**, 582.

Kanatani, H. (1969b) Induction of spawning and oocyte maturation by 1-methyladenine in starfishes. *Exp. Cell Res.* **57**, 333.

Kanatani, H. and Hiramoto, Y. (1970) Site of action of 1-methyladenine in inducing oocyte maturation in starfish. *Exp. Cell Res.* **61**, 280.

Kanatani, H. and Noumura, T. (1962) On the nature of active principles responsible for gamete-shedding in the radial nerves of starfishes. *J. Fac. Sci. Tokyo Univ., Sec. IV.* **9**, 403.

Kanatani, H., Kurokawa, T. and Nakanishi, K. (1969) Effects of various adenine derivatives on oocyte maturation and spawning in the starfish. *Biol. Bull.* **137**, 384.

Kanatani, H. and Ohguri, M. (1966) Mechanism of starfish spawning. I. Distribution of active substance responsible for maturation of oocytes and shedding of gametes. *Biol. Bull.* **131**, 104.

Kanatani, H. and Shirai, H. (1967) *In vitro* production of meiosis inducing substance by nerve extract in ovary of starfish. *Nature (London).* **216**, 284.

Kanatani, H. and Shirai, H. (1969) Mechanism of starfish spawning. II. Some aspects of action of a neural substance obtained from radial nerve. *Biol. Bull.* **137**, 297.

Kanatani, H. and Shirai, H. (1970) Mechanism of starfish spawning. III. Properties and action of meiosis-inducing substance produced in gonad under influence of gonad- stimulating substance. *Development, Growth and Differentiation.* **12**, 119.

Kanatani, H. and Shirai, H. (1971) Chemical structural requirements for induction of oocyte maturation and spawning in starfishes. *Development, Growth, and Differentiation.* (In press).

Kanatani, H., Shirai, H., Nakanishi, K. and Kurokawa, T. (1969) Isolation and identification of meiosis inducing substance in starfish *Asterias amurensis. Nature (London).* **221**, 273.

Masui, Y. (1967) Relative roles of the pituitary, follicle cells, and progesterone in the induction of oocyte maturation in *Rana pipiens. J. Exp. Zool.* **166**, 365.

Schuetz, A. W. (1967) Action of hormones on germinal vesicle breakdown in frog *(Rana pipiens)* oocytes. *J. Exp. Zool.* **166**, 347.

Schuetz, A. W. (1969a) Chemical properties and physiological actions of a starfish radial nerve factor and ovarian factor. *Gen. Comp. Endocr.* **12**, 209.

Schuetz, A. W. (1969b) Induction of oocyte shedding and meiotic maturation in *Pisaster ochraceus:* Kinetic aspects of radial nerve factor and ovarian factor induced changes. *Biol. Bull.* **137**, 524.

Schuetz, A. W. and Biggers, J. D. (1967) Regulation of germinal vesicle breakdown in starfish oocytes. *Exp. Cell Res.* **46**, 624.

Shirai, H. and Kanatani, H. (1970) Induction of oocyte maturation in calcium-free sea water by 1-methyladenine in the starfish, *Asterina pectinifera. Zool. Mag. (Tokyo).* **79**, 156.

Smith, L. D., Ecker, R. E. and Subtelny, S. (1968) *In vitro* induction of physiological maturation in *Rana pipiens* oocytes removed from ovarian follicles. *Develop. Biol.* **17**, 627.

Smith, L. D. and Ecker, R. E. (1969) Role of the oocyte nucleus in physiological maturation in *Rana pipiens. Develop. Biol.* **19**, 281.

Stevens, M. (1970) Procedures for induction of spawning and meiotic maturation of starfish oocytes by treatment with 1-methyladenine. *Exp. Cell Res.* **59**, 482.

Wilson, E. B. (1928) The cell in development and heredity. p. 1232. New York, Macmillan Co.

25

HORMONES AND FOLLICULAR FUNCTIONS

Allen W. Schuetz

I.   Introduction
II.   Experimental procedures
III.   Interruption of follicular arrest
        A.   Gonadotrophins
        B.   Steroids
        C.   Relation between ovulation and oocyte maturation
IV.   Mechanisms of hormonal interruption of follicular arrest
        A.   Site of action of hormones
        B.   Biochemical processes
        C.   Estrogens and follicular functions
V.   Steroidogenesis in amphibians
VI.   Functional maturation of oocytes *in vitro*
VII.   Structural alterations in the ovarian follicle
        A.   Membrane separation
        B.   Relation of structural changes to follicular function
VIII.   Summary
IX.   Acknowledgments
X.   References

I. INTRODUCTION

Oogenesis in most species is characterized as an intricate, progressive differentiation and growth of those sex cells (oocytes) which will give rise to the next generation. The two major components of the oocyte, nucleus and cytoplasm, both undergo highly ordered changes during the course of this process (Raven, 1961; Wischnitzer, 1966). The predominant transformations in and growth of the nucleus occur during prophase I of meiosis whereas the simultaneous cytoplasmic changes relate to the processes of vitellogenesis and oocyte growth. Oocyte differentiation, however, does not happen independently of other ovarian and body processes, or occur in all oocytes simultaneously. Oocyte changes normally occur within a structure called a follicle, and thus all aspects of oogenesis are intimately and simultaneously reflected in and related to follicular differentiation and functions. A major problem in studying follicular processes and oogenesis is the absence of reliable criteria to distinguish the various stages of follicular differentiation.

In general, two different types of association exist between the oocyte and the rest of the follicle during the course of follicle growth and differentiation. In most mammals the individual oocyte and follicle increase in size proportionately and simultaneously until oocyte growth (size) is completed. With the initiation of antrum formation and the accumulation of follicular fluid, follicle size increases enormously and relatively independent of any change in the size of the oocyte (Pincus, 1936; Blandau, 1961; Franchi, 1962; Harrison, 1962; Matthews, 1962). However, in other species including amphibians, the oocyte comprises the major physical proportion of the ovarian follicle during the entire course of follicular growth and neither antrum formation nor the accumulation of intrafollicular fluid occurs.

Regardless of the type of follicle development, both diversity and synchronization of follicular events are evident within the ovary at any particular time. Obviously, in most species, only a certain proportion of the follicle or oocyte population(s) completes development simultaneously. However, from the follicles which do develop at the same time, such as those oocytes which are ovulated on a particular day, most ova are at a stage of meiosis characteristic of the species (Austin and Walton, 1960). This clearly suggests that local synchronizing mechanisms exist at the interfollicular and intrafollicular levels. The degree and mechanisms by which intrafollicular oocyte and follicle wall maturation are synchronized is therefore of major importance.

In the oocyte, nuclear changes are of central importance to the process of meiosis and the formation of viable gametes. The nucleus of the oocyte, called the germinal vesicle, remains morphologically intact during the course of growth and differentiation and for prolonged periods after the apparent cessation of oocyte growth. With the disintegration of the germinal vesicle, chromosome condensation occurs and the meiotic process then proceeds in a manner characteristic of the species. The maintenance of the oocyte in the germinal vesicle stage is generally considered to be a state of meiotic arrest. Although meiotic arrest implies that the nucleus is involved, it has in reality been unclear whether the follicle wall, oocyte cytoplasm or nucleus, or some external factors are responsible for the arrested state. Thus it may be more proper to consider the oocyte and follicle as a unit and to designate the arrested state as one of follicular arrest.

A major impediment to understanding oogenesis concerns the nature of this follicular arrest and the mechanisms involved in its termination. It has been clear for many years that both the processes of follicular growth and the termination of follicular arrest are intimately regulated by pituitary gland secretions, presumably gonadotrophic hormones (Hisaw,1947; Everett, 1961; Schuetz, 1969). Thus, our questions concern not whether the gonadotrophic hormones initiate the intrafollicular processes but where within the follicle and by what mechanisms they act. Possible hormone tissue interactions involved in termination of follicular arrest are presented in Fig. 25.1.

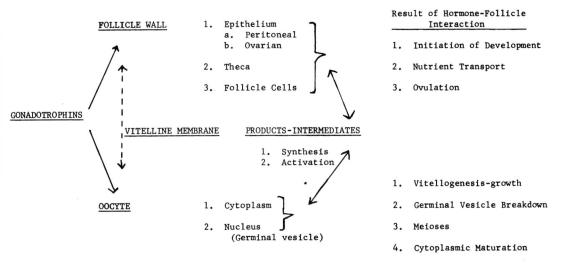

Fig. 25.1 Diagrammatic representation of possible gonadotrophic hormone-ovarian follicular component interactions involved in mediating follicle differentiation and interruption of follicular arrest.

An understanding of follicular functions at a cellular and molecular level appears central to explaining normal and abnormal follicle differentiation, hormone specificity, cytoplasmic maturation, follicular competence and atresia (Ingram, 1962). Attempts to study intrafollicular processes have, however, been fraught with many difficulties, primarily as a result of the dynamic and continuous nature of the follicular differentiation processes. The nature of the reproductive process in certain amphibians, however, presents unique characteristics which make experimental examination of some of these questions possible. These characteristics include (1) large oocytes and follicles which can be readily and individually separated from the ovary; (2) a large number of ovarian follicles in a similar state of differentiation; (3) methods for the *in vitro* culture of ovarian follicles; (4) maintenance of ovarian follicle responsiveness to hormonal stimuli under *in vitro* conditions; (5) easily recognizable alterations in follicular function; and (6) dissociation of the various components of the follicle from each other.

In general the functions of the ovary and the follicular processes have been primarily interpreted on the basis of the endocrine products secreted from the ovary, as well as the oocytes released or ovulated from the ovary. Involvement of endocrine mechanisms in regulating follicular functions at the local intraovarian level has been a matter of conjecture. Until recently the evidence for the role of steroids in regulating local follicular functions has been limited or has not been readily accepted. Considerable experimental evidence now indicates that in amphibians, steroid hormones markedly influence many aspects of follicular function.

The information presented here summarizes our present understanding of the cellular and hormonal interactions involved in three intrafollicular processes: (1) ovulation; (2) germinal vesicle breakdown; (3) follicular wall-oocyte membrane changes. These discussions are primarily concerned with the role of gonadotrophic and steroid hormones on follicular arrest in *Rana pipiens;* however, their applicability to other species appears germaine and will be discussed to some extent.

II. EXPERIMENTAL PROCEDURES

Animals.   All experiments were conducted utilizing the common leopard frog, *Rana pipiens.* Animals were purchased from commercial suppliers and maintained under conditions of artificial hibernation at 4°C in tap water. All experiments were conducted on isolated follicles individually dissected from the ovary. Because of the seasonal aspects of oogenesis in *Rana pipiens,* the gonad is normally composed of two major types of follicles, large pigmented follicles and small follicles containing unpigmented previtellogenic oocytes. The studies presented here were conducted on the large preovulatory follicles (see Fig. 25.2). Following the isolation of a pool of follicles, 15 or 20 follicles were placed into Erlenmeyer flasks containing amphibian Ringer's solution (usually 15 ml) and incubated in a Dubnoff metabolic shaker at 22°C. Incubation was normally terminated by fixation of the ovarian follicles; this was accomplished by heating the contents of the incubation flasks to the boiling point. All *in vitro* manipulations and experiments were conducted in simple salt solution (amphibian Ringer's or Holtfreter's), which usually contained penicillin G and streptomycin sulfate. Radioactively labelled and unlabelled steroids were dissolved in propylene glycol:ethanol (1:1) and added to the media. The gonadotrophic hormones utilized in these studies were triturated frog pituitary glands, commercially available preparations (PMS, HCG), or those provided by the National Institutes of Health (LH, FSH, LTH, STH).

Ovulation of oocytes from the ovarian follicle was distinguished by (1) the flattened appearance of the oocyte, (2) the presence of the irridescent vitelline membrane, and (3) the absence of the follicular membranes which contain the very prominent blood vessels (Figs. 25.2, 25.3). Ovulation can therefore be scored either by the basis of the number of oocytes or the number of follicular membrane capsules present in the incubation media.

Disintegration and disappearance of the oocyte germinal vesicle was the criterion utilized to indicate the interruption of oocyte arrest. Germinal vesicle breakdown (GVBD) was assessed following dissection of the individual oocytes or follicles. In oocytes which do not undergo nuclear disintegration, the germinal vesicle in fixed oocytes can be dissected intact as a large, white sac-like structure

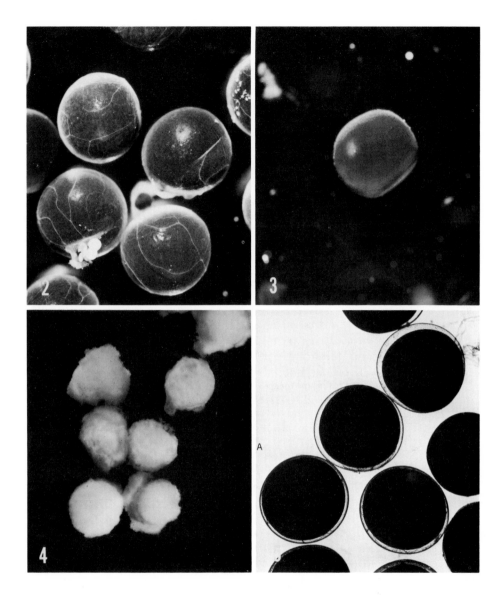

Fig. 25.2 Isolated ovarian follicles, in a state of follicular arrest, following dissection from the gonad of a hibernating frog *(Rana pipiens)*. Small previtellogenic oocytes and follicles are attached to the larger pigmented oocytes   (X 20).

Fig. 25.3   Isolated oocytes of *Rana pipiens* following ovulation from the ovarian follicle   (X 20).

Fig. 25.4   Germinal vesicles (nuclei) isolated from oocytes following fixation of ovarian follicles in follicular arrest   (X 40).

Fig. 25.5   Separation of follicular wall membranes from the oocyte following fixation of ovarian follicles exposed to progesterone and incubated for 24 hours   (X 20).

(Fig. 25.4). The germinal vesicle can also be isolated from unfixed follicles; however, greater care is required in rupturing the oocyte to prevent damage to the nucleus. Alterations in the association of the follicular membranes with the oocyte were also studied (Fig. 25.5). Subsequent to fixation of certain follicles, a clear, fluid-filled space formed between what appears to be the follicular wall tissues and the oocyte proper. This change was recorded as membranes raised or membrane separation.

III. INTERRUPTION OF FOLLICULAR ARREST

A. Gonadotrophins

Stimulation of the amphibian ovarian follicle by pituitary gonadotrophic hormone(s) is the mechanism by which follicular arrest is considered to be interrupted (Rugh, 1935a, b, 1962; Dodd, 1960). This principle applies whether the follicular processes are studied under *in vivo* or *in vitro* conditions. Ovarian follicles removed from hibernating frogs and placed *in vitro* remain in follicular arrest (no ovulation or germinal vesicle breakdown) even after several days of culture. Exceptions to this occur on rare occasions, particularly as the time of the natural breeding season approaches. Addition of amphibian pituitary gland extracts to such ovarian follicles *in vitro* results in the initiation of ovulation and oocyte maturation in a proportion of the treated follicles (Heilbrunn, Daughterty and Wilbur, 1939; Ryan and Grant, 1940). The time at which ovulation occurs *in vitro* varies considerably depending upon the season, incubation conditions and amount of hormone present (Wright, 1945).

The individual ovarian follicles rather than other ovarian tissues appear to be the only cells required for the interruption of follicular arrest by gonadotrophic hormone. Ovulation and maturation are produced by the pituitary hormones even after removal of small follicles and all extraneous tissue from the individual large follicles.

The chemical and biological properties of the gonadotrophic hormones in the amphibian pituitary have not been well established; however, they appear to be exceedingly potent and specific in their action when compared to mammalian gonadotrophins (Witschi and Chang, 1959; Hoar, 1966; Schuetz, 1971a). Alternatively, the amphibian ovarian follicles exhibit a considerable degree of selectivity *in vivo* and *in vitro* as to the types of gonadotrophins which are capable of interrupting follicular arrest (see Hisaw, 1947; Creaser and Gorbman, 1939; Dodd, 1960, for a review of the earlier literature with regard to ovulation). Numerous aspects of hormone and follicular specificity, however, remain controversial. In many cases this appears to be due to the use of impure hormones, *in vivo* interactions of injected hormones and endogenous hormones, and the utilization of animals before follicular development and differentiation are completed (reviewed by Geschwind, 1966).

Only a limited number of proteinaceous mammalian hormones have been shown to interrupt follicular arrest *in vitro*. Bergers and Li (1960) produced ovulation from ovarian fragments *(Rana pipiens)* *in vitro* with ICSH (LH) and growth hormone (STH), but not with FSH or LTH. These results were obtained, however, only after an initial *in vivo* "priming" or stimulation of the gonads with injected amphibian pituitary glands some 18 hours prior to *in vitro* assay. Schuetz (1971a) compared the activity of gonadotrophic hormones and growth hormone on both ovulation and maturation in individual isolated follicles from unprimed animals *(Rana pipiens)*. Induction of ovulation and maturation was obtained with mammalian pituitary LH or frog pituitary tissue (Table 25.1). FSH, PMS, LTH, HCG and growth hormone induced neither ovulation nor oocyte maturation. Edgren and Carter (1961) also found HCG to be relatively ineffective in stimulating ovulation *in vitro*. Furthermore a synergistic action between FSH and LH on either ovulation or maturation was not observed *in vitro* (Bergers and Li, 1960; Schuetz, 1971a). These observations suggest that both ovulation and maturation activity are present in a single gonadotrophin (LH) molecule, and presumably also in the amphibian pituitary

Table 25.1 **Ovulation and Oocyte Germinal Vesicle Breakdown Following Exposure of Isolated Ovarian Follicles *(Rana pipiens)* to Gonadotrophic Hormones or Progesterone***

Treatment	Amount	Total No. Follicles Examined	Ovulated Oocytes		Unovulated Oocytes	
			No.	No. w/GVBD	No.	No. w/GVBD
PMS	1-500 I. U.	420	0	0	420	0
HCG	1-500 I. U.	400	2	0	398	0
FSH	1-500 μg	460	7	0	453	0
LTH	1-1000 μg	220	4	0	216	0
LH	1 μg	120	1	0	119	0
	10 μg	120	3	0	117	0
	100 μg	120	42	38	78	26
	500 μg	40	38	38	2	2
Frog Pit. (equiv.)	1	120	66	64	54	21
	1/4	120	91	80	32	17
	1/10	120	88	80	29	17
Progesterone (10 μg)		120	0	0	120	120
Control		120	1	0	119	0
Vehicle Control		120	0	0	120	0

*From Schuetz, 1971a

hormone. The inactivity of some of the gonadotrophic hormones is striking in view of their inherent LH-like activity (PMS, HCG) and their steroidogenic properties (FSH, PMS, LTH, HCG) in mammals. Considerable evidence to be discussed suggests that steroid hormones produced in the follicle are responsible for the ovulation and oocyte maturation. Thus a central question remains as to whether this inactivity of the gonadotrophins is due to their inability to interact with the follicle tissues or due to the nature of the steroid which may be secreted by the follicular tissues. The actual fate of the gonadotrophic hormones during the course of ovulation and oocyte maturation induction is also unclear. Biologically active LH can be detected in incubation flasks following the induction of ovulation and maturation (Schuetz, unpublished). This suggests that complete hormone inactivation or its depletion from the medium does not occur as a result of exposure to the follicular tissues. Likewise, the physiological changes and the hormonal requirements initiated by the priming activity of the pituitary gonadotrophins remain to be elucidated.

B. Steroids

Evidence for the direct involvement of steroid hormones in interrupting follicular arrest was originally based on the demonstration that ovulation could be induced *in vitro* in the presence of steroids. Zwarenstein (1937) stated that ovulation of oocytes from ovarian fragments of *Xenopus laevis in vitro* occurred following exposure to progesterone. Steroid induction of ovulation *in vitro* was subsequently demonstrated in a variety of species including *Rana pipiens* (Bergers and Li, 1960; Wright, 1961a; Rugh, 1962).

Table 25.2 **Effect of Progesterone on Oocyte Germinal Vesicle in Isolated Ovarian Follicles of *Rana pipiens****

Treatment	Dose		Total No. of oocytes examined	Total No. of oocytes with intact germinal vesicles
Control			120	120
Vehicle control	.1	ml	120	120
Progesterone	.01	μg	120	120
	.1	μg	120	6
	1.0	μg	120	0
	10	μg	120	0
	100	μg	120	0
	1000	μg	120	60
Progesterone +	10	μg		
Versene (.1 M)	1.5	ml	120	120
Versene (.1 M)	1.5	ml	120	120

*From Schuetz, 1967a

The effects of steroids on germinal vesicle breakdown were evaluated by Schuetz (1967a) on isolated ovarian follicles. Addition of progesterone over a wide range of doses (Table 25.2) results in germinal vesicle breakdown within a 24-hour incubation period.

Follicular responsiveness to steroids is, however, dependent on or related to the structural and biological properties of the steroid hormones when either ovulation or germinal vesicle breakdown is considered. Wright (1961a) obtained ovulation from ovarian fragments *in vitro* with progestational, androgenic and certain adrenal cortical hormones but not with estrogenic steroids. Table 25.3 presents the effects of various steroids on germinal vesicle breakdown. The common steroid precursor cholesterol as well as estradiol were ineffective, whereas steroids similar to those

Table 25.3 **Effect of Various Steroids on Oocyte Germinal Vesicle Breakdown in Isolated Ovarian Follicles of *Rana pipiens****

Hormone (μg/15 ml)	Total No. of oocytes of 60 examined with intact germinal vesicles			
	0	.1	1.0	10.0
Control	60			
Vehicle control	60			
Estradiol		60	60	60
Androstenedione		30	30	30
Desoxycorticosterone	60	60	0	0
Dehydroepiandrosterone	6	60	60	30
Dehydroepiandrosterone sulfate		60	60	60
Pregnenolone		60	30	3
Pregnenolone sulfate		60	60	60
Cholesterol		60	60	60
Testosterone		60	40	1
Hormone (μg/15 ml)	0	1.	10.	100.
Control	60			
Vehicle control	60			
Corticosterone		60	38	0
Hydrocortisone		60	30	18
Androstanedione		60	60	52
Pregnanediol		60	15	0
Progesterone		0	0	0
Aldosterone		60	60	40
17a-OH-pregnenolone		60	35	15

*From Schuetz, 1967a

reported by Wright (1961) to stimulate ovulation also induced germinal vesicle breakdown. Thus the intrafollicular process of oocyte maturation and ovulation appear to be sensitive to the same types of steroid hormones. Steroid induction of germinal vesicle breakdown and ovulation *in vitro* has also been observed in the toad, *Bufo bufo* (Thornton and Evennett, 1970) and in the sturgeon, *Acipenser stellatus* (Dettlaff and Skoblina, 1969).

The efficacy of progesterone in inducing germinal vesicle breakdown is striking. Essentially all of the large oocytes undergo germinal vesicle breakdown in response to progesterone (Table 25.2, Schuetz, 1967a; Smith, Ecker and Subtelny, 1968). The capacity of progesterone to stimulate germinal vesicle breakdown, furthermore, occurs over the major part of the hibernation period and is more efficacious than pituitary stimulation. When oocyte responsiveness to progesterone actually develops is not clear since the smaller previtellogenic oocytes do not respond to the steroid.

Germinal vesicle breakdown does not appear to be an immediate response to progesterone, nor does it require the continual presence of the steroid. Both pituitary- and progesterone- induced germinal vesicle breakdown, however, are adversely affected by low temperature (Table 25.4). Nuclear disintegration begins a

Table 25.4 **Effect of Temperature on Pituitary and Progesterone Induced Ovulation and Oocyte Maturation** *In Vitro*

Treatment	Response		
22° C	Total Follicles Examined	No. Oocytes Ovulated	No. Oocytes w/GVBD
Pit. .05	100	20	75
Pit. .15	100	27	81
Prog 3 μg	100	10	99
Control	100	0	0
4° C			
Pit. .05	100	0	0
Pit. .15	100	0	0
Prog 3 μg	100	0	0
Control	100	0	0

variable number of hours (usually more than 10 hours, depending on the animal) following exposure to the steroid but less than a 30-minute exposure to the hormones is sufficient to initiate maturation (Schuetz, 1967b). A close synchronization of germinal vesicle breakdown and the succeeding maturational events also is evident in oocytes treated with steroids which initiate this process.

The actual breakdown of the germinal vesicle occurs over a period of several hours and appears to involve processes of disintegration and translocation of nuclear and cytoplasmic components. Histological and histochemical studies of this process in a variety of amphibians (Brachet, Hanocq and Gansen, 1970) and in the sturgeon (Dettlaff and Skoblina, 1969) indicate that the basal end of the germinal vesicle is initially affected and undergoes a process of vesiculation. Ultrastructural changes in oocytes of *Xenopus laevis* at the time of maturation have also been described (Gansen, 1966; Gansen and Schram, 1968).

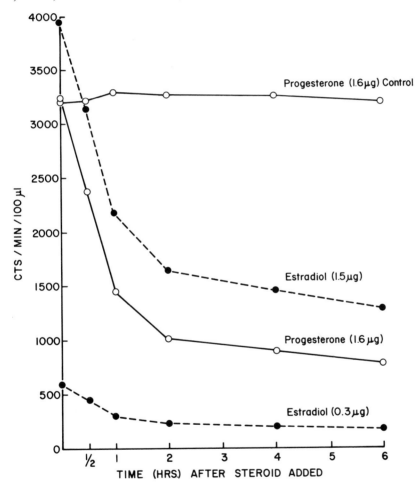

Fig. 25.6 Depletion of radioactive steroid hormones (estradiol [14]C [.19μc/1μg] progesterone [14]C [.16μc/1μg]) from media incubated *in vitro* in the presence or absence of amphibian ovarian follicles. Incubation flasks contained 100 ovarian follicles in a total volume of 10 cc of amphibian Ringer's solution. At the designated intervals a 100 μl sample of the incubation media was removed, placed in a scintillation vial and the level of radioactivity determined. Both the physiologically active (progesterone) and inactive steroids (estradiol) were depleted from the incubation media.

The inability of estrogens to stimulate ovulation and maturation as compared to progesterone is striking. Whether this was related to differences in the uptake of the steroid into the ovarian follicles was investigated with the use of radioactively labelled steroids. The results of these studies indicate (Fig. 25.6) that biologically active and inactive hormones were taken up by the ovarian follicles essentially to the same extent. Examination of the follicles after 24 hours indicated that maturation was induced in all follicles exposed to progesterone but not in the estrogen-treated follicles. Assessment of the distribution of radioactively labelled progesterone [^{14}C] within intact follicles or various components of the ovarian follicle indicate that (Table 25.5) the majority of the label was present within the oocyte cytoplasm. The

Table 25.5 **The Distribution of Progesterone [^{14}C] within Intact Ovarian Follicles and its Various Components Following** *in Vitro* **Incubation**

Structure	N	3	6	24
		cts/min/10 structures (mean ± S.D.) time (hrs) after progesterone added		
Ovarian follicles	3	21,477 ± 269	22,785 ± 557	23,582 ± 889
Follicle wall membranes	3	91 ± 14	42 ± 10	132 ± 4
Germinal vesicles	3	101 ± 28	218 ± 31	
Vitelline membranes	3			13 ± 5

Progesterone [^{14}C] (.25 μc/1.6 μg) was added to 3 Erlenmeyer flasks, each containing 100 ovarian follicles in 10 ml of amphibian Ringers' solution. At intervals designated 30 follicles from each flask were removed and washed (2 x in 15 ml of cold Ringers' solution). The 30 follicles were divided into 3 groups of 10 follicles each: Group I- follicles were fixed (boiled) and the germinal vesicle isolated before being placed in scintillation vials; Group II- whole follicles were directly transferred to scintillation vials; Group III- follicle wall membranes and vitelline membranes were dissected from unfixed follicles and then placed in scintillation vials. Scintillation fluid contained solubilizer BBS-3.

follicular membranes, vitelline membrane and germinal vesicle contained only small amounts of radioactivity at the times examined. Whether this distribution is a reflection of the major site of action of this hormone is not clear.

C. Relation Between Ovulation and Oocyte Maturation

Steroid hormones as well as gonadotrophins induce ovulation and oocyte maturation. To what extent steroids can directly substitute for the gonadotrophins, and whether ovulation and maturation are equally affected by the different hormones are questions of importance.

Under natural conditions ovulation and oocyte maturation (germinal vesicle breakdown) appear to occur at approximately the same time (Rugh, 1962). Because of this close association, however, various interpretations exist as to (1) the relationship between changes in the follicular wall and those in the oocyte, and *vice versa,* and (2) the relationship between ovulation and maturation to the stimulus initiating these follicular events. In many cases various interpretations have resulted because only a certain population of the ovary or follicles (ovulated) were studied following experimental treatment. Ryan and Grant (1940) observed that 13 of 138 ovulated oocytes contained an intact germinal vesicle some 40 hours after ovulation occurred, and concluded that ovulation is independent of maturational changes in the oocyte. Studies of Wright (1961) suggested that steroids were equally or more effective than frog pituitary homogenates in inducing ovulation, and could therefore directly substitute for the pituitary hormones. Dettlaff (1966) considered that the changes in the follicle and oocyte were not causally connected. Evidence to be discussed subsequently, however, strongly supports the hypothesis that these events are to some extent causally interrelated. Considerable evidence now indicates that these two parameters of interrupted follicular arrest *in vitro* usually are not observed simultaneously in individual follicles or when all treated follicles (ovulated and unovulated) are examined (Schuetz, 1967a, b, 1971a; Subtelny, Smith and Ecker, 1968).

The effect of various doses of progesterone and frog pituitary tissue on the aforementioned follicular responses as well as on membrane separation is presented in table 25.6. Frog pituitary homogenates ovulated more follicles but stimulated less

Table 25.6 **Ovulation and Oocyte Maturation** *In Vitro:* **Comparison of the Effects of Progesterone and Frog Pituitary Tissue***

Treatment		Response		
	No. Follicles	No. oocytes ovulated	No. oocytes w/GVBD	No. follicles w/follicular membrane raised
Progesterone - Amt.				
1 μg	100	10	100	90
3 μg	100	14	100	86
9 μg	100	14	100	86
Pituitary (Equivalent)				
1/20	100	33	80	46
3/20	100	59	79	16
9/20	100	48	64	12
Control	100	0	0	0

From Schuetz, 1971

germinal vesicle breakdown than the steroid. Essentially all unovulated oocytes in which germinal vesicle breakdown was induced also exhibited membrane separation. Maturation of oocytes was induced in all steroid treated follicles but only 14% of the oocytes were ovulated. Maturation without ovulation therefore occurred in the greatest majority of the follicles examined. This suggests that the utilization of only one of these parameters as an indication of interrupted follicular arrest leads to widely divergent results. Seasonal factors also appear to be involved in the decreased effectiveness of the pituitary hormones to initiate maturation (Table 25.6). The incidence of maturation in response to pituitary hormones in the late spring is similar to that produced by progesterone (Fig. 25.7) and again maturation without

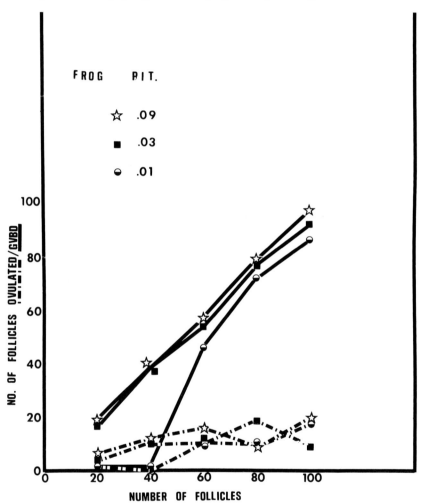

Fig. 25.7 Relationship between ovulation and maturation in individual ovarian follicles. Incubation flasks containing different numbers of ovarian follicles (20 - 100) were exposed to homogenized frog pituitary glands and incubated for 24 hours. Three doses of frog pituitary were tested and the incidence of ovulation and maturation was determined in all follicles following fixation. Maturation without simultaneous ovulation was observed in the majority of follicles regardless of the dose of the pituitary utilized. The experiment was conducted during the month of May.

simultaneous ovulation is evident. Some evidence also indicates that progesterone acts synergistically with pituitary hormones in inducing ovulation at a time when the gonads are refractory to gonadotrophic hormones alone (Wright and Flathers, 1961).

The simplest explanation for the absence of a high correlation between ovulation and maturation in individual follicles is that 2 different elements (the follicle wall and the oocyte) are responding to the hormones and their development may not be synchronized. Differences in the sensitivity of the various parameters of the follicle are not readily apparent to a particular steroid hormone. Likewise, since the induction of maturation by gonadotrophins appears to result from materials released from the follicle wall, the quantities released may be sufficient to affect only one parameter (ovulation), but insufficient to alter the oocyte germinal vesicle. The asynchrony between the two events in culture may be an expression of an asynchrony in the differentiation of these two follicular components at the time of removal from the animal, or possibly due to adverse effects of the culture system on one parameter. In any case the fact that oocytes removed from the follicle and matured with progesterone can be fertilized and develop *in vitro* suggests that oocyte differentiation is completed prior to the follicle wall (Smith, Ecker and Subtelny, 1968).

IV. MECHANISMS OF HORMONAL INTERRUPTION OF FOLLICULAR ARREST

A. Site of Action of Hormones

Where the gonadotrophic hormones act within the ovarian follicle to interrupt follicular arrest is a question of major importance; however, it is one that has been rarely considered in the past. In general, the primary locus of hormone action has been assumed to be the follicle wall since it is in this structure that alterations are required prior to ovulation. Where gonadotrophins act to initiate the process of oocyte maturation has been less clear. Dettlaff, Nikitina and Stroeva (1964) originally considered that gonadotrophic induction of germinal vesicle breakdown was the result of intraoocyte effects. Considerable evidence now suggests that the gonadotrophins initiate maturation as well as ovulation and membrane separation indirectly as a result of intermediate processes in the follicular wall. This hypothesis has resulted primarily from studies on the responsiveness to hormones of isolated oocytes and follicle tissue following microdissection (Schuetz, 1967b; Masui, 1967; Smith, Ecker and Subtelny, 1968; Dettlaff and Skoblina, 1969).

A reduction in the incidence of oocyte maturation in response to pituitary but not progesterone hormones was observed following manual removal of the follicular tissues (Schuetz, 1967b) (Table 25.7). In addition, Masui (1967) and Smith, Ecker and Subtelny (1968) found that removal of the follicle cells was rarely accomplished by manual dissection. With the complete removal of these cells by calcium deficient medium, oocyte responsiveness to pituitary hormones, but not the steroid progesterone, was lost. Masui also (1967) presented evidence that isolated follicular

Table 25.7 **Effect of Actinomycin D, Puromycin and Removal of Follicular Tissues on the Response of the Frog Oocyte Germinal Vesicle to Progesterone or Pituitary Stimulation***

Treatment	No. animals from which oocytes obtained (of 26)	Total no. oocytes examined treatment	Total no. oocytes examined w/ GVBD	% GVBD of total examined	No. animals of total responding w/ GVBD
Follicular Tissue Present					
Control (Vehicle)	26	390	0	0	0
Progesterone (10 μg)	26	390	389	99.7	26
Progesterone + Puromycin (100 μg)	20	300	30	10.0	2
Progesterone + Actinomycin D (100 μg)	22	330	329	99.7	22
Pituitary Homog. (¼ - ½ pit)	26	390	300	76.9	21
Pituitary Homog. + Puromycin (100 μg)	20	300	0	0	0
Pituitary Homog. + Actinomycin D (100 μg)	22	330	15	4.6	1
Follicular Tissue Removed					
Control (Vehicle)	22	330	0	0	0
Progesterone (10 μg)	20	300	295	98.3	20
Pituitary Homog. (¼ - ½ pit)	26	390	156	40.0	12

*From Schuetz, 1967b

tissues released a maturation-inducing substance *in vitro* following gonadotrophic stimulation. The nature of this material has not been determined. A similar reduction in oocyte responsiveness to gonadotrophins but not steroids, following removal of the follicular wall, has also been observed in the toad *Bufo bufo* (Thornton and Evennett, 1969) and in the sturgeon *Acipenser guldenstadti colchicus* (Dettlaff and Skoblina, 1969). In the sturgeon the follicle wall including the follicle cells can be separated from the oocyte without pretreatment of the follicle with enzymes or calcium free media.

These experimental data from a variety of species suggest that the gonadotrophic hormones, with respect to oocyte maturation, stimulate the release of substance(s) from one compartment of the ovarian follicle (the follicle wall) and these substances initiate changes in the second follicular compartment, the oocyte. The evidence strongly implicates the steroids as the mediators of these events.

The capacity of steroids to initiate maturation of oocytes denuded of their follicular tissues is evidence for the direct action on these substances on the oocyte (Schuetz, 1967; Masui, 1967; Smith, Ecker and Subtelny, 1968). Whether steroids initiate nuclear disintegration by a direct effect on the germinal vesicle is presently not clear. Steroids initiate oocyte cytoplasmic maturational changes without the presence of the germinal vesicle. Smith and Ecker (1969) removed the germinal vesicle from afolliculate oocytes and subsequently exposed them to progesterone. Such oocytes would respond to an artificial activation stimulus, and undergo cortical granule disintegration and vitelline membrane elevation. Changes in the oocyte cortex independent of germinal vesicle breakdown have also been obtained in sturgeon oocytes (Skoblina, 1969).

B. Biochemical Processes

Hormonal interruption of follicle arrest appears to require the stimulation of both RNA and protein synthesis. Where these events occur within the ovarian follicle, and whether they are directly involved in the process studied, is in many cases unclear. Ovulation induced by either pituitary hormones (Dettlaff, 1966; Anderson and Yatvin, 1970) or progesterone (Yatvin and Pitot, 1969) is inhibited by actinomycin D and puromycin. Oocyte maturation, whether induced by gonadotrophins or steroids, is inhibited by metabolic inhibitors of protein synthesis (Dettlaff, 1966; Schuetz, 1967b; Dettlaff and Schmerling, 1969) (Table 25.7). Steroid induced maturation, in contrast to steroid induced ovulation or pituitary induced maturation, was not inhibited by actinomycin D (Schuetz, 1967b; Smith and Ecker, 1969). These observations suggest that steroid initiation of nuclear maturation in the oocyte occurs at the site of translation rather than transcription. More definitive evidence for the involvement of RNA and protein synthesis in interrupting follicular arrest is required. Autoradiographic studies of the incorporation of RNA precursors into the ovarian follicle indicate that the follicle wall is a site of intense synthetic activity at various stages of folliculogenesis (Kessel and Panje, 1968; Anderson and Yatvin, 1970).

C. Estrogens and Follicular Functions

Considerable evidence indicates that estrogenic steroids have major and dicotomous influences (stimulatory and inhibitory) on the terminal stages of follicular differentiation and the process of follicular arrest. The stimulatory effect of estrogen on oocyte growth and vitellogenesis in amphibians is well documented (see Wallace, chapter 10) and their synthesis is presumably due to gonadotrophic hormone stimulation of steroidogenesis in the ovary (Follett, Redshaw and Nicholls, 1968). The direct involvement of estrogens on the process of follicular arrest is suggested by *in vitro* experiments. Estrogens over a wide range of doses are strikingly inactive in stimulating ovulation (Wright, 1961a), oocyte maturation (Schuetz, 1967a), or membrane separation (Schuetz, 1971b). In contrast, interruption of follicular arrest (ovulation) stimulated by gonadotrophic hormones was inhibited by estrogens (Wright, 1961b). Schuetz subsequently (1971b) demonstrated that gonadotrophin (LH or amphibian pituitary) induced oocyte maturation (Fig. 25.8)

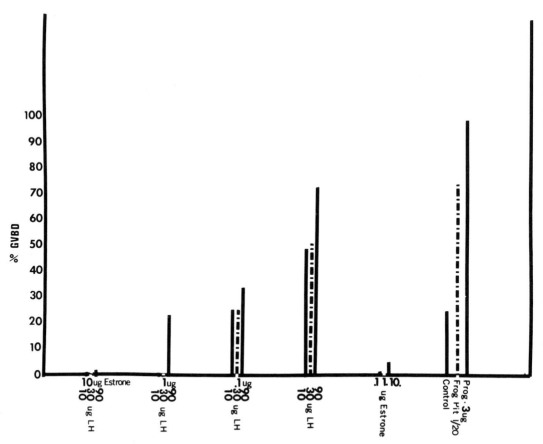

Fig. 25.8 Estrone inhibition of LH induced germinal vesicle breakdown. Ovarian follicles were incubated with varying concentrations of pituitary LH (NIH - LH - S11) and/or estrone for 24 hours and the incidence of germinal vesicle breakdown was then assessed. The results presented in the figure represent the per cent of the total number examined in which germinal vesicle breakdown occurred. Eighty follicles from four frogs (20 follicles/flask/animal) were utilized in each treatment group.

and membrane separation as well as ovulation were inhibited by estrone. Inhibition of these events by the estrogens was, furthermore, not obtained when the addition of the estrogen was delayed, suggesting the formation of intermediate substances. Inhibitory effects of estrogens were, however, dependent upon the nature of the hormone utilized to interrupt follicular arrest. Ovulation (Wright, 1961a), oocyte maturation and membrane separation all occurred when follicles were simultaneously incubated with progesterone and estrogens (Schuetz, 1971b). These data suggest that the estrogens alter the mechanism by which gonadotrophins initiate the changes in the follicle rather than by altering the capacity of the follicle and oocyte to undergo physiological alterations. The follicle wall appears to be the site of gonadotrophin inhibition by the estrogens.

Whether endogenous levels of estrogens in the ovary can exert an inhibitory influence on gonadotrophins remains to be elucidated. Their unique capacity of simultaneously exhibiting within the oocyte and follicle wall inhibitory and stimulatory influences as well as being inactive in stimulating or preventing (progesterone induced) ovulation, maturation, and membrane separation strongly suggests that the estrogens could play a major regulatory role in follicular differentiation and maintaining follicular arrest without having detrimental effects on the follicular components. In any case, the changing sensitivities of the follicle to estrogenic and progestational compounds during the course of follicle differentiation bear a striking resemblance to the events in the mammalian follicles. Whether these changing sensitivites are reflections of alterations in the types and/or amounts of steroids produced by the follicle during follicle differentiation remains to be elucidated.

V. STEROIDOGENESIS IN AMPHIBIANS

Considerable physiological and limited biochemical evidence indicates that steroids are present or synthesized in the amphibian ovary (Dodd, 1960; Forbes, 1961; Gottfried, 1964). Both estrogenic and progestational steroids appear to be present in gonadal tissue and have been identified with methods of varying specificity. Estrogenic steroids have been demonstrated in extracts of whole ovaries of *Bufo vulgaris* (Chieffi and Lupo, 1963), *Xenopus laevis* (Gallien and Chalumeau Le Foulgoc, 1960), and in the blood of *Rana esculenta* (Polzinetti-Magni, Presco, Rostozi, Bellini-Cardellini and Chieffi, 1970), and *Rana temporaria* (Cedard and Ozon, 1962). The blood estrogens appear to be ovarian in origin and the quantities secreted change during the course of oogenesis. Progesterone was isolated from ovarian extracts of *Bufo vulgaris* (Chieffi and Lupo, 1963) and identified chromatographically.

Studies on the conversion of radioactively-labelled precursors indicate the presence of steroidogenic enzymes, including those involved in the formation of progestational and estrogenic steroids, in gonad tissue of *Necturus maculosus* and *Rana pipiens* (Callard and Leathem, 1966) and *Rana temporaria* (Ozon and Breuer, 1964). Extractions of whole gonads, however, provide little information concerning the localization of the steroidogenic processes.

Histochemical studies clearly implicate the ovarian follicular tissues rather than the oocyte as being the cellular site of steroid synthesis. Jolly (1965) found that the Δ^5-3β hydroxysteroid dehydrogenase activity was localized to a greater extent in the follicular epithelium surrounding the oocyte than in the thecal cells, but not in the oocyte in the larger follicles of *Salamandra salamandra*. The general absence of steroidogenic enzymes in the oocyte and their presence in the follicle wall has also been observed in a variety of other vertebrates including teleost fish (Bara, 1965), dogfish shark (Simpson, Wright and Hunt, 1968), birds (Chieffi and Botte, 1965; Boucek and Savard, 1970; Sayler, Dawd and Wolfsen, 1970), and mammals (Pupkin, Bratt, Weisz, Lloyd and Balogh, 1966). These observations on the localization of steroidogenic enzymes are consistent with the physiological evidence presented in this review as to the importance of the follicular epithelium in mediating the gonadotrophin-induced meiotic maturation. The cellular, physiological and molecular mechanisms involved in the synthesis and secretion of the steroids within the ovarian follicle during the course of oogenesis remain to be elucidated.

VI. FUNCTIONAL MATURATION OF OOCYTES *IN VITRO*

Whether oocytes stimulated to undergo germinal vesicle breakdown under *in vitro* conditions also exhibit normal physiological maturation is a question of major significance to understanding the relationship between the oocyte and the follicle wall. Once the germinal vesicle has disintegrated a series of transformations in the nucleus (chromosomes and nucleoplasm) and cytoplasm occur which characterize the ongoing progression of oocyte development. These morphological and biochemical changes observed in amphibian follicles include: (1) formation of the meiotic spindle; (2) the various stages of meiosis including polar body formation; (3) increased contractility of the oocyte cortex (Dettlaff, 1966); (4) alterations in the distribution of ions (Morrill, 1965); (5) cortical granule alterations (Smith and Ecker, 1969); (6) extrusion of Feulgen positive bodies into the cytoplasm (Brachet, Hanocq and Gansen, 1970); (7) separation of the vitelline membrane from the oocyte following artificial activation (Smith and Ecker, 1969); (8) annulate lamellae disintegration or vesiculation (Balinsky and Devis, 1963); and (9) alteration of the cytoplasm to stimulate DNA synthesis (Gurdon, 1967; Gurdon and Speight, 1969). All of these changes have been observed or implicated following steroid-induced germinal vesicle breakdown *in vitro* in *Rana pipiens*. The most demanding criterion, however, of physiological maturation would be to demonstrate that development of matured oocytes can be effected. Ryan and Grant (1939) demonstrated that some of the oocytes of *Rana pipiens* ovulated *in vitro* by pituitary hormones and subsequently fertilized developed into normal tadpoles. Smith, Ecker and Subtelny (1968) showed that oocytes *(Rana pipiens)* removed from their follicles (some follicle cells may have been attached) and matured *in vitro* with progesterone would undergo normal development to the tadpole stage following fertilization (Table 25.8). Masui (1967) furthermore demonstrated that the follicular membranes presented no major impediment to early embryonic development. Unovulated

Table 25.8 **Development of Eggs Transferred to Foster Females***

Total number of eggs transferred	Number of experiments	Total number of eggs recovered	Number cleaved	Development, number normal at			
				Blastula	Gastrula	Neurula	Tadpole
681	7	307	229	229	225	219	201

*From Smith, Ecker and Subtelny (1968)

oocytes *(Rana pipiens)* matured *in vitro* in response to progesterone, were implanted with blastula nuclei and were observed to undergo normal cleavage to the early gastrula stages (Figs. 25.9, 25.10). These data indicate that intrafollicular oocytes

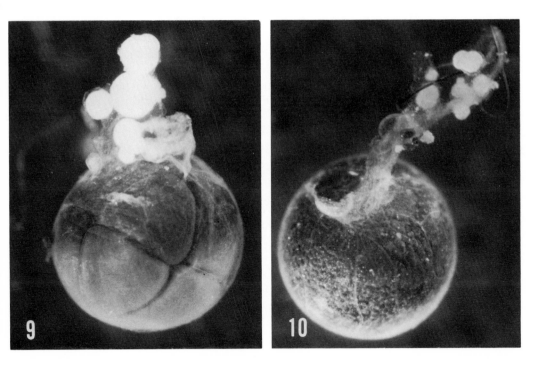

Figs. 25.9 and 25.10 Two ovarian oocytes into which blastula nuclei were injected 68 hours after the administration of progesterone. These oocytes underwent cleavage inside the follicular envelope. The ovarian epithelium may be seen still attached to the oocyte. Immature oocytes are visible in this tissue, and blood vessels may also be noted on the surface of the follicular envelope encapsulating the cleaving egg. (Fig. 25.9) a cleaving egg in the 8-cell stage. The small bleb on the left side indicates the site of nuclear implantation. (Fig. 25.10) an oocyte that has reached the blastula stage after nuclear implantation (X 50) (From Masui, 1967).

require neither ovulation, pituitary hormones, or follicular membranes for maturation to occur when progesterone is used to stimulate germinal vesicle breakdown. Differentiation of the oocyte appears therefore to have been completed in the arrested ovarian follicle at the time of the removal from the animal. When oocytes attain the capacity to undergo complete physiological maturation during the course of follicular differentiation remains to be elucidated.

In the past, induction of ovulation has been considered to be the only mechanism by which mature eggs could be obtained from the ovary. The studies discussed here clearly indicate that whether ovulation occurs is of little consequence in obtaining mature oocytes from the amphibian ovarian follicle. Steroid hormones completely duplicate the gonadotrophic hormones in their capacity to initiate normal germinal vesicle breakdown and oocyte maturation.

VII. STRUCTURAL ALTERATIONS IN THE OVARIAN FOLLICLE

A. Membrane Separation

A striking and readily apparent indication of hormonally induced alterations in the ovarian follicles is the separation of the follicle membranes from the oocyte in unovulated follicles following fixation (Fig. 25.5, Table 25.6). Alterations in the structure of the ovarian follicle of both hormone and non-hormone treated follicles are produced, however, by fixation. The data (Table 25.9) illustrate that after

Table 25.9 **Amphibian Oocyte and Follicular Dimensions Before and After Boiling: Effect of Progesterone**

Treatment	No. Oocytes	Length (mm) Animal-Vegetal Axis		Oocytes with GVBD
		Before Boiling	After Boiling	
Prop. glycol	19	1.65 ± .08*	1.81 ± .07*	0
Progesterone 1 μg	20	1.64 ± .06	1.65 ± .07	20
Progesterone 3 μg	20	1.66 ± .08	1.64 ± .06	20
Control	20	1.65 ± .09	1.85 ± 11	0

* (Mean ± S.D.)

culture for 24 hours, but before fixation, ovarian follicles have essentially the same dimensions. Following fixation, the oocyte and follicle either both increase in size or swell, but with no separation of the membrane from the oocyte, or show a separation of the follicle membranes from the oocyte with little evidence of oocyte swelling. The first case is invariably observed in unovulated non-hormone treated ovarian follicles, whereas the latter is observed in unovulated hormone treated

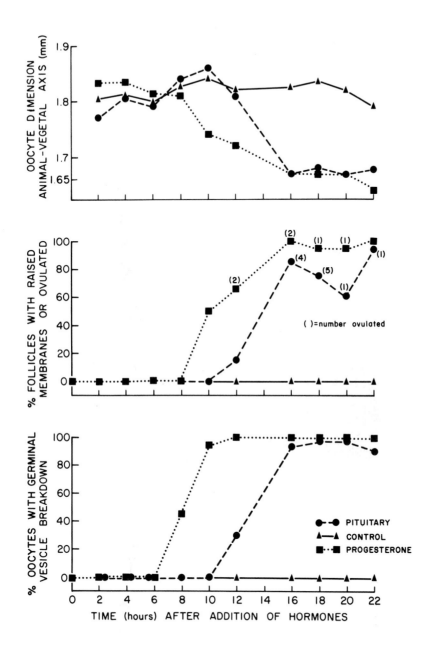

Fig. 25.11 Time course for ovarian follicular changes (germinal vesicle breakdown, follicle membrane separation, oocyte size) following fixation of ovarian follicles exposed to pituitary hormones or progesterone *in vitro*. Each point represents the mean response from a total of 20 follicles examined. All three responses were recorded from the same 20 follicles. Different flasks were examined at the different time intervals (From Schuetz, 1972).

follicles and essentially in all hormone treated follicles in which germinal vesicle breakdown does occur (Table 25.9). The swelling of non-hormone treated follicles appears to be the result of the accumulation of incubation medium within the oocyte. Figure 25.11 illustrates the time course during which membrane separation, ovulation and germinal vesicle breakdown occur following treatment with either progesterone or frog pituitary tissue. All three changes in the hormone treated follicles occurred after a several hour lag period. The events in both the pituitary and progesterone treated follicles were essentially the same except that they occurred 2-4 hours earlier in the progesterone treated follicles. This lag period in the case of the gonadotrophin treated follicles is presumably due to the time required for the formation of follicular intermediate substances.

The types of steroids and gonadotrophic hormones which initiate membrane separations are the same as those which stimulate germinal vesicle breakdown and ovulation (Schuetz, 1971a) (Table 25.1). Hormone induction of membrane separation from the oocyte is also inhibited by puromycin and cycloheximide indicating the necessity for protein synthesis in this process (Schuetz, 1972).

The actual cellular changes which are responsible for the observed separation of the membranes following fixation were investigated by microdissection of the unfixed cultured treated ovarian follicles. These results are summarized in Table

Table 25.10 **Ovulation, Oocyte Maturation and Vitelline Membrane Formation in Amphibian Ovarian Follicles: Effect of Progesterone and Frog Pituitary Hormones**

Treatment	No. animals	No. follicles	Response / 100 follicles or oocytes		
			No. ovulated	No. with GVBD*	No. of complete vitelline membranes isolated
Progesterone 3 µg	5	100	10	100	100
Frog pituitary 1/20	5	100	36	76	74
Frog pituitary 3/20	5	100	55	70	70
Control	5	100	0	0	0
Steroid Vehicle (.1 cc)	5	100	0	0	0

Twenty ovarian follicles from each animal were exposed to each treatment, incubated *in vitro* for 24 hours and then individually dissected.

*Germinal vesicle breakdown

25.10 and indicate the cellular attachments to the vitelline membrane are the major alteration produced by the hormones. In control follicles, follicle cells remain attached to the outer vitelline membrane and oocyte cortical cytoplasm remains attached to the inner surface of the vitelline membrane. Following hormone treatment (frog pituitary or progesterone), in those unovulated follicles in which germinal vesicle breakdown occurred, the vitelline membrane is isolated essentially free of follicle cells and oocyte cortical cytoplasm (Fig. 25.12). Similar changes

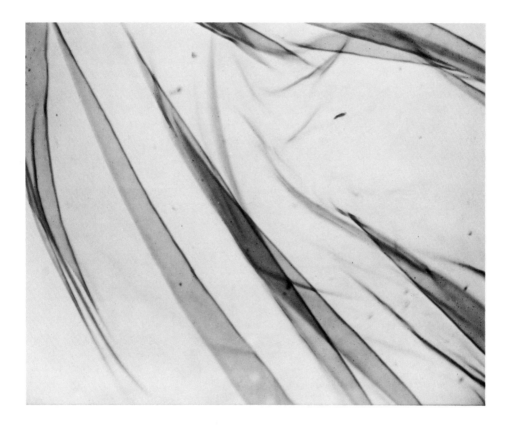

Fig. 25.12 Vitelline membrane isolated from an unovulated ovarian follicle after 24 hours of incubation with progesterone. (X 250)

presumably occur in the ovulated follicles because the vitelline membrane can also be isolated free of oocyte cortical cytoplasm from these ovulated oocytes. Thus, separation of the follicle cells from the vitelline membrane can be physically accomplished by either progesterone treatment of follicles followed by dissection, by washing follicles in calcium deficient media (Masui, 1967), or by pronase digestion (Smith and Ecker, 1969). The separation of the follicular membranes from the oocyte following fixation appears, therefore, to be the result of the interruption of the attachment of the oocyte of follicle cells to the vitelline membrane. It is thought that this detachment is the result of the retraction of macro- and micro-villus projections between the vitelline membranes and follicle cells and oocytes. Metabolic inhibitor studies furthermore indicate that hormone-induced membrane separation involves protein synthesis (Schuetz, 1972). Evidence for villus retractions has been noted before at the electron microscopic level in amphibian follicles under natural conditions (Wischnitzer, 1963; Bourne, Hope and Humphries, 1963); gonadotrophic stimulation (Gansen, 1965); and in progesterone treated follicles (Smith, Ecker and Subtelny, 1968). An interpretation of the hormone-induced structural alterations in the ovarian follicle is diagrammatically depicted in Fig. 25.13.

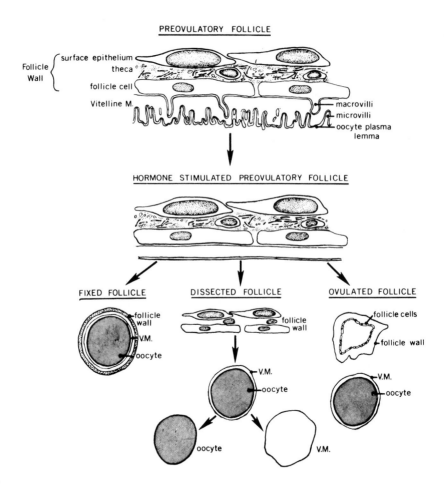

Fig. 25.13 Diagrammatic representation of the structural changes which are considered to occur in the preovulatory amphibian ovarian follicle following pituitary or progesterone treatment. The effects of the pituitary are presumably mediated by intermediates produced in the follicle wall.

B. Relation of Structural Changes to Follicular Functions

Rugh (1935b) originally observed that the follicle cells were retained within the ovarian follicle wall, rather than attached to the oocyte following ovulation. How this separation was accomplished was unknown but the localized rupture of the ovarian follicle wall was considered to be the primary follicular event required for ovulation. The present data appear to indicate that structural changes over the entire surface of the ovarian follicle (separation of follicle cells) as well as a localized rupture are both necessary before ovulation can be accomplished. Although detachment of only the follicle cells from the outer surface of the vitelline membrane would appear to be necessary for ovulation to occur, the detachment of the oocyte from the vitelline membrane also occurs at the same time. The membrane separation is probably an important preliminary for vitelline membrane elevation following fertilization or artifical activation. The separation is also the probable explanation for the observation that unovulated progesterone treated follicles can be activated by pricking (Masui, 1967; Subtelny, Ecker and Smith, 1968).

The instances of membrane separation in the absence of ovulation noted in these studies are striking and support the idea that follicular rupture and membrane separation, although both induced by the same hormones, are separate events (Fig. 25.13). The actual rupture of the amphibian follicle is restricted to the area of the peritoneal epithelium and appears to involve synthetic and catalytic processes. Rugh (1935a) originally suggested that a localized enzymatic disintegration of the follicle may be the primary cause for follicle rupture. Recent work in mammals (Espey, 1967) indicates that disintegration of collagenous materials occurs in the area of follicle rupture. Similar changes in the amphibian ovarian follicle have also been described by Anderson and Yatvin (1970). Evidence suggests that a "collagenase"-like enzyme may be responsible for these changes (Espey and Rondell, 1968; Rondell, 1969). A striking consequence of experimentally stimulating membrane separation and maturation without ovulation is that a mature oocyte is entrapped within the follicle. The ultimate significance of this is that oocyte atresia may result, not particularly as a result of abnormal or degenerative processes in an abnormal oocyte, but merely as a result of the asynchronous timing which exists between oocyte maturation and the follicle's capacity to be ovulated. This is strongly supported by the data of Masui (1967) and Smith, Ecker and Subtelny (1968) which indicate that oocytes matured with steroid undergo normal development following fertilization or implantation of nuclei. Alternatively, the villus projections into the vitelline membrane have been implicated in the pinocytosis and the transfer and accumulation of nutrients into the oocyte from the body (Gansen, 1966). The interruption of these processes by the retraction of the villus projections may also markedly interfere with oocyte viability and lead to the oocyte atresia. In any case, a variety of cellular and biochemical events occur within the ovarian follicles in response to a single hormone. The synchronization of the biochemical and structural events appears to be required for normal follicular and oocyte development whereas asynchrony probably leads to abnormal processes.

VIII. SUMMARY

Evidence discussed here clearly indicates that the endocrine regulation of the processes of follicular arrest in amphibians can be studied under *in vitro* conditions. Gonadotrophic hormones act on the differentiated ovarian follicle to initiate three closely synchronized events: ovulation, separation of the follicle cells and oocyte from the vitelline membrane, and germinal vesicle breakdown in the oocyte. There is little evidence to indicate, however, that the gonadotrophins are directly involved in initiating any of these follicular occurrences. Steroid hormones, depending upon their structure, stimulate all three of these follicular changes, and these three events furthermore exhibit the same types of specificity in their responsiveness to the steroid hormones. Progestational, adrenal cortical and androgenic steroids can initiate these follicular changes whereas estrogenic steroids are ineffective. Estrogenic steroids, in contrast, act to inhibit these three follicular events when gonadotrophins are utilized to initiate them, but not when incubated in the presence of stimulatory steroids (progesterone). Estrogens, however, do have stimulatory effects on follicular growth and differentiation primarily in mediating vitellogenesis. The ovarian follicle wall rather than the oocyte appears to be the site of gonadotrophic hormone stimulation of intermediate processes which are responsible for the induction of the three structural changes discussed. In particular, the follicle cells appear to be the cellular site involved in the mediation of gonadotrophin induced oocyte maturation and for determining gonadotrophin specificity. Evidence is discussed which indicates that steroidogenesis occurs in the amphibian ovarian follicle and is restricted to the somatic tissue (follicle wall). The occurrence of the three events within individual follicles is to some extent asynchronous and the significance of these asynchronies to normal and abnormal follicular processes of ovulation and atresia are discussed. Normal physiological maturation of the oocyte, however, can be achieved by steroids alone in either the presence or the absence of follicle wall or simultaneous ovulation.

Thus it appears that the closely synchronized events within the preovulatory ovarian follicle may have a common basis for their induction and that this is necessary to release the oocyte from the follicle and to prepare the oocyte for fertilization. The evidence suggests that the synthesis and/or release of steroid(s) from the follicle wall, and its subsequent local action within the two compartments of the ovarian follicle (follicle wall and oocyte) provides a simple means of accomplishing the synchronization of these events with a single mediator.

IX. ACKNOWLEDGMENTS

Financial support for these studies provided by the National Institutes of Child Health and Human Development (Research Grant 5-RO1-HD-03797), the Ford Foundation, and the National Cancer Institute (5TI-CA-5079) is gratefully acknowledged.

X. REFERENCES

Anderson, J. W. and Yatvin, M. B. (1970) Metabolic and ultrastructural changes in the frog ovarian follicle in response to pituitary stimulation. *J. Cell. Biol.* **46**, 3.

Austin, C. R. and Walton, A. (1960) Fertilization. *In* "Marshall's Physiology of Reproduction". (A. S. Parkes, ed.), Vol. 1, pt. 2, p. 310. Boston, Little Brown.

Balinsky, B. I. and Devis, R. J. (1963) Origin and differentiation of cytoplasmic structures in the oocyte of *Xenopus laevis. Acta Embryol. Morph. Exp.* **6**, 55.

Bara, G. (1965) Histochemical localization of Δ^5-3β-hydroxysteroid dehydrogenase in the ovaries of a teleost fish, (*Scomber scomber* L.). *Gen. Comp. Endocrinol.* **5**, 284.

Bergers, A. C. J. and Li, C. H. (1960) Amphibian ovulation *in vitro* induced by mammalian pituitary hormones and progesterone. *Endocrinology.* **66**, 255.

Blandau, R. J. (1961) Biology of eggs and implantation. *In* "Sex and Internal Secretions". (W. C. Young, ed.), Vol. II, p. 797. Baltimore, Williams and Wilkins.

Boucek, R. J. and Savard, K. (1970) Steroid formation in the avian ovary *in vitro (Gallus domesticus).Gen. Comp. Endocrinol.* **15**, 6.

Brachet, J., Hanocq, F. and Gansen, Van P. (1970) A cytochemical and ultrastructural analysis of *in vitro* maturation in amphibian oocytes. *Develop Biol.* **21**, 157.

Callard, I. P. and Leathem, J. (1966) Steroid synthesis by amphibian ovarian tissue. *Gen. Comp. Endocrinol.* **7**, 80.

Cedard, L. and Ozon, R. (1962) Teneur en oestrogenes du sang de la Grenoville rousse (*Rana temporaria* L.). *Compt. Rend. Soc. Biol.* **156**, 1805.

Chieffi, G. and Lupo, C. (1963) Identification of sex hormones in the ovarian extracts of *Torpedo marmorata* and *Bufo vulgaris. Gen. Comp. Endocrinol.* **3**, 149.

Chieffi, G. and Botte, Y. (1965) The distribution of some enzymes involved in the steroidogenesis of hen's ovary. *Experientia.* **21**, 16.

Creaser, C. W. and Gorbman, A. (1939) Species specificity of the gonadotrophic factors in vertebrates. *Quart. Rev. Biol.* **14**, 311.

Dettlaff, T. A. (1966) Action of actinomycin and puromycin upon frog oocyte maturation. *J. Embryol. Exp. Morph.* **16**, 183.

Dettlaff, T. A., Nikitina, L. A. and Stroeva, O. G. (1964) The role of the germinal vesicle in oocyte maturation in anurans as revealed by the removal and transplantation of nuclei. *J. Embryol. Exp. Morph.* **12**, 851.

Dettlaff, T. A. and Schmerling, Zh. G. (1968) Puromycin-sensitive phases and protein synthesis in the period of oocyte maturation in *Rana temporaria. C. R. Acad. Sci. USSR (Russian).* **181**, 501.

Dettlaff, T. A. and Skoblina, M. N. (1969) The role of germinal vesicle in the process of oocyte maturation in *Anura* and *Acipenseridae. Annales d'Embryologie et de Morphogenese.* Suppl. **1**, 133.

Dodd, J. M. (1960) Gonadal and gonadotrophic hormones in lower vertebrates. *In* "Marshall's Physiology of Reproduction". (A. S. Parkes, ed.), Vol. 1, pt. 2, p. 417. Boston, Little Brown.

Edgren, R. A. and Carter, D. L. (1961) Blockade of progesterone-induced *in vitro* ovulation of *Rana pipiens* with chorionic gonadotropin. *Gen. Comp. Endocrinol.* **1**, 492.

Espey, L. L. (1967) Ultrastructure of the apex of the rabbit Graafian follicle during the ovulatory process. *Endocrinology.* **81**, 267.

Espey, L. L. and Rondell, P. (1968) Collagenolytic activity in the rabbit and sow Graafian follicle during ovulation. *Am. J. Physiol.* **214**, 326.

Everett, J. W. (1961) The mammalian female reproductive cycle and its controlling mechanisms. *In* "Sex and Internal Secretions". (W. C. Young, ed.), Vol. I, p. 497. Baltimore, Williams and Wilkins.

Follett, B. K., Redshaw, M. R. and Nicholls, T. J. (1968) The vitellogenic response in the South African clawed toad (*Xenopus laevis* Daudin).*J. Cell. Physiol.* Suppl. **72**, 91.

Forbes, Thomas R. (1961) Endocrinology of reproduction in cold-blooded vertebrates. *In* "Sex and Internal Secretions". (W. C. Young, ed.), Vol. II, p. 1035. Baltimore, Williams and Wilkins.

Franchi, L. L. (1962) The structure of the ovary. B. Vertebrates. *In* "The Ovary". (S. Zuckerman, ed.), Vol. 1, p. 121. New York, Academic Press.

Gallien, L. and Chalumeau-Le Fouloc, M. T. (1960) Mise en evidence de steroides oestrogenes dans l'ovaire juvenile de *Xenopus laevis* (Daudin) et cycle des oestrogens au cours de la ponte. *Compt. Rendus.* **251**, 460.

Gansen, Van P. (1966) Ultrastructure comparee du cytoplasme peripherique des oocytes murs et des oeufs vierges de *Xenopus laevis* (Batracien anoure). *J. Embryol. Exp. Morph.* **15**, 355.

Gansen, Van P. and Schram, A. (1968) Ultrastructure et cytochimie ultrastructurale de la vesicule germinative et du cytoplasme perinucleaire de l'oocyte mur de *Xenopus laevis*. *J. Embryol. Exp. Morph.* **20**, 375.

Geschwind, I. I. (1966) Species specificity of anterior pituitary hormones. *In* "The Pituitary Gland". (G. W. Harris and B. T. Donovan, eds.), Vol. 2, p. 589. Berkeley, University of California Press.

Gottfried, H. (1964) The occurrence and biological significance of steroids in lower vertebrates. A review. *Steroids.* **3**, 219.

Gurdon, J. B. (1967) On the origin and persistence of a cytoplasmic state inducing nuclear DNA synthesis in frogs' eggs. *Proc. Nat. Acad. Sci. U.S.A.* **58**, 545.

Gurdon, J. B. and Speight, V. A. (1969) The appearance of cytoplasmic DNA polymerase activity during the maturation of amphibian oocytes into eggs. *Exp. Cell. Res.* **55**, 253.

Harrison, R. J. (1962) The structure of the ovary. C. Mammals. *In* "The Ovary". (S. Zuckerman, ed.), Vol. 1, p. 143. New York, Academic Press.

Heilbrunn, L. V., Daughterty, K. and Wilbur, K. M. (1939) Initiation of maturation in the frog egg. *Physiol. Zool.* **12**, 97.

Hisaw, F. L. (1947) Development of the Graafian follicle and ovulation. *Physiol. Rev.* **27**, 95.

Hoar, W. S. (1966) Hormonal activities of the Pars Distalis in cyclostomes, fish and amphibia. *In* "The Pituitary Gland". (G. W. Harris and B. T. Donovan, eds.), Vol. 1, p. 242. Berkeley, University of California Press.

Hope, J., Humphries, A. A. Jr., and Bourne, G. H. (1963) Ultrastructural studies in developing oocytes of the salamander *Triturus viridescens*. I. The relationship between follicle cells and developing oocytes. *J. Ultrastruct. Res.* **9**, 302.

Ingram, D. L. (1962) Atresia. *In* "The Ovary". (S. Zuckerman, ed.), Vol. 1, p. 247. New York, Academic Press.

Jolly, M. J. (1965) Mise en evidence histochimique d'une Δ^5-3β-hydroxysteroide- dehydrogenase dans l'ovaire de l'urodele *Salamandra salamandra* (L.) a differents stades du cycle sexual. *C. R. Acad. Sci.* **261**, 1569.

Jorgensen, C. B. and Vijayakumar, S. (1970) Annual oviduct cycle and its control in the toad *Bufo bufo* (L.). *Gen. Comp. Endocrinol.* **14**, 404.

Kessel, R. G. and Panje, W. R. (1968) Organization and activity in the pre- and post-ovulatory follicle of *Necturus maculosus*. *J. Cell Biol.* **39**, 1.

Masui, Y. (1967) Relative roles of the pituitary, follicle cells and progesterone in the induction of oocyte maturation in *Rana pipiens*. *J. Exp. Zool.* **166**, 365.

Matthews, L. H. (1962) The structure of the ovary. A. Invertebrates. *In* "The Ovary". (S. Zuckerman, ed.), Vol. 1, p. 89. New York, Academic Press.

Morrill, G. A. (1965) Water and electrolyte changes in amphibian eggs at ovulation. *Exp. Cell Res.* **40**, 664.

Ozon, R. and Breuer, H. (1964) Untersuchungen uber den Stoffwechsel von Steroidhormonen bie Vertebraten. IV. Aromatisierung von Testosteron zu Ostrogenen im ovar des Frosches *(Rana temporaria).* *Hoppe-Seylers Z. Physiol. Chem.* **337**, 61.

Pincus, G. (1936) The Eggs of Mammals. New York, Macmillan Company.

Polzonetti-Magni, A., Lupo di Prisco, C., Rastogi, R. K., Bellini-Cardellini, L. and Chieffi, G. (1970) Estrogens in the plasma of females of *Rana esculenta* during the annual cycle and following ovariectomy. *Gen. Comp. Endocrinol.* **14**, 212.

Pupkin, K. A., Bratt, H., Weisz, J., Lloyd, C. W. and Balogh, K. (1966) Dehydrogenases in the rat ovary. I. A histochemical study of Δ^5-3β- and 20 α-hydroxysteroid dehydrogenases and enzymes of carbohydrate oxidation during the estrous cycle. *Endocrinology.* **79**, 316.

Raven, C. P. (1961) Oogenesis: The Storage of Developmental Information. Oxford, Pergamon Press.

Rondell, P. (1970) Follicular processes in ovulation. *Fed. Proc.* **29**, 1875.

Rugh, R. (1935a) Ovulation in the frog. I. Pituitary relations to induced ovulation. *J. Exp. Zool.* **71**, 149.

Rugh, R. (1935b) Ovulation in the frog. II. Follicular rupture to fertilization. *J. Exp. Zool.* **71**, 163.

Rugh, R. (1962) Experimental Embryology. Minneapolis, Burgess Press.

Ryan, F. J. and Grant, R. (1940) The stimulus for maturation and for ovulation of the frog's egg. *Physiol. Zool.* **13**, 383.

Sayler, A., Dawd, A. J. and Wolfsen, A. (1970) Influence of photoperiod on the localization of Δ^5-3β hydroxysteroid dehydrogenase in the ovaries of maturing Japanese quail. *Gen. Comp. Endocrinol.* **15**, 20.

Schuetz, A. W. (1967a) Effect of steroids on the germinal vesicle of oocytes of the frog *(Rana pipiens)* in vitro. *Proc. Soc. Exp. Biol. Med.* **124**, 1307.

Schuetz, A. W. (1967b) Action of hormones on germinal vesicle breakdown in frog *(Rana pipiens)* oocytes. *J. Exp. Zool.* **166**, 347.

Schuetz, A. W. (1969) Oogenesis: Processes and their regulation. *In* "Advances in Reproductive Physiology". (A. McLaren, ed.), Vol. 4, p. 99. London, Logos Press.

Schuetz, A. W. (1971a) *In vitro* induction of ovulation and oocyte maturation in *Rana pipiens* ovarian follicles. Effects of steroidal and nonsteroidal hormones. *J. Exp. Zool.* (In press).

Schuetz, A. W. (1971b) Estrogens and ovarian follicular functions in *Rana pipiens. Gen. Comp. Endocrinol.* (In press).

Schuetz, A. W. (1972) Induction of structural alterations in the preovulatory amphibian ovarian follicle by hormones. *Biol. Reprod.* (In press).

Simpson, T. H., Wright, R. S. and Hunt, S. V. (1968) Steroid biosynthesis *in vitro* by the component tissues of the ovary of dogfish *(Scyliorhinus caniculus). J. Endocrinol.* **42**, 519.

Skoblina, M. N. (1969) Independence of the cortex maturation from germinal vesicle material during the maturation of amphibian and sturgeon oocytes. *Exp. Cell. Res.* **55**, 142.

Smith, L. D., Ecker, R. E. and Subtelny, S. (1966) The initiation of protein synthesis in eggs of *Rana pipiens. Proc. Nat. Acad. Sci. U.S.A.* **56**, 1724.

Smith, L. D., Ecker, R. E. and Subtelny, S. (1968) *In vitro* induction of physiological maturation in *Rana pipiens* oocytes removed from their ovarian follicles. *Develop. Biol.* **17**, 627.

Smith, L. D. and Ecker, R. E. (1969) Role of the oocyte nucleus in physiological maturation in *Rana pipiens. Develop. Biol.* **19**, 281.

Subtelny, S., Smith, L. D. and Ecker, R. E. (1968) Maturation of ovarian frog eggs without ovulation. *J. Exp. Zool.* **168**, 39.

Thornton, V. F. and Evennett, P. T. (1970) Endocrine control of oocyte maturation and oviducal jelly release in the toad *Bufo bufo* (L.) .*Gen. Comp. Endocrinol.* **13**, 268.

Wischnitzer, S. (1963) The ultrastructure of the layers of developing yolk forming oocytes from *Triturus viridescens. Z. Zellforsch.* **60**, 452.

Wischnitzer, S. (1966) The ultrastructure of the cytoplasm of the developing amphibian egg. *Adv. Morph.* **5**, 131.

Witschi, E. and Chang, C. Y. (1959) Amphibian ovulation and spermiation. *In* "Comparative Endocrinology". (A. Gorbman, ed.), p. 149. New York, John Wiley.

Wright, P. A. (1945) Factors affecting *in vitro* ovulation in the frog. *J. Exp. Zool.* **110**, 565.

Wright, P. A. (1961a) Induction of ovulation *in vitro* in *Rana pipiens* with steroids. *Gen. Comp. Endocrinol.* **1**, 20.

Wright, P. A. (1961b) Influence of estrogens on induction of ovulation *in vitro* in *Rana pipiens. Gen. Comp. Endocrinol.* **1**, 381.

Wright, P. A. and Flathers, A. R. (1961) Facilitation of pituitary induced frog ovulation by progesterone in early fall. *Proc. Soc. Exp. Biol. Med.* **106**, 346.

Yatvin, M. B. and Pitot, H. C. (1969) Effect of the inhibition of RNA synthesis on *in vitro* ovulation by frog ovaries. *Nature (London).* **223**, 62.

Zwarenstein, H. (1937) Experimental induction of ovulation with progesterone. *Nature (London).* **139**, 112.

26

INTERACTION BETWEEN OOCYTES AND FOLLICULAR CELLS

A. V. Nalbandov

I.   Introduction
II. Experimental methods and results
III. Discussion
IV. Acknowledgments
V.   References

I. INTRODUCTION

This paper is based on research by a team of my co-workers who include Drs. M. E. El-Fouly, Stanislawa Stoklosowa, and Miss Mary Nekola whose individual contributions will be identified in the body of the paper. They set out to answer the question of what role, if any, the follicular ovum plays in the luteinization process which occurs after the egg is shed at ovulation. Perusal of the literature provided very little information on the subject but did contain hints that there may be a controlling relationship between the ovum and the cellular lining of the follicle. Thus, Falck (1959) showed that whole follicles (containing the ovum) when transplanted to the anterior chamber of the eye of rats did not luteinize, while similarly transplanted walls of follicles (without oocytes) were transformed morphologically into luteal cells resembling those present in corpora lutea. In a whole series of papers, Channing (see, for instance, 1970) showed that granulosa cells of horse, pig, monkey, and woman, when cultured *in vitro*, were transformed morphologically into what she calls "lutein cells" which are capable of producing progestin. It should be noted that in harvesting the granulosa cells for culture, the chances are overwhelming that the ovum is lost or destroyed and is thus not present in her cultures. These two studies hint that the presence of the ovum somehow prevents luteinization of cells lining the follicle and when absent allows follicular cells to synthesize progestin.

This problem was explored by direct experimental means and a three-pronged attack was initiated. First, it was decided to assess what would happen if the ovum were removed from the tertiary follicle by careful suction. This technique was perfected by Dr. El-Fouly. Second, the work of Channing was essentially repeated except that a conscious effort was made to incorporate many or few oocytes into the tissue culture of follicular cells. This study by M. Nekola is now in progress (Nekola and Nalbandov, 1971). Lastly, the work of Dr. Stoklosowa, who has incubated whole follicles in organ culture and studied their fate, will be described.

II. EXPERIMENTAL METHODS AND RESULTS

The technique of ovectomy, (removal of the ovum from tertiary follicles), is described by El-Fouly, Cook, Nekola and Nalbandov (1970). It involves the insertion of a fine needle through the stroma of a rabbit or pig ovary into an antral follicle and, by mild suction, the removal of the ovum from the cumulus oophorus. The success of the operation is always judged by the examination of the aspirated fluid and cellular debris under the microscope to verify the presence of the ovum and to estimate the quantity of granulosa cells removed. (If too much of the granulosa layer is removed the follicle usually degenerates.) To provide controls for ovectomized follicles two operations are performed, in one of which a needle is inserted and the liquor follicli is allowed to leak out; in the other the needle is inserted and the contents of the antrum are mildly agitated on the assumption that in many cases the ovum will be torn loose from the cumulus but will remain in the antrum. In both of these control operations the follicle reforms in the majority of cases, appears normal on microscopic examination (Figs. 26.1-26.3), and may be induced to ovulate normally with appropriate hormonal treatment. Such control follicles never become luteinized.

The story is drastically different in the case of ovectomy. It was found that 87% of the ovectomized follicles (as of this time many hundreds), are transformed into transient luteal structures which morphologically and endocrinologically are indistinguishable from corpora lutea of pseudopregnant rabbits. These structures are not solid like normally induced corpora lutea but have a hollow center and they last for only 6-8 days after which they degenerate (Figs. 26.1-26.3).

While this paper is not concerned with the endocrine problems involved, a few remarks seem in order. It should be noted that ovectomy or simple puncturing of follicles may be performed on a single ovary. Thus, an ovary may contain side by side normal untreated follicles, ovectomized follicles which have luteinized, and punctured follicles which have reformed. This fact suggests that the initial luteinization process resulting from ovectomy does not depend on any central hormonal control system because if it did, even the nonovectomized follicles should show morphological and endocrinological changes. The luteinization process, thus, seems to be due only to one fact: the removal of the ovum. It was also found by Cropper, El-Fouly, Cook and Nalbandov (1970) that, if one injects as little as $2\mu g$ of purified LH into the empty antrum immediately after ovectomy, it is possible to transform the transient luteinized, hollow ovectomized follicle into a solid corpus luteum which lasts as long as it would in pseudopregnant animals. The endocrinology of these findings has many interesting ramifications and it is being investigated further.

Figures 26.1-26.4 show the morphological transformation which occurs after follicular puncture or ovectomy and Table 26.1 supports the contention that the temporary structures produced by ovectomy are endocrinologically competent. Attention is also called to the fact that mere puncture of the follicle which allows the follicular fluid to leak out but usually does not damage the egg, causes such punctured follicles to produce significantly more progesterone than is produced by untreated follicles from estrous rabbits, but not as much as is synthesized by ovectomized follicles (Table 26.1).

Table 26.1 **Progestin Concentration in Ovarian Venous Plasma**

Group	Days post treatment	Progestin (ng/ml)	
		20a-OH	Progesterone
Ovectomy	3	497** ± 40 (6)[a]	245** ± 48 (7)
Puncture	3	945 ±137 (6)	72 ± 11 (6)
Ovectomy	8	1167 ± 77 (2)	31 ± 19 (2)
Pseudopregnant	3	70 ± 11 (3)	214 ± 47 (3)
Sham operated	3	425 ± 1 (2)	NM (2)
Estrus	—	118 ± 81 (2)	NM (2)

NM = not measurable (less that 25 ng/ sample).

[a] Mean ± standard error (number of animals).

** $p < 0.01$ for ovectomy (3 days) vs. puncture for both steroids.

Data from: El-Fouly *et al.* (1970) *Endocrinology* 87: 288-293.

The phenomenon discussed above is not restricted to rabbits. Preliminary but adequate data show that ovectomy also leads to follicular luteinization in pigs although the endocrinology of the process in pigs seems to be more involved than it is in rabbits.

Next, the work was extended to tissue culture of follicular cells. Here Miss

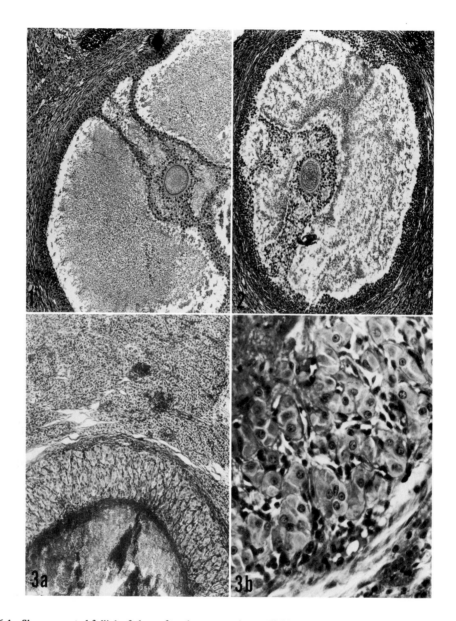

Fig. 26.1 Sham-operated follicle, 3 days after sham operation   (X 70).

Fig. 26.2 Punctured follicle 3 days after puncturing. Note that the antrum is refilled with fluid and that there is no evidence of luteinization of follicle wall   (X 70).

Fig. 26.3 *Left:* ovectomized follicle 3 days after ovectomy. Note the heavily luteinized follicle wall   (X 70).
Right: enlarged portion of picture on left (X 270), showing that the cells strongly resemble the morphology of the lutein tissue of the normal corpus luteum.

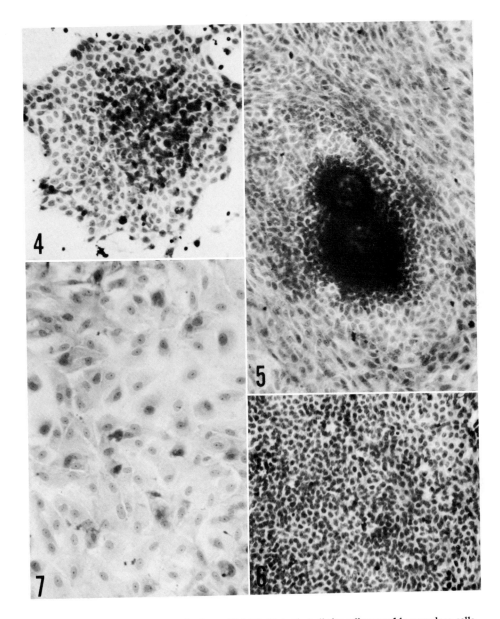

Fig. 26.4   Follicle cells after 18-24 hrs of culture (X 140). Note that all the cells resemble granulosa cells.

Fig. 26.5   Two oocytes and surrounding cells after 3 days of culture (X 140). Note that the cells close to oocyte are morphologically different from those on the periphery.

Figs. 26.6 and 26.7   Compare Figs. 26.6 and 26.7 and note distinct difference in type of cell population. Fig. 26.6 was taken after 5 days of culture and Fig. 26.7 after 8 days (both X 140). The cells in Fig. 26.6 resemble granulosa cells; those in Fig. 26.7 resemble lutein cells.

Nekola used rat ovaries which were dispersed by the use of pronase and shaking procedures which eventually provided her with cultures, some of which contained practically no oocytes while others contained many. (For details of the technique see Nekola and Nalbandov, 1971.) The numbers of eggs present in the cultures has no significance for the present discussion because the experiment was originally designed to study the effect of the number of eggs on the kind of hormones which predominate in the culture medium. (This work is now in progress.) The interesting and significant aspect of this work lies in the fact that when cultures containing ova were stained and examined under the microscope a distinct dichotomy of cell types was consistently noted. Thus, if one examines cover slips after the third day of incubation one notices that the morphological appearance of the cells immediately surrounding ova is totally different from the cells which are farther away from the oocytes (Figs. 26.4-26.7). The cells in close proximity to oocytes obviously resemble epithelial cells and are reminiscent of granulosa cells, while those further removed are fibroblastic in appearance and, while they can not be said to be typical lutein cells, they resemble them much more than do the cells close to the ovum. The difference between the two types of cells becomes strikingly obvious by comparing Figures 26.6 and 26.7 which are taken at the same magnification. These morphological differences do not exist on day 1 of culture (Fig. 26.4) which strongly suggests that the two different types of cells seen on later days of culture were derived from an originally homogeneous population of cells all of which resemble epithelial cells on day one of culture.

There is one other aspect of these studies that deserves passing comment. It is noted (Fig. 26.5) that the growth of granulosa-type cells immediately around the ovum is frequently much more prolific than the growth of these cells or of the luteal-type cells farther removed from the oocytes. This finding may imply that the ovum in some way stimulates growth of cells in its immediate vicinity.

The strong inference that the differentiation of follicular cells in tissue culture is in some way affected by the ovum, is strengthened by the work of Dr. Stoklosowa. In her studies she employs organ culture of whole follicles, both small and large, which she removes surgically and painstakingly from rats in estrus. The general course of events is approximately as follows. After one and occasionally two days of culture the ovum appears to remain normal or at least one can say that there are no obvious degenerative changes in its membrane or in the nucleus or nucleolus (Fig. 26.8). It is also noted that both the granulosa and the theca layers show no obvious morphological changes during the first two days of organ culture. However, after 3 or more days of culture the oocyte shows obvious degenerative changes including a breakdown of the membrane and of the nucleus. These changes occur in both small follicles lacking an antrum as well as tertiary follicles with an antrum (Figs. 26.9, 26.10, 26.11). When the egg begins to disintegrate the follicular cells also change morphologically and, in fact, those farthest away from the oocyte and closest to the theca show the most far-reaching changes. The cells which after 1 or 2 days of culture were obviously granulosa cells, now in the presence of a degenerated or degenerating ovum resemble lutein cells. We attribute this morphological change in follicular cells to the progressive decay of the ovum.

III. DISCUSSION

We have presented 3 lines of evidence, one derived from *in vivo* studies involving removal of the ovum from follicles and two lines involving experiments with *in vitro* culture of granulosa cells in the presence or absence of ova, and organ culture of whole follicles. All of the observations strongly suggest that the oocyte plays a major role in the differentiation of follicular cells. At present we are assuming that the ovum must be producing a luteostatic substance which prevents the luteinization of granulosa cells.

It is also concluded from all the lines of evidence presented, that this substance exerts its effect at the local level in that in culture the cells closest to the ovum retain their original morphological integrity while those at some distance from the oocyte do not. In the *in vivo* work involving ovectomy, the ovum may be removed from one single follicle in an ovary and only the ovectomized follicle will luteinize while the neighboring follicles will remain normal. This fact further shows that the effect of ovectomy is manifested purely locally and never involves other follicles in spite of their very close proximity to the ovectomized follicle. Even if all follicles on one ovary are ovectomized this does not affect the follicles in the contralateral ovary which remain normal and persist for prolonged periods of time. It is postulated that the luteostatic substance must also be present in the follicular fluid. Indeed, preliminary experiments show that the follicular fluid from pig follicles does contain a substance which reduces the rate of progesterone synthesis by formed corpora lutea in rabbits but so far the experimental procedures have not been designed to test its luteolytic ability. The chemical nature of this substance remains unknown but because it is not present in pig blood plasma, the substance may be restricted to follicular fluid which, in chemical composition, is very similar to plasma.

An alternative explanation can be proposed which postulates that instead of producing a luteostatic substance which is then transferred to the follicular fluid, the ovum metabolizes and thus removes a luteotrophic substance, perhaps a hormone, which may be present in follicular fluid. This idea does not seem probable on the basis of Nekola's experiments in which no follicular fluid is present in the incubation medium and in which only the cells farthest away from the ovum show signs of luteinization while the cells surrounding the oocyte resemble granulosa cells. This observation seems to suggest that something diffuses out of the ovum to maintain the morphological integrity of granulosa cells. Obviously, much more work is needed before this and many other problems raised by these findings can find a satisfactory explanation.

It has been pointed out that luteinization of the follicles may sometimes occur in the presence of the ovum, as it does, for instance, in cases when massive doses of LH are injected into rats. Here the follicle luteinizes without ovulating and the egg becomes entrapped in the lutein tissue. This fact is not considered contradictory to

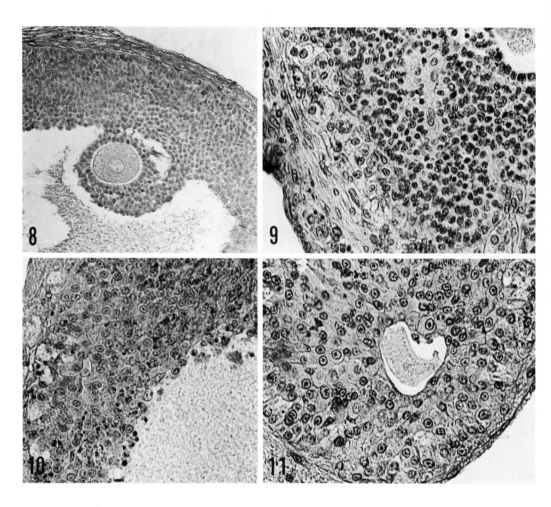

Fig. 26.8 Tertiary follicle after 1 day of culture (X 230). Note that the ovum appears normal and that there is no sign of luteinization.

Fig. 26.9 Large follicle after 3 days of culture (X 290). The oocyte (upper right) has degenerated and the theca and some of the granulosa cells have begun to luteinize.

Fig. 26.10 Wall of a tertiary follicle with a degenerated oocyte and luteinization after 3 days of culture (X 290).

Fig. 26.11 Small follicle with degenerated oocyte and extensive luteinization of the wall after 6 days of culture (X 290).

our hypothesis because it is obvious that the action of LH is not normal in that it does not cause ovulation, as it should, but causes luteinization. It appears possible that the abnormally high dose of LH has a deleterious effect on the ovum and the subsequent luteinization of the follicle may be due to the presence of a defective ovum.

Equally unexplained remains the fact that the follicles of many species begin to secrete progesterone prior to ovulation and prior to any obvious morphological changes in the granulosa and theca layers. (In mares and in women some luteinized cells are present in the pre-ovulatory follicle). The ability of the follicle to make progesterone prior to ovulation implies, in the light of the present discussion, that shortly before ovulation the oocytes lose some of their ability to restrain luteinization (as in women and mares), or to prevent progesterone synthesis (as in rabbits, guinea pigs, and many other species). Attention is called to the data in Table 1. which show that mere puncture of the follicle which apparently does not damage the egg and which allows the punctured follicles to reform without luteinizing, does increase their ability to synthesize progesterone very significantly. This raises the question of the degree to which either ovarian or hypophysical hormones are able to modify the ability of oocytes to regulate the morphology of follicular cells and to control synthetic pathways. This problem deserves attention, but has received little.

The work in progress in our laboratory has shown that it is not only possible to extirpate the ovum from follicles but it is feasible to inject hormones directly into single follicles and cause them to ovulate normally. These approaches provide useful techniques by which the effects of hormones on oocytes can be studied *in vivo*.

Finally, it should be pointed out that the role of the ovum on the endocrinology and morphology of follicular cells is not a phenomenon that is restricted to rabbits but has also been demonstrated in the domestic pig. It appears reasonable to assume that similar effects will be found in other species of mammals and that the phenomenon may be universal.

IV. ACKNOWLEDGMENTS

These studies were supported, in part, by NIH grant HD 3043 and contract 69-2135.

V. REFERENCES

Channing, C. (1970) Influences of *in vivo* and *in vitro* hormonal environment upon luteinization of granulosa cells in tissue culture. *In* "Recent Progr. Hormone Res." (E. B. Astwood, ed.), Vol. 26, p. 589. New York, Academic Press.

Cropper, M., El-Fouly, M., Cook, B., and Nalbandov, A. V. (1970) Corpus luteum formation of ovectomized follicles treated with LH. *Proc. Endocrine Soc. 52nd Meeting*. St. Louis, p. 80.

El-Fouly, M., Cook, B., Nekola, M., and Nalbandov, A. V. (1970) Role of the ovum in follicular luteinization. *Endocrinology.* **87**, 288.

Falck, B. (1959) Site of production of oestrogen in rat ovary as studied in micro-transplants. *Acta Physiol. Scand.* Suppl. **163**, 101.

Nekola, M. and Nalbandov, A. V. (1971) Morphological changes of rat follicular cells as influenced by oocytes. *Biol. Reprod.* **4**, (In press.).

INDEXES

SUBJECT INDEX

acid phosphatase, 88, 89, 111
  in follicle cells, 103
  in oocytes, 97-103
actinomycin D, 147, 175, 199, 234, 356,
  402, 404, 495
adenine, derivatives and oocyte maturation
  of, 459-476
adenosine ribohydrolase, 471
alkaline phosphatase, in primordial germ
  cells, 302
allopatric species, 328
α-amanitin, 195, 196, 198, 201, 202
amphibian, nuclear structure during oogenesis
  of, 119-138
androgenetic, 281
aneuploid, 280
  autosomal, 291-292
annulate lamellae, in oocytes, 21-23, 29
antimycin A, 356
Balbiani body, 17
blastocyst, 242, 448
  centriole in, 54
  delayed yolk of, 52
  parthenogenetic, 281, 283, 284
blastula, 232, 233
  genetic expression in, 2, 130, 136
  nuclei, 499
Call-Exner bodies, in follicles of human and
  rabbit, 23-25
capacitation, 444
cell cycle, G$_2$-phase, 362, 368, 369
  S-phase. See DNA synthesis.
centrifugal stratification of mouse oocytes,
  83
centriolar satellite, 55, 56, 66, 84, 93
centrioles, 296
  during oogenesis, 9, 29, 54, 93
  in oocyte, 55, 57, 111
centromere. See kinetochore.
Chiari-Frommel syndrome, 447
cholesterol, 487
chromatin, 196
  association with histones, 142
  control of transcription, 157
  during mouse oocyte maturation, 68-72
  in cat cells, 280
  in primordial germ cells, 9
chromosomal mosaics, 292-293

chromosome, 55, 56, 69, 72, 83, 215, 234,
  448
  lampbrush, 121, 123-126, 130-133, 137,
  138, 171, 227, 401
  maternal, 1
  paternal, 1
  proteins of, 141-158
colchicine, 75, 286
"collagenase"-like enzyme, 480
conjugation, 5
  see also fertilization
corpora lutea, 514
cortical granules, 27, 29, 93, 100, 103, 498
culture media, amphibian Ringer's, 229, 482
  Eagle's medium, 303, 306, 379, 381
  Ham's F10, 441, 445
  Holtfreter's, 482
  Krebs-Ringer, 66, 89, 242
  199, 441
  Waymouth's medium, 303
cyanide, 356
cycloheximide, 356, 502
cytochrome oxidase, 219
cytoplasmic bridges, 14, 15
cytoplasmic inheritance, 221-223
desmosome, 96, 97, 103
deuterostome, 130, 131, 137
DNA, 121, 125, 126, 130-138, 142, 147, 171,
  194, 196, 197, 202
  chloroplast, 215
  cytoplasmic, 215-233, 234
  interaction with histones, 153
  in Xenopus ovary, 170
DNAase, 121
DNA polymerase, 154-156
DNA synthesis, 150, 232, 362, 365, 368, 369,
  374, 498
endocrine, control of oogenesis, 2
endoplasmic reticulum, 53, 315
  association with Golgi, 93
  association with mitochondria, 60-62
  during oocyte maturation, 27, 75
  during oogenesis, 21
  formation of acid hydrolases, 111
energy metabolism. See metabolic activity.
estradiol, 407, 487, 489
estradiol-17β, 343, 344, 460
estrogen, 343, 344, 345, 356, 375, 392, 496-497

estrous cycle, 361, 362, 365
estrus, 366, 378, 385, 518
1-ethyladenine, 467, 468, 469
exocytosis, 112
Fallopian tube, 5, 35, 381, 425, 430
fat body, 322, 323
fertilization, 1, 2, 9, 29, 60, 228, 279,
    286, 293, 332, 393, 425, 498
    of *in vitro* matured oocytes, 405, 420
fluoride, 356
follicle, *see also* Graafian follicle.
    organ culture of, 518
follicle cells, 262, 271, 306, 317, 324, 352,
    393, 415, 448
    acid phosphatase localization, 103
    DNA of, 217
    horseradish peroxidase localization, 107
    influence on oocyte development, 248-249
    interaction with oocytes, 513-522
    RNA of, 171
    ultrastructural changes during oogenesis,
       9-38, 88, 93, 97
follicle stimulating hormone (FSH), 248, 375,
    378, 392, 397, 400, 441, 444, 445,
    482, 485, 486
follicular fluid, 418, 445, 448, 515, 519
    formation of, 23
follicular function and hormones, 479-506
follicular growth, in culture, 314
    in mouse ovary, 361-375
    organ culture of mouse ovary, 377-393
    ultrastructural changes in, 15-25
folliculogenesis, 2
gap junction, 97
gastrulation, 151, 219, 325
gene, 2
    ribosomal, 168, 170-175, 177-179, 181-183,
       186-187, 193-212
genetic information, utilization of, 129-138
genital ridges, 6, 7, 321, 322
germarial cystocytes, 254-263
germinal vesicle, 68, 79, 172, 227, 402, 413,
    414, 415, 459, 472
    breakdown, 228, 441, 463, 464-470, 482-506
    DNA of, 170
glycogen, 48, 50, 55
Golgi complex, 55
    change during oocyte maturation, 27, 29, 79
    in follicle cells, 23
    in oocyte during oogenesis, 21, 93-113
    in primordial germ cell, 9

gonadogenesis, 2
gonadotrophins. *See specific hormone.*
gonad stimulating substance, 460-464
Graafian follicle, 88, 93, 111, 112, 384,
    398, 401, 408
    maturation in organ culture, 377-393
granulosa cells, 361, 362, 363, 368, 393,
    401-402, 513-522
gynogenesis, 282, 283
haploid development, 280-283
hermaphrodite, 336, 397
heterochromatin, 66-84
heteroploidy, in mammalian egg, 279-298
histones, 142-154
    changes in development of, 151-152
    characterization of, 143-146
    degradation of, 150-151
    fractions of, 146-149
    interaction between DNA and, 153-154
    synthesis of, 149-150
horseradish peroxidase, 88, 89, 103-107, 111
human chorionic gonadotrophin (HCG), 174,
    328, 341, 343, 344, 351, 354, 355,
    378, 381, 388, 390, 392, 408, 415,
    418, 427, 429, 430, 441, 444, 448,
    482, 485, 486
human menopausal gonadotrophin (HMG), 444,
    445, 448
hyaluronidase, induction of parthenogenesis,
    279, 280, 448
Δ^5-3β hydroxysteroid dehydrogenase,
    498
insulin, 355
intercellular junctions, 25
interconnecting canals, 256
iodoacetate, 356
kinetochore, 56, 65-84
lampbrush chromosome. *See* chromosome.
laparoscopy, 444, 445, 448
lipovitellin, 341, 348, 349, 379
liquor folliculi. *See* follicular fluid.
LTH, 482, 485, 486
luteinization, 2, 513, 514, 515, 518
luteinizing hormone (LH), 248, 375, 392, 397,
    408, 418, 482, 485, 514, 519
luteostatic substance, 519
lysosomes, 88, 97, 103, 107, 111, 317
male pronucleus growth factor, 407, 408
maturation of oocyte *in vitro*, 25, 27,
    29, 32, 66-84, 418-422, 439-448
maturation of oocyte related to ovulation,

413-432
meiosis inducing substance, 460, 470
  see also l-methyladenine
menstrual cycle, 420, 427, 441, 445
menstrual period, last (LMP), 445, 447
metabolic activity of oocyte, 17, 93, 111,
  112, 217, 219, 241-249, 393, 415
metaphase I, 75-79, 383, 388, 390, 400, 401,
  414, 418
metaphase II, 383, 388, 390, 400, 401, 414,
  418, 420
metestrus, 366
l-methyladenine, 2, 460, 465, 467, 468,
  471-472
l-methyladenosine, 467, 468, 469, 471
l-methyladenosine ribohydrolase, 470-471
microtubules, 55, 56, 57, 68, 72, 75, 79,
  83, 256
microvilli, 17, 27, 32, 103, 111, 315, 393
mitochondria, 58, 59, 61, 216, 219, 313, 315
  and permeability changes in, 245
  association with endoplasmic reticulum during
  oogenesis, 60
  DNA in, 215-223
  during maturation, 27, 79
  during oogenesis, 21, 57-62, 103
  genetic content of, 2
  in primordial germ cell, 9, 303
  origin of embryonic, 223
  protein synthesis in, 220
  RNA in, 219-223
mitochondrial DNA, 168, 170, 215-223
mitochondrial RNA, 168, 219-223
morula, 54, 441
  centriole in, 55
  histones in, 151
  mitochondria of, 60
neurula, 322, 325
non-disjunction, 266-267, 288-292
nucleolar genes, 121-123
nucleolus, 119-138, 194, 227
nucleus, changes during oogenesis, 9-12
nurse cells, 132, 138, 242, 244, 248
  mutations influencing development of,
  267-268
oligomycin, 356
ovarian factor, 460
  see also meiosis-inducing substance
ovary, organ culture of, 307, 310, 311, 314,
  315, 317, 318, 378-380

tissue culture of, 305-307
ovectomy, 514, 515, 519
oviduct. *See* Fallopian tube.
ovulation, 32, 35, 60, 174, 365, 377, 415,
  482, 484, 486, 496, 521
  of immature stages, 422-425
paracentrosome, 56-57
parthenogenesis, 228, 393, 404, 445, 448
  and heteroploidy in the mammalian egg,
  279-298
pheromone, 461
phosvitin, 341, 348, 349, 357
pinocytosis, 88, 111, 112, 349, 356
pituitary, 229, 485, 491
pituitary gonadotropins, 459, 460, 484-493,
  495
pituitary homogenate, 459
polar bodies, 35, 151, 401, 441, 444, 448,
  474, 475, 498
  first, 29, 32, 79, 83, 285, 383, 388
  second, 32, 281, 283, 284, 285
pregnancy, 365, 441
pregnant mares serum gonadotrophin (PMSG),
  378, 379, 388, 390, 392, 415, 430, 482,
  485, 486
preimplantation period, centrioles during, 55
  heteroploidy during, 280
  metabolic activity during, 241-249
  mitochondria during, 60
  yolk as energy source during, 53, 55
primordial germ cell, 321-324
  migration of, 6-9
  movement of, 301, 302, 303, 305
progesterone, 229, 408, 459, 460, 472, 473,
  487, 488, 489, 493, 495, 497, 515, 521
progestin, 513
prolactin, 408
pronase, 503, 518
protein synthesis, 495, 502
  during oocyte maturation, 227-236
protein uptake by oocytes, 339-357
protostoma, 130, 131, 132, 137
pseudogastrulation, 234-235
puromycin, 79, 149, 356, 473-475, 495-502
radial nerve factor, 460
  see also gonad stimulating substance
ribonucleoprotein particle
  42S, 179, 181-182
  80S, 177, 179
ribosomes, 23, 27, 53, 72, 79, 83, 174, 175,

317
rifampicin, 199, 200
RNA, 130, 132, 133
  associated with chromosomes, 156-157
  in cytoplasm during maturation, 27
  in ovary of frog, 177-179, 181-183
RNAase, 121, 134, 149, 199
RNA-DNA hybridization, 130, 133, 134
RNA (heterogeneous), 131, 136, 171
RNA polymerase, 123, 126, 152, 154-156, 194,
  195-202
RNA synthesis, 131, 137, 152, 171-175, 227,
  495
  regulation of, in oocyte of *Xenopus laevis*,
  193-212
mRNA, 130, 136, 137, 169
  maternal, 130, 171-172
rRNA, 227
  5S, 170, 174-175, 177-179, 181-183, 186
  18S, 123, 170, 172, 175, 177-179, 181-183,
  186, 203, 211, 267
  20S, 123, 126
  28S, 123, 170, 172, 177-179, 181-183, 186,
  203, 267
  30S, 123, 126
  40S, 121
  regulation of synthesis of, 193-212
tRNA, 227
  4S, 168, 170, 172, 177, 203, 205
sperm (spermatozoa, spermatozoon, sperms),
  1, 2, 5, 9, 143, 223, 255, 282, 293, 425,
  441, 444, 445
  capacitated, 311, 312, 405
  treatment with ionizing radiation, 283
spermatocyte, 280
  of *Bombyx mori*, 256, 259
  of crayfish, 12
  of dipteran, 130
  of *Drosophila*, 138
spermatids, 56, 72
spermatogonia, 303
steroid hormones, effect on follicle, 486-506
steroidogenesis, in amphibians, 497-498
STH, 482, 485
synaptonemal complex, 242, 243
triploidy, 286-288
vincristine, 75
vitellogenesis, 336, 341-357
vitellogenin, incorporation by ovary, 345-349
  synthesis and turnover, 343-345

yolk, composition of, 50, 341
  formation of, 339-357
  in oocytes, 2, 47-55, 112, 216, 227,
  324, 325
yolk plate, 48, 49, 51, 216-217, 341, 342,
  347, 348, 349, 357
yolk sac, 302
zona pellucida, 23, 25, 35, 97, 103, 107,
  306, 308, 309, 311, 315, 393, 418
zygoid, 280, 283, 292

SPECIES INDEX

Acipenser
  A. guldenstadti colchicus
    germinal vesicle breakdown in, 495
  A. stellatus
    germinal vesicle breakdown in, 196
Amanita phalloides, 196
armadillo, oocyte maturation in, 416
  origin of primordial germ cells in, 301
Aspergillus niger
  mitochondrial RNA of, 220
Amblystoma
  RNA in germinal vesicle of, 172
ape, parthenogenetic development of, 278
Apodemus
  pre-reduction in, 288
bat, oocyte maturation in, 461
Bombyx mori
  mutant
    small egg (sm, 3-41.8), 270
  spermatocyte of, 255, 259
  sterile mutations in, 254
  vitellogenesis in, 270
Bufo
  B. bufo
    germinal vesicle breakdown in, 488,
    495
  B. vulgaris
    estrogens in ovary of, 497
    progesterone in ovary of, 497
cat, oocyte maturation in, 417, 422
  sex chromatin in cells of, 278
Chinese hamster, fibroblast of, 75
cow, histones of, 143, 148
  mitochondria in oocyte of, 60, 245
  oocyte maturation in, 392, 401, 402, 417
crayfish, spermatocyte of, 12
deer mouse, yolk in egg of, 50
dog, oocyte maturation in, 417, 431
Drosophila, 295
  genetic control of oogenesis of, 253-271
  mutants of:
    abnormal oocyte, 268
    almondex, 271
    crossover suppressor, 264-266
    deep orange, 271
    female sterile, 258, 260, 261
    female sterile (2)E, 267-268
    Fs(2)D, 262

    fused, 258, 260, 261
    morula, 267-268
    rudimentary, 271
    singed 36a, 267-268
    stubbloid 105 deficiency, 264
    suppressor of (Hairy wing)[2],
    267-268
    tiny, 271
  spermatocyte of, 138
  D. melanogaster
    synaptonemal complex in, 254
Engystomops
  duration of lampbrush chromosome stage in,
  131, 132
Ericulus
  oocyte maturation in, 416, 431
Escherichia coli
  chromosome of, 136
  mitochondrial RNA of, 220
  RNA polymerase of, 196-199
ferret, oocyte maturation in, 417
  parthenogenesis in, 279
  yolk in oocyte of, 53
goat, mitochondria in oocyte of, 60
guinea pig, Golgi complex in oocyte of, 21,
  27, 93-97
  mitochondria in oocyte of, 245
  oocyte maturation in, 417
  parthenogenetic development of, 278
  pinocytosis in oocyte of, 356
  yolk in oocyte of, 50
Habrobracon juglandis
  microtubules in ovarian cystocyte of, 256
hamster, annulate lamellae in oocyte of, 21
  cytoplasmic bridges in oocyte of, 14
  Golgi complex in oocyte of, 27
  mitochondria in oocyte of, 245
  oocyte maturation in, 417
  yolk in oocyte of, 50
horse, granulosa cells of, 513, 521
  oocyte maturation in, 417, 431
human (man, woman)
  annulate lamellae in oocyte of, 21, 29
  color blindness in, 288
  cortical granules in oocyte of, 29
  culture of ovaries of, 378
  cytoplasmic bridges during oogenesis in, 14
  fertilization *in vitro* of, 444-445

first polar body of, 32
Golgi complex in oocyte of, 21, 27
granulosa cells of, 513, 521
mitochondria in oocyte of, 245
oocyte maturation in, 417, 439-444
oogenesis in, 3
parthenogenetic development of, 278, 282,
283
primordial germ cells of, 301, 302
triploidy in, 285
Hyalophora cecropia
vitellogenesis in, 270
Lithodermium undulatum
paracentrosome of, 57
Locusta migratoria
RNA synthesis on lampbrush chromosomes of,
131
man. *See* human
mole, oocyte maturation in, 416
monkey, Golgi complex in oocytes of, 21
granulosa cells of, 513
Macaca mulatta, 38
mitochondria in oocytes of, 245
oocyte maturation in, 417
pig-tail monkey, oocyte of, 52
yolk in oocyte of, 53
spider monkey, oocyte of, 61, 62
mouse, centriolar satellite of, 55
cortical granules in oocyte of, 29
cytoplasmic bridges during oogenesis of, 14
enzymic changes during development of, 247
first polar body of, 29
follicle growth in ovary of, 361-375,
377-396
Golgi complex in oocyte of, 93-97
haploidy in, 278-279, 281
kinetochore of egg in, 66-84
metabolism of egg in, 241-249
mitochondria in oocyte of, 57, 245
movement of primordial germ cells in,
302-303
nucleolus in oocyte of, 29
oocyte maturation in, 416-432, 441
origin of primordial germ cells in, 301, 302
ovarian culture of, 310-311, 314-315,
317-318, 378
ribosomes in ovulated egg of, 53
spermatocyte of, 72
tabby gene of, 288
yolk in egg of, 55

Necturus
genome of, 137
N. maculosus
steroidogenesis in, 497
Neurospora
cytoplasmic inheritance in, 221
mitochondria DNA of, 220
Notophthalmus (Triturus) viridescens. See
T. viridescens.
opossum, origin of primordial germ cells in,
301
Orzyias latipes, 473
pea, histones of, 143, 148
pig, granulosa cells of, 513, 521
oocyte maturation in, 392, 401, 402
ovectomy of, 514, 515
tubal oocyte of, 425
pigeon, spermatocyte of, 66
rabbit, cytoplasmic bridges during oogenesis
of, 14
first polar body of, 32
genital ridges of, 7
Golgi complex in oocyte of, 21, 27, 93-97
metabolism of egg in, 241
nucleolus in oocyte of, 29
oocyte maturation *in vitro* of, 402,
416, 441
ovectomy of, 514, 515
parthenogenesis in, 279, 281, 282
primordial germ cells of, 7, 8
resumption of meiosis of, 400-402
yolk in oocyte of, 50, 53
Rana
R. esculanta
estrogens in ovary of, 497
R. pipiens
amino acid pool in oocyte of, 229
follicular function of, 482-506
mitochondrial DNA of, 218
oocyte maturation in, 459, 472, 473
protein synthesis in oocyte of, 228-236
R. temporaria, 340, 341
estrogens in ovary of, 497
steroidogenesis in, 497
rat, culture of follicle cells and oocytes
of, 518
cytoplasmic bridges during oogenesis in, 14
first polar body of, 29
follicle of, 513
follicular growth in, 374

Golgi complex in oocyte of, 27
heteroploidy in, 248
mitochondria in oocyte of, 245
morula of, 54
oocyte maturation of, 416
origin of primordial germ cells of, 301, 302
ovarian culture in, 380
parthenogenesis in, 243
proliferation of oogonia of, 12
yolk in oocyte of, 50, 55
Salamandra salamandra
Δ^5-3β hydroxysteroid dehydrogenase
in follicle of, 498
sea urchin, cortical granules in egg of, 27
cytoplasmic DNA of, 217
cytoplasmic RNA of, 170
expression of oocyte RNA of, 168
histones in, 151, 172
Lytechinus pictus
mitochondrial DNA of, 218
sperm of, 143
sheep, mitochondria in eggs of, 60
oocyte maturation of, 417
ovarian culture in, 378-380
yolk in oocyte of, 53
shrew, oocyte maturation in, 416, 422
snail, histones during spermatogenesis and
fertilization of, 151
Spisula solidissima
fertilization in, 28
starfish, oocyte maturation and adenine
derivatives of, 459-476
Aphelasterias japonica, 464
Asterias
A. amurensis, 460, 464, 466
A. forbesi, 461, 464, 466, 467
A. rubens, 474
Asterina pectinifera, 461, 462, 463, 464,
466, 467, 471, 472, 473, 474
Astropecten
A. aurantiacus, 466, 474
A. scoparius, 464
Ceramaster placenta, 466
Coscinasterias acutispina, 464
Leptasterias hexactis, 466
Luidia quinaria, 464, 474
Marthasterias glacialis, 466, 467
Mediaster aequalis, 466
Patiria miniata, 464, 466, 467, 471
Pisaster

P. brevispinus, 466
P. giganteus, 466
P. ochraceus, 460, 461, 464, 466
Pycnopodia helianthoides, 466
Triturus
T. cristatus
lampbrush chromosomes and RNA synthesis
in, 171
T. viridescens
DNA in oocytes of, 170
duration of lampbrush stage in, 131
follicular structure of, 349, 350
oogenesis in, 121-138
Urechis
kinetochore in egg of, 66, 75
mitochondrial DNA of, 218, 220
vole, oocyte maturation in, 417
woman. *See* human.
Xenopus
X. laevis, 322, 326-327, 329
anucleoate mutant of, 171, 187
duration of lampbrush stage in, 131-136
estrogens in ovary of, 497
genetic expression during oogenesis in,
130
gonadal effect on germ cells of, 321-337
mitochondrial DNA and RNA of, 218,
219-23
oogenesis of, 121-138
ovulation of, 486
Oxford nuclear marker of, 328, 332
RNA synthesis during oogenesis in,
167-187, 193-212
ultrastructure during maturation of
oocyte in, 489
"yolk DNA" of, 217
yolk formation in, 341-357
X. l. laevis, 325
X. l. victorianus, 325
X. mulleri, 222
X. tropicalis, 326-327, 328, 329
yeast, cytoplasmic inheritance in, 221
mitochondrial DNA of, 220

AUTHOR INDEX

Adams, E. C., 9, 15, 17, 21, 27, 50, 55, 88, 89, 302, 356, 393, 448

Adler, A., 154

Aggarwal, S. K., 254, 270

Ahren, K., 248

Akai, H., 255, 259, 265

Alfert, M., 27, 53

Allen, E., 12, 417, 427, 439

Allfrey, V. G., 149, 150, 152, 153, 156, 168, 171, 175, 233

Alvarez, B., 113

Amoroso, E. C., 281

Anderson, E., 53, 57, 88, 93, 97, 111, 112, 245, 356, 357

Anderson, J. W., 495, 505

Anderson, L. M., 112

Anderson, P. J., 89

Ando, T., 154

Andre, J., 62

Arms, K., 232, 234

Arvy, L., 88

Ashwell, M., 220

Askonas, B. A., 181

Attardi, B., 171

Attardi, G., 171

Auerbach, S., 248

Austin, C. R., 27, 50, 53, 278, 279, 281, 284, 291, 381, 325, 480

Awdeh, Z., 181

Baca, M., 17, 21, 25, 27, 29, 32, 417

Bacsich, P., 278

Bahn, E., 271

Bajer, A., 66

Baker, R. D., 416, 417

Baker, T. G., 9, 17, 21, 60, 138, 377, 379, 401, 420

Balbiani, E. G., 93

Balfour-Lynn, S., 282

Balinsky, B. T., 47, 498

Ballentine, R., 341

Balogh, K., 498

Baltimore, D., 181

Baltus, E., 170, 216, 217, 234

Banon, P., 88

Bara, C., 498

Baramki, T. A., 278, 286, 417, 420

Barka, T., 89

Barnett, W. E., 221

Bartalos, N., 278, 286

Bartley, J., 150

Bautz, E. K. F., 194, 200

Bavister, B. D., 417, 420, 445

Beams, H. W., 17, 89, 97, 112, 113, 356

Beato, M., 375

Beatty, R. A., 278, 279, 281, 282, 283, 284, 286, 291, 294, 295

Beatty, R. B., 121, 172, 183

Beaumont, H. M., 12, 17

Becker, Y., 174

Bekhor, I., 154, 157

Bell, J. H., 21, 32

Bell, P. R., 216

Bell, W. J., 270

Bellair, J. T., 158

Bellairs, R., 112

Bellini-Cardellini, L., 497

Beltermann, R., 35

Benjamin, W. B., 155

Berg, P., 196

Berger, E. R., 217

Bergers, A. C. J., 484, 485, 486

Bernardi, G., 222

Berry, R. O., 417

Bienstbach, F., 375

Bier, K., 132, 133, 138

Biggers, J. D., 97, 112, 241, 242, 243, 244, 245, 246, 248, 249, 378, 379, 390, 415, 416, 420, 460, 461, 464

Birnstiel, M. L., 119, 123, 170, 172, 183, 193, 194, 208

Bjorkman, N., 97

Blackler, A. W., 321, 322, 323, 324, 325, 328

Blair, D. G., 218

Blanchette, E. J., 15, 21

Bland, L. J., 417, 427, 439

Blandau, R. J., 9, 14, 15, 29, 32, 89, 302, 303, 306, 314, 391, 398, 416, 418, 480

Bloch, D. P., 151

Bodenstein, D., 258, 269, 270, 271

Boell, E. J., 219

Boiron, M., 174

Boling, J. L., 416, 418

Bomsel-Helmreich, O., 284, 285, 286

Bonner, J., 142, 144, 145, 146, 148, 149, 153, 154, 155, 156, 157

Borghese, E., 398

Borst, P., 219, 220
Borum, K., 14
Borun, T. W., 149, 150
Böstrom, H., 23
Botte, Y., 498
Botticelli, C. R., 460
Boucek, R. J., 498
Bounoure, L., 6
Bourgeois, S., 208
Brachet, J., 170, 216, 217, 228, 234, 474, 489, 498
Brackett, B. G., 445
Bradbury, J. T., 392
Bradbury, S., 351
Braden, A. W. H., 279, 284, 291, 425
Brambell, F. W. R., 6, 9, 14, 35, 301
Branca, A., 35, 416, 417
Brandes, D., 88
Bratt, H., 498
Breed, W. G., 417, 425
Bresciani, F., 375
Breuer, H., 497
Briggs, R. W., 228, 234, 236
Brinkley, B. R., 66, 72, 75, 83
Brinster, R. L., 243, 244, 247, 248, 448
Britten, R. J., 130, 133, 136, 137
Brown, A. C., 341
Brown, D. D., 119, 123, 168, 170, 171, 172, 174, 175, 177, 183, 186, 193, 194, 197, 204, 208, 211, 217, 218, 219, 220, 227
Brown, D. H., 221
Brown, E. H., 254, 258
Brown, H., 57, 60, 62
Brown, S. W., 266
Brownlee, G. G., 174, 177
Bruce, H. M., 281
Brutlag, D., 155, 157
Bryan, J., 466
Buck, C. A., 221
Buck, R. C., 66
Burckhard, C., 416
Burgess, R. R., 194, 200
Burke, J. F., 112
Burkl, W., 35
Burnett, R. C., 258
Burr, H. S., 416
Bustin, M., 144, 146
Butler, G. C., 142, 147
Calarco, P., 57, 62, 72
Call, E., 23

Callahan, P. X., 147
Callan, H. G., 119, 121, 131, 137, 171
Callard, I. P., 497
Carnevali, F., 222
Carr, D. H., 286
Carter, D. L., 485
Cassens, C. A., 260
Cassidy, J. D., 254, 255
Cedard, L., 497
Celestino da Costa, A., 6
Chaet, A. B., 460, 461
Chalkley, R. G., 142, 145, 146, 150, 155,
Chalumeau-Le Foulgoc, M. T., 345, 497
Chambers, R., 472
Chambon, P., 196
Champagne, M., 151
Chandley, A. C., 441
Chang, L. M. S., 158
Chang, C. Y., 484
Chang, M. C., 279, 282, 391, 393, 400, 408, 420, 425, 427
Channing, C., 513, 514
Chappell, J. B., 245
Chase, J. W., 219
Chieffi, C., 497, 498
Chiquoine, A. D., 6, 302
Cho, W. K., 378, 379, 390
Church, R. B., 113, 157
Clarke, J. R., 417, 425
Cleland, S., 79
Clewe, T. H., 391, 417, 427
Coe, W. R., 416
Coen, D., 222
Coggeshall, R., 88
Cognetti, G., 151
Cohen, A. I., 50
Cohn, Z. A., 111
Cole, R. D., 142, 144, 146, 149, 156
Cole, H. H., 417, 431
Comb, D. G., 156, 174
Condon, W., 57, 97
Cons, J. M., 113, 357
Cook, B., 514
Cook, E. A., 217
Coppo, A., 194, 199, 200
Corner, C. W., 417
Cornette, J. C., 248
Cotran, R. S., 89
Coulon, E. M., 66
Court Brown, W. M., 267

Cox, R. A., 168
Craig, S. P., 170
Crampton, C. F., 143
Creaser, C. W., 484
Crippa, M., 126, 130, 131, 133, 136, 137, 168, 172, 234
Croes-Buth, S., 392
Crone, M., 14
Crooke, A. C., 378
Cropper, M., 514
Cross, P. C., 422, 448
Cummings, M. R., 267, 270
Dahmus, M., 142, 155, 157
Danziger, S., 302
Dapples, C. C., 267
Darlington, C. D., 66, 266
Darnell, J. E., 172, 174, 178
Daughterty, K., 484
Davidson, E. H., 2, 126, 129, 130, 131, 133, 134, 135, 136, 137, 194, 227, 234, 253, 168, 172, 175
Davidson, O. W., 113
Dawd, A. J., 498
David, I. B., 119, 168, 170, 175, 193, 194, 197, 204, 208, 216, 217, 220, 221, 234
Dawson, A. B., 417
Day, F. T., 417, 431
Deane, H. W., 88
Debeyre, A., 6
DeDuve, C., 88, 112
De Filippes, F. M., 157
De Fonbrune, P., 416
de Jongh, S. E., 394
Delage, Y., 472
De Lange, R. J., 144, 148, 149
Dempsey, E. W., 278, 427
Denis, H., 234
De Recondo, A. M., 158
Dettlaff, T. A., 229, 474, 488, 489, 491, 493, 495
Deutsch, J., 222
de Vellis, J., 112
Devis, R. J., 498
de Vitry, F., 234
de Winiwarter, H., 12
Dick, D. A. T., 351
Dick, E. G., 351
Dickmann, Z., 425
Dienstbach, F., 375
Di Mauro, E., 194, 199, 200

Dingman, W. C., 151, 155
Dixon, G. H., 154
Dodd, J. M., 484, 497
Doira, H., 270
Dominitz, R., 111
Donachie, W. D., 258
Donahue, R. P., 66, 83, 97, 112, 242, 243, 244, 247, 248, 402, 415, 416, 417, 420, 422, 427, 429, 439, 441, 448
Doorme, J., 416
Dott, H. M., 88
Doty, P., 142, 156
Doty, J., 196
Dounce, A. L., 150
Droller, M. J., 112
Dumont, J. N., 112, 325, 341, 343, 347, 349, 350, 351
Duncan, G. W., 248
Dunn, J. J., 194, 200
Dyer, R. F., 14
Dziuk, P. J., 417, 422, 425
Ecker, R. E., 228, 229, 230, 231, 232, 233, 234, 235, 459, 460, 472, 473, 474, 488, 491, 493, 495, 498, 499, 503, 504, 505
Edelman, M., 220
Edgren, R. A., 485
Edstrom, J. E., 171
Edwards, R. G., 38, 281, 284, 378, 379, 381, 391, 400, 415, 416, 417, 418, 419, 420, 422, 441, 444, 445, 448
El-Fouly, M., 513, 514, 515
Elgin, S. C. R., 142, 144
Enders, A. C., 50, 53, 55, 57, 60
Enders, R. K., 417, 431
Endo, Y., 27
Engle, E. T., 416, 425, 427
Enzmann, E. V., 243, 400, 416, 418, 425, 427, 441
Epifanova, O. I., 375
Epler, J. L., 221
Epstein, C. J., 247, 248
Espey, L. L., 505
Evans, D., 119, 170, 193
Evans, H. M., 12, 417, 431
Evennett, P. T., 488, 495
Everett, J. W., 481
Everett, N. B., 6, 301, 302, 323
Exner, S., 23
Faber, M., 373
Fainstat, T., 378, 392

Falck, B., 513
Falk, G. J., 271
Fambrough, D., 142, 144, 145, 146, 148, 149, 155
Fan, H., 253
Farquhar, M. G., 111
Fasman, G. D., 154
Fawcett, D. W., 14, 21, 103
Fechheimer, N. S., 278
Fedorko, M. E., 111
Felsenfeld, G., 153
Fichot, O., 158
Ficq, A., 170, 216, 217
Finke, E. H., 68
Fischberg, M., 183, 279, 286, 291, 328
Fiume, L., 196
Flathers, A. R., 493
Flax, M. H., 53
Follett, B. K., 343, 345, 349, 496
Foote, W. D., 379, 392, 401, 417, 418, 420
Forbes, Thomas, R., 497
Ford, E. H. R., 72
Ford, P. J., 168, 171, 174
Forget, B. G., 174
Forrester, S., 147, 155
Fowler, R. E., 378, 379, 381, 415
Franchi, L. L., 9, 12, 14, 17, 21, 60, 138, 393, 401, 480
Frenster, J. H., 155
Friedgood, H. B., 416, 417
Friend, D. S., 111
Fritz, H. I., 378, 379, 390
Froehner, S. C., 142, 144
Frost, J. K., 88
Fujimura, F., 142, 155
Furlan, M., 150, 156
Fuss, A., 301
Galey, F., 112
Galibert, F., 174
Gall, J. G., 68, 119, 121, 122, 123, 125, 126, 131, 171, 172, 174, 193, 227, 253
Gallien, L., 345, 497
Gallwitz, D., 149, 233
Gansen, Van. P., 489, 498, 504, 505
Gates, A. H., 378, 381, 416, 418, 419, 422
Gay, H., 72
Geiduschek, E. P., 194
Gellhorn, A., 155
Gerard, M., 400, 402, 404, 405
Gershey, E. L., 150, 152

Geschwind, I. I., 484
Getz, M. J., 157
Giese, A., 50
Gilmour, R. S., 157
Gitlin, G., 341
Glass, L. E., 113, 341, 347, 357
Glick, D., 121
Goldsmith, L., 154
Gonatas, N. K., 66, 79
Gondos, B., 8, 9, 10, 14
Gorbman, A., 113, 484
Gothie, S., 416
Gottfried, H., 497
Gowen, J. W., 265, 266
Graham, C. F., 279, 291, 448
Graham, R. C., Jr., 89
Grant, P., 340, 341
Grant, R., 484, 491, 498
Green, E., 234
Greenawalt, J. W., 220
Greenaway, P. J., 148
Greenberg, J. R., 172
Gregg, J. R., 341
Grell, E. H., 263, 269
Greller, E. A., 112
Gresson, R. A. R., 416
Gros, F., 171
Gross, P. R., 130, 133, 136, 151, 168, 170, 1?, 228, 234
Gulyas, B., 52, 53
Gurdon, J. B., 142, 168, 171, 174, 183, 186, 232, 236, 329, 332, 498
Guthrie, M. J., 416
Gutierrez, R. M., 153
Hadek, R., 15, 23, 50, 53
Hahn, W. E., 113
Hamberger, L. A., 248
Hamilton, W. J., 417, 431
Hamkalo, B. A., 121, 137
Hammond, J., 391
Hancock, J. L., 422
Hanocq, F., 489, 498
Hanocq-Quertier, J., 170, 216, 217, 234
Hargitt, G. T., 6, 301
Harrison, R. J., 480
Hartman, C. G., 417
Hartmann, N. R., 368
Haselkorn, R., 194
Haslett, G. W., 150
Hawley, E. S., 220

Hayashi, H., 144, 147
Hayashi, M., 444
Heape, W., 416
Hedberg, E., 27
Heidger, P. M., 103
Heilbrunn, L. V., 484
Heinrich, J. J., 113
Hennen, S., 234
Henneguy, M. G., 93
Hepler, P. K., 79
Hersh, R. T., 2
Hertig, A. T., 9, 15, 17, 21, 27, 50, 55, 88, 89, 302, 356, 393, 448
Hess, O., 119, 138
Hew, H. Y., 151
Hewlett, J., 260
Hilgartner, C. A., 150, 155
Hinton, C. W., 265
Hiramoto, Y., 471, 472
Hisaw, F. L., 392, 460, 481, 484
Hnilica, L., 147, 151, 153
Hoadley, L., 417
Hoar, W. S., 484
Hodes, M. E., 158
Hollenberg, C. P., 220
Holtzman, E., 111
Hope, J., 17, 21, 504
Hough, B. R., 132, 133, 134, 135, 172, 227, 234
Houssais, J. F., 171
Howk, R., 156, 158
Hruban, Z., 103
Huang, A. S., 181
Huang, M. I. H., 171
Huang, P. C., 154, 157
Huang, R. C., 142, 143, 145, 153, 154, 155, 156, 157, 158
Huberman, J., 142, 155
Huebner, E., 112
Humphrey, R. R., 234, 324
Hunt, J. A., 171
Hunt, S. V., 498
Hunter, A. L., 151, 170
Hunter, R. H. F., 417, 418, 419
Hyman, L. H., 130
Ickowicz, R., 150
Iizuka, R., 417
Ikegami, S., 460
Inglis, C. J., 154
Ingram, D. L., 35, 112, 377, 391, 393, 481
Inoue, S., 66

Ioannou, J. M., 14
Ishikawa, K., 144, 147
Iwai, K., 144, 147
Iwamatsu, T., 473
Izawa, M., 168, 171
Jacob, H., 222
Jacob, J., 267
Jacob, S. T., 196, 197
Jacques, P. J., 112
Jagiello, G. M., 416, 417, 418, 441
Jared, D. W., 217, 227, 343, 344, 345, 346, 347, 348, 349, 350, 352, 353, 355, 356, 357
Jarett, L., 68
Jeffers, K. R., 416
Jensen, R., 142, 150, 155
Jergil, B., 154
Jericijo, M., 150, 156
Johns, E. W., 145, 147, 150, 155
Jokelainen, P. T., 66
Joklik, W. K., 174
Jolly, M. J., 498
Jones, E. C., 361, 373, 374
Jones, H. W., 417, 420
Jones, K., 172, 194
Jones, O. W., 196
Journey, L. J., 75
Julien, J., 174
Julin, C., 416
Justus, J. T., 228, 234
Kabat, D., 150
Kajiwara, K., 150
Kanatani, H., 460, 461, 462, 463, 464, 466, 467, 470, 471, 473
Karasaki, S., 341, 342
Karnicki, J., 417, 418, 441
Karnovsky, M. T., 89
Kaufman, B. P., 72
Kedes, L., 130, 151, 172
Kelley, D. E., 175
Kemp, N. E., 341
Kennedy, J. F., 417, 427, 439, 441, 448
Kessel, R. G., 23, 113, 495
Kidston, M., 253
Kim, S., 152
King, R. C., 254, 255, 257, 258, 259, 260, 261, 263, 264, 265, 266, 267, 268, 269, 270, 271
King, T. J., 228, 234, 236
Kingery, H. M., 6, 12, 301
Kingsbury, B. F., 9
Kinkade, J. M., 146

Kirby, K. S., 217
Kirkham, W. B., 416
Kleiman, L., 142, 155
Kleinsmith, L. J., 153, 156
Kung, G. M., 154, 157
Klug, W. S., 267, 268, 269
Knigge, K. M., 112
Knight, E., 174, 178
Koch, E. A., 254, 255, 257, 258, 260, 261, 265, 266
Koga, H., 15
Kohne, D. E., 133, 136
Koike, K., 220
Koskimies, O., 248
Kowallik, K., 57
Krafka, J., 278
Kramen, M. A., 245, 246
Kramer, F. R., 126, 131, 133, 136, 137, 168, 172
Krarup, T., 373
Krauskopf, C., 53
Krishan, A., 66
Krohn, P. L., 361, 373, 374
Kroon, A. M., 219
Kuechler, E., 172
Kuhlmann, W., 416
Kuntzel, H., 220
Kunz, W., 131, 133, 138
Kurokawa, T., 460, 464, 466, 467
Labrie, F., 171
Lamb, D. C., 147
Lamberti, A., 194, 199, 200
Lams, H., 416, 417
Langan, T. A., 153, 156
Lanzavecchia, G., 15
Larsen, C. L., 174
Latham, H., 174
La Torre, J. L., 341
Laycock, D. G., 171
Leak, L. V., 103, 112
Leathem, J. H., 112, 377, 497
Lee, C. S., 137
Lejeune, J., 267
Lelong, J. C., 174
Leng, M., 153
Levy, E., 14, 392
Li, C. H., 484, 485, 486
Liau, M. C., 172
Lieberman, I., 150
Lima de Faria, A., 66
Lindsay, D. T., 151

Lindsley, D. L., 263, 269
Ling, V., 154
Lingrel, J. B., 158
Lipner, H., 404
Littau, V. C., 149, 152, 233
Littauer, U. Z., 220
Littna, E., 168, 172, 174, 175, 177, 183, 211, 219, 227
Lloyd, C. W., 498
Lloyd, L., 119, 121
Lobel, B. L., 88
Lockard, R., 158
Lockshin, R. A., 253
Loeb, J. A., 158
Loening, V. E., 172, 194
Loewenstein, J. E., 50
London, I., 171
Long, J. A., 416
Longley, W. H., 417, 422
Longo, F. J., 53
Lostroh, A. J., 392
Luck, D. J. L., 220, 222
Luck, J. M., 145
Luft, J. H., 68
Lupo, C., 497
Luykx, P., 66, 75
Macgregor, H. C., 119, 121, 123, 170, 171, 19?, 253
Mackler, B., 60
Maheshwari, N., 156
Makimo, S., 416
Mancini, R. E., 113
Mandel, P., 196
Mandl, A. M., 9, 12, 14, 138, 361, 417, 419
Mangioni, C., 15
Mansani, F. E., 21
Manton, I., 57
Mantsavinos, R., 158
Marcaud, L., 171
Marcus, P., 253
Marino, P., 194, 199, 200
Mark, E. L., 416
Martinovitch, P. N., 378, 398
Marushige, K., 142, 151, 154, 155, 157
Marzluff, W., 150, 152
Mascia, M., 157
Mastroianni, L. Jr., 17, 21, 27, 29, 32, 35, 38, 5?, 89, 393, 416, 417, 441, 445
Masui, Y., 229, 351, 459, 460, 493, 495, 498, 499, 503, 505

Matley, D. L., 401
Matthews, L. H., 480
Matthey, R., 288
Mauleon, P., 377, 392
Maunsbach, A. B., 112
Mazen, A., 151
McCarthy, B. J., 157
McCarty, K., 150, 152
McConnaughy, R. A., 460
McConnell, D. J., 157
McFeely, R., 286
McGaughey, R. W., 416, 427
McIntosh, J. R., 79
McKay, D. G., 302
McKay, M., 150
Mead, J. F., 112
Meisler, M. H., 153
Menkin, M. F., 444, 448
Merriam, R. W., 232, 234
Meyer, G. F., 265
Meyer, J. E., 392
Mikamo, K., 285, 291
Miller, J. W., 72
Miller, O. L. Jr., 121, 172, 183
Millington, P. F., 341
Mintz, B., 6, 302
Mirsky, A. E., 126, 130, 131, 133, 136, 137, 149, 152, 153, 168, 170, 171, 172, 175, 194, 233
Mishell, D. R. Jr., 21, 32
Misra, D. N., 137
Mittwoch, U., 278
Miura, A., 147
Moffit, J. G., 417
Mole-Bajer, J., 66
Monier, R., 174
Monoulou, J. C., 221
Monroy, A., 228
Moore, D. H., 112
Moore, J. E. S., 417
Moore, R. W., 247
Moore, S., 143
Morgan, T. H., 265
Moricard, R., 416
Morrill, G. A., 498
Moser, F., 27
Moses, M. J., 12
Motomura, A., 15
Motta, P., 23
Muchmore, J., 150
Mueller, G. C., 149, 150, 233

Muhlethaler, K., 216
Munro, H. N., 196, 197
Murray, K., 147, 148
Muta, T., 15
Nakamura, T., 416
Nakanishi, K., 460, 464, 466, 467
Nalbandov, A. V., 420, 514, 518
Narayan, S. K., 172
Nass, M. M. K., 221
Nowicka, J., 448
Noyes, R. W., 391, 427
Neal, P., 379
Nebel, B. R., 66
Neelin, J. M., 142, 147
Nekola, M., 513, 514, 518
Nelson, B. L., 350, 352, 353, 355, 356
Nelsen, O. E., 301
Netter, P., 222
Newell, Q. U, 417, 427, 439
Newman, H. H., 416
Nicholls, T. J., 345, 496
Nicolaieff, A., 222
Nikitina, L. A., 229, 493
Nieuwkoop, P. D., 6
Noll, H., 220
Noriega, C., 417, 441
Novikoff, A. B., 88, 111
Nowicka, J., 279, 291
Noyes, R. W., 416, 417, 425
Oakberg, E. F., 17
Odeblad, E., 23
Odor, D. L., 9, 14, 15, 29, 32, 89, 314, 417
Ohba, Y., 147
Ohguri, M., 460
Ohlenbusch, H., 142, 155
Ohno, S., 14, 278, 341, 416
Okamoto, T., 14
Olins, A. L., 153
Olins, D. E., 147, 153, 154
Olivera, B., 142, 155
Onozato, H., 23
Osborn, M., 148, 155
Otsuki, Y., 270
Oura, C., 9, 32
Ozaki, H., 151
Ozdzenski, W., 302
Ozon, R., 497
Paesi, F. J. A., 392
Paik, W. J., 152
Pan, M. L., 270

Panje, W. R., 495
Panyim, S., 142, 145, 146, 150
Paoletti, R., 143
Pardue, M. L., 68
Parkes, A. S., 281
Parnas, H., 171
Paul, J., 157
Pearson, O. P., 416, 417, 422, 431
Pedersen, T., 361, 362, 368, 371, 373, 374, 392,
Penman, S., 168, 253
Penrose, L. S., 289
Perkowska, E., 170, 193
Perry, R. P., 172, 175
Peters, H., 14, 362, 371, 392
Petrochilo, E., 222
Petrucci, D., 444
Pharris, B. B., 248
Phillips, D. M. P., 150
Piko, L., 217, 218, 284
Pincus, G., 243, 278, 279, 282, 400, 416, 417,
   418, 425, 427, 441, 480
Pinzino, C. J., 156
Piperno, G., 222
Pitot, H. C., 495
Pogo, A. O., 152, 153
Pogo, B. G. T., 152, 153
Polani, P. E., 289
Polge, C., 417, 418, 419, 420, 425
Polzinetti-Magni, A., 497
Pool, W. R., 404
Porter, K. R., 32, 35, 112
Pratt, H., 168
Pratt, J. P., 417, 427, 439
Presco, C., 497
Prescott, D. M., 150, 258
Pressman, B. C., 245
Pupkin, K. A., 498
Purdy, J. M., 445
Race, R. R., 290
Rall, S. C., 144, 146
Ralph, R. K., 175
Ramuz, M., 196
Rapola, J., 248
Rasmussen, P. S., 146
Rauh, W., 301
Raven, C. P., 3, 17, 112, 479
Redshaw, M. R., 343, 345, 496
Reeder, R. H., 170, 194, 197, 204, 208, 211, 221
Reich, E., 222
Reid, B. R., 149, 156

Rein, G., 416, 417
Renger, H. C., 220
Renninger, D. R., 15, 32
Renninger, D. F., 417
Reynolds, E. S., 68
Rhoades, M. M., 3
Ribbert, D., 133, 138
Rich, A., 172
Richardson, K. C., 68
Rifkin, M. R., 220
Riggs, A. D., 208
Ritossa, F. M., 267
Robbins, E., 66, 79, 149, 150
Robertson, J. E., 416, 417
Robinson, A., 417
Rocha, G., 445
Rock, J., 444, 448
Roeder, R. G., 196, 197
Rogers, M. E., 126, 172, 174
Rondell, P., 505
Roodyn, D. B., 220, 221
Rosenbaum, R. M., 88
Rosset, R., 174
Rostozi, R. K., 497
Roth, T. F., 112
Rowlands, I. W., 378
Rubaschkin, W., 301, 417
Rubinstein, L., 248
Ruby, J. R., 14
Rugh, R., 484, 486, 490, 505
Rumery, R. E., 302, 303, 306, 310, 398
Runner, M. N., 430
Ruska, H., 112
Russell, E. S., 302
Russell, L. B., 278, 289
Rutter, W. J., 196, 197
Ryan, F. J., 484, 491, 498
Ryan, R. J., 417, 418, 441
Ryle, M., 378
Ryser, H. J. P., 111
Saacke, R. G., 21, 60, 62
Sadgopal, A., 150
Saglik, S., 278
Sainmont, G., 12
Sajdel, E. M., 196, 197
Salb, J., 253
Sandler, L., 269
Sang, J. H., 269
Sanger, R., 290
San Lin, R. I., 112

Sarkar, N., 156, 174
Satake, K., 146
Sato, H., 466
Saunders, B., 417, 441
Saunders, G. F., 157
Sayler, A., 498
Schaffhausen, B., 154
Scharff, M. D., 149
Schectman, A. M., 112
Scherrer, K., 171
Schindler, A. M., 285, 291
Schjeide, O. A., 112, 356
Schlafke, S. J., 50, 53, 55, 57, 60
Schmerling, Zh. G., 495
Schroeder, T. E., 255
Schuetz, A. W., 229, 390, 459, 460, 461, 464,
   474, 481, 484, 485, 486, 487, 488, 491, 493,
   494, 495, 496, 497, 501, 502, 504
Schultz, R. J., 295
Schwartz, M., 218
Seaman, F., 150
Seifart, H. K., 196
Seitz, H. M., 445
Selvig, S. E., 170
Senger, P. L., 21, 60, 62
Shapiro, H., 282, 343
Shapiro, K., 153
Sharp, D., 57, 97
Shaver, E. L., 286
Shaw, L., 142, 145, 155, 156
Sheehan, J. R., 17
Shelton, M., 374
Shettles, L. B., 444
Shih, T. Y., 157
Shin, W. Y., 111
Shirai, H., 460, 461, 462, 464, 466, 467, 470
Signoret, J., 234
Silverman, L., 121
Simkins, C. S., 6, 301
Simons, D., 417
Simpson, T. H., 498
Sirlin, J. L., 183, 267
Skalko, R. G., 14
Skoblina, M. N., 474, 488, 489, 493, 495
Slater, E. C., 247
Slonimski, P. P., 222
Smart, J. E., 142, 144, 155
Smith, C. W. 247, 248
Smith, E. L., 144, 148, 149
Smith, J., 137

Smith, K. D., 157
Smith, L. D., 168, 228, 229, 230, 231, 232, 233,
   234, 235, 322, 459, 460, 472, 473, 474, 488,
   491, 493, 495, 498, 499, 503, 504, 505
Smith, M. M., 157
Smith, P. A., 254, 257, 258, 265, 266
Snow, M. H. L., 137
Snyder, L., 194, 199, 200
Sobotta, J., 416
Soderwall, A. L., 416, 418
Soeiro, H. C., 174, 178
Sonnenblick, B. P., 253
Sotelo, J. R., 32, 35
Spalding, J., 150
Sparvoli, E., 72
Spaulding, J. F., 417
Speight, V. A., 498
Spiegel, R. B., 195
Spiegelman, S., 267
Spirin, A. S., 234
Sporn, M. D., 151, 155
Stavnezer, J., 158
Stavy, L., 168
Stedman, E., 142
Stedman, E., 142
Stefanini, M., 9, 32
Stegner, H. E., 17, 21, 60
Stein, W. H., 143
Steinert, G., 474
Stellwagen, R. H., 142, 144, 146
Steptoe, P. C., 417, 418, 420, 444, 445, 448
Stern, C., 285
Stern, S., 241, 247, 248, 416
Stevely, W. S., 153
Stevens, M., 460, 466, 473
Stirpe, F., 196
Stocken, L. A., 153
Stosch, H. A. von., 57
Straus, W., 111
Strauss, F., 416, 431
Stroeva, D. G., 229, 493
Stubblefield, E., 66, 72, 75, 83
Subtelny, S., 168, 228, 229, 230, 459, 460, 488,
   491, 493, 495, 498, 499, 504, 505
Summers, W. C., 195
Suzuki, H., 208
Suzuki, S., 38, 417
Swain, E., 301
Swanson, C. P., 125
Swanson, R. F., 220

Swezy, O., 12
Swift, H., 103
Szabo, P. L., 50
Szollosi, D., 15, 27, 53, 55, 56, 62, 72, 93
Takata, K., 341
Takei, S., 150
Talbert, A. J., 243, 419
Tandler, C. J., 341
Tardini, A., 21
Tarkowski, A. K., 279, 291, 448
Tecce, G., 222
Telfer, W. H., 112, 270
Teplitz, R., 14
Thibault, C., 278, 279, 281, 282, 285, 286, 379,
    392, 400, 402, 404, 405, 408, 417, 418, 420
Thiel-Bartosh, E., 35
Thomas, C. A. Jr., 121, 137, 182
Thomas, D. Y., 222
Thomson, A., 417
Thomson, J. L., 243, 244
Thornton, V. F., 488, 495
Thuring, R. W. J., 220
Tibbits, F. D., 401
Tidwell, T., 152
Tocchini-Valentini, G. P., 194, 199, 200
Tozer, F., 417
Travers, A. A., 194, 195, 200
Tsatsaris, B., 416
Tsvetikov, A. N., 146
Turpin, R., 267
Tyler, A., 217, 218, 228, 234
Uehlinger, V., 142
Umana, R., 150
Urist, M. R., 356
Uzzell, T., 295
Van Beneden, E., 416
Van Bruggen, E. F. J., 220
Van de Kerckhoue, D., 398
Van den Bergh, S. G., 247
Van der Stricht, O., 93, 416, 417, 431
Van der Stricht, R., 417
Vanneman, A. S., 301
Vasquez-Nin, G. H., 35
Venable, J. H., 88
Vendrely, C., 216
Vendrely, R., 216
Venini, M. A., 398
Verma, I. M., 220
Vickers, A. D., 284, 286
Vidali, G., 147, 150, 152

Villar, O., 113
Vinograd, J., 217, 218
Vitali-Mazza, L., 21
von Hippel, P. H., 153
Waddington, C. H., 50
Waldeyer, W., 12
Wales, R. G., 245
Wallace, H., 123, 183
Wallace, R. A., 112, 217, 227, 325, 341, 343
    344, 345, 346, 347, 348, 349, 350, 351,
    352, 353, 355, 356, 357, 496
Walton, A., 391, 480
Wang, T. V., 155, 156, 158
Ward, M. C., 417
Ward, R. T., 62
Warner, J. R., 175, 178
Warrick, E., 306, 398
Wartenberg, H., 17, 21, 60, 347
Watson, J. D., 175
Watson, M., 68
Wattiaux, J. M., 267, 268, 269
Wattiaux, R., 88, 112
Weakley, B. S., 9, 14, 15, 17, 21, 27, 50, 53
Weber, C. S., 123, 170, 175, 183
Weber, R., 219
Weber, K., 148, 155
Wegienka, E. A., 247, 248
Weiss, S. B., 156
Weissman, S. M., 174
Weisz, J., 498
Whaley, A., 75
White, B. J., 302, 303
Whitten, W. K., 242
Whittingham, D. G., 97, 112, 242, 243, 244, 2·
    381, 393, 396, 415, 416
Widholm, S., 142, 155
Wilbur, K. M., 484
Wilkie, D., 220, 221, 222
Williams, P. C., 378, 392
Williamson, A. R., 181
Williamson, R., 174, 177
Wilson, E. B., 2, 93, 129, 228, 472
Wilson, E. D., 378, 379, 381
Wilson, L. P., 269
Wischnitzer, S., 15, 57, 79, 349, 350, 479, 50·
Witkowska, A., 279, 291, 448
Wittek, M., 341
Witschi, E., 6, 235, 302, 322, 484
Wolff, E. K., 306
Wolfsen, A., 498

Wolstenholme, D. R., 219

Wood, D. D., 220

Woolam, D. H. M., 72

Work, W. S., 220

Wright, P. A., 484, 486, 487, 488, 491, 493,
496, 497

Wright, R. S., 498

Wyburn, G. M., 278

Yamada, E., 15

Yamamoto, K., 23

Yamate, A. M., 391, 416, 427

Yarger, R. J., 263

Yatvin, M. B., 495, 505

Young, W. C., 377, 416, 418

Yuncken, C., 441

Zachariae, F., 23

Zajdela, J., 171

Zalokar, M., 269

Zamboni, L., 9, 14, 15, 17, 21, 23, 25, 27, 29,
32, 35, 47, 53, 55, 89, 416, 417, 418, 425

Zarrow, M. X., 378, 381

Zohary, D., 266

Zondek, B., 306

Zubay, G., 142, 156

Zuckerman, S., 3, 6, 12, 14, 323, 361

Zwarenstein, H., 343, 486

Zylber, E., 168